XVIII International School on Nuclear Physics, Neutron Physics and Applications 2009

Journal of Physics: Conference Series Volume 205

Varna, Bulgaria
21-27 September 2009

ISBN: 978-1-61738-272-7

Printed from e-media with permission by:

Curran Associates, Inc.
57 Morehouse Lane
Red Hook, NY 12571

Some format issues inherent in the e-media version may also appear in this print version.

Printed by Curran Associates, Inc. (2010)

For permission requests, please contact the Institute of Physics
at the address below.

Institute of Physics
Dirac House, Temple Back
Bristol BS1 6BE UK

Phone: 44 1 17 929 7481
Fax: 44 1 17 920 0979

techtracking@iop.org

Additional copies of this publication are available from:

Curran Associates, Inc.
57 Morehouse Lane
Red Hook, NY 12571 USA
Phone: 845-758-0400
Fax: 845-758-2634
Email: curran@proceedings.com
Web: www.proceedings.com

TABLE OF CONTENTS

PREFACE

The present volume contains the lectures and short talks given at the XVIII International School on Nuclear Physics, Neutron Physics and Applications. The School was held from 21 to 27 September 2009 in Hotel "Lilia" located on 'Golden Sands' (Zlatni Pyasaci) Resort Complex on the Black Sea coast, near Varna, Bulgaria. The School was organized by Institute for Nuclear Research and Nuclear Energy of Bulgarian Academy of Sciences. Co-organizer of the School was Bulgarian Nuclear Regulatory Agency. The event was sponsored by National Science Fund of Bulgaria.

According to the long-standing tradition the School has carried every second year since 1973. The School content has been restructured according to our new enlarged international links and today it is more similar to international conference than to classical nuclear physics school. This new image attracts a lot of young scientists and students from many countries. This year – 2009, we had pleasure to welcome more than 50 distinguished sciences as lecturers. Additionally, 14 young colleagues received the opportunity to present a short contribution. The program ranges from recent achievements in nuclear structure and reactions to the hot problems of application of nuclear methods, reactor physics and nuclear safety. The 94 participants enjoyed the scientific presentations and discussions as well as the relaxing atmosphere at the beach and during the pleasant evenings.

The main topics have been the following:

- Nuclear excitations at various energies.
- Nuclei at high angular moments and temperature.
- Structure and reactions far from stability
- Symmetries and collective phenomena
- Methods for lifetime measurements
- Astrophysical aspects of nuclear structure
- Neutron nuclear physics
- Nuclear data
- Advanced methods in nuclear waste treatment
- Nuclear methods for applications

Several colleagues contributed to the organization of the School. We would like to thank to them and especially to the Scientific Secretary of the School Dr. Dimitar Tonev, the member of Organizing Committee Prof. Sevdalina Dimitrova, and Dr. Dimitar Tarpanov for their cordiality and high level assistance.

Sofia, 12 of November 2009

Co-chair persons of the
Organizing Committee

Prof. Dr. Sc. Ch. Stoyanov

Prof. Dr. Sc. N. Janeva

SPORT

XVIII International School on Nuclear Physics, Neutron Physics and Applications

September 21 - 27, 2009
Varna, Bulgaria

Organised by:

Institute for Nuclear Research and Nuclear Energy,
Sofia, Bulgaria

Supported by:

Bulgarian Nuclear Regulatory Agency,

National Science Fund of Bulgaria

International Advisory Committee

A. Bracco *(University of Milan)*	**A. Covello** *(U. Napoli)*	**A. Polanski** *(Dubna)*
C. Lister *(ANL)*	**F. Gunsing** *(CEA Saclay)*	**F. Kaeppeler** *(FZK, Karlsruhe)*
G. De Angelis *(LNL)*	**J. Blomgren** *(Uppsala University)*	**J. Draayer** *(SURA)*
J. Stamenov *(INRNE, Sofia)*	**L. Kostov** *(NRA, Bulgaria)*	**L. Leal** *(ORNL)*
L. Tassan-Got *(Universite Paris-Sud)*	**M. Harakeh** *(KVI)*	**M. Kirchbach** *(UASLP, Mexico)*
M. Ploszajczak *(GANIL)*	**N. Pietralla** *(TU, Darmstadt)*	**Nguyen Van Giai** *(IPN - Orsay)*
N. Zamfir *(NIPNE-Bucharest)*	**O. Bouland** *(CEA/LANL)*	**P.-H. Heenen** *(ULB, Bruxelles)*
P. Rulhusen *(IRMM-JRC)*	**R. Capote Noy** *(IAEA, Vienna)*	**R. Casten** *(Yale)*
R. Julin *(JYI)*	**S. Gales** *(IN2P3,France)*	**V. Shvetsov** *(JINR, Dubna)*
V. Voronov *(BLTP, Dubna)*	**V. Zelevinsky** *(MSU)*	**Y. Penionzhevich** *(JINR,Dubna)*

Local Organizing Committee

Ch. Stoyanov	*co-chairperson*	nuclear physics
N. Janeva	*co-chairperson*	neutron physics, applications
D. Tonev	*scientific secretary*	
S. Dimitrova		

List of participants

1. Anton **Antonov**,
   INRNE-BAS, 72, Tzarigradsko chaussee, Blvd.
   BG – 1784 , Sofia, Bulgaria,
   aantonov@inrne.bas.bg
2. Marilena **Avrigeanu**,
   NIPNE-Bucharest, Str. Atomistilor no.407, P.O. BOX MG-6,
   Bucharest – Magurele, Romania,
   m_avrig@yahoo.com
3. Evgeny **Balbutsev**,
   JINR, Dubna, Joliot-Curie 6, 141980 Dubna, Moscow region, Russia,
   balbuts@thsun1.jinr.ru
4. Maria **Barbaro** ,
   U. Torino, Corso Massimo D'Azeglio 60/H, Italy,
   barbaro@to.infn.it
5. Davide **Bianco**,
   U. Napoli Federico II, Corso Umberto I, 80138 Napoli, Italy,
   bianco@na.infn.it
6. Andrey **Blazhev**,
   U. Köln, Zülpicher Straße 77, 50937 Cologne, Germany,
   a.blazhev@ikp.uni-koeln.de
7. Angela **Bonaccorso**,
   INFN, Sez. di Pisa, Largo Pontecorvo 3, 56127 Pisa, Italy,
   bonac@df.unipi.it
8. Dennis **Bonatsos**,
   N.C.S.R. Demokritos, 153 10 - Agia Paraskevi Athens P.O. 60228,
   Greece,
   bonat@inp.demokritos.gr
9. Ruxandra **Borcea**,
   IRMM, JRC, Retieseweg 111, B-2440 Geel, Belgium,
   ruxandra.borcea@ec.europa.eu
10. Slobodan **Brant**,
    U. Zagreb, p.p. 407, Trg maršala Tita 14 HR-10002, Zagreb,
    Croatia,
    brant@polaris.phy.hr
11. Aldo **Covello**,
    U. Napoli Federico II, Corso Umberto I, 80138 Napoli, Italy,
    covello@na.infn.it
12. Raquel **Crespo**,
    IST- Lisbon, Centro de Física Nuclear, Av. Prof. Gama Pinto, 2,
    1699, Portugal,
    raquel.crespo@tagus.ist.utl.pt
13. Jozsef **Cseh**,
    MTA ATOMKI, H-4001 Debrecen, PO Box 51, Hungary,
    cseh@atomki.hu
14. Dominique **Curien**,
    IPHC-DRS UdS, 23 rue du Loess BP 28 - 67037 Strasbourg Cedex 2,
    France ,
    dominique.curien@ires.in2p3.fr

15. Miroslav **Danchev**,
U. Sofia, 5 James Bourchier Blvd., 1164 Sofia, Bulgaria,
miroslav@phys.uni-sofia.bg
16. Antoaneta **Damyanova**,
U. Sofia, 5 James Bourchier Blvd., 1164 Sofia, Bulgaria,
adamyanova@phys.uni-sofia.bg
17. Judit **Darai**,
U. Debrecen, Pf. 105, 4010 Debrecen, Hungary
darai@atomki.hu
18. Alick **Deacon**,
U. of Manchester, Oxford Road Manchester, M13 9PL, England
alick.deacon@manchester.ac.uk
19. Isabelle **Deloncle**,
CSNSM, Orsay Bâtiments 104 et 108, 91405 Orsay CAMPUS, France
isabelle.deloncle@csnsm.in2p3.fr
20. Arnoldas **Deltuva**,
U. Lisbon, P-1649-003 Lisbon, Portugal,
deltuva@cii.fc.ul.pt
21. Sevdalina **Dimitrova**,
INRNE-BAS, 72, Tzarigradsko chaussee, Blvd.
BG – 1784 , Sofia, Bulgaria,
sevdim@inrne.bas.bg
22. Alessia **Di Pietro**,
INFN, via S. Sofia 62, I-95125, Catania, Italy
Dipietro@lns.infn.it
23. Jerry **Draayer**,
DPA-LSU, Baton Rouge, LA 70803, USA
draayer@sura.org
24. Enrico **Fioretto**,
INFN- LNL, Viale dell'Università 2, 35020 Legnaro (PADOVA), Italy
fioretto@lnl.infn.it
25. Christoph **Fransen**,
U. Köln, Zülpicher Straße 77, 50937 Cologne, Germany,
fransen@ikp.uni-koeln.de
26. Matthias **Fritzsche**,
TU Darmstadt, D-64289 Darmstadt, Germany
fritzsche@ikp.tu-darmstadt.de
27. Mitko **Gaidarov,**
INRNE-BAS, 72, Tzarigradsko chaussee, Blvd.
BG – 1784 , Sofia, Bulgaria,
gaidarov@inrne.bas.bg
28. Huben **Ganev**,
INRNE-BAS, 72, Tzarigradsko chaussee, Blvd.
BG – 1784 , Sofia, Bulgaria,
huben@inrne.bas.bg
29. Vladimir **Garistov**,
INRNE-BAS, 72, Tzarigradsko chaussee, Blvd.
BG – 1784 , Sofia, Bulgaria,
garistov@inrne.bas.bg

30. Eduardo **Garrido**,
IEM-CSIC, Serrano 123, E-28006 Madrid, Spain,
e.garrido@iem.cfmac.csic.es
31. Ana **Georgieva**,
INRNE-BAS, 72, Tzarigradsko chaussee, Blvd.
BG – 1784 , Sofia, Bulgaria,
anageorg@inrne.bas.bg
32. Panagiotis **Georgoudis**,
N.C.S.R. Demokritos, 153 10 - Agia Paraskevi Athens P.O. 60228, Greece,
potoulis@gmail.com
33. Ivaylo **Ivanov**,
INRNE-BAS, 72, Tzarigradsko chaussee, Blvd.
BG – 1784 , Sofia, Bulgaria,
ivanov@inrne.bas.bg
34. Martin **Ivanov**,
INRNE-BAS, 72, Tzarigradsko chaussee, Blvd.
BG – 1784 , Sofia, Bulgaria,
martin@inrne.bas.bg
35. Natalia **Janeva**,
INRNE-BAS, 72, Tzarigradsko chaussee, Blvd.
BG – 1784 , Sofia, Bulgaria,
pripesho@inrne.bas.bg
36. Eric **Johnson**,
Florida State University, 32306 FL, USA
grogache@fsu.edu
37. Ari **Jokinen**,
U. Jyväskylä, PL 35, 40014 Jyväskylän yliopisto, Finland
jokinen@jyu.fi
38. Dimitar **Kadrev**,
INRNE-BAS, 72, Tzarigradsko chaussee, Blvd.
BG – 1784 , Sofia, Bulgaria,
kadrev@inrne.bas.bg
39. Doycho **Karagyozov**,
U. Sofia, 5 James Bourchier Blvd., 1164 Sofia, Bulgaria,
dkaragyozov@phys.uni-sofia.bg
40. Sofia **Karampagia**,
N.C.S.R. Demokritos, 153 10 - Agia Paraskevi Athens P.O. 60228, Greece,
skaram@central.ntua.gr
41. Yury **Kopach**,
JINR, Dubna, Joliot-Curie 6, 141980 Dubna, Moscow region, Russia,
kopatch@nf.jinr.ru
42. Amel **Korichi**,
CSNSM, Orsay Bâtiments 104 et 108, 91405 Orsay CAMPUS, France
amel.korichi@csnsm.in2p3.fr
43. Andreas **Krugmann**,
TU Darmstadt, D-64289 Darmstadt, Germany
krugmann@ikp.tu-darmstadt.de

44. Galina **Krumova**,
U. of Ruse, 8 Studentska Str., 7017 Ruse, Bulgaria,
gal@ru.acad.bg
45. Todorka **Kutsarova**,
INRNE-BAS, 72, Tzarigradsko chaussee, Blvd.
BG – 1784 , Sofia, Bulgaria,
dorak@inrne.bas.bg
46. Hristo **Laftchiev**,
INRNE-BAS, 72, Tzarigradsko chaussee, Blvd.
BG – 1784 , Sofia, Bulgaria,
laftchiev@gmail.com
47. Silvia **Leoni**,
University of Milano and INFN, Via Celoria 16, 20133 Milano, Italy
silvia.leoni@mi.infn.it
48. Gèsa **Lévai**,
MTA ATOMKI, H-4001 Debrecen, PO Box 51, Hungary,
levai@atomki.hu
49. Jean **Libert**,
IPN - Orsay, bat. 100, 91406 Orsay, France,
libert@ipno.in2p3.fr
50. Nicola **Lo Iudice**,
U. Napoli Federico II, Corso Umberto I, 80138 Napoli, Italy,
nicola.loiudice@na.infn.it
51. Jie **Meng**,
SPSKLNPT-Peking U., 100871 Beijing, China
mengj@pku.edu.cn
52. Zhanna **Mezentseva**,
JINR, Dubna, Joliot-Curie 6, 141980 Dubna, Moscow region, Russia,
mzv@nf.jinr.ru
53. Nikolay **Minkov**,
INRNE-BAS, 72, Tzarigradsko chaussee, Blvd.
BG – 1784 , Sofia, Bulgaria,
nminkov@inrne.bas.bg
54. Ani **Minkova**,
INRNE-BAS, 72, Tzarigradsko chaussee, Blvd.
BG – 1784 , Sofia, Bulgaria,
animin@phys.uni-sofia.bg
55. Avazbek **Nasirov**,
JINR, Dubna, Joliot-Curie 6, 141980 Dubna, Moscow region, Russia,
nasirov@jinr.ru
56. **Nguyen** Van Giai,
IPN - Orsay, bat. 100, 91406 Orsay, France,
nguyen@ipno.in2p3.fr
57. José Roberto **Oliveira**
Instituto de Fısica da Universidade de Sao Paulo, SP, Brazil
zero@if.usp.br
58. Pavel **Petkov**,
INRNE-BAS, 72, Tzarigradsko chaussee, Blvd.
BG – 1784 , Sofia, Bulgaria,
pp@inrne.bas.bg

59. Norbert **Pietralla**,
TU Darmstadt, D-64289 Darmstadt, Germany
pietralla@ikp.tu-darmstadt.de
60. Gabriela **Popa**,
Ohio U., USA
popag@ohio.edu
61. Georgi **Rainovski**,
U. Sofia, 5 James Bourchier Blvd., 1164 Sofia, Bulgaria,
rig@phys.uni-sofia.bg
62. Francesco **Recchia**,
INFN- Sezione di Padova, Padova, Italy,
recchia@pd.infn.it.
63. Peter **Ring**,
TU Munich, 85748 Garching, Germany,
ring@ph.tum.de
64. Ivan **Ruskov**,
INRNE-BAS, 72, Tzarigradsko chaussee, Blvd.
BG – 1784 , Sofia, Bulgaria,
ivan@inrne.bas.bg
65. Pedro **Sarriguren**,
IEM-CSIC, Serrano 123, E-28006 Madrid, Spain,
sarriguren@iem.cfmac.csic.es
66. Abdulrahman **Scheik Obeid**,
TU Darmstadt, D-64289 Darmstadt, Germany
scheikh@ikp.tu-darmstadt.de
67. Valentina **Semkova**,
INRNE-BAS, 72, Tzarigradsko chaussee, Blvd.
BG – 1784 , Sofia, Bulgaria,
valents@inrne.bas.bg
68. Valery **Shvetsov**,
JINR, Dubna, Joliot-Curie 6, 141980 Dubna, Moscow region, Russia,
shv@nf.jinr.ru
69. George **Souliotis**,
Cyclotron Institute, Texas A&M University, USA.
soulioti@comp.tamu.edu.
70. Paul **Stevenson**,
U. Surrey, Guildford, Surrey, GU2 7XH, UK
p.stevenson@surrey.ac.uk
71. Mario **Stoitsov**,
ORNL, USA
stoitsov@ornl.gov
72. Chavdar **Stoyanov**,
INRNE-BAS, 72, Tzarigradsko chaussee, Blvd.
BG – 1784 , Sofia, Bulgaria,
stoyanov@inrne.bas.bg
73. Suzana **Szilner,**
Ruder Boskovic Institute, HR-10 001 Zagreb, Croatia
szilner@irb.hr

74. Dimitar **Tarpanov**,
INRNE-BAS, 72, Tzarigradsko chaussee, Blvd.
BG – 1784 , Sofia, Bulgaria,
dimitert@yahoo.co.uk
75. Laurent **Tassan-Got**,
IPN - Orsay, 91406 Orsay, France,
tassango@ipno.in2p3.fr
76. Dimitar **Tonev**,
INRNE-BAS, 72, Tzarigradsko chaussee, Blvd.
BG – 1784 , Sofia, Bulgaria,
dimitar.tonev@lnl.infn.it
77. Rositsa **Topchiyska**,
U. Sofia, 5 James Bourchier Blvd., 1164 Sofia, Bulgaria,
rtopchiyska@phys.uni-sofia.bg
78. Aurora **Tumino**,
INFN, via S. Sofia 62, I-95125, Catania, Italy
tumino@lns.infn.it
79. Calin **Ur**,
INFN and U. Padova, Italy
calin.ur@pd.infn.it
80. Eric **Voutier**,
LPSC / IN2P3 / CNRS – UJF – INP, 53 rue des Martyrs, 38026
Grenoble cedex, France,
voutier@lpsc.in2p3.fr
81. Shaun **Wyngaardt**,
U. Stellenbosch, Stellenbosch, 7600, South Africa,
shaunmw@sun.ac.za
82. Volker **Werner**,
U. Yale P.O. Box 208120, New Haven, CT 06520, USA
volker.werner@yale.edu
83. Orlin **Yordanov**,
INRNE-BAS, 72, Tzarigradsko chaussee, Blvd.
BG – 1784 , Sofia, Bulgaria,
orliny@inrne.bas.bg
84. Petar **Zhivkov**,
INRNE-BAS, 72, Tzarigradsko chaussee, Blvd.
BG – 1784 , Sofia, Bulgaria,
petar.zhivkov@gmail.com

All papers published in this volume of *Journal of Physics: Conference Series* have been peer reviewed through processes administered by the proceedings Editors. Reviews were conducted by expert referees to the professional and scientific standards expected of a proceedings journal published by IOP Publishing

Collective motion in the frame of phase space moments (Nuclear Scissors)

E B Balbutsev [1]

[1]Joint Institute for Nuclear Research, 141980 Dubna, Moscow Region, Russia

E-mail: `balbuts@thsun1.jinr.ru`

Abstract. We consider the phase space moments (or Wigner Function Moments (WFM)) method, which is developed to describe the collective motion. The method is generalized to take into account pair correlations. Its connection with RPA and Green's function method is analyzed in the simple model, the Harmonic Oscillator with Quadrupole–Quadrupole (HO+QQ) interaction. Possibilities of WFM method are demonstrated on an example of the nuclear scissors mode.

Keywords: nuclear scissors, Wigner function moments.

1. Introduction

The method of phase space moments, or Wigner function moments method was suggested in the Bogoliubov Laboratory of Theoretical Physics about 30 years ago. It was successfully applied to study isoscalar and isovector giant multipole resonances and low-lying collective modes of rotating and nonrotating nuclei with various realistic forces [1]. The method is developed and modified continually. For example, the exact relation between the RPA and WFM variables and the respective dynamical equations is found. The very close connection of the WFM method with the Green's Function (GF) one is established [2]. Last time the method is generalized to take into account pair correlations [3].

All these points will be discussed on an example of simple model HO+QQ, which allows one to obtain the majority of results in the analytical form. The object of the demonstration will be the scissors mode - very interesting and rather curious collective excitation predicted

33 years ago. Its experimental discovery has initiated a cascade of theoretical studies. An excellent review was given by D. Zawischa [4]. Very briefly the situation can be described in the following way. All microscopic calculations with effective forces reproduce experimental data with respect to the position and the strength of the scissors mode. However, the situation is more obscure in regard to simple phenomenological models whose aim is to explain the physics of the phenomenon and to interpret it in the most simple and transparent terms. So it will be interesting to compare the possibilities, advantages and disadvantages of various methods in the description of all subtleties of this mode.

2. Scissors mode by WFM, RPA and Green's function methods

The basis of the method is the TDHF equation for the one-body density matrix $\rho^\tau(\mathbf{r}_1, \mathbf{r}_2, t)$:

$$i\hbar\frac{\partial\hat{\rho}^\tau}{\partial t} = \left[\hat{H}^\tau, \hat{\rho}^\tau\right], \tag{1}$$

where $\hat{H}^\tau$ is the one-body self-consistent mean field Hamiltonian and τ is an isotopic spin index. It is convenient to modify equation (1) introducing the Wigner transform of the density matrix

$$f^\tau(\mathbf{r}, \mathbf{p}, t) = \int d^3s\ \exp(-i\mathbf{p}\cdot\mathbf{s}/\hbar)\rho^\tau(\mathbf{r} + \frac{\mathbf{s}}{2}, \mathbf{r} - \frac{\mathbf{s}}{2}, t) \tag{2}$$

and of the Hamiltonian $H_W^\tau(\mathbf{r}, \mathbf{p})$. Then one arrives [5] at

$$\frac{\partial f^\tau}{\partial t} = \frac{2}{\hbar}\sin\left\{\frac{\hbar}{2}\left[(\nabla)^H\cdot(\nabla^p)^f - (\nabla^p)^H\cdot(\nabla)^f\right]\right\} H_W^\tau f^\tau, \tag{3}$$

where the upper index on the bracket stands for the function on which the operator in these brackets acts. It is shown in [1], that by integrating equation (3) over the phase space $\{\mathbf{p}, \mathbf{r}\}$ with the weights $x_{i_1}x_{i_2}\ldots x_{i_k}p_{i_{k+1}}\ldots p_{i_{n-1}}p_{i_n}$, where k runs from 0 to n, one can obtain a closed finite set of dynamical equations for Cartesian tensors of the rank n. Taking linear combinations of these equations one is able to represent them through irreducible tensors, which play the role of collective variables of the problem. However, it is more convenient to derive the dynamical equations directly for irreducible tensors using the technique of tensor products [6]. For this it is necessary to rewrite the Wigner function equation (3) in terms of cyclic variables

$$r_{+1} = -\frac{1}{\sqrt{2}}(x_1 + ix_2)\ , \quad r_0 = x_3\ , \quad r_{-1} = \frac{1}{\sqrt{2}}(x_1 - ix_2)$$

and take its moments with various tensor products of r_α and p_α.

The microscopic Hamiltonian of the model is

$$H = \sum_{i=1}^{A}(\frac{\hat{\mathbf{p}}_i^2}{2m} + \frac{1}{2}m\omega^2\mathbf{r}_i^2) + \bar{\kappa}\sum_{\mu=-2}^{2}(-1)^\mu\sum_i^Z\sum_j^N q_{2-\mu}(\mathbf{r}_i)q_{2\mu}(\mathbf{r}_j)$$

$$+\frac{1}{2}\kappa\sum_{\mu=-2}^{2}(-1)^{\mu}\{\sum_{i\neq j}^{Z}q_{2-\mu}(\mathbf{r}_i)q_{2\mu}(\mathbf{r}_j)+\sum_{i\neq j}^{N}q_{2-\mu}(\mathbf{r}_i)q_{2\mu}(\mathbf{r}_j)\}, \tag{4}$$

where $q_{2\mu}=\sqrt{16\pi/5}\,r^2Y_{2\mu}$ is a quadrupole operator and N, Z are the numbers of neutrons and protons. Substituting spherical functions by tensor products $r^2Y_{2\mu}=\sqrt{\frac{15}{8\pi}}r^2_{2\mu}$, where $r^2_{\lambda\mu}\equiv\{r\otimes r\}_{\lambda\mu}=\sum_{\sigma,\nu}C^{\lambda\mu}_{1\sigma,1\nu}r_\sigma r_\nu$ and $C^{\lambda\mu}_{1\sigma,1\nu}$ is the Clebsch-Gordan coefficient, one can write the mean field Hamiltonian as

$$\hat{H}^\tau=\frac{\hat{p}^2}{2m}+\frac{1}{2}m\,\omega^2r^2+6\sum_\mu(-1)^\mu Z^\tau_{2-\mu}r^2_{2\mu}, \tag{5}$$

where $Z^{\rm n}_{2\mu}=\kappa R^{\rm n}_{2\mu}+\bar{\kappa}R^{\rm p}_{2\mu}$, $Z^{\rm p}_{2\mu}=\kappa R^{\rm p}_{2\mu}+\bar{\kappa}R^{\rm n}_{2\mu}$, $R^\tau_{\lambda\mu}(t)=\int d\{\mathbf{p},\mathbf{r}\}r^2_{\lambda\mu}f^\tau(\mathbf{r},\mathbf{p},t)$ and $\int d\{\mathbf{p},\mathbf{r}\}\equiv 2(2\pi\hbar)^{-3}\int d^3p\int d^3r$. Integration of equation (3) with the weights $r^2_{\lambda\mu}$, $(rp)_{\lambda\mu}\equiv\{r\otimes p\}_{\lambda\mu}$ and $p^2_{\lambda\mu}$ yields the following set of equations [7]:

$$\frac{d}{dt}R^\tau_{\lambda\mu}-\frac{2}{m}L^\tau_{\lambda\mu}=0, \qquad \lambda=0,2,$$
$$\frac{d}{dt}L^\tau_{\lambda\mu}-\frac{1}{m}P^\tau_{\lambda\mu}+m\,\omega^2R^\tau_{\lambda\mu}-12\sqrt{5}\sum_{j=0}^{2}\sqrt{2j+1}\{^{11j}_{2\lambda1}\}(Z^\tau_2R^\tau_j)_{\lambda\mu}=0, \qquad \lambda=0,1,2,$$
$$\frac{d}{dt}P^\tau_{\lambda\mu}+2m\,\omega^2L^\tau_{\lambda\mu}-24\sqrt{5}\sum_{j=0}^{2}\sqrt{2j+1}\{^{11j}_{2\lambda1}\}(Z^\tau_2L^\tau_j)_{\lambda\mu}=0, \qquad \lambda=0,2, \tag{6}$$

where $\{^{11j}_{2\lambda1}\}$ is the $6j$-symbol and the following notation is introduced

$$P^\tau_{\lambda\mu}(t)=\int d\{\mathbf{p},\mathbf{r}\}p^2_{\lambda\mu}f^\tau(\mathbf{r},\mathbf{p},t),\quad L^\tau_{\lambda\mu}(t)=\int d\{\mathbf{p},\mathbf{r}\}(rp)_{\lambda\mu}f^\tau(\mathbf{r},\mathbf{p},t). \tag{7}$$

By definition $q_{2\mu}=\sqrt{6}r^2_{2\mu}$, the quadrupole moment $Q^\tau_{2\mu}=\sqrt{6}R^\tau_{2\mu}$, $R^\tau_{00}=-Q^\tau_{00}/\sqrt{3}$ with $Q^\tau_{00}=N_\tau<r^2>$ being the mean square radius of neutrons or protons. The tensor $L^\tau_{1\nu}$ is connected with angular momentum by the relations $L^\tau_{10}=\frac{i}{\sqrt{2}}I^\tau_3$, $L^\tau_{1\pm1}=\frac{1}{2}(I^\tau_2\mp iI^\tau_1)$.

We rewrite equations (6) in terms of isoscalar and isovector variables $R_{\lambda\mu}=R^{\rm n}_{\lambda\mu}+R^{\rm p}_{\lambda\mu}$, $\bar{R}_{\lambda\mu}=R^{\rm n}_{\lambda\mu}-R^{\rm p}_{\lambda\mu}$ (and so on) with the isoscalar $\kappa_0=(\kappa+\bar{\kappa})/2$ and isovector $\kappa_1=(\kappa-\bar{\kappa})/2$ strength constants. Equations will be simplified to get the results in analytical form. Writing all variables as a sum of their equilibrium value plus a small deviation $R_{\lambda\mu}(t)=R^{eq}_{\lambda\mu}+\mathcal{R}_{\lambda\mu}(t)$, $P_{\lambda\mu}(t)=P^{eq}_{\lambda\mu}+\mathcal{P}_{\lambda\mu}(t)$, $L_{\lambda\mu}(t)=L^{eq}_{\lambda\mu}+\mathcal{L}_{\lambda\mu}(t)$, $\bar{R}_{\lambda\mu}(t)=\bar{R}^{eq}_{\lambda\mu}+\bar{\mathcal{R}}_{\lambda\mu}(t)$, $\bar{P}_{\lambda\mu}(t)=\bar{P}^{eq}_{\lambda\mu}+\bar{\mathcal{P}}_{\lambda\mu}(t)$, $\bar{L}_{\lambda\mu}(t)=\bar{L}^{eq}_{\lambda\mu}+\bar{\mathcal{L}}_{\lambda\mu}(t)$, we linearize the equations of motion in $\mathcal{R}_{\lambda\mu}$, $\mathcal{P}_{\lambda\mu}$, $\mathcal{L}_{\lambda\mu}$ and $\bar{\mathcal{R}}_{\lambda\mu}$, $\bar{\mathcal{P}}_{\lambda\mu}$, $\bar{\mathcal{L}}_{\lambda\mu}$ (small-amplitude approximation). Further, the equilibrium deformation and mean square radius of neutrons are supposed to be equal to that of protons, i.e. $\bar{R}^{eq}_{20}=\bar{R}^{eq}_{00}=0$. Due to this approximation the equations for isoscalar and isovector systems are decoupled, every set of equations being split into five independent subsets with $\mu=0,\pm1,\pm2$.

The scissors mode is described by the isovector subset with $\mu = 1$ Supposing, as usual, $\kappa_1 = \alpha\kappa_0$, and taking for κ_0 the self-consistent value $\kappa_0 = -\frac{m\bar{\omega}^2}{4Q_{00}}$, where $\bar{\omega}^2 = \frac{\omega^2}{1+\frac{2}{3}\delta}$ [7] with the standard definition of the deformation parameter $Q_{20} = (4/3)Q_{00}\delta$, we find the characteristic equation whose solutions are

$$\Omega_{\pm}^2 = \bar{\omega}^2(2-\alpha)(1+\delta/3) \pm \sqrt{\bar{\omega}^4(2-\alpha)^2(1+\delta/3)^2 - 4\bar{\omega}^4(1-\alpha)\delta^2}. \quad (8)$$

The high-lying solution Ω_+ gives the frequency Ω_{iv} of the $\mu = 1$ branch of the IsoVector GQR (IVGQR). The low-lying solution Ω_- gives the frequency Ω_{sc} of the scissors mode.

A direct way of calculating the reduced transition probabilities is provided by the theory of the linear response of a system to a weak external field

$$\hat{W}(t) = \hat{W}\exp(-i\Omega t) + \hat{W}^\dagger \exp(i\Omega t). \quad (9)$$

For magnetic excitations $\hat{W} = \hat{W}_{1\mu}^{\mathrm{p}} = \sum_{s=1}^{Z} \hat{w}_{1\mu}(s)$, $\hat{w}_{1\mu} = \sqrt{\frac{3}{2\pi}}\frac{\mu_N}{i\hbar}(r\hat{p})_{1\mu}$, $\mu_N = \frac{e\hbar}{2mc}$ and

$$B(M1)_\nu = \frac{1-\alpha}{8\pi}\frac{m\bar{\omega}^2}{\hbar}Q_{00}\delta^2 \frac{\Omega_\nu^2 - 2(1+\delta/3)\bar{\omega}^2}{\Omega_\nu[\Omega_\nu^2 - \bar{\omega}^2(2-\alpha)(1+\delta/3)]}\mu_N^2. \quad (10)$$

For electric excitations $\hat{W} = \hat{W}_{2\mu}^{\mathrm{p}} = \sum_{s=1}^{Z} \hat{w}_{2\mu}(s)$, $\hat{w}_{2\mu} = e\,r^2 Y_{2\mu} = e\sqrt{\frac{15}{8\pi}}r_{2\mu}^2$ and

$$B(E2)_\nu = \frac{e^2\hbar}{m}\frac{5}{16\pi}Q_{00}\frac{(1+\delta/3)\Omega_\nu^2 - 2(\bar{\omega}\delta)^2}{\Omega_\nu[\Omega_\nu^2 - \bar{\omega}^2(2-\alpha)(1+\delta/3)]}. \quad (11)$$

The **RPA approach** gives exactly the same results for energy and transition probabilities, therefore one can suspect very close relation between WFM and RPA. Really, this relation can be established by considering the linear response of the system to a weak external time-dependent field (9). In this case WFM collective variables can be written as linear combinations of corresponding Transition Matrix Elements (TME) [2], for example,

$$\delta R_{\lambda\mu}^\tau(t) \equiv \mathcal{R}_{\lambda\mu}^\tau(t) = \sum_\nu (<0|\hat{R}_{\lambda\mu}^\tau|\nu> c_\nu - <\nu|\hat{R}_{\lambda\mu}^\tau|0> \bar{c}_\nu)e^{-i\Omega t} + c.c. \quad (12)$$

where $\hat{R}_{\lambda\mu} = \sum_{s=1}^{A}\{r_s \otimes r_s\}_{\lambda\mu}$. Inserting such expressions for $\delta R_{\lambda\mu}$, $\delta L_{\lambda\mu}$ and $\delta P_{\lambda\mu}$ into (6) we can derive equations, which can be called as dynamical equations for corresponding transition matrix elements. Exactly the same dynamical equations for TME can be derived in RPA by combining RPA equations [5] in accordance with the definition of matrix elements:

$$<0|\hat{F}^\tau|\nu> = \sum_{mi}(f_{im}^\tau X_{mi}^{\tau,\nu} + f_{mi}^\tau Y_{mi}^{\tau,\nu}).$$

So, there exists one-to-one correspondence between the set of dynamical equations for WFM variables and the set of dynamical equations for transition matrix elements. On the other hand the TME equations are just linear combinations of the RPA equations. Therefore we can conclude that RPA and WFM approaches are equivalent in all aspects.

Solving the Vlasov equation (which is obtained by conserving on the right hand side of equation (3) only the first term of the sin-function expansion) by the **Green's function method** with the mean field (5) one derives [8, 2] the integral equation for the Wigner function. Taking the proper phase space moments of this equation one reproduces the (linearized) dynamical equations (6), that indicates on the very close relationship of two methods.

3. Pairing

Pair correlations can be taken into account by working with time dependent HFB equations. The detailed form of the TDHFB equations is

$$\begin{aligned} i\hbar\dot{\hat\rho} &= \hat h\hat\rho - \hat\rho\hat h - \hat\Delta\hat\kappa^\dagger + \hat\kappa\hat\Delta^\dagger, & -i\hbar\dot{\hat\kappa} &= -\hat h\hat\kappa - \hat\kappa\hat h^* + \hat\Delta - \hat\Delta\hat\rho^* - \hat\rho\hat\Delta, \\ -i\hbar\dot{\hat\rho}^* &= \hat h^*\hat\rho^* - \hat\rho^*\hat h^* - \hat\Delta^\dagger\hat\kappa + \hat\kappa^\dagger\hat\Delta, & -i\hbar\dot{\hat\kappa}^\dagger &= \hat h^*\hat\kappa^\dagger + \hat\kappa^\dagger\hat h - \hat\Delta^\dagger + \hat\Delta^\dagger\hat\rho + \hat\rho^*\hat\Delta^\dagger, \end{aligned} \quad (13)$$

where $\hat\rho$ and $\hat\kappa$ are normal and abnormal density matrices respectively, $\hat\Delta$ is the pairing gap. We work with the Wigner transformation [5] of these equations. For example, the first one reads:

$$\begin{aligned} i\hbar\dot f &= i\hbar\{h, f\} - \Delta\kappa^* + \kappa\Delta^* - \frac{i\hbar}{2}\{\Delta, \kappa^*\} + \frac{i\hbar}{2}\{\kappa, \Delta^*\} \\ &\quad - \frac{\hbar^2}{8}[\{\{\kappa, \Delta^*\}\} - \{\{\Delta, \kappa^*\}\}] + ..., \end{aligned} \quad (14)$$

where functions h, f, Δ, and κ are the Wigner transforms of $\hat h$, $\hat\rho$, $\hat\Delta$, and $\hat\kappa$, respectively, $\{f, g\}$ is the Poisson bracket of functions $f(\mathbf{r}, \mathbf{p})$ and $g(\mathbf{r}, \mathbf{p})$; the dots stand for terms proportional to higher powers of $\hbar$.

To study the quadrupole collective motion with $K^\pi = 1^+$ in axially symmetric nuclei, it is necessary to calculate moments of Eq. (14) (+ three other equations) with the weight functions

$$W = \quad xz, \quad p_xp_z, \quad zp_x + xp_z \equiv \hat L, \quad \text{and} \quad zp_x - xp_z \equiv \hat I_y.$$

This procedure yields 16 equation for collective variables $\int d(\mathbf{p}, \mathbf{r})Wf$ and $\int d(\mathbf{p}, \mathbf{r})W\kappa$. However, due to symmetry considerations 8 of them turn out trivial. Applying the approximation $\delta\kappa_+(\mathbf{r}, \mathbf{p}) \ll \delta\kappa_-(\mathbf{r}, \mathbf{p})$ one is able to reduce the problem to a set of only six dynamical equations (strictly speaking 12 ones: 6 for protons and 6 for neutrons). Making the standard approximation to decouple isovector and isoscalar subsets, we find that the isovector subset has two integrals of motion allowing one to reduce the eigenvalue problem to a quadratic equation. Its two solutions

$$E_\pm^2 = \mathcal{D}_\omega \pm \sqrt{\mathcal{D}_\omega^2 - [8\Delta\tilde{\Delta}\epsilon^2 + 4(\hbar\bar{\omega})^4\delta^2](1-\alpha) + 24\alpha\kappa_0\tilde{\Delta}k_0\hbar^4/m^2} \tag{15}$$

describe the energy E_+ of the IVGQR and the energy E_- of the scissors mode. Here $\epsilon^2 = \hbar^2\bar{\omega}^2(1+\frac{\delta}{3})$, $k_0 = 2\int d(\mathbf{p},\mathbf{r})\kappa^0(\mathbf{r},\mathbf{p})$, $\mathcal{D}_\omega = 2\Delta\tilde{\Delta} + \epsilon^2(2-\alpha)$, $2\tilde{\Delta} = |V_0|I_{pp}^{\kappa\Delta}$,

$$I_{pp}^{\kappa\Delta}(\mathbf{r},p) = \frac{r_p^3}{\sqrt{\pi}\hbar^3}e^{-\gamma p^2}\int \kappa^0(\mathbf{r},p')\left[\phi_0(2\gamma pp') - 4\gamma^2 p'^4\phi_2(2\gamma pp')\right]e^{-\gamma p'^2}p'^2 dp', \tag{16}$$

$$\gamma = r_p^2/4\hbar^2, \quad \phi_0(x) = \frac{1}{x}sh(x), \quad \phi_2(x) = \frac{1}{x^3}[(1+\frac{3}{x^2})sh(x) - \frac{3}{x}ch(x)],$$

r_p and V_0 are parameters of the pair interaction $v(|\mathbf{p}-\mathbf{p}'|) = -|V_0|(r_p\sqrt{\pi})^3 e^{\gamma|\mathbf{p}-\mathbf{p}'|^2}$. It is worth noting that contrary to the case without pairing [7] the energy E_- does not go to zero for deformation $\delta = 0$. The calculation of transition probabilities shows that this mode of a spherical nucleus can be excited by an electric field and it is not excited by a magnetic field.

The transition probabilities are calculated with the help of linear response theory:

$$B(E2)_\nu = 2|<\nu|\hat{F}_{21}^{\mathrm{p}}|0>|^2 = \frac{e^2\hbar^2}{m}\frac{5}{4\pi}Q_{00}^{\mathrm{p}}\frac{(1+\delta/3)(E_\nu^2 - 4\Delta\tilde{\Delta}) - 2(\hbar\bar{\omega}\delta)^2}{E_\nu[E_\nu^2 - \mathcal{D}_\omega]}, \tag{17}$$

$$B(M1)_\nu = 2|<\nu|\hat{F}_{11}^{\mathrm{p}}|0>|^2 = \frac{m\bar{\omega}^2}{4\pi}(1-\alpha)Q_{00}^{\mathrm{p}}\delta^2\frac{E_\nu^2 - 4\Delta\tilde{\Delta} - 2\epsilon^2}{E_\nu[E_\nu^2 - \mathcal{D}_\omega]}\mu_N^2. \tag{18}$$

It is necessary to notice that the transition probability, as given by formula (18), has the experimentally observed quadratic deformation dependence.

In the **isoscalar** case we have two solutions

$$E_\pm^2 = 2\Delta\tilde{\Delta} + \epsilon^2 \pm \sqrt{(2\Delta\tilde{\Delta} + \epsilon^2)^2 + 24\kappa_0\tilde{\Delta}k_0\hbar^4/m^2}, \tag{19}$$

$E_+ \equiv E_{is}$ being the energy of the isoscalar giant quadrupole resonance and $E_- \equiv E_{ISL}$ being the energy of the IsoScalar Low-Lying Excitation (ISLLE). It is important to note that the very existence of the ISLLE relies on two factors: 1) pair correlations and 2) quantum correction.

Results of our calculations [3] for most of the nuclei where the scissors mode has been observed are presented in Figure 1. As it is seen, the inclusion of pairing considerably improves the agreement between the theory and experiment.

4. Conclusion

The connection of the WFM method with the GF one and RPA is discussed. It is shown that WFM and GF methods are very close to one another. Contrary to the RPA, both work in phase space with no need to introduce a single particle basis and yield identical sets of dynamical equations for the moments. To show the analytical equivalence between the WFM and RPA methods one needs to introduce the dynamical equations for the transition matrix elements. They can be derived either from the RPA equations for the amplitudes X_{kq}, Y_{kq} or from the WFM dynamical equations for the moments.

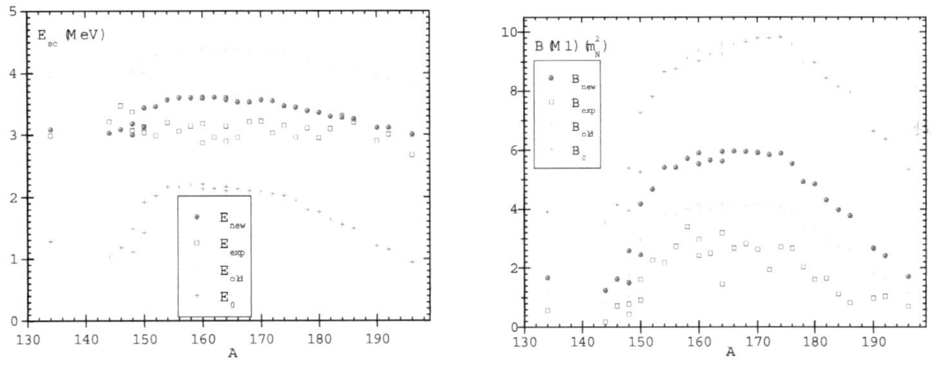

Figure 1. Energies and transition probabilities of the scissors mode as a function of the mass number A. E, B_{new}: continuity equation is fulfilled, E, B_{old}: continuity equation is violated, E, B_0: theory without pairing.

The analytical formulae reproduce very well the experimentally observed deformation dependence of the energy and $B(M1)$ factor of scissors. The results of calculations are in reasonable agreement with experimental data, pair correlations being extremely important. The joint action of pairing and quantum corrections leads to the appearance of a low-lying isoscalar excitation.

The study of low-lying modes and scissors on the basis of normal and anomalous densities, calculated within a microscopic approach including spin degrees of freedom, will be the subject of the future work.

References

[1] Balbutsev E B 1991 *Sov. J. Part. Nucl.* **22** 159

[2] Balbutsev E B and Schuck P 2007 *Annals of Physics* **322** 489

Journal of Physics: Conference Series **205** (2010) 012001 doi:10.1088/1742-6596/205/1/012001

[3] Balbutsev E B, Malov L A, Schuck P and Urban M 2009 *Physics of Atomic Nuclei* **72** 1305

[4] Zawischa D 1998 *J. Phys.* G **24** 683

[5] Ring P and Schuck P 1980 *The Nuclear Many-Body Problem* (Berlin: Springer–Verlag)

[6] Varshalovitch D A, Moskalev A N and Khersonski V K 1975 *Quantum Theory of Angular Momentum* (Leningrad: Nauka)

[7] Balbutsev E B and Schuck P 2003 *Nucl. Phys.* A **720** 293
Balbutsev E B and Schuck P 2003 *Nucl. Phys.* A **728** 471

[8] Kohl H, Schuck P and Stringari S 1986 *Nucl. Phys.* A **459** 256

An upgraded version of an importance sampling algorithm for large scale shell model calculations

Bianco D[1], Andreozzi F[1], Lo Iudice N[1], Porrino A[1], Dimitrova S[2]

[1]Universitá di Napoli Federico II, Dipartimento Scienze Fisiche, Monte S. Angelo, via Cintia, 80126 Napoli, Italy

[2]Institute of Nuclear Research and Nuclear Energy, Sofia, Bulgaria

E-mail: loiudice@na.infn.it

Abstract. An importance sampling iterative algorithm, developed few years ago, for generating exact eigensolutions of large matrices is upgraded so as to allow large scale shell model calculations in the uncoupled m-scheme. By exploiting the sparsity properties of the Hamiltonian matrix and projecting out effectively the good angular momentum, the new importance sampling allows to reduce drastically the sizes of the matrices while keeping full control of the accuracy of the eigensolutions. Illustrative numerical examples are presented.

1. Introduction

Even after the advent of powerful computing facilities, standard diagonalization methods are not suitable for large scale shell model(SM) calculations. They require in fact N^3 operations, N denoting the dimensions of the matrix [1]. On the other hand, if one is interested in few nuclear eigenstates, as it is often the case, one may resort to alternative methods.

Most of them are based on the Lanczos [2] algorithm and are especially efficient for determining the low lying eigenvalues [3]. Its numerical implementation, however, deserves special care. In fact, the generated states are mathematically but not numerically orthogonal. The error so induced propagates rapidly with the number of iterations and may yield "ghost" or spurious states, like the multiple appearance of the same state or the intrusion of states of unwanted angular momenta. Several codes based on Lanczos are now available. Among them, widely adopted is *Antoine* [4], which takes full advantage of the sparse nature of the nuclear Hamiltonian when used in the $m-$scheme.

An alternative approach, that circumvents the direct diagonalization of H, is the shell model Monte Carlo (SMMC) method [5, 6, 7]. This has to deal with the famous *sign problem* which has not found yet a clear-cut solution. Even with this limitation, SMMC has proved to be a powerful and efficient tool for studying ground-state and thermal properties of medium-mass nuclei as well as electroweak nuclear properties. SMMC evaluates expectation values and strength functions, but does not give explicit eigenvectors.

The quantum Monte Carlo diagonalization (QMCD) method combines the two approaches [8, 9]. It is basically an importance sampling method which generates stochastically by a MC technique a truncated basis for the direct diagonalization of the many-body Hamiltonian. On the other hand, the basis states so generated are not orthogonal and form in general a redundant set. Moreover, they do not have the spin as good quantum number. Specific procedures have been developed to obviate these shortcomings. QMCD has been adopted extensively and with success for systematics in the (s, d) and (f, p) nuclear regions.

An approximate method, based on the density matrix renormalization group (DMRG) [10], was adopted to achieve a drastic truncation of the shell model space in nuclear systems [11]. The method came out to work more effectively in the coupled scheme and, in this form, was used to compute the low energy spectra of nuclei in the (f, p) shell, like ^{48}Cr [12, 13].

Few years ago, we developed a method [14, 15] that faces directly the diagonalization of the Hamiltonian. It is an iterative algorithm which generates a selected set of eigenvectors of a large matrix and is extremely simple to be implemented. Is is also endowed with an importance sampling that yields a drastic reduction of the space and allows to extrapolate to asymptotic values.

The method was implemented in the coupled j-scheme only, which is too time consuming in the construction of the Hamiltonian matrix. On the other hand, in the m-scheme one has to deal with matrices whose sizes become prohibitively large as the number of shells increases. In order to circumvent this problem, we have modified the sampling procedure [16] according to some prescriptions, inherent the method, that allow a drastic cut of the dimensions of the Hamiltonian matrix. How this is done will be illustrated shortly here.

2. The algorithm

Let us consider a symmetric matrix $A = \{a_{ij}\}$ representing a self-adjoint operator $\hat{A}$ in an orthonormal basis $\{| 1\rangle, | 2\rangle, \ldots, |i >, \ldots, | N\rangle\}$. Let us suppose, for simplicity, that the basis states are ordered according to some criterion, for instance increasing unperturbed energies, when the operator $\hat{A}$ is the nuclear shell model Hamiltonian.

The algorithm consists of several iteration loops. The first loop goes through the following steps:

1a) Consider the lowest n_0 basis states $|i >$ $(n_0 << N)$, construct and diagonalize the $n_0 \times n_0$ submatrix $A_0 = (a_{ij})$,

1b) select the lowest v eigenvalues $\lambda_1^{(0)}, \ldots, \lambda_k^{(0)}, \ldots, \lambda_v^{(0)}$ and the corresponding eigenvectors

$$| \varphi_k^{(0)}\rangle = \sum_{i=1}^{n_0} c_i^{(0)} \mid i\rangle\rangle, \tag{1}$$

1c) consider now the subspace spanned by the v eigenvectors $| \varphi_k^{(0)}\rangle$ plus the basis states $|i >$ $(i = n_0 + 1, n_0')$, and construct the new submatrix

$$A_1 = \begin{pmatrix} \Lambda_0 & B_0 \\ B_0^{(T)} & A_0^c \end{pmatrix}, \tag{2}$$

where Λ_0 is the v-dimensional diagonal matrix, composed of the eigenvalues $\lambda_k^{(0)}$ $(k = 1, v)$, $A_0^c = \{a_{ij}\}$ $(i, j = n_0 + 1, n_0')$ is a submatrix of dimensions $n_1 = n_0' - n_0$, B_0 and its transpose are off-diagonal blocks composed of the matrix elements $b_{kj}^{(0)} = \langle\varphi_k^{(0)} \mid \hat{A} \mid j\rangle$ $(k = 1, v), (j = n_0 + 1, n_0')$,

1d) diagonalize A_1 and extract the new lowest v eigenvalues $\lambda_k^{(1)}$ $(k = 1, v)$ and the corresponding eigenvectors $| \varphi_k^{(1)}\rangle$,

1e) consider now the new subspace spanned by the new lowest v eigenvectors $| \varphi_k^{(1)}\rangle$ plus the basis states $|i >$ $(i = n_0' + 1, n_0'')$, construct the new submatrix A_2 just as done in point 1c) for A_1 and, after its diagonalization, extract the new lowest v eigenvalues $\lambda_k^{(2)}$ $(k = 1, v)$ and the corresponding eigenvectors $| \varphi_k^{(2)}\rangle$.

The just outlined procedure is iterated until the full basis is exhausted. This zero approximation loop yields the approximate v eigenvalues and eigenvectors

$$E_k^{(1)} \equiv \lambda_k^{(N)}, \qquad | \psi_k^{(1)}\rangle \equiv | \varphi_k^{(N)}\rangle = \sum_{i=1}^{N} c_k^{(N)}(i) \mid i\rangle. \tag{3}$$

These eigensolutions are the new entries for a new iteration. More specifically, we consider the basis composed of the eigenvectors $\psi_1^{(k)}$ plus the original basis states $\{| j\rangle\}$. Since the vectors $\psi_1^{(k)}$ are linear combinations of the $| j\rangle$ states, this new basis is no longer orthonormal and may be even redundant. We have therefore to solve an eigenvalue problem of general form and resort to the Choleski decomposition method. With this modifications, the subsequent iteration loops proceed as the first one and generate a sequence of v vectors $\psi_1^{(k)}, \ldots, \psi_v^{(k)}$ that converge to the exact eigensolutions [14]

3. Importance sampling

The outlined algorithm, though of simple implementation, requires the storage of at least one eigenvector. Since for many complex systems the dimensions of the Hamiltonian matrix become prohibitively large, one must rely on some importance sampling that allows to truncate the space by selecting only the basis states relevant to the exact eigensolutions. A notable example is the stochastic diagonalization method [17], which samples the basis states relevant to the ground state through a combination of plane (Jacobi) rotations and matrix inflation.

A similar sampling procedure can be implemented in the framework of our diagonalization process. Exploiting the fact that the algorithm yields quite accurate solutions already in the first approximation loop, we have devised a sampling which makes use of the first loop only, accordingly modified:

1a) Turn the v-dimensional principal submatrix $\{a_{ij}\}\,(i, j = 1, v)$ into the diagonal form Λ_v with eigenvalues $\lambda_1, \lambda_2, \ldots, \lambda_v$.

1b) For $j = v + 1, \ldots, N$, diagonalize the $v + 1$-dimensional matrix

$$A = \begin{pmatrix} \Lambda_v & \vec{b}_j \\ \vec{b}_j^T & a_{jj} \end{pmatrix}, \tag{4}$$

where $\vec{b}_j = \{b_{1j}, b_{2j}, \cdot, b_{vj}\}$.

1c) Select the lowest v eigenvalues $\lambda'_i, (i = 1, v)$ and accept the new state only if

$$\sum_{i=1,v} | \lambda'_i - \lambda_i | > \epsilon \tag{5}$$

Otherwise restart from point 1b) with a new j. The outcome of this procedure is that the selected states span a $n_s(< N)$ dimensional space, so that the subsequent refinement loops iterate only on the smaller set of n_s basis vectors. The number of operations is thereby reduced by a factor n_s/N.

4. Importance sampling algorithm (ISA) in the m-scheme

The model space under consideration is spanned by the states $|j >= |\alpha_1, \alpha_2, \ldots, \alpha_v, M >$, where $M = m_1 + \ldots + m_i + \ldots + m_v$ is the total magnetic quantum number of a system of v valence nucleons and $\alpha_i = \{a_i m_i\}$ $(a_i = \{n_i l_i j_i\})$ denote the single particle quantum numbers.

Let us consider the modified Hamiltonian

$$H_J = H + c[\mathbf{J}^2 - J(J + 1)]^2, \tag{6}$$

where H is a two-body Hamiltonian of general form, $\mathbf{J}$ the total spin operator, and c a positive constant.

We select a set of configurations $\{a_1^{n_1}, a_2^{n_2}, \dots a_i^{n_i} \dots a_v^{n_v}\}$, where n_i is the occupation number of the s.p. configurations a_i. They define a subspace M_0. Under the action of $\mathbf{J}^2$, the states $|j, N>_0 = |\alpha_1^{n_1}, \alpha_2^{n_2}, \dots \alpha_i^{n_i} \dots \alpha_v^{n_v}>$, belonging to a given configuration, transform among themselves in virtue of the commutation relations $[\mathbf{J}, j_i^2] = 0$.

For fixed J, we diagonalize the Hamiltonian matrix H_J in M_0 by using, for instance, the algorithm outlined in Sect. 2. If the constant c is chosen so as to push up in energy the eigenvalues with total spin $J' \neq J$, the diagonalization algorithm generates v lowest eigenvalues $\lambda_1^{(0)}, \dots, \lambda_v^{(0)}$ and eigenvectors $\varphi_1^{(0)}, \dots, \varphi_v^{(0)}$. These have a good J in virtue of the rotational invariant Hamiltonian and the mentioned invariance of each configuration $\{a_1^{n_1}, a_2^{n_2}, \dots a_i^{n_i} \dots a_v^{n_v}\}$ with respect to $\mathbf{J}^2$.

The eigenvectors $\varphi_1^{(0)}, \dots, \varphi_v^{(0)}$ are coupled by H only to those basis states of the complementary subspace $M_0^{(C)}$ that differ from the ones in M_0 by at most two single particle states or, in the second quantization language, two particle-two hole $(2p - 2h)$ states. Let us denote them by $|j>_2$. We now consider the subspace spanned by the eigenvectors $\varphi_1^{(0)}, \dots, \varphi_v^{(0)}$ plus the basis states $|j>_2$ and adopt the algorithm to diagonalize H_J in this subspace. A new set of lowest eigenvalues $\lambda_1^{(1)}, \dots, \lambda_v^{(1)}$ and eigenvectors $\varphi_1^{(1)}, \dots, \varphi_v^{(1)}$ is generated.

These new eigenvectors are coupled only to the basis states, to be denoted by $|j>_4$, differing from the original $|j>_0$ by at most $4p - 4h$. The diagonalization of H_J in the new subspace, spanned by $\varphi_1^{(1)}, \dots, \varphi_v^{(1)}$ plus $\{|j>_4\}$, generates a new set of v lowest eigensolutions of good J.

By iterating the procedure just outlined, we cover eventually the full space and the set of v eigensolutions tend to the exact ones.

4.1. Upgraded importance sampling

Even after having exploited the sparsity of the Hamiltonian matrix, the diagonalization procedure might become prohibitively lengthy as the dimensions of the space increase. It is therefore useful, if not necessary, to resort to the sampling procedure. This has been simplified as follows. Let us fix a sequence of positive small numbers ϵ_i of decreasing values $\epsilon_0 > \epsilon_1 > \dots > \epsilon_i > \dots > \epsilon_n$. Once the diagonalization of H_J in the initial subspace M_0 yields the lowest v eigenvalues $\lambda_1^{(0)}, \dots, \lambda_v^{(0)}$ with the corresponding eigenvectors $\varphi_1^{(0)}, \dots, \varphi_v^{(0)}$, we pick only the $(2p, 2h)$ basis states $|j>$ that fulfill the condition

$$\frac{b_0^2(kj)}{a_{jj} - \lambda(0)_k} > \epsilon_0 \tag{7}$$

where $k = 1, v$ and a_{jj} $(j = n_0 + 1, N,)$ is the diagonal matrix element corresponding to the state $|j>$ coupled to $\varphi_k^{(0)}$ through the matrix element $b_0(kj) = <j|H_J|\varphi_k^{(0)}>$.

We then adopt the diagonalization algorithm to diagonalize H_J in the subspace spanned by the v states $\{\varphi_k^{(0)}\}$ plus the basis states $\{|j>\}$ selected according to the sampling condition (7). New eigenvalues $\lambda_1^{(1)}, \dots, \lambda_v^{(1)}$ and eigenvectors $\varphi_1^{(1)}, \dots, \varphi_v^{(1)}$ are generated. These eigenfunctions are now linear combinations of the original $(0p-0h)$ states of the subspace M_0 plus the $(2p-2h)$ states selected according to the sampling condition (7). We now explore again the subspace complementary to M_0 and select all the states $|j>$ that fulfill the updated condition

$$\frac{b_1^2(kj)}{a_{jj} - \lambda_k^{(1)}} > \epsilon_1 \tag{8}$$

where $k = 1, v$ and $b_1(kj) = <j|H_J|\varphi_k^{(1)}>$ $(j = n_0 + 1, N)$.

The diagonalization algorithm applied to the new subspace spanned by the v states $\{\varphi_k^{(1)}\}$ plus the set of states $\{|j>\}$ yields new eigensolutions $\lambda_1^{(2)}, \ldots, \lambda_v^{(2)}$ and eigenvectors $\varphi_1^{(2)}, \ldots, \varphi_v^{(2)}$. These updated eigenvectors include $2p-2h$ states not admitted in the first run (since $\epsilon_0 > \epsilon_1$) plus $(4p-4h)$ states. We now iterate the sampling procedure with decreasing values of ϵ_i, obtaining more and more accurate eigensolutions, until we exhaust the full basis. For $\epsilon_n \to 0$, the sampling yields the exact eigensolutions.

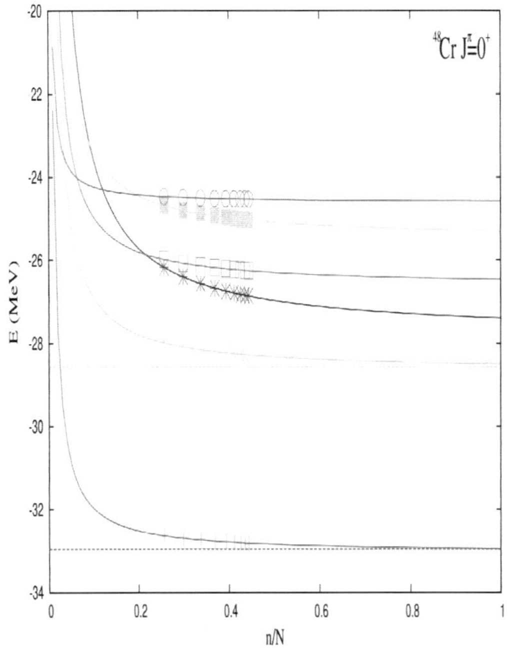

Figure 1. Convergence properties of the lowest $J^\pi = 0^+$ eigenvalues in ^{48}Cr

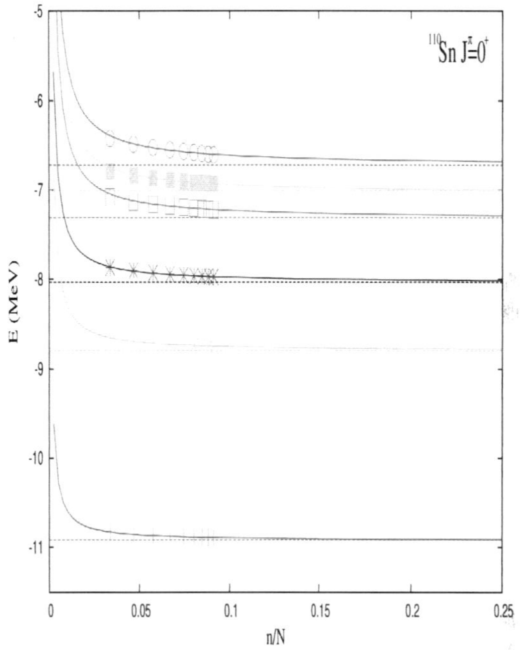

Figure 2. Convergence properties of the lowest $J^\pi = 0^+$ eigenvalues in ^{110}Sn

5. Numerical implementation

We applied the sampling algorithm to the semi-magic ^{110}Sn and the N=Z even-even ^{48}Cr. The model spaces are:

1) $P \equiv \{2d5/2, 1g7/2, 2d3/2, 3s1/2, 1h11/2\}$ for the valence neutrons of ^{110}Sn,
2)$P \equiv \{1f7/2, 1f5/2, 2p3/2, 2p1/2\}$ for the 4 valence protons and neutrons of ^{48}Cr.

For ^{110}Sn we used the single particle energies (in MeV) [20] $\epsilon_{2d5/2} = 0$, $\epsilon_{1g7/2} = 0.08$, $\epsilon_{3s1/2} = 2.45$, $\epsilon_{2d3/2} = 2.55$, $\epsilon_{1h11/2} = 3.20$. As for ^{48}Cr, we used the same energies adopted in Ref.[19], namely $\epsilon_{1f7/2} = 0$, $\epsilon_{2p3/2} = 2.0$, $\epsilon_{2p1/2} = 4.0$, $\epsilon_{1f5/2} = 6.5$.

We adopted a realistic effective interaction deduced from the CD-Bonn potential [18] for ^{110}Sn, and used the KB3 interaction [19] for ^{48}Cr.

The first task consisted in choosing the initial subspace M_0. Its dimensions n_0 increase with the number of eigenstates of good J we intend to generate. To yield six eigenstates, as in our case, the space dimensions came out to be of the order $n_0 \sim 100$. Having chosen M_0, we applied the iterative sampling procedure (7) for a given ϵ. At each iteration, the sampling condition (7) picks up new states that differ from the $0p-0h$ states of the original subspace M_0 by an increasing number of $p-h$ states. We applied the sampling procedure with decreasing values of

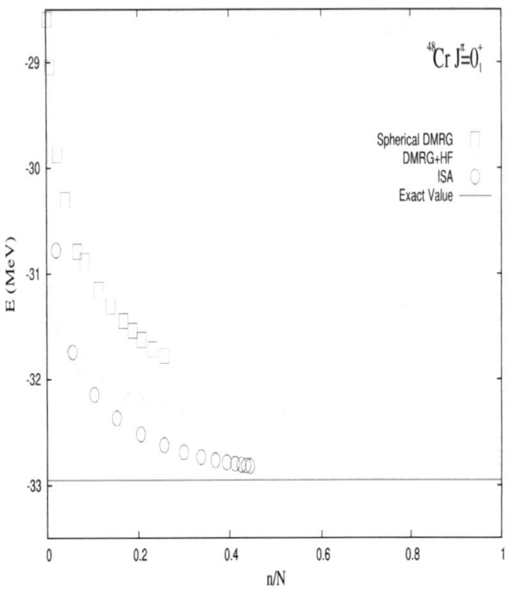

Figure 3. Convergence of the lowest $J = 0$ level of ^{48}Cr in the present approach and in the DMRG

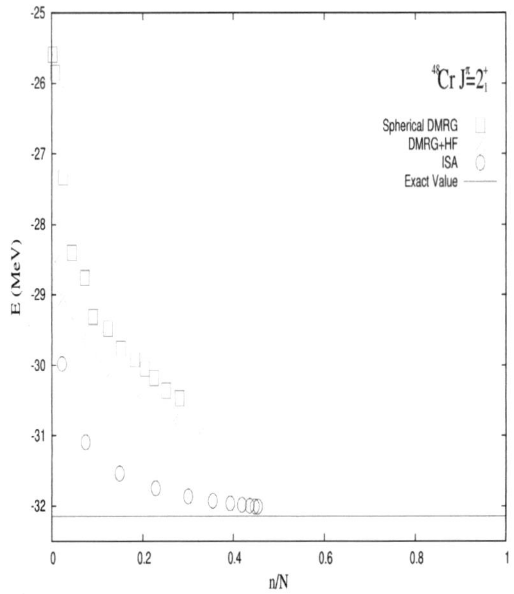

Figure 4. Convergence of the lowest $J = 2$ level of ^{48}Cr in the present approach and in the DMRG

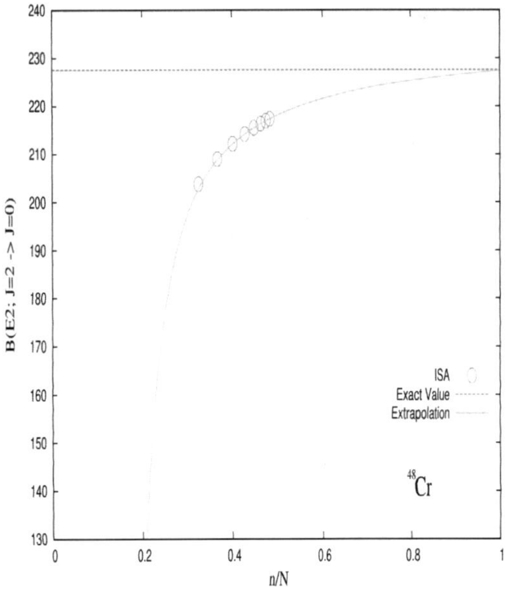

Figure 5. Convergence properties of the $E2$ strength in ^{48}Cr

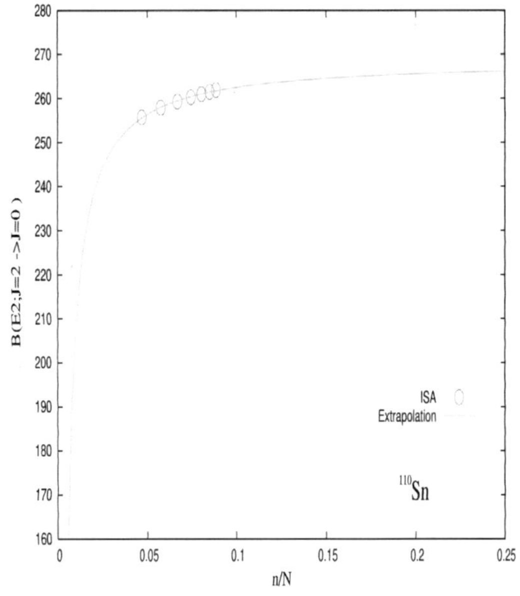

Figure 6. Convergence properties of the $E2$ strength in ^{110}Sn

ϵ_i. Each value of ϵ determines uniquely the dimension n_i of the Hamiltonian matrix to be the diagonalized.

Figs. 1 and 2 plot the lowest six $J = 0$ eigenvalues of ^{48}Cr and ^{110}Sn versus the dimensions n of the matrices resulting from the corresponding decreasing values of the sampling parameter ϵ.

As shown in the plots, all six eigenvalues approach closely the exact values for relatively small n-dimensional subspaces, especially in ^{110}Sn (Fig. 2).

The convergence is more rapid than in other approaches like the DMRG. This is illustrated for the lowest $J = 0$ and $J = 2$ levels of ^{48}Cr in Figs. 3 and 4. It is to be pointed out, however, that the DMRG seems to be more effective in the j-coupled scheme [12, 13]. Figs. 5 ad 6 show that a fast convergence is achieved also for the the $E2$ transitions strengths.

The smooth behavior of energies and $E2$ strengths versus n is to be noticed. This allows to extrapolate the computed data to asymptotic values. These are shown to practically coincide with the exact quantities.

6. Concluding remarks

According to the present results, the m-scheme implementation of the importance sampling iterative algorithm for diagonalizing large matrices looks very promising. The m-scheme combined with an effective J-projection of the states allow to take advantage of the sparsity of the Hamiltonian matrix thereby reducing drastically the sizes of the matrix. The J-projection allows also to generate a number of low-lying eigenstates for each J value that is appreciably larger than in other approaches.

A further crucial reduction is achieved by the importance sampling properly upgraded with respect to the original formulation. The importance sampling enhances greatly the convergence of the iterative process and makes possible an extrapolation to asymptotic values for energies and transition strengths that practically coincide with the exact values.

A high accuracy of the results is achieved also for the wavefunctions, as illustrated by the convergence properties of the $E2$ transition strengths. Some illustrative examples show that the convergence rate is appreciably faster than in other approaches. In virtue of these features, we are confident that the method can be applied to Hamiltonian matrices of very large dimensions. This hope is supported by preliminary calculations.

References

[1] Wilkinson J H 1965 *The Algebraic Eigenvalue Problem* (Oxford: University Press)
[2] Golub G H and Van Loan C F 1996 *Matrix Computations* (Baltimore:John Hopkins University Press)
[3] Caurier E *et al* 2005 *Rev. Mod. Phys.* **77** 427
[4] Caurier E and Nowacki F 1999 *Acta Physica Polonica* **30** 705
[5] White J A , Koonin S E and Dean D J 2000 *Phys. Rev. C* **61** 034303
[6] Koonin S E Dean D J and Langanke K 1997 *Phys. Rep.* **278** 1
[7] Koonin S E Dean D J and Langanke K 1997 *Annu. Rev. Nucl. Part. Sc.* **47** 463
[8] Otsuka T, Honma M and Mizusaki T 1998 *Phys. Rev. Lett.* **81** 1588
[9] Otsuka T *et al* 2001 *Prog. Part. Nucl. Phys.* **47** 319
[10] White S R 1993 *Phys. Rev. B* **48** 10345
[11] Dukesky J and Pittel S 2001 *Phys. Rev. C* **63** R061303
[12] Pittel S and Sandulesku N 2006 *Phys. Rev. C* **73** 014301
[13] Thakur B, Pittel S, Sandulescu N 2008 *Phys.Rev.* C **78** 041303
[14] Andreozzi Porrino A and Lo Iudice N 2002 *J. Phys. A: Math. Gen.* **35** L61
[15] Andreozzi F, Lo Iudice N and Porrino A 2003 *J. Phys. : Nucl. Part. Phys.* **29** 2319
[16] Andreozzi F, Bianco D, Lo Iudice N and Porrino to be submitted for publication.
[17] Davidson E R 1975 *J. Computational Phys.* **17** 87
[18] Machleidt R 1989 *Adv. Nucl. Phys.* **19** 189
[19] Caurier E, Zuker A P, Poves A and Martinez-Pinedo G 1994 *Phys. Rev. C* **50** 225
[20] Dikmen E, Öztürk O, Valliere M 2009 *J. Phys. : Nucl. Part. Phys.* **36** 045102

Journal of Physics: Conference Series **205** (2010) 012003 doi:10.1088/1742-6596/205/1/012003

$^{10-11}$Li and $^{13-14}$Be Studied by Projectile Fragmentation and pp-RPA

G Blanchon[(1,2,3)], **A Bonaccorso**[(1)], **D M Brink**[(4)], **and N Vinh Mau**[(5)]

[(1)] Istituto Nazionale di Fisica Nucleare, Sez. di Pisa, Largo Pontecorvo 3, 56127 Pisa, Italy.
[(2)] Scuola di Dottorato G. Galilei, Università di Pisa, Pisa, Italy.
[(3)] Present Address: CEA/ DAM, BP 12- 91680 Bruyeres-le-Chatel, France.
[(4)] Department of Theoretical Physics, 1 Keble Road, Oxford OX1 3NP, U. K.
[(5)] Institut de Physique Nucléaire, IN2P3-CNRS, F-91406, Orsay Cedex, France.

E-mail: bonac@df.unipi.it

Abstract. The level ordering in the unbound nuclei ^{10}Li and ^{13}Be is established on a firm basis using a time dependent projectile fragmentation model, comparing to experimental data and using structure inputs obtained from a semi-phenomenological core-vibration coupling model of two-neutron halo nuclei. The information on the shell ordering corresponds to the understanding of the neutron-core interaction. This is a building block of any three body model of borromean nuclei. As a consistency test we show that the energy spectra of some Beryllium isotopes obtained by using pp-RPA predict the same shell ordering as extracted from the reaction model vs. data analysis and that the well known structure of ^{11}Li is reproduced.

1. Introduction

Projectile fragmentation is used to study two neutron halo projectiles as in Refs.[1]-[17]. The two-neutron emission is supposed to be due to the nuclear interaction with the target. The mechanism is such that one neutron is stripped by the target, and thus taken away from the elastic channel: it cannot therefore be measured in coincidence with the core. The other neutron is left behind, for example in a resonance state, which then decays. For a decaying s-state this mechanism has been described by the sudden approximation in Ref.[2] under the hypothesis that while the first neutron is stripped and not detected, the second neutron is emitted at large impact parameters with no final state interaction with the target. This method leads to a transition probability independent of the impact parameter. The second neutron however re-interacts with the core which, for example, is going to be ^{9}Li in the case of the ^{11}Li projectile, while it will be ^{12}Be in the case of the projectile fragmentation of ^{14}Be, since both ^{10}Li and ^{13}Be are not bound. Experiments with a ^{14}B projectile [3] have also been performed, in which the n-^{12}Be relative energy spectra have been reconstructed by coincidence measurements. In such a nucleus the valence neutron is weakly bound, while the valence proton is strongly bound. Thus the neutron will probably be emitted in the first step and then re-scattered by the core minus one proton nucleus. The projectile-target distances at which this kind of mechanism would be relevant are probably not so large to neglect the effect of the neutron-target interaction.

2. Projectile Fragmentation

We calculate breakup by a semiclassical approximation in which the dynamics is controlled by the three potentials describing nucleon-core, nucleon-target, and core-target interactions. The collision is described in terms of only the three-body variables of nucleon coordinate, projectile coordinate, and target coordinate. The projectile-target relative motion is treated semiclassically by using a trajectory of the center of the projectile relative to the center of the target $\mathbf{R}(t) = \mathbf{b}_c + vt\hat{\mathbf{z}}$ with constant velocity v in the z direction and impact parameter $\mathbf{b}_c$ in the xy plane. The formalism is then valid for incident energies above the Coulomb barrier. To first order such inelastic-like excitations can be described along this trajectory by the time dependent perturbation amplitude [18, 19] for a transition from a nucleon state ψ_i bound in the projectile, to a final continuum state ψ_f, such that $A_{fi} = \frac{1}{i\hbar}\int_{-\infty}^{\infty} dt \langle \psi_f(\mathbf{r},t)|V(\mathbf{r},\mathbf{R}(t))|\psi_i(\mathbf{r},t)\rangle$, where the interaction responsible for the neutron transition $V(\mathbf{r},\mathbf{R}(t))$, moves past on the constant velocity path $\mathbf{R}$(t). The radial part $\phi_i(\mathbf{r})$ of the single particle initial state wave function $\psi_i(\mathbf{r},t)$ is calculated in a potential $V_{WS}(r)$ which is fixed in space and chosen such as to reproduce the experimental valence particle separation energy. In the special case of exotic nuclei the traditional approach to inelastic excitations needs to be modified. For example the final state can be eigenstate of a potential $U(r)$ (cf.Eq. (3)) modified with respect to V_{WS} because some other particle is emitted during the reaction process as discussed in the Introduction. The final state interaction might also have an imaginary part which would take into account the coupling between a continuum state and an excited core. The rôle of the target represented by V is just to perturb the initial bound state wave function and to allow the transition to the continuum by transferring some momentum to the neutron. Then it is enough to choose a simplified form of the interaction, such as a delta-function potential $V(r) = v_2\delta(x)\delta(y)\delta(z)$. The value of the strength $v_2 \equiv$ [MeV fm^3] used in the calculation is taken equal to the volume integral of the appropriate neutron-target interaction. The coordinate system and other details of the calculations can be found in Ref.[4].

The probabilities for different processes can be represented in terms of the amplitude as $dP/d\xi = \sum |A_{fi}|^2 \delta(\xi - \xi_f)$ where ξ can be momentum, energy or any other variable for which a differential cross section is measured. Direct one-particle re-arrangment reactions of the peripheral type in presence of strong core-target absorption can be described by an equation like [19, 20] $\frac{d\sigma_{-n}}{d\xi_f} = C^2S\int d\mathbf{b_c}\frac{dP_{bup}(b_c)}{d\xi_f}P_{ct}(b_c)$, (see Eq. (2.3) of [20]) and C^{2}S is the spectroscopic factor for the initial single particle state. The core survival probability is defined in terms of a S-matrix function of the core-target distance of closest approach b_c. A simple parameterisation is $P_{ct}(b_c) = |S_{ct}|^2 = e^{(-\ln 2 exp[(R_s - b_c)/a])}$. The strong absorption radius R$_s \approx 1.4(A_p^{1/3} + A_t^{1/3})$ fm is defined as the distance of closest approach for a trajectory that is 50% absorbed from the elastic channel and a=0.6 fm is a diffuseness parameter.

Due of the strong core absorption these calculations are performed using the asymptotic form of the initial and final state wave functions. Introducing the quantization condition[19] the probability spectrum reads

$$\frac{dP_{in}}{d\varepsilon_f} = \frac{2}{\pi}\frac{v_2^2}{\hbar^2 v^2}C_i^2\frac{m}{\hbar^2 k}\frac{1}{2l_i+1}\Sigma_{m_i,m_f}|\sin(\delta+\nu)|^2|I_{m_i,m_f}|^2. \quad (1)$$

The generalization including spin is given in Appendix B of Ref.[4] where also the form factor $|I_{m_i,m_f}|$ is given explicitly. $|\sin(\delta+\nu)|^2 = |1-\bar{S}|^2$ where $\bar{S} = e^{2i(\delta+\nu)}$ is an off-the-energy-shell S-matrix representing the final state interaction of the neutron with the projectile core. It depends on a phase which is the sum of δ, the free particle n-core phase shift, plus ν the phase of the matrix element $|I|$. An analytical estimate value of $|I|^2$, for an s to s transition, is

$$|I|^2 = \frac{1}{b_c^4}\frac{2\pi b_c}{\sqrt{\gamma^2+k^2}}e^{-2\gamma b_c}\exp\left(-\gamma b_c\frac{\gamma^2+k^2}{4\bar{k}^2}\right)$$
$$= \frac{1}{b_c^3}\frac{2\pi}{\sqrt{\gamma^2+k^2}}\exp\left(-2\gamma b_c\left(1+\frac{q}{4\bar{k}}\right)\right), \qquad (2)$$

where $\bar{k} = mv/\hbar$ and $q = (\gamma^2 + k^2)/2\bar{k}$.

The above analytical expression is accurate to within 10% for impact parameters around the strong absorption radius and for neutron-core energies less that 1.5MeV. The agreement improves for larger impact parameters.

On the other hand, the sudden approximation has been followed in Ref.[2] leading to $\frac{d\sigma}{d\varepsilon_f} \sim \frac{1}{k}\frac{|\sin(\delta+\beta)|^2}{(\gamma^2+k^2)}$, with $\beta = arctan(k/\gamma)$. It is clear that while in the sudden approach the initial and final state overlap is taken in the whole coordinate space, irrespective of the target and of the beam velocity, in our method the overlap of the initial and final wave functions decreases with the core-target impact parameter according to the form factor $|I|$.

3. Structure and Reactions of $^{13-14}$Be

Table 1. LHS: Scattering length of the 2s continuum state, energies and widths of the p- and d-resonances in ^{10}Li (top) and ^{13}Be (bottom) and corresponding strength parameters for the δV potential.

	ε_{res} MeV	Γ_j MeV	a_s fm^{-1}	α_l MeV
$2s_{1/2}$			-17.2	-10.0
$1p_{1/2}$	0.63	0.35		3.3
$1d_{5/2}$	1.55	0.18		-9.8
$2s_{1/2}$			-0.8	8.0
$1p_{1/2}$	0.67	0.28		8.34
$1d_{5/2}$	2.0	0.40		-2.36

Uncertainties in the interpretation of experimental results as compared to structure calculations were at the origin of our motivations to try to understand whether the neutron-^{12}Be relative energy spectra obtained from fragmentation of ^{14}Be or ^{14}B would show differences predictable in a theoretical model. If differences will be found in the experimental results with ^{14}B and ^{14}Be beams they could be due to an interplay between structure and reaction effects. Here we discuss only the ^{14}Be case. Some results are also given for the $^{10-11}$Li systems [5], which being already well understood are used as test cases.

The ground state of ^{14}Be has spin $J^\pi = 0^+$. In a simple model assuming two neutrons added to a ^{12}Be core in its ground state the wave function is: $|^{14}Be >= [b_1(2s_{1/2})^2 + b_2(1p_{1/2})^2 + b_3(1d_{5/2})^2] \otimes |^{12}Be, 0^+ >$. Then the bound neutron can be in a 2s, $1p_{1/2}$ or $1d_{5/2}$ state. However the situation is much more complicated [12]-[15] and in particular the calculations of Ref. [16] show that there is a large component $(2s_{1/2}, 1d_{5/2}) \otimes |^{12}Be, 2^+ >$ with the core in its low energy 2^+ state which can modify the neutron distribution.

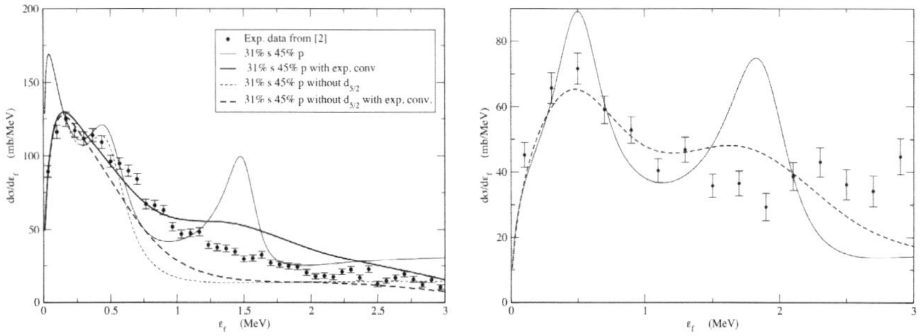

Figure 1. LHS: Sum of all transitions from the s and p initial states, weighted by the known C^2S, for the reaction ^{11}Li+^{12}C → n+^{9}Li+X . Full thick line is the folding of the calculated spectrum with the experimental resolution curve. RHS: Sum of all transitions from the s initial state for the reaction ^{14}Be+^{12}C → n+^{12}Be+X. Dashed line is the folding of the calculated spectrum with the experimental resolution curve.

Table 2. Theoretical and experimental values of S_{2n} (MeV) in ^{10}Be and ^{12}Be, $\langle r^2\rangle^{1/2}$ (fm), $\langle\lambda^2\rangle^{1/2}$ (fm) and $\langle\rho^2\rangle^{1/2}$ (fm) in ^{12}Be.

	$S_{2n}(^{10}\text{Be})$	$S_{2n}(^{12}\text{Be})$	$\langle r^2\rangle^{1/2}$	$\langle\lambda^2\rangle^{1/2}$	$\langle\rho^2\rangle^{1/2}$
Theory	8.50	3.60	2.77	4.10	4.89
Exp.	8.48	3.67±0.01	2.59±0.06		

Table 3. Theoretical and experimental values of S_{2n} (MeV) in ^{12}Be and ^{14}Be, $\langle r^2\rangle^{1/2}$ (fm), $\langle\lambda^2\rangle^{1/2}$ (fm) and $\langle\rho^2\rangle^{1/2}$ (fm) in ^{14}Be in the two cases of non-inversion (first line) and inversion (second line) of the 2s and $1p_{1/2}$ shells.

	$\epsilon(1p_{1/2})$	$\epsilon(2s)$	$S_{2n}(^{12}\text{Be})$	$S_{2n}(^{14}\text{Be})$	$\langle r^2\rangle^{1/2}$	$\langle\lambda^2\rangle^{1/2}$	$\langle\rho^2\rangle^{1/2}$
Theory	-3.15	0.09	2.91	0.51	3.45	5.45	8.45
	0.48	-3.15	3.67	1.33	2.96	4.27	4.72
Exp.	-	-	3.67±0.01	1.34 ± 0.11	3.1±0.4	4.5±1.0	5.4±1.0

Table 4. Main RPA amplitudes of ^{12}Be ground state.

X_{ab}				$X_{\alpha\beta}$
$(1p_{1/2})^2$	$(1p_{1/2}\ 2p_{1/2})$	$(2s)^2$	$(1d_{5/2})^2$	$(1p_{3/2})^2$
0.76	0.31	0.50	0.43	0.57

To describe the valence neutron in the ^{13}Be continuum we assume that the single neutron hamiltonian with respect to ^{12}Be has the form $h = t + U + iW$ where t is the kinetic energy and $U(r) = V_{WS} + \delta V$ is the real part of the neutron-core interaction. For the time being the imaginary part is taken equal to zero. V_{WS} is again a Woods-Saxon potential plus spin-orbit whose parameters are given in Ref. [4], and δV is a correction:

$$\delta V(r) = 16\alpha e^{2(r-R)/a}/(1 + e^{(r-R)/a})^4 \tag{3}$$

which originates from particle-vibration couplings [13] relevant only for the low energy spectrum. Values for the parameter α are given in Table 1. The above form is suggested by a calculation of

Journal of Physics: Conference Series **205** (2010) 012003 doi:10.1088/1742-6596/205/1/012003

Table 5. Main RPA amplitudes for ^{14}Be ground state without (first line) and with (second line) inversion of 2s-1p$_{1/2}$ shells.

$\epsilon(1p_{1/2})$	$\epsilon(2s)$	X_{ab}			$X_{\alpha\beta}$	
		$(2s)^2$	$(1d_{5/2})^2$		$(1p_{3/2})^2$	$(1p_{1/2})^2$
-3.15	0.09	-0.93	-0.49		0.32	0.36
		$(1d_{5/2})^2$	$(1p_{1/2})^2$	$(1p_{1/2}\ 2p_{1/2})$	$(1p_{3/2})^2$	$(2s)^2$
0.48	-3.15	0.69	-0.73	0.52	-0.61	0.45

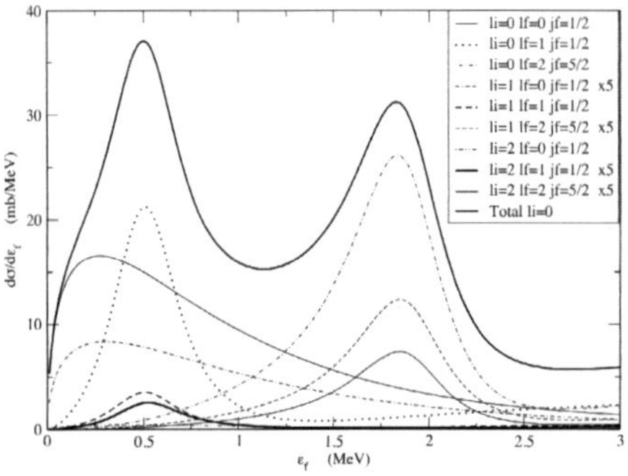

Figure 2. Individual transitions from the bound s,p,d components of the initial wave function to the unbound s,p,d states. Only the contribution from the initial s-orbital are then summed to give the full curve.

such couplings using Bohr and Mottelson collective model of the transition amplitudes between zero and one phonon states. Therefore our structure model is not a simple *single-particle in a potential* model but contains in it the full complexity of single-particle vs. collective couplings. However the fact that such a complexity can be put in a form like Eq.(3) is an added value to our approach. If simple fittings of experimental data will be obtained, then the parameters of a semi-phenomenological potential can be deduced and linked to a more microscopic model. To give some examples of a possible comparison with available data, we show in Fig. 1 the experimental points from H. Simon et al. [9] for the reactions ^{11}Li+^{12}C $\rightarrow$ n+^{10}Li +X and ^{14}Be+^{12}C $\rightarrow$ n+^{12}Be+X at 250 A.MeV and our calculations. More details in the legenda and in Refs. [4, 5].

Results obtained including the s, p and d initial states in ^{14}Be leading to unbound ^{13}Be states are shown in Fig.2. Each curve corresponds to just one transition as indicated. The solid curve is the sum of all transitions from the s-bound state which gives the dominant contribution to the transition probability due to the long tail of the wave function. To make them visible some curves have been multiplied by a factor of five as indicated in the legenda.

In Fig.3 we compare the s to s cross sections from our model with the results of the sudden formula used in [2]. It is clear that the sudden model is accurate only for the *true* virtual state of ^{10}Li. The s-threshold state in ^{13}Be has a very small scattering length and it is probably a complicated state coupled to an excited core.

On the other hand with the s and p single particle states of the LHS of Table 3, pp-RPA calculations have been performed. The details of the formalism, preliminary results and comparison with experimental data are presented in Tables 2,3,4 from [21] and [22], using two hypothesis for the level ordering. The best results are obtained for the s(bound)-p(unbound) inversion case. From the same tables it is also interesting to look at the RPA amplitudes, given in Table 5 which are closely related to the spectroscopic factors and give a hint on the occupation

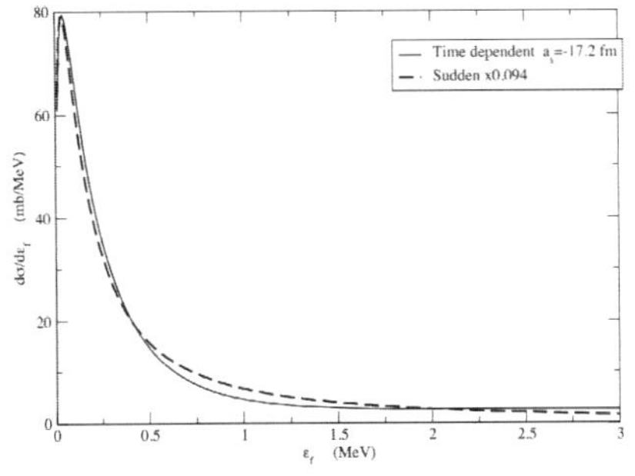

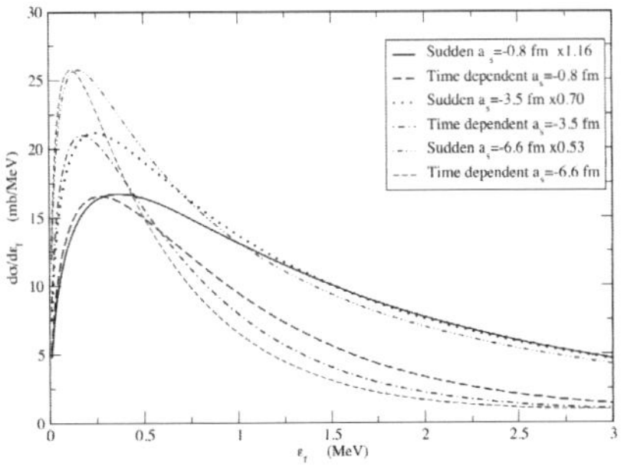

Figure 3. LHS: n-^{9}Li relative energy spectrum; comparison for the s to s transition of the fragmentation method result (solid curve) with that according to the sudden formula (dashed curve). Both calculations use the exact phase shifts. The sudden method is very accurate for a virtual state. RHS: n-^{12}Be relative energy spectrum; comparison of fragmentation method for an s to s transition with results of the sudden formula using the same optical model phase shifts corresponding to scattering lengths as indicated. The sudden formula is not accurate in this case due to the small scattering lengths. For each case we give in the legenda the normalization factor between the two calculations.

probabilities of the orbitals. It is evident that the mixing of s-p-d states is very strong in both ^{12}Be and ^{14}Be and that these nuclei have a *dynamical* nature far from the traditional shell ordering and closure concepts. Furthermore, notice that ^{13}Be is the first nucleus in which a s-p inversion is understood to happen across threshold.

Finally there is the question of the s-state in the continuum clearly seen in various experiments and necessary to interpret the ^{13}Be spectrum. As we have discussed in Ref.[4] and also found in [9] this state has a small scattering length thus implying a weak interaction in the $l = 0$ channel between the neutron and ^{12}Be. It does not appear to be important in the final ^{14}Be level scheme and in Ref.[22] it is interpreted as a phantom of the s-bound state near threshold in ^{13}Be.

4. Conclusions and Outlook

The field of Rare Isotopes Studies is very active, growing steadily and rapidly. Some recent achievements in projectile fragmentation theory for elastic breakup have been presented here together with new structure results of $^{13-14}$Be. From the structure point of view, in the search for the dripline position, a very important role is played by the study of nuclei unstable by neutron emission. This is one of the most important subjects which need to be adressed and developed further in the near future and for which some suggestions of possible structure and reaction formalisms have been presented.

References

[1] Marqués F M *et al* 2001 *Phys. Rev.* C **64** 061301(R); Orr N 2003 *Prog. Theor. Phys. Suppl.* **146** 201
[2] Bertsch G F , Hencken K and Esbensen H 1998 *Phys. Rev.* C **57** 1366
[3] Lecouey J L 2004 *Few Body Syst* **34** 21
[4] Blanchon G, Bonaccorso A, Brink D M, García-Camacho A and Vinh Mau N 2007 *Nucl. Phys.* A**784** 49
[5] Blanchon G, Bonaccorso A, Brink D M and Vinh Mau N 2007 *Nucl. Phys.* A **791** 303
[6] Blanchon G, Bonaccorso A and Vinh Mau N 2004 *Nucl. Phys.* A **739** 259
[7] Thoennessen M *et al* 1999 *Phys. Rev.* C **59** 111; 1999 *Phys. Rev.* C **60** 027303
[8] Labiche M *et al* 2001 *Phys. Rev. Lett.* **86** 600
[9] Simon H *et al* 2007 *Nucl. Phys.* A **791** 267
[10] Korsheninnikov A A *et al* 1995 *Phys. Lett.* B **343** 53
[11] Kondo Y it et al 2007 *RIKEN Accel. Prog. Rep.* **40** 27
[12] Thompson I J and Zhukov M V 1996 *Phys. Rev.* C **53** 708
[13] Vinh Mau N and Pacheco J C 1996 *Nucl. Phys.* A **607** 163
[14] Pacheco J C and Vinh Mau N 2002 *Phys. Rev.* C **65** 044004
[15] Descouvemont P 1994 P̂hys. Lett. B **331** 71 ; 1995 *Phys. Rev.* C **52** 704
[16] Tarutina T, Thompson I J, Tostevin J A 2004 *Nucl. Phys.* A **733** 53
[17] Labiche M, Marques F M, Sorlin O and Vinh Mau N 1999 *Phys. Rev.* C **60** 027303
[18] Alder K and Winther A 1975 *Electromagnetic Excitation* North-Holland
[19] Bonaccorso A and Brink D M 1988 *Phys. Rev.* C **38** 1776; 1991 *Phys. Rev.* C **43** 299; 1991 *Phys. Rev.* C **44** 1559
[20] Bonaccorso A 1999 *Phys. Rev.* C **60** 054604
[21] Blanchon G 2008 *Nuclear Reactions and Structure For Nuclei Far from Stability* Ph.D. thesis, University of Pisa, unpublished. http://www.infn.it/thesis/
[22] Blanchon G, Vinh Mau N and Bonaccorso A 2009 *in preparation*

Journal of Physics: Conference Series **205** (2010) 012004 doi:10.1088/1742-6596/205/1/012004

Realistic low-momentum effective interactions and nuclear structure near double closed shells

A Covello[1,2], L Coraggio[2], A Gargano[2] and N Itaco[1,2]

[1]Dipartimento di Scienze Fisiche, Università di Napoli Federico II,
Complesso Universitario di Monte S. Angelo, I-80126 Napoli, Italy
[2]Istituto Nazionale di Fisica Nucleare,
Complesso Universitario di Monte S. Angelo, I-80126 Napoli, Italy

E-mail: covello@na.infn.it

Abstract. We report on a shell-model study of nuclei with four valence nucleons in the ^{132}Sn and ^{208}Pb regions. This aims at investigating to what extent the striking similarity existing between the low-energy properties of ^{134}Sb and ^{210}Bi persists when adding a pair of identical particles. We employ realistic low-momentum effective interactions derived from the CD-Bonn nucleon-nucleon potential through use of the $V_{\rm low-k}$ approach. The calculated results are in very good agreement with the available experimental data and emphasize the persistence of a close resemblance between the spectroscopy of the two regions when moving away from the one proton-one neutron systems.

1. Introduction

A fundamental problem of nuclear physics is to understand the properties of nuclei starting from the forces among nucleons. Nowadays, there are two main lines of attack to attain this goal. The first one comprises the so-called *ab initio* calculations in which nuclear properties, such as binding and excitation energies, are calculated directly from first principles of quantum mechanics using an appropriate computational scheme. To this category belong the Green's function Montecarlo Method [1], the no-core shell-model [2], and the coupled cluster method [3]. Clearly, all *ab initio* calculations need huge amount of computational resources and are therefore currently limited to light nuclei.

The second line of attack consists of using the shell model with two-nucleon effective interactions derived microscopically from the bare nucleon-nucleon (NN) potential. In this kind of calculations, which are generally referred to as realistic shell-model calculations, only the valence nucleons are treated as active particles while the core polarization effects are taken into account perturbatively in the derivation of the effective interaction $V_{\rm eff}$. As we shall discuss later, these effects are of key importance for an accurate description of nuclear properties. Of course, this approach allows to perform calculations for medium- and heavy-mass nuclei which are far beyond the reach of *ab initio* calculations.

The two main ingredients of realistic shell-model calculations are the NN potential V_{NN} and the many-body methods for deriving $V_{\rm eff}$. As regards the first point, there are several high-quality potentials which fit equally well (χ^2/datum $\approx$ 1) the NN scattering data up to the inelastic threshold. This means that their on-shell properties are essentially identical, namely they are phase-shift equivalent. This may raise the question of how much nuclear structure results may depend on the choice of the NN potential one starts with. We shall come back to this point in the next section.

As regards the derivation of $V_{\rm eff}$, a main difficulty is the strong short-range repulsion contained in all modern NN potentials. As is well known, the traditional way to circumvent this problem is the Brueckner G-matrix method. However, a few years ago a new approach to the renormalization of V_{NN} has been proposed [4], which has proved to be an advantageous alternative to the use of the G-matrix [5]. This consists in deriving from V_{NN} a low-momentum potential, $V_{\rm low-k}$, that preserves the deuteron binding energy and scattering phase shifts of V_{NN} up to a certain cutoff momentum Λ. Since $V_{\rm low-k}$ is a smooth potential, it can be used directly to derive $V_{\rm eff}$.

Since its introduction in 2002, we have used this approach to derive realistic low-momentum effective interactions for shell-model calculations in various mass regions. In this paper, we first give an outline of the theoretical framework in which these calculations have been performed, referring the reader to [5] for a detailed description. Then we present and discuss some recent results for nuclei with four valence nucleons in the ^{132}Sn and ^{208}Pb regions, focusing attention on the similarity between the spectroscopy of these two regions. A short summary is given in the last section.

2. Outline of theoretical framework

The shell-model effective interaction $V_{\rm eff}$ is defined, as usual, in the following way. In principle, one should solve a nuclear many-body Schrödinger equation of the form

$$H\Psi_i = E_i\Psi_i, \tag{1}$$

with $H = T + V_{NN}$, where T denotes the kinetic energy. This full-space many-body problem is reduced to a smaller model-space problem of the form

$$PH_{\rm eff}P\Psi_i = P(H_0 + V_{\rm eff})P\Psi_i = E_iP\Psi_i. \tag{2}$$

Here $H_0 = T + U$ is the unperturbed Hamiltonian, U being an auxiliary potential introduced to define a convenient single-particle basis, and P denotes the projection operator onto the chosen model space,

$$P = \sum_{i=1}^{d} |\psi_i\rangle\langle\psi_i|, \tag{3}$$

d being the dimension of the model space and $|\psi_i\rangle$ the eigenfunctions of H_0. The effective interaction $V_{\rm eff}$ operates only within the model space P. In operator form it can be schematically written as

$$V_{\rm eff} = \hat{Q} - \hat{Q}' \int \hat{Q} + \hat{Q}' \int \hat{Q} \int \hat{Q} - \hat{Q}' \int \hat{Q} \int \hat{Q} \int \hat{Q} + \; ... \;\; , \tag{4}$$

where $\hat{Q}$, usually referred to as the $\hat{Q}$-box, is a sum of irreducible valence-linked diagrams, and the integral sign represents a generalized folding operation. $\hat{Q}'$ is obtained from $\hat{Q}$ by removing terms of first order in the interaction. Once the $\hat{Q}$-box is calculated, the folded-diagram series of equation (4) can be summed up to all orders by iteration methods, as, for instance, the Lee-Suzuki one [6].

As pointed out in the Introduction, we "smooth out" the strong repulsive core contained in V_{NN} by constructing a low-momentum potential $V_{\rm low-k}$. This is achieved by integrating out the high-momentum modes of V_{NN} down to a cutoff momentum Λ. This integration is carried out with the requirement that the deuteron binding energy and phase shifts of V_{NN} up to Λ are preserved by $V_{\rm low-k}$. Once the $V_{\rm low-k}$ is obtained, we use it, plus the Coulomb force for protons, as input interaction for the calculation of the matrix elements of the shell-model effective interaction within the framework of the folded-diagram method outlined above. The calculation of $\hat{Q}$, which is in principle an infinite sum of irreducible diagrams, can only be made approximately by selecting certain classes of diagrams. In our calculations we have included in $\hat{Q}$ all the 1-body and 2-body diagrams up to second order in $V_{\rm low-k}$.

A detailed description of the derivation of $V_{\rm low-k}$ from V_{NN} as well as a discussion of its main features can be found in [4, 5]. However, we should mention here that the use of $V_{\rm low-k}$ largely reduces the ambiguity in the choice of V_{NN}. In fact, we have verified [5] that shell-model effective interactions derived from phase-shift equivalent NN potentials through the $V_{\rm low-k}$ approach lead to very similar results. In other words, $V_{\rm low-k}$ gives an approximately unique representation of the NN potential.

3. Calculations and results

We start by giving some details of our shell-model calculations and then present and discuss the results obtained for the four-valence particle nuclei ^{136}I, ^{136}Sb and ^{136}Sn in ^{132}Sn region and their counterparts in ^{208}Pb region, ^{212}At, ^{212}Bi and ^{212}Pb. As mentioned in the Introduction, our interest centers on the close resemblance between the spectroscopy of these two regions. In this connection, it is worth noting that the existence of this resemblance was pointed out long ago in [7], where it was noticed that every single-particle proton or neutron state in ^{132}Sn region, characterized by quantum numbers (nlj), has its counterpart around ^{208}Pb with quantum numbers $(nl+1j+1)$. It has been, however, the substantial progress in gaining information on neutron-rich nuclei around ^{132}Sn that has made this issue the focus of recent studies [8, 9, 10, 11].

For the nuclei in ^{132}Sn region we assume that the valence protons occupy the five levels $0g_{7/2}$, $1d_{5/2}$, $1d_{3/2}$, $2s_{1/2}$, and $0h_{11/2}$ of the 50-82 shell, while for the neutrons the model space includes the six levels $0h_{9/2}$, $1f_{7/2}$, $1f_{5/2}$, $2p_{3/2}$, $2p_{1/2}$, and $0i_{13/2}$ of the 82-126 shell. Similarly, in ^{208}Pb region we take as model space for the valence protons the six levels of the 82-126 shell and let the valence neutrons occupy the seven levels $1g_{9/2}$, $0i_{11/2}$, $0j_{15/2}$, $2d_{5/2}$, $3s_{1/2}$, $1g_{7/2}$, and $2d_{3/2}$ of the 126-184 shell. As regards the choice of the single-proton and single-neutron energies, they have been taken from experiment. The adopted values are reported in [12] and [13] for the ^{132}Sn and ^{208}Pb regions, respectively.

For both regions the two-body effective interaction is derived from the CD-Bonn NN potential [14] renormalized by means of the $V_{\rm low-k}$ approach, as outlined in the previous section. As in our previous studies (see [11] and references therein), the cutoff momentum Λ is given the value 2.2 fm^{-1}. The computation of the diagrams included in the $\hat{Q}$-box is performed within the harmonic-oscillator basis using intermediate states composed of all possible hole states and particle states restricted to the five proton and neutron shells above the Fermi surface. The oscillator parameter is 7.88 MeV for the $A = 132$ region and 6.88 MeV for the $A = 208$ region, as obtained from the expression $\hbar\omega = 45A^{-1/3} - 25A^{-2/3}$. The shell-model calculations have been performed by using the OXBASH computer code [15].

We now present and discuss the results of our study. To start with, in table 1 we compare the calculated excitation energies with the available experimental ones [16, 17, 18] in ^{136}Sb, ^{136}I, ^{212}Bi and ^{212}At. We see that the agreement between experiment and theory is very good, the discrepancies being below 50 keV for the large majority of states.

To give emphasis to the resemblance between the spectroscopy of the ^{132}Sn and ^{208}Pb regions, we show in figure 1 the calculated energies of ^{136}Sb and ^{136}I for the lowest states with $J^\pi = 0^-$ to $J^\pi = 7^-$ and in figure 2 the energies of the $J^\pi = 0^-$ to $J^\pi = 9^-$ states in ^{212}Bi and ^{212}At. These curves may be considered the evolution of the $\pi g_{7/2}\nu f_{7/2}$ and $\pi h_{9/2}\nu g_{9/2}$ multiplets in the one proton-one neutron nuclei ^{134}Sb and ^{210}Bi, respectively, which we also report in figures 1 and 2 for the sake of comparison. In fact, this allows direct visualization of how the structure of proton-neutron multiplets is affected when adding two identical particles. A detailed discussion

Table 1. Experimental and calculated excitation energies (in MeV) for ^{136}Sb, ^{136}I, ^{212}Bi, and ^{212}At.

Nucleus	J^π	$E_{\rm expt}$	$E_{\rm calc}$
^{136}Sb	1^-	0.0	0.0
	2^-	0.053	0.021
	4^-	0.226	0.179
	6^-	0.278	0.210
^{136}I	1^-	0.0	0.0
	2^-	0.087	0.101
	3^-	0.222	0.203
^{212}Bi	0^-	0.239	0.174
	1^-	0.0	0.0
	2^-	0.115	0.141
	3^-	0.213	0.214
	4^-	0.251	0.258
^{212}At	1^-	0.0	0.0
	2^-	0.160	0.196
	3^-	0.206	0.227
	5^-	0.275	0.296
	8^-	0.328	0.338
	9^-	0.223	0.301

of the proton-neutron multiplets in ^{210}Bi and ^{134}Sb can be found in [13, 19], where we have also made an analysis of the various contributions to our effective interaction. This has given evidence of the crucial role played by the core-polarization effects, in particular those arising from $1p$-$1h$ excitations.

As regards the structure of the states in ^{136}Sb and ^{136}I, they are dominated by the $\pi g_{7/2}(\nu f_{7/2})^3$ and $\pi(g_{7/2})^3\nu f_{7/2}$ configurations, respectively. Similarly, for ^{212}Bi and ^{212}At the states reported in figure 2 are dominated by the $\pi h_{9/2}(\nu g_{9/2})^3$and $\pi(h_{9/2})^3\nu g_{9/2}$ configurations, respectively. For all four nuclei, however, there is a significant configuration mixing. This is particularly large for ^{212}Bi, with a percentage of components other than dominant one going from 25 to 52% .

¿From comparison of figures 1 and 2 we see that the curves for the four-valence-particle nuclei all have a similar shape, but they differ from that of the multiplets in ^{134}Sb an ^{210}Bi. The main new features are an overall flattening of the staggering and a rather sharp energy increase of the 0^- state with respect to the 1^- one. These are a manifestation of the dominant role played by pairing correlations, which means that the two additional protons or neutrons in both A = 136 and 212 systems are prevalently coupled to zero angular momentum. However, the wave functions of the states reported in figures 1 and 2 also contain components with the pair coupled to $J \neq 0$, which have a larger weight when a neutron rather than a proton pair is added. As a matter of fact, it turns out that the proton-proton pairing component of our effective interaction plays a more significant role than that of the neutron-neutron one. As is shown in [20], this is largely due to the $1p$-$1h$ core excitations.

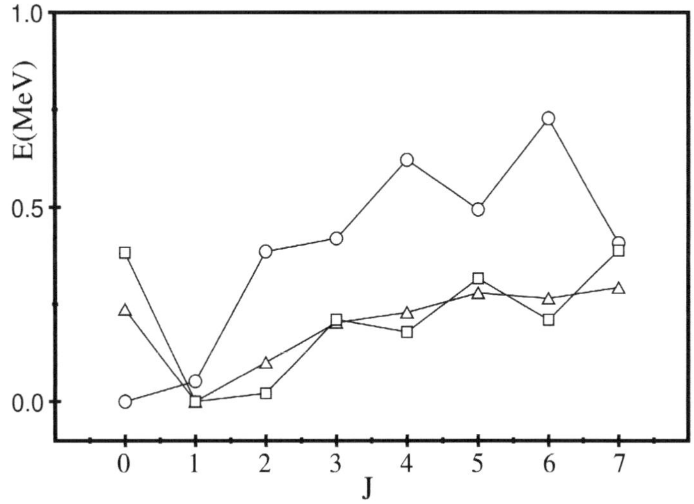

Figure 1. Proton-neutron multiplets in ^{134}Sb (circles), ^{136}Sb (squares) and ^{136}I (triangles).

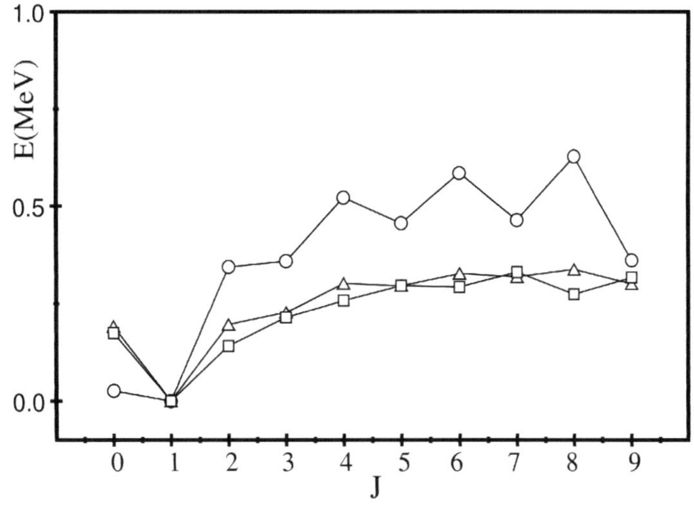

Figure 2. Proton-neutron multiplets in ^{210}Bi (circles), ^{212}Bi (squares) and ^{212}At (triangles).

As regards the effects of components with the pair coupled to $J \neq 0$, we have verified that their presence tends to produce staggering with phase opposite to that observed in one proton-one neutron systems. This is evident for ^{136}Sb but not for ^{212}Bi. The explanation resides in the fact that in ^{212}Bi this effect is quenched by configuration mixing, which, as mentioned before, we have found to be more relevant for this nucleus. As regards ^{136}I and ^{212}At, we don't find evidence of significant staggering, the curves in figures 1 and 2 being almost flat but for the lowest angular momenta. This is because, as mentioned above, the wave functions of the considered states in these nuclei contain a small percentage of nonzero-coupled proton pair components, giving rise to a staggering that is largely washed out by the configuration mixing.

Let us now turn to the four-valence-neutron nuclei ^{136}Sn and ^{212}Pb. In figure 3 we report the calculated and experimental [16] energies of the lowest states in the latter together with our predictions for the former. We see that the calculated curve for ^{136}Sn shows just the same behavior as that relative to its counterpart nucleus ^{212}Pb, the latter reproducing quite well the experimental one. For both nuclei, the considered states receive significant contributions from configurations other than the dominant one. In fact, we find that the percentage of the $(\nu f_{7/2})^4$ configuration for the states of ^{136}Sn and of the $(\nu g_{9/2})^4$ configuration for the states of ^{212}Pb ranges from 65 to 83% and from 51 to 75%, respectively.

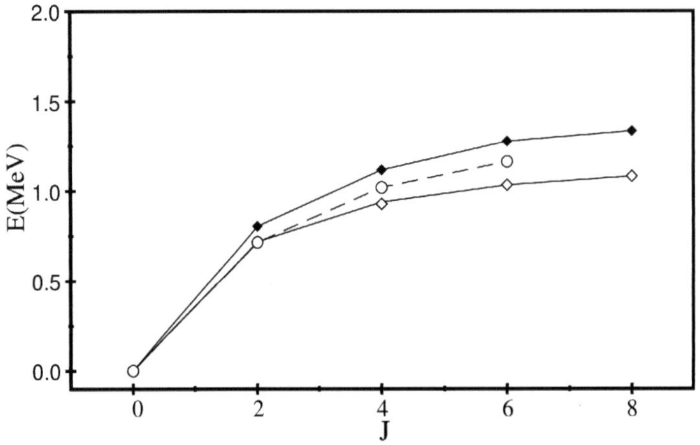

Figure 3. Energies of the low-lying states in ^{136}Sn and ^{212}Pb. The theoretical results are represented by open diamonds (^{212}Pb) and open circles (^{136}Sn) while the experimental data for ^{212}Pb by black diamonds.

4. Summary

We have presented here the results of a comparative shell-model study of four-valence-particle nuclei in the ^{132}Sn and ^{208}Pb regions. In our calculations we have employed realistic low-momentum effective interactions derived from the CD-Bonn NN potential without the use of any adjustable parameters.

Our results for all the considered nuclei are in very good agreement with the available experimental data and show evidence that a close resemblance between the spectroscopy of these two regions persists when moving away from the immediate neighbors of doubly magic ^{132}Sn and ^{208}Pb. This makes us confident in the soundness of our predictions for both the missing states of the proton-neutron multiplets in ^{136}Sb, ^{136}I, ^{212}Bi and ^{212}At and the low-energy spectrum of exotic ^{136}Sn. We hope that they may be verified in a not too distant future.

References

[1] Pieper S 2008 *Proc. Int. School of Physics "E. Fermi", Course CLXIX*, ed A Covello, F Iachello, R A Ricci and G Maino (Amsterdam: IOS Press) p 111 and references therein
[2] Navrátil P, Quaglioni S, Stetcu I and Barrett B 2009 *J. Phys.* G **36** 083101 and references therein
[3] Hagen G, Dean D J, Hjorth-Jensen M, Papenbrock T and Schwenk A 2007 *Phys. Rev.* C **76** 044305 and references therein
[4] Bogner S, Kuo T T S, Coraggio L, Covello A and Itaco N 2002 *Phys. Rev.* C **65** 051301(R)
[5] Coraggio L, Covello A, Gargano A, Itaco N and Kuo T T S 2009 *Prog. Part. Nucl. Phys.* **62** 135
[6] Suzuki K and Lee S Y 1980 *Prog. Theor. Phys.* **64** 2091
[7] Blomqvist J 1981 *Proc. 4th Int. Conference on Nuclei Far from Stability (Helsingor 1981)* CERN Report 81-09 (Geneva: CERN) p 536
[8] Fornal B *et al.* 2001 *Phys. Rev* C **63** 024322

[9] Korgul *et al.* 2007 *Eur. Phys. J.* A **32** 25
[10] Isakov V I *et al.* 2007 *Phys. At. Nucl.* **70** 818
[11] Covello A, Coraggio L, Gargano A and Itaco N 2009 *Acta Phys. Pol.* B **40** 401
[12] Coraggio L, Covello A, Gargano A and Itaco N 2005 *Phys. Rev.* C **72** 057302
[13] Coraggio L, Covello A, Gargano A an Itaco N, 2007 *Phys. Rev.* C **76** 061303(R)
[14] Machleidt R 2001 *Phys. Rev.* C **63** 024001
[15] Brown A B, Etchegoyen A and Rae D M 1988 *The Computer Code OXBASH* MSU-NSCL Report No. 524
[16] Data extracted using the NNDC On-line Dtata Service from the ENSDF database, version of October 1, 2009
[17] Urban W *et al.* 2006 *Eur. Phys. J.* A **27** 257
[18] Simpson G S *et al.* 2007 Phys. Rev. C **76** 041303(R)
[19] Coraggio L, Covello A, Gargano A an Itaco N, 2006 *Phys. Rev.* C **73** 031302(R)
[20] Covello A, Coraggio L, Gargano A and Itaco N 2009 *Phys. Rev.* C **80** 021305(R)

Journal of Physics: Conference Series **205** (2010) 012005 doi:10.1088/1742-6596/205/1/012005

Higher Tamm-Dankoff approximation for rotational nuclear states

H Laftchiev[1], J Libert[2], P Quentin[3] and Ha Thuy Long [4].

[1]INRNE - Bulgarian Academy of Sciences, 72 Tzarigradsko chaussee, blvd.-1784 Sofia, Bulgaria.
[2]IPN-Orsay, CNRS-IN2P3 Universit Paris XI, 15 rue Georges Clmenceau F-91406 Orsay France.
[3]CENBG CNRS IN2P3-Universit Bordeaux I, F-33175 Gradignan cedex France.
[4]Vietnam National University at Hanoi, 334 Nguyen Trai, Hano, Vietnam.

Abstract. Within the Higher Tamm Dancoff Approximation (HTDA), single particle states associated with a mean field of the Hartree-Fock type are used into shell model like calculations. Recently, this approach has been used for ground state calculations in magic nuclei and in the $N = Z$ mass region, as well as to study properties of some important isomeric states. The Cranking version of this formalism (Cr. HTDA) represents an attempt to reproduce rotational bands up to high spins in heavy nuclei. In this context, through the use of the mean field part of HTDA, many deformation and rotation effects (as intruder orbitals and effects of the time reversal symmetry breaking) are included in the configuration space. This work discusses the Cr. HTDA results obtained for two Pb isotopes (in order to test the approach), when using up to 2 particles - 2 holes excitations and provides tests for enlargement of the configuration space by 4 particle - 4 holes excitations. It is focused on the configuration space properties, improvement of truncation schemes, computational problems and saturation behavior of some physical quantities.

1. Introduction

The study of nuclear structure, when using effective phenomenological nucleon-nucleon forces is a well developped theoretical approach and has already given many and impressive experimentally verified results. When using microscopic approaches, nuclear phenomena could be described quite precisely with significant predictive power. The combination of microscopic approach and effective phenomenological nucleon-nucleon interactions allows to model nucleus behavior in mostly any state. As far as, superdeformed bands represent an example of two extremes: extreme in deformation and in some cases extreme in very high spin, they are suitable for test of modelling theories. Such extreme states allow streching the theoretical approaches to their limits. As such a test, one studies in this article the present performance of the cranked Higher Tamm-Dankoff Approximation (Cr. HTDA) theory in a domain, where Cr. HFB studies already show at the same time important successes, as well as some deficiency related to non-exact conservation of the particle number.

The HTDA theory, was described and developed in Ref. [1, 2] and used in some specific cases [3, 4, 5, 6] of nuclear phenomena. The specific use for rotational bands description is hindered by exeptional strong configuration space requirements, in the domain of superdeformed heavy nuclei, where a lot of Cr. HFB ([7, 8, 9, 10, 11] or Cr. RHB [12] calculations where performed.

Nevertheless, such an approach represents a unique possiblity to test effective phenomenological nucleon-nucleon forces when particle number is treated exactly and in a natural way does not allow spurious transition phenomena.

Without using projection methods, the Cr. HTDA could avoid configuration mixing between neighboring nuclei and thus could describe some fine shell structure driven phenomena as band crossings or band terminations. Nevertheless, a configuration space of sufficient size is required. We will discuss here some conditions able to assess the space size adapted to the problem. The testing calculations presented here concern two Pb isotopes - ^{192}Pb and ^{194}Pb.

2. The Cr. HTDA essentials

The HTDA method, has been fully described in Ref. [1, 2]. Let us recall the three main steps upon which it relies and some particularities of the interactions used in the current study:

i) A n-particle -n hole Slater determinant basis set is built on a Hartree-Fock "vacuum" solution associated with the one-body density matrix ρ^0 and the corresponding selfconsistent HF Hamiltonian $h^0\left(\rho^0\right)$. In the present work, this mean field is built on the ground of the Skyrme SkM* parametrization of the effective Skyrme nucleon-nucleon interaction [13].

ii) To mock the exact residual interaction we consider a δ pairing interaction and deducing its contributions of the "mean field" type leads us to the residual interaction V_0^{res} and therefore to the Hamiltonian $H_0^{HTDA} = h^0(\rho^0) + V_0^{res}$. The corresponding stationary Schroedinger equation is solved. In practice, only the state with the lowest energy is of physical interest here, (namely Ψ). It is extracted using standard Lanczos algorithm and formally writes :

$$|\Psi_0\rangle = |\phi_0\rangle + \sum_{i=\forall\, 1p1h} \chi_1^i \left|\phi_1^i\right\rangle + \sum_{j=\forall\, 2p2h} \chi_2^j \left|\phi_2^j\right\rangle + \sum_{k=\forall\, 3p3h} \chi_3^k \left|\phi_3^k\right\rangle + ..., \tag{1}$$

where the many-body ground-state wave-function is described by the Slater determinant $|\phi_0\rangle$ built with the neutron or proton lowest energy single particle (sp) states. When promoting a nucleon from a hole state φ_a to a particle state φ_a' neglecting the mean-filed changes, one gets a new determinant $\left|\phi_1^i\right\rangle$ which corresponds to the particular 1 particle -1 hole excitation (1p1h) associated with the exchange of φ_a by φ_a'. In this way, one built the many-body basis of Slater determinants $\left|\phi_2^j\right\rangle$, $\left|\phi_3^k\right\rangle$ ..., corresponding to 2 particles -2 holes (2p2h), 3 particles -3 holes (3p3h) ... excitations of the reference quasivacuum Slater determinant $|\phi_0\rangle$. Within this space, the N (or Z) body wavefunction is finally represented by the expansion coefficients χ_n^m. The basis of Slater determinant $|\phi_n^m\rangle$ is of course infinite and has to be truncated to be handled by computers. That point will be discussed in the next chapters. Note that here, like in most HFB or HF plus BCS calculations, the considered excitations do not mix different charge states.

iii) From the correlated wavefunction $\Psi_{\alpha=0}$ written as the combination of Slater determinants Eq. (1), one deduces a new one-body density :

$$(\rho_{\alpha=1}^{corr})_{ij} = \left\langle \Psi_{\alpha=0} \left| a_j^+ a_i \right| \Psi_{\alpha=0} \right\rangle \tag{2}$$

which defines through Skyrme functional a new one body Hamiltonian $h^0(\rho_{\alpha=1}^{corr})$, whose solutions define a new set of n particle - n hole states (step i). The HTDA Hamiltonian writes in this new space $H_{\alpha=1}^{HTDA} = h^0(\rho_{\alpha=1}^{corr}) + V_{\alpha=1}^{res}$ from which the new correlated wave function (step ii) is extracted as $\Psi_{\alpha=1}$, the process being pursued increasing the index α up to convergence with requirements for the lowest energy or for the decomposition numbers χ_n^j of the wave function Ψ.

2.1. The Cranked HTDA Hamiltonian

These basic principles have been kept in the present routhian approach, in which a linear constraint on the component J_x of the angular momentum is added, writing therefore:

$$\widehat{H} = \widehat{H}_{Skyrme} + \widehat{H}_{Coulomb} + \widehat{H}_{pair} - \Omega\widehat{J}_x \tag{3}$$

where $\widehat{H}_{pair}$ is a zero-range volume δ pairing interaction, and where Ω is the angular velocity i.e. the Lagrange multiplier associated with the dynamical constraint $\widehat{J}_x$ - the x-axis component of the angular momentum.

Equation (1) is used to write the HTDA Hamiltonian matrix of in the *npnh* representation :

$$H_{ij} = \left\langle nPnH_i|\widehat{H}|mPmH_j\right\rangle = \left\langle \phi_n^i|\widehat{H}|\phi_m^j\right\rangle = \begin{pmatrix} & 0P0H & 1P1H & 2P2H & ... \\ 0P0H & H_{00} & H_{01} & H_{02} & ... \\ 1P1H & H_{10} & H_{11} & H_{12} & ... \\ 2P2H & H_{20} & H_{21} & H_{22} & ... \\ ... & ... & ... & ... & ... \end{pmatrix} \tag{4}$$

One neglects the small polarization changes of the mean field caused by the *npnh* excitations, when n is a small number. This is done in analogy with the Koopmans approximations used for odd systems. Thus, mean field contributions vanish in the non diagonal terms of (4). The non-diagonal matrix elements are only due to the residual interaction defined as :

$$\widehat{V}^{res} = \widehat{V} - \widehat{V}_{HF}, \tag{5}$$

where $\widehat{V}$ is the two-body interaction including all interactions and $\widehat{V}_{HF}$ is the one-body reduction of the Skyrme and Coulomb interactions. The non-diagonal matrix elements of $\widehat{H}$ are thus given by :

$$H_{ij} = \left\langle \phi_n^i|\widehat{V}^{res}|\phi_m^j\right\rangle. \tag{6}$$

The only terms which are due to the mean field belong to the diagonal matrix elements H_{ii} which are written as:

$$\left\langle \phi_n^i|\widehat{H}_{Skyrme} + \widehat{H}_{Coulomb} - \Omega\widehat{J}_x|\phi_n^j\right\rangle = H_{00} + \delta_{ij}(\sum_p e_p^i - \sum_h e_h^i) \tag{7}$$

Here e_h and e_p are the energies of the hole and particle sp-states defining ϕ_n^i and $H_{00} = \langle\phi_0|K + V_{HF}|\phi_0\rangle$ is the total energy of the quasivacuum Slater determinant.

2.2. The zero-range pairing force

In general in both channels - particle-hole or particle-particle, one could use the same interaction or different ones, but it is difficult to find at this stage of the research, a computationally doable approach with realistic interactions which works on both channels. Therefore, in this study one uses a volume zero-range pairing force as residual interaction (Ref. [14]) - similar to that employed in static HTDA calculation of Ref. [1, 2]), limited to act only on the states whose energies are in the vicinity of the Fermi energy λ. To avoid any artificial sharp cutoff energy dependence (due to the appearance or disappearance of some single particle state into the window upon varying any continuous parameter like Ω), it is customary to introduce a smoothing factor $f(e_i)$ defined by a cutoff parameter X (in the present case $X = 4MeV$) and a smoothing parameter μ (here, $\mu = 0.5MeV$) and written as :

$$f^2(e_i) = \frac{\left(1 + exp\left(-\frac{X}{\mu}\right)\right)}{\left(1 + exp\left(\frac{(e_i-\lambda)-X}{\mu}\right)\right)}. \tag{8}$$

The matrix element of the interaction takes then the form:

$$\widetilde{V}_{ijkl}^{pair} = V_0 \left\langle ij\right| \left| \frac{1 - \overrightarrow{\sigma}_1 . \overrightarrow{\sigma}_2}{4} \delta(\overrightarrow{r}_1 - \overrightarrow{r}_2) \right| \widetilde{kl} \right\rangle f(e_i) f(e_j) f(e_k) f(e_l), \tag{9}$$

where V_0 is the pairing strength for a given isospin, and where σ_i are the usual Pauli matrices. Finally, V_0 values will be discussed hereafter in section 4. We take into account only the sp-states whose Hartree-Fock energies belong to the window:

$$\lambda - (X + 5\mu/2) < e_i < \lambda + (X + 5\mu/2), \tag{10}$$

where λ stands for the Fermi energy.

3. Symmetries

Similarly to Ref. [8, 15] we preserve the parity and signature symmetries in the HF cranking Hamiltonian. The corresponding sp-spectrum (and the n particles - n holes deduced Slater determinants) are also parity-signature symmetric. The selection rules for the zero-range volume pairing interaction when using sp-states with good parity-signature symmetries are such that:

$$V_{ijkl}^{pair} = 0, \; when, \; s_i s_j s_k s_l \neq 1 \, or \, \pi_i \pi_j \pi_k \pi_l \neq 1,$$

where s_i, s_j, s_k, s_l and π_i, π_j, π_k, π_l are signatures and parities of the sp states with indexes i, j, k, and l respectively.

When taking these rules into account, it is straightforward to verify through the Wick's theorem from Eqs. (6), (7) and (9), that the Hamiltonian matrix become block-diagonal in four blocks of states, corresponding to the four possible couples of parity and signature values. This allows us to diagonalize separately in four blocks for each isospin. Of course, the final result to be considered will lie in the block which gives the lowest total energy $\left\langle \widehat{H} \right\rangle = \left\langle \Psi | \widehat{H} | \Psi \right\rangle$. Similarly, the sp-density matrix is also block diagonal and hence one performs the routhian calculations separately each block.

4. Truncations and parametrization of the force

In such a shell-model like problem, the corner stone is clearly to determine the size of the *npnh* basis which will give a convincing accuracy and convergence of solutions. This depends on the the size of the restricted sp-space for the pairing force action, on the rank of excitations included in the configuration space and, an eventual energy cut-off of the configurations of a given rank. Some important remarks can be made in order to determine the "sufficient" conditions for convenient and still tractable size of the basis. Upon increasing the configuration space, one must study the trend of some important physical quantities which "saturate", which may be interpreted as an indication that this configuration space is not enough. Further, considerations over convergence and boundary conditions for such quantities must be taken into account in order to determine the right recipe for basis truncations and the parametrization of the residual interaction.

Here, one is concerned by the reproduction of the pairing properties of superdeformed rotational bands in the $A = 190$ mass region. In general, one can adjust the pairing strength parameters V_0 for neutrons and protons (named here V_n and V_p) in order to obtain meaningful

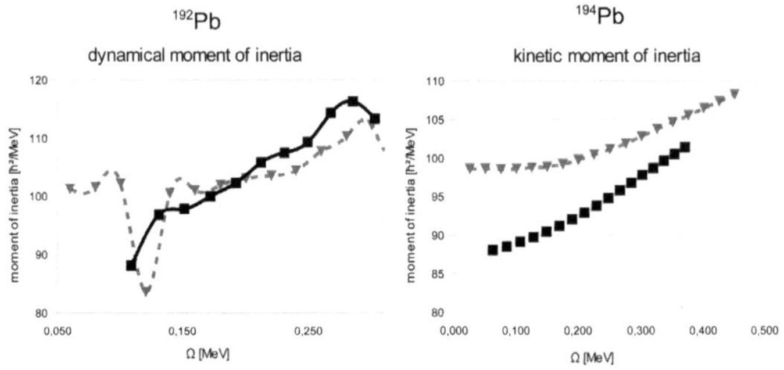

Figure 1. HTDA dynamical moments of inertia of ^{192}Pb (red, triangles), HTDA kinetic moments of inertia of ^{194}Pb (red, dashed line, triangles), experimental values (squares) versus angular velocity.

values for the kinetic moment of inertia at low angular velocities. But, when vibrational modes are not explicitely included in the formalism, one must substract about 15% of such a value in order to obtain the experimentaly deduced moments of inertia when taking them into account considerations (see Ref. [16]) for vibrational collectivity. For simplicity, everywhere in this study, one consider $V_n = V_p$, but computations for non-equal V_n and V_p have been also performed [17].

Figure 1 displays the Cr. HTDA results for the^{192}Pb Yrast SD band kinetic moments of inertia, compared to the data, using the following truncation conditions: only 0p0h, 1p1h and 2p2h excitations are included in the configuration space, their energy is restricted to be below $2X + 5\mu/2$ value. Although, good convergence is obtained, an important band crossing at low angular momentum is reproduced and the band termination is also in accord with the data; the slope of the moment of inertia is too weak and one could imagine that such phenomenon is related with the size of the truncated configuration space. The same is true for the slope of the kinetic moment of inertia of SD-1 band of the ^{194}Pb isotope. In order to understand the reason for the weak slopes , one tests first the response of the system, when changing the pairing interaction strength within large boundaries. It is given on figure 3 for the ^{194}Pb Yrast SD band, where a measure of spreading (S_d) of the occupation numbers around the Fermi level, defined by $S_d = tr(\hat{\rho}(1-\hat{\rho}))^{\frac{1}{2}}$ (which is equivalent in the BCS notation to $S_d = \sum_i (u_i v_i)$, see e.g. Ref. [17]) is plotted against the pairing strength.

Figure 2 displays the Cr. HTDA results as obtained from static calculations ($\Omega = 0$). When $S_d \to 0$ then one is performing Cr. HTDA without correlations. When one increases the pairing strength the amount of correlations (measured by S_d) gradually increases, upon arriving at some level, where it tends to increase only marginally. Something limits its further expansion. It is our contentio that this limits its further expansion. One supposes that this is due to the restricted character of the configuration space. As a matter of fact, it is limited in terms of energy and rank of the p-h excitations.

On figure 3, at low angular velocity ($\Omega = 50KeV$), one shows the progression of the angular momentum versus the change of X. When $X \to 0$ one must clearly obtain the Cr. HF limit. Upon increasing the cut-off parameter, the pairing correlation effects are increasing and are expressed also by increasing. This may be expressed also as an increase of counter-rotating intrinsic currents (see Ref. [18, 19]) and thus they gradually decrease the angular momentum, the kinetic moment of inertia and minimize the collective kinetic rotational energy for a given angular velocity. Some shell effects related to the nature of new included sp-states in the

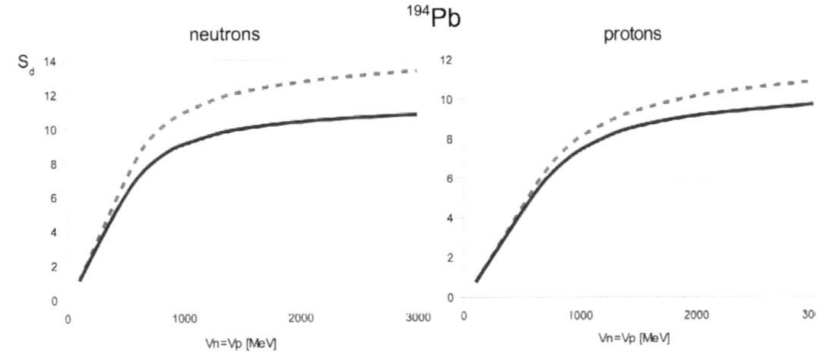

Figure 2. Spreading of the occupation numbers around the Fermi level (S_d, dimensionless) versus pairing strength $V_n = V_p$ (in MeV). The blue continuous line corresponds to Cr. HTDA calculations for up to 2p2h excitations, the red dashed - adds highly truncated (as described in the body text) version of the 4p4h space.

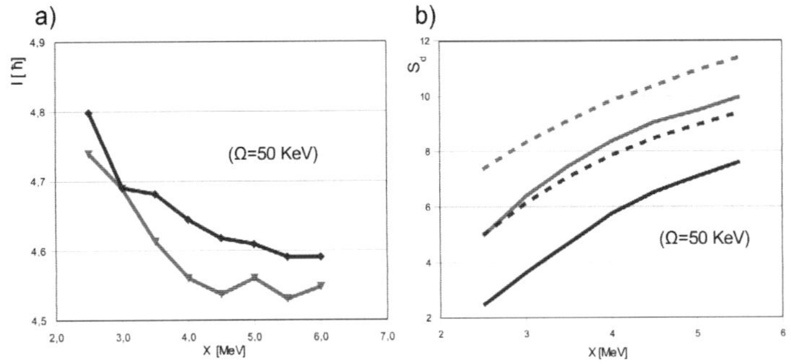

Figure 3. Angular momentum I - (side a) and spreading S_d - (side b) around the Fermi level versus the cut-off parameter X . Red curves and full triangles on side a (resp. blue curves and diamonds on side (a)) stands for $V_n = V_p = 1400MeV$ (resp. $V_n = V_p = 700MeV$) On side (b) full lines (resp. dashed lines) indicate proton (resp. neutron) distribution.

window could change slightly the smoothness of such progression, but not the trend. When one increases too much the pairing interaction window, the intrinsic vortical current generation begins to saturate and the shell effects could even counter the decreasing trend as it is shown at this figure for higher X. This behaviour on figure 3 is more prononced for the curve with the larger value of the pairing strength $V_n = V_p = 1400MeV$ as compared with $V_n = V_p = 700MeV$, because when $X = 4MeV$ such a pairing strength already lies deep in the saturation zone as seen on figure 2. As far as, the type of excitations is limited to 2p2h, one is unable to obtain less saturated curves, especially using pairing strengths, suitable to reproduce the experimental I. As an example, for $X = 4MeV$ and $\mu = 0.5MeV$, one must choose a very high pairing strength ($1400MeV$) in order to obtain meaningful value of I at low spin. Then, the only way to get pairing strengths, which at the same time correspond to non-saturated stretches of the curves on figure 2 and give the right I, is to increase the rank of excitations included in the configuration space. Especially interesting is the addition of 4p4h excitations to the configuration space, because when the residual interaction favors pairing correlations, the excitations of pair rank represent biggest share in the wave function decomposition mix in equation 1. On figure 2 one has also reported the change, produced over the spreading (S_d) by adding 4p4h excitations under a special selection/truncation scheme discussed below. As it is seen from figure 2 the addition of 4p4h excitations allows to obtain the the same amount that with $V_n = V_p = 1400MeV$, but with lower values of pairing strength. This time, a range of $V_n = V_p$ between 900 and 1000 MeV is sufficient to achieve the same amount of correlations (and spin). That is mostly out of the saturation zone. When computational difficulties will be resolved as discussed below, one will include more 4p4h and possibly all of the 3p3h excitations in the configuration space and thus one will obtain realistic pairing strengths probably far below the saturation region. At the limit of a full configuration space one would obtain pairing strengths very close to the BCS ones.

The recipe to select the most "important" 4p4h excitations included in the space of the present calculations is described below. At any Cr. HTDA iteration with 0p0h, 1p1h, 2p2h, 4p4h excitations a faster and preliminary iteration with only up to 2p2h excitations is performed, where the 200 2p2h excitations with highest probability (χ^2) in the decomposition (1) are isolated. Next, the genuine Cr. HTDA iteration with 0p0h, 1p1h, 2p2h, 4p4h excitations is performed, where the 4p4h excitations used are only those, which are based on the selected 200 high probable 2p2h excitations. This is done by exciting 2 more particles and holes on these particular configurations. They are moreover limited in energy to up to $2X + 5\mu/2$ energy level, which is the same as those for the 2p2h configurations and it is in agreement with the definition of the sp-window. In principle, the bulk of 4p4h excitations lies above this limit, but the computational limits restrict us to such schemes.

As already said, to extract a few eigenstates and eigenvalues in such matrices, the Lanczos algorithm has been used with the numerical code written by B.N. Parlett and D.S. Scott [20] available as an open source.

5. Computational background

A good computing speed is requested to be able to use an appropriate configuration space. The difference from existing large scale shell model calculations is that the configuration space could be smaller in our case, due to the fact that for stable heavy nuclei $N > Z$. One could, thus, neglect neutron-proton pairing correlations as already mentioned. But, by imposing meaningful value for the pairing interaction window - $X = 4\,MeV$ and $\mu = 0.5\,MeV$ one could obtain up to 60 sp-states in the so-defined "valence" space for one nucleon type. This exceeds by far the usual large shell model valence space numbers, where Wick theorem are performed for much more restricted spaces. In such Cr. HTDA window, only parity-signature symmetries are imposed and thus configuration space for up to $10^6 - 10^7$ could be obtained. Here, the problem with the computation time does not lie in the Lanczos procedure process, as it is the case for large scale shell model calculations. In the Cr. HTDA method, the dimension of the energy matrix n is smaller, and eigenvectors for matrices of corresponding dimension could be easily found, taking into account their sparse nature, in a Lanczos algorithm. Here, the time for building the matrix is critical, as far as many Wick theorem calculations must be performed quickly (about $10^{12} - 10^{14}$ times) when Hamiltonian is changed at any iteration. Such problem could be solved by identifying and thus excluding quickly the non-zero elements of H_{ij}. On figure 4 the dependence of time of matrix construction and eigen value finding is displayed as a function of the size of the matrix n.

One uses in our case a binary description of any excitation, as an up to 64-bit bitstring is assigned to describe the particle part of any type of excitation and another 64-bit bitstring for the hole part of the excitation. For example the excitation defined below on the left side:

$$|A>= a^+_{i_1} a^+_{i_2} a_{i_3} a_{i_3} |HFcore> \rightarrow \{(00001000010000)_{particles}, (0000000100100)_{holes}\}, \quad (11)$$

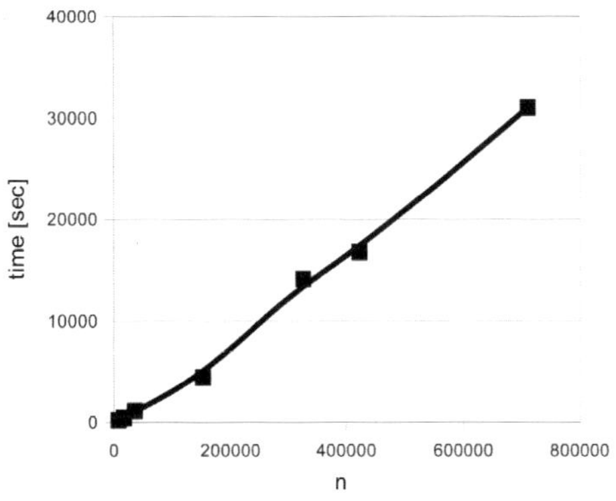

Figure 4. The time necessary to build the matrix and find the lowest eigensolution versus different dimensions (n) of the matrix

is represented by the couple of bitstrings on right. Here, any bit set to 1 represents a specific particle or hole sp state excited in the considered configuration. The bits corresponding to other sp states are set to 0. In such a way excitations in a pairing action window of up to 128 sp-states could be described. Let us consider matrix elements of equation (6)of the type

$$\sum_{ijkl} V_{ijkl} < HFcore|a^+_{i_1}a^+_{i_2}...a_{i_{n+1}}a_{i_{n+2}}...|\ a^+_i a^+_j a_l a_k\ |a^+_{j_1}a^+_{j_2}...a_{j_{n+1}}a_{j_{n+2}}...|HFcore> \tag{12}$$

A 2-body operator can change only 4 states/bits of the description of the right quantum state. The Hamming distance between two bitstrings representing the states $|A>$ and $|B>$ is defined as the sum of the numbers of particle and hole excitations necessary to transform one state into the other. This allows us to write in simple terms the selection rules for having non vanishing matrix elements for a p-body interaction : the Hamming distance between the l.h.s and the r.h.s of the matrix element should be less than $2p$. For example, if one looks for $< A|V^{res}|B >$ calculation, where $|A>$ is described by the couple of bitstrings in Eq. (11) and $|B>$ for instance as:

$$|B>= a^+_{j_1}a^+_{j_2}a_{j_3}a_{j_3}|HFcore> \rightarrow \{(00000000000011)_{particles}, (0001100000000)_{holes}\}, \tag{13}$$

then the bitwise exclusive (XOR) operation between bitstrings, which describe $|A>$ and $|B>$ gives:

$$\{(00001000010011)_{particles}, (0001100100100)_{holes}\}, \tag{14}$$

where the ones take places of sp-excited states which differ in $|A>$ and $|B>$. In this particular case, one sees that $|A>$ and $|B>$ differ by 8 excitations and thus it requires at least a 4-body V^{res} operator. Then one can simply skip the calculation of this matrix element and go the next. The faster the Hamming distance computation is, the faster the H_{ij} construction becomes and less truncation limits of the configuration space are needed for a given total computing time. One uses, here, the Donald Knuth algorithm (Ref. [21]) for quick finding the Hamming distance for any i and $j > i$ (the H_{ij}matrix is symmetric and real) which could in principle compute one Hamming distance for 10-60 clock cycles, when one programs it in a high level language. But, recently the NSA of US obliged indirectly microprocessors vendors to add to

Journal of Physics: Conference Series **205** (2010) 012005 doi:10.1088/1742-6596/205/1/012005

mass market microprocessor's instruction sets (used so frequently in NSA supercomputers), instructions that allow hardware realizations of Hamming distance calculation for bitsets. These SIMD instructions (instruction POPCNT in the SSE4a instruction set of the x86 processors) allow assessing 30 bitstrings per clock cycle, thus promising much higher speed of calculation of our matrix elements. Using them together with other similar techniques of calculations, the inclusion of all 3p3h and 4p4h excitations into our configuration space should be possible. It is expected that such an enlargment of our configuration mixing basis should be able to cure the deficiencies of the moments of inertia slopes obtained up to now with the Cr. HTDA method.

6. Conclusions and perspectives

The Cr. HTDA calculations, when performed with sufficiently large configuration spaces and less truncations, should constitute a powerful tool to study theoretically rotational bands. In the same framework, pairing or (low multipolarity) multipole-multipole forces could be taken into account. Such task requires some advanced computational techniques. When applied, they are expected to yield improved results for rotational or mixed vibrational-rotational bands. It has been the object of this contribution to explore the potential consequences of using such tecniques to describe these bands.

Acknowledgements

One of authors was supported by contract F-1502/05 of Bulgarian NSF in order to achieve part of the work and to participate in the Varna Nuclear Physics school. Three of us were supported also by PICS agreement 4847 (2009) CNRS-IN2P3/BAS.

References

[1] Pillet N, Quentin P and Libert J 2002 *Nucl. Phys.* A **697** 141.
[2] Pillet N, PhD Thesis 2000, Université Bordeaux I, France.
[3] Ha T L, Ph. D. Thesis 2004 Université Bordeaux I, France.
[4] Pillet N, Berger J-F and Caurier E 2008 *Phys. Rev.* C **78**, 024305
[5] Bonneau L, Quentin P and Sieja K 2007 *Phys. Rev.* C **76**, 014304
[6] Naidja H, Ph. D. Thesis 2009 Universit Sciences et Technologies - Bordeaux I
[7] Terasaki J, Heenen P-H, Dobaczewski P and Flocard H 1995 *Nucl. Phys.* A **593** 1.
[8] Gall B, Bonche P, Dobaczewski J et al. 1994 *Z. Phys.* A **348** 183.
[9] Peru S, PHD thesis 1997 Université Paris VII France.
[10] Bender M, Bonche P, Duguet T and Heenen P-H 2004 *Phys.Rev.* C **69**, 064303
[11] Valor A, Egido J L and Robledo L M 1997 *Phys. Lett.* B **393** 249.
[12] Afanasjev A V, Konig J and Ring P 1999 *Phys. Rev.* C **60** 051303.
[13] Bartel J, Quentin P, Brack M, Guet C and Haakansson H B 1982 *Nucl. Phys.* A **386** 79.
[14] Krieger S J, Bonche P, Flocard H and Quentin P 1990 *Nucl. Phys.* A **517** 275.
[15] Goodman A L 1974 *Nucl. Phys.* A **265** 466.
[16] Libert J, Girod M and Delaroche J-P 1999 *Phys. Rev.* C **60** 054301.
[17] Libert J, Laftchiev H, Quentin P 2003 Contribution to 22nd International Workshop on Nuclear Theory, 16-21, Rila Mountains, Heron Press-Sofia
[18] Durand M, Schuck P, Kunz J 1985 *Nucl.Phys.* A **439** 263
[19] Laftchiev H, Samsoen D, Quentin P and Mikhailov I N 2003 *Phys. Rev.* C **67** 014301.
[20] Parlett B N and Scott D N 1979 *Math.Comp.* **217** 33-145.
[21] Knuth D E, *The Art of Computer Programming*, Addison-Wesley Professional

Journal of Physics: Conference Series **205** (2010) 012006 doi:10.1088/1742-6596/205/1/012006

Ab Initio Open Core Shell Model for Nuclear Structure

J P Draayer, T Dytrych, and K D Sviratcheva

Department of Physics and Astronomy, Louisiana State University, Baton Rouge, LA 70803, USA

E-mail: draayer@lsu.edu

Abstract. The SU(3) symmetry-adapted version of the No-Core Shell Model (NCSM), which reduces to the Elliott SU(3) Model in its $0\hbar\omega$ limit, is described and shown to be effective in providing an efficient description of low-lying eigenstates of ^{12}C and ^{16}O. A symmetry-guided framework is suggested based on our recent findings of low-spin and high-deformation dominance in realistic NCSM results. This holds promise to significantly enhance the reach of *ab initio* shell models.

1. Significance of an *ab initio* symmetry-adapted open-core shell model

The dual challenge of understanding the strong interaction that precludes perturbative treatments, and the complexities of the quantum many-particle nuclear system that admit, for example, correlated phenomena and cluster sub-structures, places extraordinary demands on both physics and computational science. We aim to achieve significantly enhanced reach beyond that of the current no-core shell model (NCSM) by exploiting our knowledge of dominant symmetries, first of the interaction itself and then of those that emerge as a result of the many-body dynamics.

Ab initio approaches, which build upon a 'first principles' foundation, afford tremendous impact on advancing the frontiers of nuclear physics in multiple arenas, bridging from fundamental interactions and quantum chromodynamics (QCD) considerations to the physics of many-nucleon systems, including astrophysics and cosmology applications. Such approaches are crucial for achieving predictive power from fundamental input, and hence are vital for advancing energy applications as well as for gaining an understanding of nucleosynthesis and related nuclear reactions that are key to understanding the structure of atomic nuclei at a fundamental level.

It is important to acknowledge the impact the current *ab initio* NCSM (or open-core shell model, OCSM, for low N_{max} cutoff parameter) has already had on achieving a better understanding of the structure of light nuclei, from the deuteron up through ^{16}O and even for some selected cases beyond (e.g., see [1, 2] for NCSM and cluster NCSM/RGM results). However, it is also important to understand the limitations of the NCSM approach as currently implemented. Specifically, the combinatorial growth in the size of the basis or 'model spaces' that one encounters as a function of the N_{max} total $\hbar\Omega$ excitations, and the associated computational complexity and storage requirements this implies, means extending its reach relies on the availability of ever-larger and more powerful HPC resources. Unfortunately, the N_{max} values that can currently be accommodated, even if extended by 2 or 4 units to 10 or $12\hbar\Omega$ for 'p-shell'

Journal of Physics: Conference Series **205** (2010) 012006 doi:10.1088/1742-6596/205/1/012006

nuclei, fall short of what is required to take into account the presence of collective modes reflected in enhanced transition rates and cluster configurations. We aim to overcome this limitation, and achieve, for the first time, a unified microscopic description of low-lying states that reflect the close interplay of particle-hole modes, α clustering and collective rotational dynamics.

The symmetry-adapted OCSM (SA-OCSM) [3] advances the NCSM concept by recognizing that special symmetries often dominate the dynamics, and this means that physically relevant eigenstates can be described by a relatively few collective basis states that correspond to a special linear combination of a large number of NCSM basis states. (The SA-OCSM states can be obtained through a unitary transformation from the m-scheme basis used in the conventional shell model.) Hence, within the SA-OCSM framework, the growth of the model space can be managed by winnowing the model space to only physically relevant states as determined through symmetry considerations. The underlying concept of this framework is illustrated in our proof-of-principle study [3, 4] that exploits symplectic Sp(3, $\mathbb{R}$) symmetry and its SU(3) subgroup symmetry in an analysis of large-scale nuclear physics applications for ^{12}C and ^{16}O. What one learns from the outcome of these studies is that typically a small fraction of the full model space (Fig. 1), several orders of magnitude less than that of the corresponding NSCM approach, suffices to represent most of the physics – typically 90% or more as measured by projecting NCSM results onto a symmetry-adapted equivalent basis and noting that only a small subset of the full space contributes to the low-dynamics.

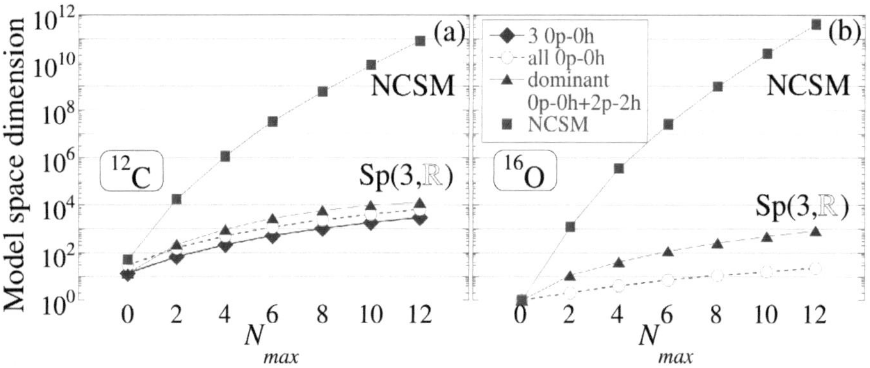

Figure 1. NCSM space dimension as a function of the $N_{\max}$ model space compared to that of the Sp(3, $\mathbb{R}$) subspace of the most dominant symplectic basis states: (a) $J = 0, 2$, and 4 for ^{12}C, and (b) $J = 0$ for ^{16}O.

The significance of the symplectic Sp(3, $\mathbb{R}$) symmetry for a microscopic description of a quantum many-body system of interacting particles [5, 6] follows from the physical relevance of its 21 generators[1]. Specifically, the generators are directly related to the particle momentum ($p_{s\alpha}$) and coordinate ($q_{s\beta}$) operators and constructed as $\sum_s p_{s\alpha}p_{s\beta}$, $\sum_s(q_{s\alpha}p_{s\beta} \pm q_{s\beta}p_{s\alpha})$, and $\sum_s q_{s\alpha}q_{s\beta}$ with α, $\beta = x$, y, and z for the 3 spatial directions and $s = 1, ..., A$ running over the nucleons. These generators realize important observables, such as the many-particle kinetic energy $\sum_{s,\alpha} p^2_{s\alpha}/2m$, the mass quadrupole moment and angular momentum operators, together with multi-shell collective vibrations and vorticity degrees of freedom for a description of rotational dynamics in a continuous range from irrotational to rigid rotor flows.

[1] While in its quantal realization the Sp(3, $\mathbb{R}$) symmetry is known to realize the famous Bohr-Mottelson collective model (1975 Physics Nobel Prize), it is interesting to note that in its classical realization the symplectic Sp(3, $\mathbb{R}$) symmetry underpins the dynamics of rotating bodies and, for example, the Sp(3, $\mathbb{R}$) model has been used to describe the rotation of deformed stars and galaxies [7].

The symplectic generators can be written as SU(3) tensor operators in terms of the harmonic oscillator raising ($b^{\dagger(1\,0)}$) and lowering ($b^{(0\,1)}$) operators,

$$\begin{aligned} A^{(2\,0)}_{\mathcal{L}M} &= \frac{1}{\sqrt{2}}\sum_{n=1}^{A}\left[b_n^{\dagger(1\,0)} \times b_n^{\dagger(1\,0)}\right]^{(2\,0)}_{\mathcal{L}M} \\ C^{(1\,1)}_{\mathcal{L}M} &= \sqrt{2}\sum_{n=1}^{A}\left[b_n^{\dagger(1\,0)} \times b_n^{(0\,1)}\right]^{(1\,1)}_{\mathcal{L}M} \\ H^{(0\,0)}_{00} &= \sqrt{3}\sum_{n=1}^{A}\left[b_n^{\dagger(1\,0)} \times b_n^{(0\,1)}\right]^{(0\,0)}_{00} + \frac{3}{2}A \end{aligned} \qquad (1)$$

with $B^{(0\,2)}_{\mathcal{L}M} = (-)^{\mathcal{L}-M}(A^{(2\,0)}_{\mathcal{L}-M})^{\dagger}$. The eight operators $C^{(1\,1)}_{\mathcal{L}M}$ generate the SU(3) subgroup of $\mathrm{Sp}(3,\mathbb{R})$ and are related to the angular momentum operator L_{1q} and the Elliott algebraic quadrupole moment tensor $\mathcal{Q}^a_{2q}$ as follows,

$$\begin{aligned} C^{(1\,1)}_{1q} &= L_{1q}, \quad q = 0, \pm 1 \\ C^{(1\,1)}_{2q} &= \frac{1}{\sqrt{3}}\mathcal{Q}^a_{2q}, \quad q = 0, \pm 1, \pm 2. \end{aligned} \qquad (2)$$

The deformation-related $(\lambda\,\mu)$ set of quantum numbers labels SU(3) irreducible representations, irreps. Consequently, SU(3)-symmetric states (and hence symplectic basis states that are built on these) bring forward important information about nuclear shapes and deformation in terms of $(\lambda\,\mu)$, for example, $(0\,0)$, $(\lambda\,0)$ and $(0\,\mu)$ describe spherical, prolate and oblate shapes, respectively.

A basis of a symplectic irrep is constructed by acting with symmetrically coupled polynomials in the symplectic raising operators, $A^{(2\,0)}$, on a set of basis states of a symplectic bandhead, $|\sigma; S_\sigma\rangle$, which is a $\mathrm{Sp}(3,\mathbb{R})$ lowest-weight state,

$$|\sigma n\rho\omega\kappa(LS_\sigma)JM_J\rangle = \left[\left[A^{(2\,0)} \times A^{(2\,0)} \ldots \times A^{(2\,0)}\right]^{n} \times |\sigma; S_\sigma\rangle\right]^{\rho\omega}_{\kappa(LS_\sigma)JM_J}, \qquad (3)$$

where $\sigma \equiv N_\sigma\,(\lambda_\sigma\,\mu_\sigma)$ labels $\mathrm{Sp}(3,\mathbb{R})$ irreps, $n \equiv N_n\,(\lambda_n\,\mu_n)$, $\omega \equiv N_\omega\,(\lambda_\omega\,\mu_\omega)$, and $N_\omega = N_\sigma + N_n$ is the total number of oscillator quanta related to the eigenvalue, $N_\omega\hbar\Omega$, of the three-dimensional harmonic oscillator Hamiltonian. The symplectic structure divides the full space into $\mathrm{Sp}(3,\mathbb{R})$ irreps (multi-shell "vertical" cones). The structure accommodates particle-hole (p-h) configurations in a natural way and in this way spans the entire space. The $\mathrm{Sp}(3,\mathbb{R})$ symmetry provides for a further organization of the SU(3) basis of the SA-OCSM, which like the latter, also spans the entire space.

2. Symmetry-guided framework

The symmetry-guided framework of the SA-OCSM utilizes a natural preference towards low-spin and high-deformation dominance as revealed in realistic NCSM wavefunctions. Specifically, our recent study [8] showed that for the $6\hbar\Omega$ NCSM results for ^{12}C using the effective N^{3}LO interaction (and similarly for other interaction choices, such as JISP16) with $\hbar\Omega$=15 MeV (Fig. 2), proton (neutron) spin values S_π $(S_\nu) = 0$ and 1 are sufficient to describe about 99% of the converged $J = 0^+_{\rm gs}$, 2^+_1, and 4^+_1 ^{12}C NCSM eigenvectors. The residual 1% involves time/memory consuming higher-spin and consequently less relevant configurations. Hence, retaining only the relevant proton (neutron) spin values allows one, with the same computer resources, to accommodate the full basis within the selected spin spaces up through higher

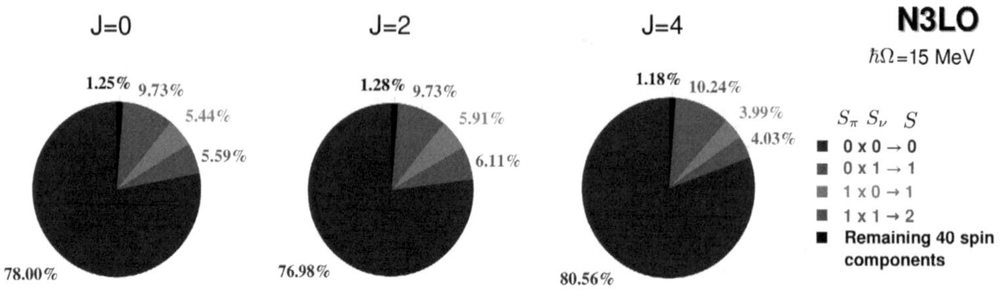

Figure 2. Intrinsic spin structure of $J = 0^+_{\rm gs}, 2^+_1$, and 4^+_1 NCSM states in ^{12}C obtained using N^{3}LO effective interactions in the $N_{\rm max} = 6$ model space with $\hbar\Omega = 15$ MeV. Contributions from high S_π (S_ν) proton (neutron) spin values are found negligible.

$\hbar\Omega$ values. Furthermore, in a proton-neutron formalism, the spin quantum numbers dictate the spatial symmetry because the fermion statistics requires overall antisymmetry within each sector. Hence, low-spin implies maximum spatial symmetry, which is where one finds the SU(3) configurations of maximum spatial deformation (see, e.g., Fig. 3) [4] .

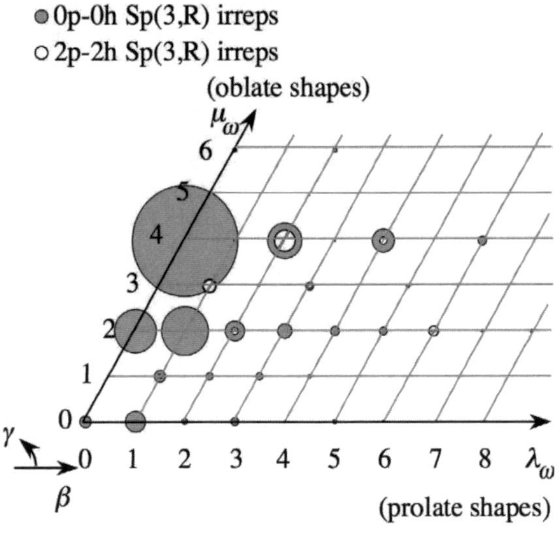

Figure 3. Probabilities (specified by the area of the circles) for the symplectic states which make up the most important 0p-0h (blue circles) and $2\hbar\Omega$ 2p-2h (empty circles) symplectic irreps, within the 0^+ ground state in ^{12}C, calculated by NCSM, $\hbar\Omega = 15$ MeV. The Sp(3, $\mathbb{R}$) states are grouped according to their $(\lambda_\omega\ \mu_\omega)$ SU(3) symmetry, which is mapped onto the $(\beta\,\gamma)$ shape variables of the collective model.

Winnowing considerations of this type illustrate that the SA-OCSM offers a systematic framework for down-selecting to physically relevant and manageable subspaces associated with full NCSM based on spin and deformation selection, which are complementary and mutually reinforcing. This is illustrated in Fig. 4, where a manageable (e.g, $N_{\rm max} = 6$) NCSM many-body space is schematically shown (a) together with an associated manageable space within the symmetry-guided framework that underpins the SA-OCSM (b). In the SA-OCSM structured approach, the space is first separated into proton and neutron subspaces and then, in keeping with the LS coupling foundation of the SA-OCSM, this is further organized according to the total proton spin (S_ν) and neutron spin (S_π) coupled to total spin (S). This decomposition enables one to trim the space to the dominant low-spin configurations.

These spin spaces can be further organized into SU(3) structures, each of which realizes a

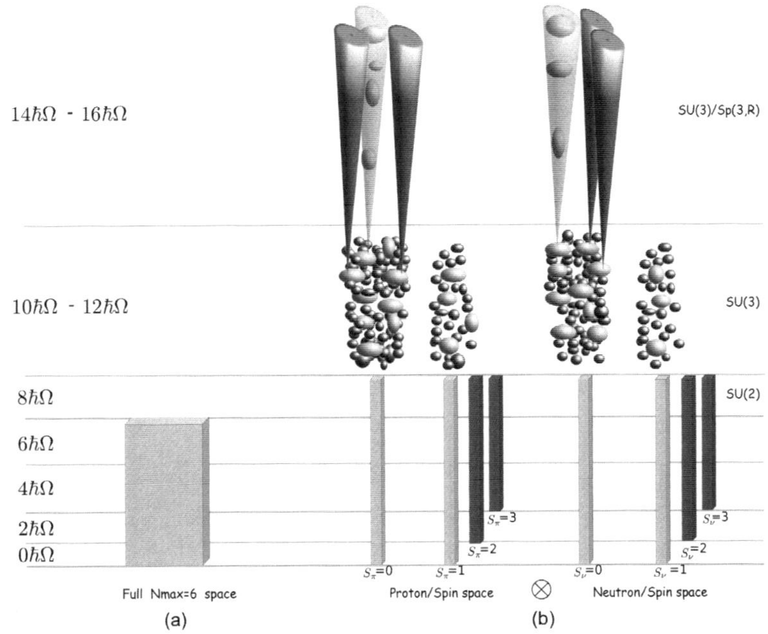

Figure 4. Schematic illustration of a manageable many-body basis of (a) NCSM ($N_{\max} = 6$) and (b) SA-OCSM with the symmetry-guided extension of high-$\hbar\Omega$ configurations. Green (light) color denotes physically significant configurations and grey (dark) color denotes physically irrelevant configurations.

nuclear shape deformation (represented as 'ellipsoids' in Fig. 4). Since typically configurations of maximum spatial deformation dominate, only a percentage of the proton(neutron)/spin subspace need be further considered (green 'ellipsoids') to accommodate the physically most significant configurations, including particle-hole excitations that are important for a description of clustering modes, in higher $\hbar\Omega$ model spaces (e.g., $N_{\max} = 10 - 12$ in Fig. 4). And beyond these, the Sp(3, $\mathbb{R}$) symplectic symmetry can guide an extension of model spaces to incorporate even higher $\hbar\Omega$ values (e.g., $N_{\max} = 14 - 16$ in Fig. 4) [3] required to gain convergence of the lowest bound 0^+ states in light nuclei (green cones in Fig. 4).

An illustrative example for the symmetry-guided framework of the SA-OCSM is shown in Fig. 5 for the case of the $N \sim Z$ pf-nucleus ^{64}Ge of astrophysical significance. The figure shows the combinatorial (near-exponential) growth of the full NCSM space (blue squares). Employing spin considerations, the full space can be reduced roughly by an order of magnitude while achieving considerable decrease in computing intensity because low-spin configurations typically have a simpler structure than for higher ones. In the example of Fig. 5, we chose to retain only proton (neutron) spin values $S_\pi = 0$ and 1 ($S_\nu = 0$ and 1) coupled to total spin $S = 0$, 1 and 2 (no restrictions, red diamonds). Further reductions of several order of magnitude can be achieved by selecting SU(3) proton (neutron) modes according to their shape deformation with the most deformed structures playing the foremost role in nuclear collectivity. As for the spin, the selected proton SU(3) configurations are coupled to the selected neutron configurations to yield the total number of many-body states (yellow, green and purple triangles). Beyond these winnowing considerations, only reduced matrix elements of the associated highest-weight states need to be calculated and stored because the associated coupling coefficients can be computed 'on the fly'.

For shell models that employ restricted model spaces a suitable Hamiltonian renormalization,

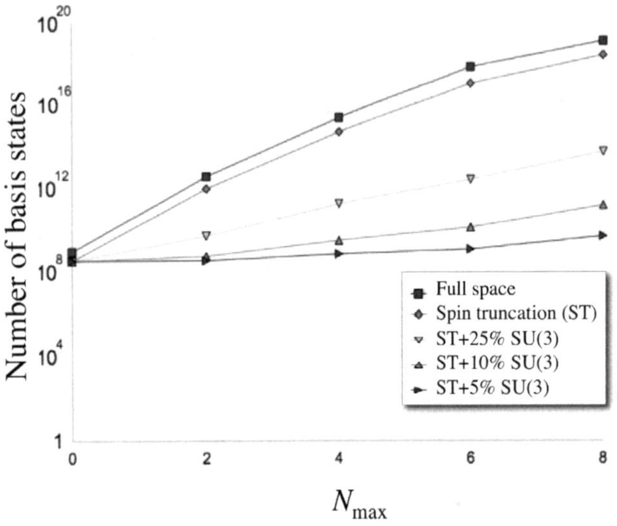

Figure 5. Dimensions of various model spaces for $^{64}_{32}Ge_{32}$: conventional NCSM model spaces (blue squares) and reduced model space dimensions when only proton (neutron) S_π (S_ν) = 0 and 1 values coupled to total spin 0, 1 and 2 are considered (red diamonds) and when beyond this only 25%, 10% and 5% of the most deformed proton (neutron) SU(3) configurations are retained (yellow, green and purple triangles, respectively) as a function of $N_{\rm max}$.

which is properly derived from realistic nucleon-nucleon interactions tied to QCD, is required in order to preserve predictive capabilities. A novel approach, the Similarity Renormalization Group (SRG) [9], which was recently adopted in nuclear physics and continues to enjoy successes in, e.g, condensed matter and high energy physics, decouples spaces that are relevant for the regime of nuclear dynamics from highly-energetic irrelevant configurations. Moreover, reflecting the symmetries in play for the SA-OCSM, we suggest a SU(3)-based SRG renormalization [10], which is suitable for SU(3)-adapted shell model calculations. This is because the SRG in a SU(3) basis results in a unitarily transformed NN interaction that respects the SU(3) symmetry, that is, a (near-)diagonal Hamiltonian in the basis used. In addition, while an important but challenging part of the SRG approaches is to properly account for the many-body forces induced during the renormalization (in order to preserve the unitarity of the SRG transformations), the use of symmetries, e.g. highest-weight states and reduced matrix elements, plays a crucial role in addressing this issue.

In short, the SA-OCSM builds upon a proton-neutron formalism in a LS coupling scheme with the proton/neutron spaces organized into subspaces of definite spin and further into spatial SU(3) representations that key to deformation, and for even higher $\hbar\Omega$ model spaces, into symplectic $\mathrm{Sp}(3,\mathbb{R})\supset\mathrm{SU}(3)$ structures. The symmetry-guided framework utilizes the physical relevance at each level of the structured space, namely, low-spin and high-deformation dominance together with most important patterns of symplectic excitations. This approach will open up an entire region of the periodic table to investigation with *ab initio* methods with forefront predictive capabilites.

Supported by the U.S. National Science Foundation (PHY-0500291 & OCI-0904874), and the Southeastern Universities Research Association (http:/www.sura.org).

References

[1] Navrátil P, Vary J and Barrett B 2000 *Phys. Rev. Lett.* **62** 054311
[2] Navrátil P, Quaglioni S, Stetcu I and Barrett B 2009 *J. Phys. G: Nucl. Part.* **36** 083101
[3] Dytrych T, Sviratcheva K, Bahri C, Draayer J and Vary J 2007 *Phys. Rev. Lett.* **98** 162503
[4] Dytrych T, Sviratcheva K D, Draayer J P, Bahri C and Vary J P 2008 *J. Phys. G: Nucl. Part. Phys.* **35** 123101
[5] Rosensteel G and Rowe D 1977 *Phys. Rev. Lett.* **38** 10
[6] Rowe D 1985 *Reports on Progr. in Phys.* **48** 1419
[7] Rosensteel G 1993 *Astrophys. J.* **416** 291

[8] Dytrych T, Draayer J P, Sviratcheva K D, Bahri C and Vary J P 2009 *"Nuclear Structure And Dynamics '09": Proceedings of the International Conference, Dubrovnik, Croatia, 4-8 May 2009* vol 1165 ed Milin M, Niksic T, Szilner S and Vretenar D (AIP Conference Proceedings)
[9] Bogner S, Furnstahl R and Perry R 2007 *Phys. Rev. C* **75** 061001(R)
[10] Sviratcheva K, Dytrych T and Draayer J 2009 *to be submitted*

Journal of Physics: Conference Series **205** (2010) 012007 doi:10.1088/1742-6596/205/1/012007

Microscopic study of low energy collective states in even-even nuclei: a prospective analysis of dynamical corrections to vibrational mass parameters.

J Libert[1], J-P Delaroche[2], M Girod[2], H Goutte[2], S Hilaire[2], S Péru[2], N Pillet[2] and GF Bertsch[3]

[1] Institut de Physique Nucléaire, IN2P3-CNRS/Université Paris-Sud,91406 Orsay cedex, France.

[2] CEA, DAM, DIF, F-91297 Arpajon France.

[3] Department of Physics and Institute of Nuclear Theory, Box 351560 University of Washington Seattle, WA 98915 USA.

E-mail: `libert@ipno.in2p3.fr`

Abstract. A systematic study of low energy nuclear structure at normal deformation has been carried out using the Generator Coordinate Method mapped onto a 5-Dimensional Collective quadrupole Hamiltonian (5DCH) by using the Gaussian Overlap Approximation (GCM-GOA). The collective space is spanned by Hartree Fock Bogoliubov (HFB) states under axial and triaxial quadrupole constraints deduced with the D1S Gogny force. In addition, our 5DCH includes the Thouless-Valatin dynamical corrections to its rotational kinetic terms. The work is described in detail elsewhere together with the corresponding comparisons with experimental data when it is available. Many properties show a satisfactory agreement with experiment, but there is an almost systematic overestimation of vibrational band head energies, which is the subject of the present paper. We show here the performance of the theory on related observables, and propose improvements of the theory to address the problem of the vibrational band heads. An important reason for the deficient is the treatment of vibrational inertial parameters. In particular, the theory needs to include the dynamical Thouless-Valatin corrections to the vibrational terms in the 5DCH. In present work, on the aim of a simple formula grounded by known symmetry rules within the 5DCH, these dynamical TV corrections are roughly estimated, allowing us to handle their possible effects in term of spectroscopic properties, to present a guess of the next model improvment, and to isolate some areas in the chart where states $I_n^\pi = 0_2^+$, 2_2^+ or 2_3^+ seem to have important components out of the scope of a pure collective quadrupole approach.

1. Introduction.

Until recently, the extensions beyond mean field needed to describe dynamical properties such as excitation energies have not been carried out in a systematic way on the whole chart of nuclides. However, a large scale study on the ground of a microscopical quadrupole 5- Dimensional Collective Hamiltonian (5DCH) has been recently proposed [1]. It relies upon quadrupole Constrained Hartree-Fock-Bogoliubov (CHFB) mean field deduced from the D1S Gogny force. Under the Gaussian Overlap Approximation, the Generator Coordinate Method is employed to reduce the so-called Griffin-Hill-Wheeler equation to a Shrödinger form. Local approximations as Inglis Belyaev cranking formulas and the self-consistent cranking model are then involved to evaluate the ingredients of the deduced collective 5DCH. As a result, a large set of benchmarks have been obtained in this context, concerning ground state (gs) properties, and low energy quadrupole collective state spectroscopy up to $I^{\pi} = 6^{+}$ at Normal Deformation (ND), in even-even nuclei with proton numbers Z=10 to 110, from dripline to dripline and neutron number $N \leqslant 200$. This connected work is described in great details in [1] together with available comparisons with experimental data.

The aim of the present paper, is to report about a preliminary and prospective study about vibrational mass parameters within this 5DCH approach. As a matter of fact, the 5DCH in the state of art defined by [1] takes into account microscopically Thouless-Valatin (TV) dynamical corrections in its rotational terms. This is not the case in its vibrational term. The implementation of an Adiabatic Time Dependant HFB evaluation of the vibrational kinetic tensor remains a quite heavy task, in particular numerically for large scale calculations. This implementation, in the spirit defined for instance in [2], is presently in progress but not yet achieved. However, as it is the case for the rotational term, the TV correction is expected to increase mass parameters and therefore to decrease vibrational quasi-β and quasi-γ bandhead energies. This expected effect is therefore consistent with an observed and almost general overestimation in of $I_n^{\pi} = 0_2^{+}$ and $\left(2_2^{+} \text{ or } 2_3^{+}\right)$ state excitation energies in [1].

In previous versions of the 5DCH belonging to the same context, in particular approaches of Super Deformation (SD) in the $Hg-Pb$ [3] and the Actinide [4] regions, rotational TV corrections where approximately taken into account by a constant 4/3 scaling factor applied to Inglis Belyaev moments of inertia. That has been shown to be in general case a quite realistic assumption. This somewhat crude input has been overcrossed in [1] , but we adopt now a similar strategy for the vibrational term. However, as it will be shown, the precise quantitative information obtained about TV effects on rotational term, through known symmetry relations in the 5DCH, may drive a somewhat more sophisticated scaling of the vibrational mass parameters.

2. Brief review of formalism.

We limit ourselves to a fast overview of the 5DCH method to fix the notation. Any detail can be found in [1] and references therein. In the second subsection, the method employed to renormalize vibrational mass parameters is examined in full detail.

2.1. Overview on the 5DCH approach.

2.1.1. CHFB equations. The present 5DCH approach relies first of all upon the resolution of static and dynamic CHFB equations within the triaxial set of symmetries, namely :

$$\delta\left\langle \Phi^{\omega\dot{q}}\left(q_0,q_2\right)\left|\widehat{H}-\lambda_0\widehat{Q}_0-\lambda_2\widehat{Q}_2-\lambda_Z\widehat{Z}-\lambda_N\widehat{N}-\left(\omega\widehat{I}_k\right)-\left(\dot{q}\widehat{\mathcal{P}}\right)\right|\Phi^{\omega\dot{q}}\left(q_0,q_2\right)\right\rangle=0, \quad (1)$$

-where $\Phi^{\omega\dot{q}}\left(q_0,q_2\right)$ is the Bogoliubov vacuum of quasiparticle, the corresponding total energy being named $E_{CHFB}^{\omega\dot{q}}\left(q_0,q_2\right)$.

-where $\widehat{H}$ is the two-body hamiltonian deduced from the Gogny effective interaction, here under its most usual parametrisation D1S [5]-[6].

-where Lagrange multipliers λ_m for static constraints are fixed by the number of particles of each type, ($\left\langle\widehat{Z}\right\rangle = Z$, $\left\langle\widehat{N}\right\rangle = N$) and the quadrupole deformations ($\left\langle\widehat{Q}_0\right\rangle = q_0$, $\left\langle\widehat{Q}_2\right\rangle = q_2$).

-where the optional dynamical constraints (see section 2.1.3) refer to ω the angular velocity and $\widehat{I}_k$ the component on the axe k of the angular momentum, and, similarly, $\dot{q}$ the velocity in deformation variables and $\widehat{\mathcal{P}}$ the Baranger Veneroni impulsion operator for the adiabatic

Journal of Physics: Conference Series **205** (2010) 012007 doi:10.1088/1742-6596/205/1/012007

collective motion [7], i.e. $\widehat{\mathcal{P}} = i\hbar\left[\frac{d\mathcal{R}_0}{dq}, \mathcal{R}_0\right]$, $\mathcal{R}_0$ being the generalized density matrix associated with the static Bogoliubov state $\Phi^{00}(q)$.

As it is usual, quadrupole moments (q_0, q_2) are replaced by the dimensionless variables (β, γ) defined by $\beta = \sqrt{5\pi}\sqrt{q_0^2 + 3q_2^2} \,/\, \left(3A^{\frac{5}{3}}\, r_0^2\right)$ and $\gamma = \arctan\left(\sqrt{3}q_2/q_0\right)$ with $A = N + Z$ and $r_0 = 1.2\ fm$. Associated cartesian coordinates $(a_0 = \beta\cos\gamma,\ a_2 = \beta\sin\gamma)$ are also in use.

The CHFB equations are solved by expansion of single particle states onto a triaxial harmonic oscillator basis [8] [9]. Improvments for present large scale calculations are provided in [1].

2.1.2. The main 5DCH equation. Under the Gaussian Overlap Approximation, the Griffin-Hill-Wheeler equation [10] [11] for the five components of the quadrupole tensor in a space spanned by the CHFB states Φ^{00} (and their rotational transforms) is reduced to a Schrödinger form (see a detailed discussion about the GCM-GOA approach in [12]). The deduced collective Hamiltonian $\widehat{\mathcal{H}}_{\mathrm{col}\,l}$ written in the intrinsic axes system is referred by three Euler angles (through operators $\widehat{I}_k$) and the two quadrupole deformation variables (a_0, a_2). It takes the form of the so-called Bohr Hamiltonian in its most general version [13], i.e.

$$\widehat{\mathcal{H}}_{\mathrm{col}\,l} = \frac{1}{2}\sum_{k=1}^{3}\frac{\widehat{I}_k^2}{\mathcal{J}_k} + \sum_{m,n=0\ and\ 2} D^{-\frac{1}{2}}\frac{\partial}{\partial a_m}D^{\frac{1}{2}}\left(B_{mn}\right)^{-1}\frac{\partial}{\partial a_n} + V, \tag{2}$$

where $D(a_0, a_2)$ is the metric and where the potential part $V\ (a_0, a_2)$ writes [12] $V = E^{00}_{CHFB} - \Delta V^{ZPE}$.

The six kinetic functions of the deformation (a_0, a_2), i.e. the three moments of inertia $\mathcal{J}_k$ along principal axes k, the three mass parameters B_{00}, $B_{02} = B_{20}$ and B_{22} together with the Zero Point Energy correction ΔV^{ZPE} are evaluated under local approximations as explained in section 2.1.3.

A numerical resolution of the collective Schrödinger equation is implemented, following methods explained in [3]. The collective wavefunction Ψ_n^{IM} associated with the n^{th}eigenvalue with angular momentum I (with projections and M and K on the third axis of the laboratory and intrinsic frame respectively is finaly such that :

$$\widehat{\mathcal{H}}_{\mathrm{col}\,l}\left|\Psi_n^{IM}\right\rangle = E_n^I\left|\Psi_n^{IM}\right\rangle \quad and \quad \left|\Psi_n^{IM}\right\rangle = \sum_{\substack{0\leqslant K\leqslant I \\ K\ even;K\neq 0\ for\ I\ odd}} g_n^{IK}(a_0, a_2)\left|IMK\right\rangle \tag{3}$$

where $|IMK\rangle$ is a standard normalized combination of rotational Wigner matrices.

2.1.3. Evaluation of $\mathcal{J}_k$, B_{mn} and ΔV^{ZPE} functions. A selfconsistent cranking calculation is implemented for each deformation point. In this context, moments of inertia are calculated as :

$$\mathcal{J}_k = \mathcal{J}_k^{TV} = \lim_{\omega\to 0}\frac{\left\langle\Phi^{\omega\ 0}\right|\widehat{I}_k\left|\Phi^{\omega\ 0}\right\rangle}{\omega}, \tag{4}$$

which has been shown to be equivalent to the Thouless-Valatin inertia [14]. A fixed value of ω, namely $\omega = 0.002$ MeV, is employed to approximate the limit numerically.

The static CHFB calculation of $\Phi^{0\ 0}(a_0, a_2)$ leads us with its related quasiparticle (qp) states $|\mu\rangle$, $|\nu\rangle$, ... and qp energies E_μ, E_v, ... That allows to refer also in present work to standard Inglis-Belyaev inertia [15] [16] labelled as $\mathcal{J}_k^{Inglis}$, calculated as for instance in [3].

In an Adiabatic Time Dependant HFB frame, Thouless Valatin vibrational mass parameters would be similarly defined [2] [3] [7] as a limit at low velocity of the type :

$$B_q^{TV} = \lim_{\dot{q}\to 0} \frac{\left\langle \Phi^{0\ \dot{q}} \left| \mathcal{P} \right| \Phi^{0\ \dot{q}} \right\rangle}{\dot{q}}. \tag{5}$$

Such calculation being not yet achieved, standard cranking formulas are employed. Cranking mass parameters are deduced within a linear response theory very similar to those involved for Inglis-Belyaev moment of inertia. Here, they are therefore also labelled as "Inglis" mass parameters. Defining in standard way the moment $\mathcal{M}_k^{ij}(q)$ for the Bogoliubov state $\Phi^{0\ 0}(q_0, q_2)$, mass parameters and ZPE correction are evaluated as [12] [1] :

$$B_{ij}^{Inglis}(q) = \frac{1}{2}\frac{\mathcal{M}_{-3}^{ij}(q)}{\left(\mathcal{M}_{-1}^{ij}(q)\right)^2}.; \quad \Delta V^{ZPE}(q) = \frac{1}{4}\sum_{i,j}\frac{\mathcal{M}_{-2}^{ij}(q)}{\mathcal{M}_{-3}^{ij}(q)}. \tag{6}$$

2.2. Renormalisation of vibrational mass parameters.

The symmetry rules within the quadrupole 5DCH (attached with both rotational invariance and intrinsic axes relabelling) have been studied in great detail long time ago in the pioneering work of ref. [13] and have been put by these authors under an elegant and compact form. With slight differences in notations, their conclusions for the kinetic part can be summarized as follows :

The kinetic term of the 5DCH is fully described in the most general case by two functions of deformation variables (a_0, a_2) named here B_{Rot} and B_{Vib}. These two functions, which must be analytic in (a_0, a_2) are defined in the whole collective space and connected to usual moments of inertia $\mathcal{J}_k$ and vibrational mass parameters B_{mn} by :

$$\begin{aligned} \mathcal{J}_k(\beta,\gamma) &= 4(\beta\sin\gamma_k)^2 B_{Rot}(\beta,\gamma_k) \\ B_{22}(\beta,\gamma) &= B_{Vib}(\beta,\gamma_3) \\ with\ \ \gamma_k &= \gamma - k\frac{2\pi}{3} \end{aligned} \tag{7}$$

Microscopic calculations providing moments of inertia and mass parameters in the first sextant ($0 \leqslant \gamma \leqslant \pi/3$), functions B_{Rot} and B_{Vib} are finally defined on this ground in the whole collective space,

i) using the subsidiary relations :

$$\begin{aligned} B_{00}(\beta,\gamma) &= \frac{2}{3}B_{Vib}(\beta,\gamma_1) + \frac{2}{3}B_{Vib}(\beta,\gamma_2) - \frac{1}{3}B_{Vib}(\beta,\gamma_3) \\ B_{02}(\beta,\gamma) &= \left(B_{Vib}(\beta,\gamma_1) - B_{Vib}(\beta,\gamma_2)\right)/\sqrt{3} \end{aligned} \tag{8}$$

ii) knowing they must be even functions of a_2 and therefore invariant in the change $\gamma \to -\gamma$.

iii) and knowing they must be equal to each other on the axis $a_2 = 0$ and therefore have to verify:

$$B_{Rot}(\beta,\gamma=0) = B_{Vib}(\beta,\gamma=0)\ ; B_{Rot}(\beta,\gamma=\pi) = B_{Vib}(\beta,\gamma=\pi). \tag{9}$$

Conditions eq.9 constitue a quite weak constraint, because the B_{Rot} function is related to moment of inertia (eq.7) through a factor $\sin^2\gamma$ which vanishes identically on the axis. However, we

Journal of Physics: Conference Series **205** (2010) 012007 doi:10.1088/1742-6596/205/1/012007

have tryed to exploit them to renormalize our Inglis-Belyaev mass parameters using a simple prescription, which is defined by the following steps :

i) For each nuclear system, and for each deformation β, microscopical functions B are decomposed in two parts, an even ($A^{\oplus}$)and an odd ($A^{\ominus}$) part in the change $\gamma \rightarrow (\pi - \gamma)$, (i.e. also $a_0 \rightarrow -a_0$). We define therefore :

$$A^{\oplus}(\beta,\gamma) = \frac{1}{2}\left(B(\beta,\gamma) + B(\beta,\pi-\gamma)\right) \; ; A^{\ominus}(\beta,\gamma) = \frac{1}{2}\left(B(\beta,\gamma) - B(\beta,\pi-\gamma)\right) \qquad (10)$$

Due to their a_0-parity property, functions A can be easily scaled to have a given value along the $a_2 = 0$ axis.

ii) At the issue of our microscopical calculations functions B_{Rot}^{Inglis} and B_{Rot}^{TV} are known. The effect of TV dynamical correction on rotational part are now approximed at each deformation β on applying scaling factors $\alpha^{\oplus}(\beta)$ and $\alpha^{\ominus}(\beta)$ to the corresponding Inglis Belyaev functions $A_{Rot}^{Inglis\oplus}(\beta,\gamma)$ and $A_{Rot}^{Inglis\ominus}(\beta,\gamma)$. In practice, α^i factors are determined through a least square process on the numerical mesh $\{\beta_m, \gamma_n\}$, on minimizing

$$\forall\beta_m,\ \{i = \oplus,\ominus\},\ \delta\sum_n \left|\alpha^i(\beta_m)\ A_{Rot}^{Inglis\ i}(\beta_m,\gamma_n) - A_{Rot}^{TV\ i}(\beta_m,\gamma_n)\right|^2 = 0 \qquad (11)$$

That of course involves points of the mesh with $\gamma \neq 0$ and $\gamma \neq \pi$ only, where the B_{Rot}^{TV} function and its deduced parts $A_{Rot}^{TV\ k}$ are completely defined. This first step of the prescription can be shown to be very satisfactorying, on comparing true TV moments of inertia (which are kept in the 5DCH calculations) and those deduced (using eq.7 after inverting eq.10) from the Inglis scaled functions $A_{Rot}^{RnTV\ i} = \alpha^i(\beta)\ A_{Rot}^{Inglis\ i}(\beta,\gamma)$.

An example is displayed in the left hand side (a) of figure 1 where the two sets of moments of inertia as function of the triaxiality γ are compared for the ^{110}Ru nucleus at $\beta = 0.3$. Scaling factors α^i are found to be very smooth functions of β varying typically between 1.0 and 3/2. That means that the overall behaviour of the B_{Rot}^{TV} function is mainly contained in the B_{Rot}^{Inglis} one, and that these two functions can be related each other with a very good accuracy by a weak and regular scaling transformation.

iii) The last step of the prescription consists in applying the same "scaling" on vibrational function B_{Vib}^{Inglis}, securing thus the condition on axis $a_2 = 0$. Thus, on the ground of this minimum of information trick, one defines the new vibrational functions :

$$A_{Vib}^{RnTV\ i}(\beta,\gamma) = \alpha^i(\beta)\ A_{Vib}^{Inglis\ i}(\beta,\gamma),\ with\ \{i = \oplus,\ominus\} \qquad (12)$$

Deduced kinetic functions B_{Vib}^{RnTV} and other related functions B are displayed for our example (i.e. ^{110}Ru at $\beta = 0.3$) on the right hand side (b) of figure 1. As shown in there, the condition on axis $a_2 = 0$ (eq. 9) is pretty well verified for the fully coherent description with B_{Rot}^{Inglis} and B_{Vib}^{Inglis}, but is broken if only rotational kinetic terms includes the dynamical TV contribution as it the case in the systematic calculation of [1]. Finally, the ad hoc scaled function B_{Vib}^{RnTV} verifies the equalities on axis to B_{Vib}^{TV}.

3. Results.

3.1. The low energy spectrum of ^{110}Ru.

We have taken ^{110}Ru (Z=44,N=66) nucleus as an example. It is therefore convenient to see on this perticuliar case the differences in spectrum induced by various hypotheses employed as

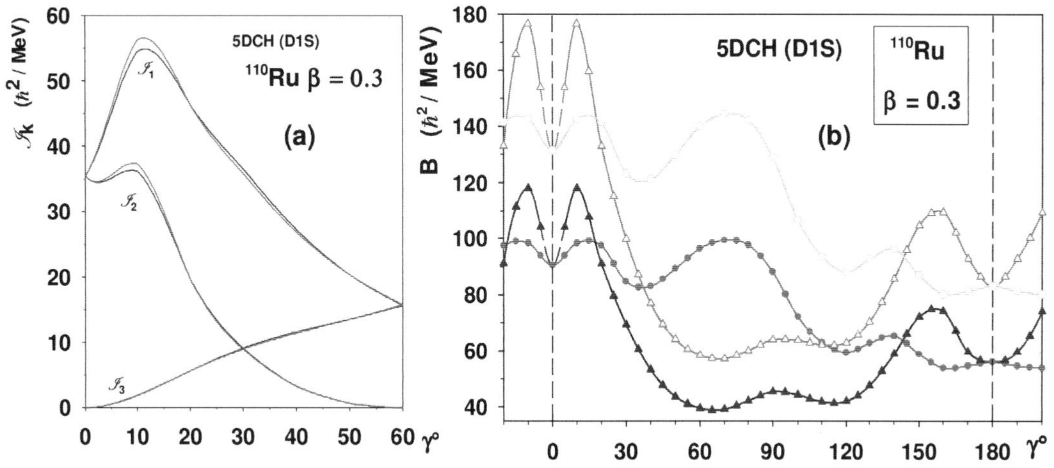

Figure 1. On side (a), for the ^{110}Ru nucleus, at deformation $\beta = 0.3$, Thouless-Valatin moments of inertia $\mathcal{J}^{TV}_{k=1,2,3}$ (red lines) are compared as function of the triaxiality γ with corresponding "scaled" (see text) Inglis-Belyaev moments of inertia (blue lines). On side (b)B^{Inglis}_{Rot} (full triangles, blue line), B^{TV}_{Rot} (open triangles, pink line), B^{Inglis}_{Vib} (full dots, red line) and B^{RnTV}_{Vib} (open dots, green line) -see text for definitions- are displayed as functions of the triaxilality γ.

input of the 5DCH. This nucleus is a triaxial-$\gamma-$soft nucleus -see its Potential Energy Surface in figure 2-. As shown in this figure, a significative progress in the description of in gs- quasi-β and quasi-γ bands is obtained, step by step, from the pure Inglis-Belyaev calculation of the kinetic tensor " $\left(\mathcal{J}^{Inglis}_k, B^{Inglis}_{mn}\right)$ " , to the calculation including exact TV corrections for the rotational term only " $\left(\mathcal{J}^{TV}_k, B^{Inglis}_{mn}\right)$ " -i.e. those obtained in [1]-, and finally to the present calculation including also the approximate estimation of the TV vibrational term labelled here " $\left(\mathcal{J}^{TV}_k, B^{RnTV}_{mn}\right)$ ". It is noticeable that in such a soft system, a correction of vibrational term affects not only the band head energies but also the rotational sequence (in particular the gs band). In this nucleus, there is no available experimental data for the quasi-β band head energy. It is predicted here at 0.950 MeV. In this nucleus, as it is the most general case in the chart, the level $I^{\pi}_n = 2^+_2$, with more than 80% of the norm belonging to the K=2 part of the space (in any of these calculations) is clearly the quasi-γ band head. That seems to be consistent with the experimental spectrum.

3.2. Vibrational states in the Gd (Z=64) isotopic chain.

Now we consider another example taken in the so-called $N = 90$ $\gamma-$soft-rigid prolate transition in the rare earths, namely in the Gd ($Z = 64$) isotopic chain : In figure 3, the evolution in the gs band energies and of $I^{\pi}_n = 0^+_2$, 2^+_2 and 2^+_3 state energies are displayed as a function of the neutron number N for the $\left(\mathcal{J}^{TV}_k, B^{Inglis}_{mn}\right)$ and the $\left(\mathcal{J}^{TV}_k, B^{RnTV}_{mn}\right)$ calculations and for the experiment. For these nuclei, in the gs band, the renormalization of vibrational part of the kinetic tensor is without any significative consequences. As discussed in [1], and specialy in the gs band, the vicinity of the shell closure for N=82 induces some discrepancies with experiment. For all these nuclei, in both calculations (with the noticeable exception of the $N = 90$ ^{154}Gd), the $I^{\pi}_n = 2^+_2$ has the properties a γ-band head with more than 75% of the norm in the $K = 2$ part of the space. An inversion with the $I^{\pi}_n = 2^+_3$ occurs at $N = 90$ where the two states are

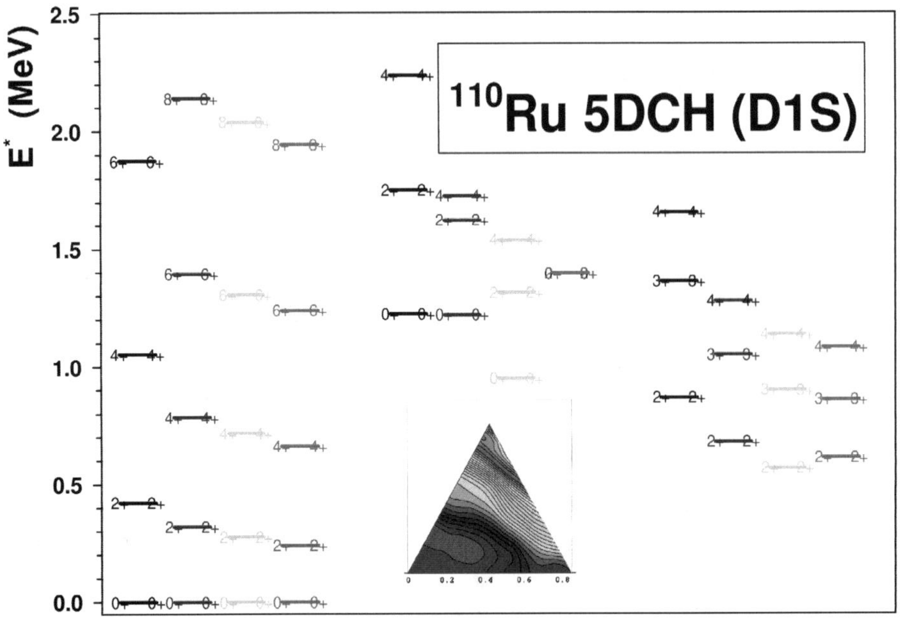

Figure 2. For the ^{110}Ru nucleus, the experimental low energy spectrum (in red color) for the gsband (on the left), the quasi-β band -NB the 0+ band head is unknown- (on center) and the quasi-γ band (on the right) is compared with three theoretical spectra deduced respectively with kinetic functions ($\mathcal{J}_k^{Inglis}$, B_{mn}^{Inglis})-levels in black color-, ($\mathcal{J}_k^{TV}$, B_{mn}^{Inglis}) -levels in blue color-, and ($\mathcal{J}_k^{TV}$, B_{mn}^{RnTV}) -levels in green color-. Experimental data are extracted from the compilation [17]. Potential energy surface is also displayed in a sextant. Isolines are separated by 1 MeV.

very close in energy. The $N = 90$ γ−soft-rigid prolate transition is well localized : The gs deformation $(\langle\beta\rangle, \langle\gamma\rangle)$ is $(0.17, 25°)$ at $N = 86$, reaches $(0.34, 12°)$ at $N = 90$, and then follows a regular behaviour to more prolate rigid system, reaching $(0.37, 4°)$ at $N = 100$. As shown, the improvment of the description on renormalizing vibrational kinetic term is noticeable but is not sufficent to lower enough the band head energies in the most deformed Gd nuclei. Thus, if around N=90 in Gd nuclei, a quadrupole collective space seems to be rich enough to describe the low energy collective modes, -this is confirmed by a very good agreement with experiment obtained for E0 and E2 transition probabilities in ^{154}Gd- , it is highly probable that other degrees of freedom take some importance with increasing deformation and have therefore to be also taken into account in the future.

3.3. An overview of results in the whole chart.
A statistic approach has been performed in a large area of the chart ($18 \leqslant Z \leqslant 98$). Nuclei included in the panel are those i) for wich the experimental energy of the considered state ($I_n^\pi = 0_2^+$, 2_2^+ or 2_3^+) is known in the Brookheaven nuclear data table [17] and ii) and those for which the calculated spectrocopy has been validated. (some nuclei at or close the doubly shell closure are not considered -see the discussion of this point in [1]-. Each of these excluded nuclei corresponds to an empty point in the map displayed in figure 5. Results of this statistical analysis are displayed in figure 4. One can see in this figure the almost systematic overestimation of excitation energies for states $I_n^\pi = 0_2^+$, 2_2^+ and 2_3^+ respectively, obtained in the standard calculation referred as $\left(\mathcal{J}_k^{TV}, B_{mn}^{Inglis}\right)$. As a result, clouds of points obtained in the

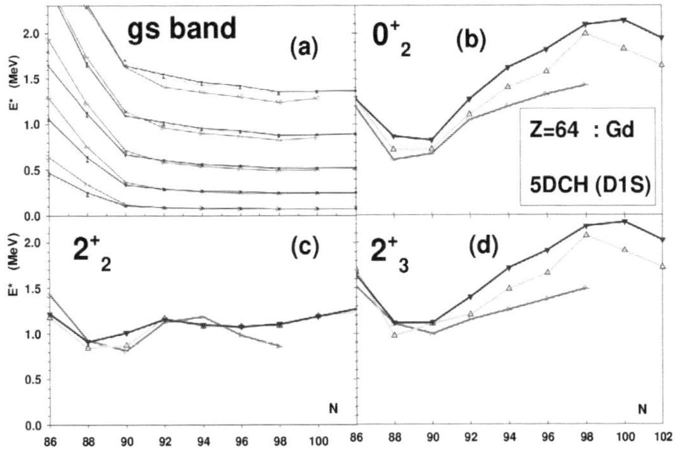

Figure 3. Evolution in gs band energies (a) and of $I_n^\pi = 0_2^+$ (b), 2_2^+ (c) and 2_3^+ (d) state energies as function of the neutron number for the Gd isotopic chain. Experimental values, extracted from [17], are joined by red lines. Blue lines with full triangle down correspond to the standard calculation $\left(\mathcal{J}_k^{TV}, B_{mn}^{Inglis}\right)$, whereas green lines with open triangle up correspond to the present calculation with renormalized vibrational mass parameters i.e. $\left(\mathcal{J}_k^{TV}, B_{mn}^{RnTV}\right)$.

$\left(\mathcal{J}_k^{TV}, B_{mn}^{RnTV}\right)$ approach are closer to the line of "perfection" (defined by $E^{Th} = E^{Exp}$) and more symmetrically distributed around it. Of course, there is also a certain amount of points quite far from this line in any case. Experimental states have been considered here only for their spin and order in the spectrum. Clearly a certain amount of them does not belong (at all or for a part) to quadrupole collective modes.

To localize and have in hand discrepancies between experimental and theoretical values obtained with the present $\left(\mathcal{J}_k^{TV}, B_{mn}^{RnTV}\right)$ calculation, a map in the plan (N,Z) of the ratio E^{Th}/E^{Exp} for the state $I_n^\pi = 0_2^+$ is produced in figure 5. A simple overview on this map shows a quite satisfactorying agreement. However, apart ponctual and isolated problems, apart defaults directly connected with double shell closures, a quite large and noticeable area of descrepancies experience-theory appears in the well deformed actides (Points of the map with $90 \leqslant Z \leqslant 98$, $134 \leqslant N \leqslant 152$) where, even with such renormalized mass parameter, the 0_2^+ state energy is overestimated by at least 30%. In this region, these results seems to indicate, that the experimental 0_2^+ state is not the quasi$-\beta$ band head (a quasi$-\beta$ band head could exists at higher energy), or that this state is a more complicate mixing, involving other degrees of freedom. Due to the systematic character of the descrepancy in the region, these degrees of freedom are probably of higher collective multipole type. Further studies will be devoted to understand more deeply these facts.

4. Conclusions.

Clearly, the function B^{Vib} including its TV contributions should be evaluated microscopically instead of being simply rescaled over the Inglis-Belyaev function. However, it is believable that the present renormalized kinetic function B_{RnTV}^{Vib}, which exhibits the general behaviour of the B_{Inglis}^{Vib} one, drives a lowering of vibrational band heads which has at least the good order of magnitude. Under this assumption, it has been shown that a global quantitative improvment

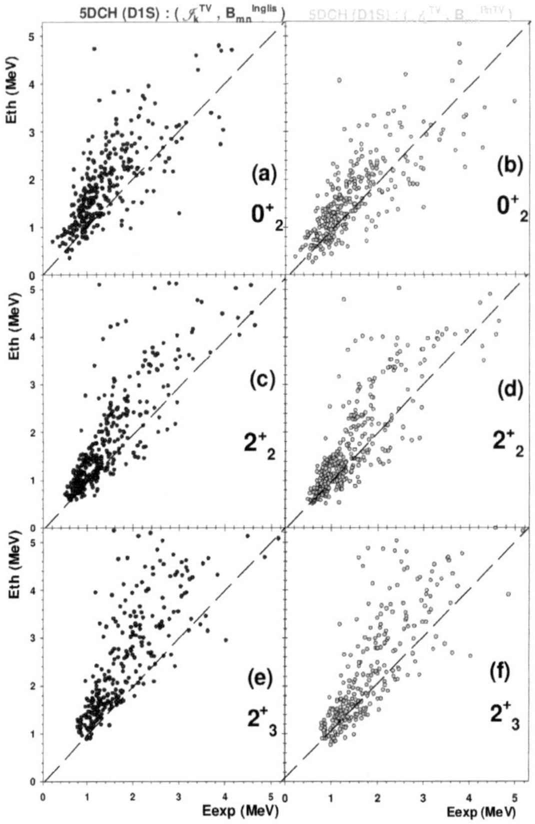

Figure 4. States $I_n^\pi = 0_2^+$ (a,b), 2_2^+ (c,d) and 2_3^+ (e,f) for even-even nuclei with $20 \leqslant Z \leqslant 100$ are displayed as function of their theoretical energies (vertical axis) and experimental ones (horizontal axis) On the left (a,c,e) with blue points (resp right (b,d,f) with green points) theoretical values are issueing from the standard calculation with $\left(\mathcal{J}_k^{TV}, B_{mn}^{Inglis}\right)$ (resp the calculation with renormalized vibrational mass parameters with $\left(\mathcal{J}_k^{TV}, B_{mn}^{RnTV}\right)$. In each plot, as a guide for eyes, the dashed line corresponds to $E^{Th} = E^{Exp}$.

of the theory can be expected at next step, precisely on evaluating within an Adiabatic Time Dependant HFB approach these vibrational TV contributions. As known, beyond the pure quadrupole collective space considered here, a microscopic theory considering single particle modes is clearly needed in the vicinity of shell closures. But moreover, it has been shown here, that for some well deformed nuclei , (for instance in the actinide region), a convincing description of $I_n^\pi = 0_2^+$ states presumably implies to take also into account collective degrees of freedom of higher multipoles.

Acknowledgments
To participate as invited speaker at the XVIII International Workshop in Varna, one of us (J.L.) has beneficiated of a full support by the PICS agreement N°4847 (2009) between Bulgarian Academy of Science-University of Sofia and french CNRS-IN2P3.

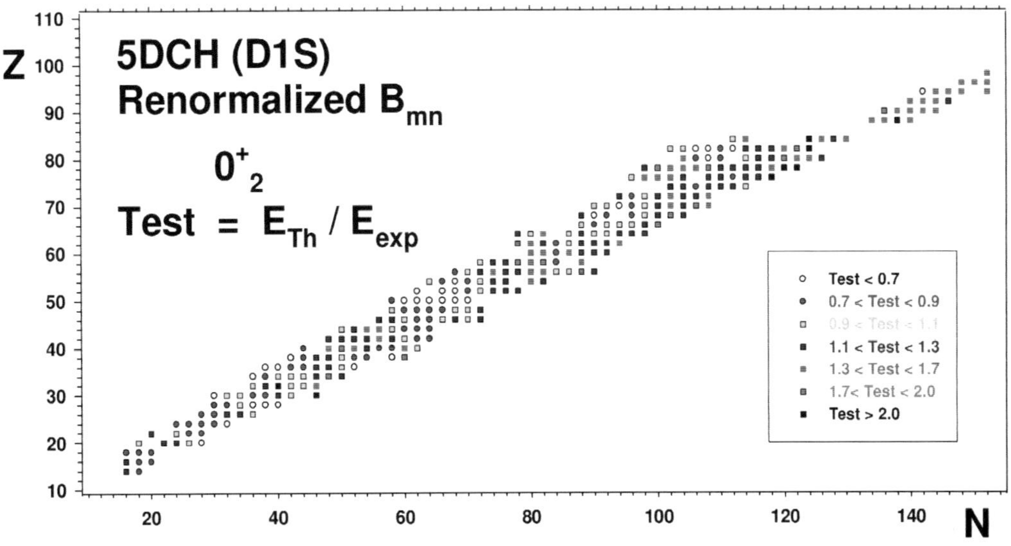

Figure 5. Map (N, Z) of the ratio Test=$E_{Th}/E_{\rm exp}$, where E_{Th} is the theoretical energy of the $I^{\pi}_{n} = 0^{+}_{2}$ state deduced with renormalized vibrational mass parameters B^{RnTV}_{mn} (see text for definition) and $E_{\rm exp}$ is the corresponding experimental energy extracted from the compilation [17].

References

[1] Delaroche J-P, Girod M, Libert J, Goutte H, Hilaire S, Peru S, Pillet N and Bertsch GF 2009 submitted to *Phys. Rev.* C, *Preprint* arXiv:0910.2940.
[2] Yuldashbaeva E Kh, Libert J, Quentin P and Girod M 1999 *Phys. Lett.* B **461** 1.
[3] Libert J, Girod M, and Delaroche J-P 1999 *Phys. Rev.* C **60** 054301.
[4] Delaroche J-P, Girod M, Goutte H and Libert J 2006 *Nucl.Phys.* A **771** 103.
[5] Dechargé J and Gogny D 1980 *Phys. Rev.* C **21** 1568.
[6] Berger J-F, Girod M and Gogny D 1991 *Comp. Phys. Comm.* **63** 365.
[7] Baranger M and Vénéroni M 1978 *Ann. Phys. N.Y.***114** 123.
[8] Girod M and Grammaticos B 1983 *Phys. Rev.* C **27** 2317.
[9] Girod M, Delaroche J-P, Berger J-F and Libert J 1994 *Phys. Lett.* B **325** 1.
[10] Hill D L and Wheeler J A 1953 *Phys. Rev.* **89** 1102.
[11] Griffin J J and Wheeler J A 1957 *Phys. Rev.* 108, 311.
[12] Girod M and Grammaticos B 1979 *Nucl. Phys.* A **330** 40.
[13] Kumar K and Baranger M 1967 *Nucl. Phys.* A **92** 608.
[14] Thouless D J and Valatin J G 1962 *Nucl. Phys.* **31** 211.
[15] Inglis D R 1956 *Phys. Rev.* **103** 1786.
[16] Belyaev S T 1961 *Nucl. Phys.* **24** 322.
[17] Brookhaven database, http://www.nndc.bnl.gov/.

Journal of Physics: Conference Series **205** (2010) 012008 doi:10.1088/1742-6596/205/1/012008

Role of multiphonon configurations in nuclear spectra and responses: a new approach

Lo Iudice N[1], Andreozzi F[1], Knapp F[2] Porrino A[1] Kvasil J[2]

[1]Dipartimento di Scienze Fisiche Università di Napoli Federico II and INFN, Sezione di Napoli, Napoli Italy

[2]Institute of Particle and Nuclear Physics, Faculty of Mathematics and Physics, Charles University, V Holesovickach 2, 180 000 Prague 8, Czech Republic

E-mail: loiudice@na.infn.it

Abstract. A basis of multiphonon states is generated by constructing and solving iteratively a set of equations of motion. Represented in this basis, the Hamiltonian has a simple structure and can be easily brought to diagonal form. The method is adopted to compute the energy levels and the electromagnetic responses in ^{16}O, chosen as test ground because of its complex structure.

1. Introduction

The role of the multiphonon configurations in determining the properties of the low-lying nuclear spectra and in shaping the high-energy giant resonances is becoming more and more manifest. In fact, the experimental evidence of multiphonon collective mode at low-energy and of multiple giant resonance at high energy has grown rapidly in recent years [1, 2, 3, 4]. Moreover, multiphonon configurations affect crucially the width and damping of the giant resonances [5].

Tamm-Dancoff (TDA) and random-phase approximations (RPA) can only describe the global features of collective modes and explain in part their fine structure by accounting for their decay to single-particle excitations (Landau damping). These approaches, however, are not suited to the study of the anharmonicities of the many low-lying multiphonon spectra and of the high-energy multiple resonances. They are not even able to account for the full fragmentation of the giant resonances.

It is therefore necessary to go from the particle-hole ($p-h$) to a larger space that includes more complex $np-nh$ configurations and account properly for the couplings between them.

Several theoretical approaches have been devoted to the achievement of such a task. A notable one is the nuclear field theory [6], where the $p-h$ space is coupled to $2p-2h$ configurations and to the continuum. Such a method is clearly well suited for describing the damping of the giant resonances.

Of considerable interest is also a method that extends Migdal's theory and exploits the Green function techniques to enlarge the configuration space beyond the TDA and RPA [7]. The method was applied with considerable success to the study of collective modes and giant resonances.

The most widely adopted approach is the quasiparticle-phonon model (QPM) [8], that effectively diagonalizes a separable Hamiltonian of general form in a space spanned by states with a number of phonons up to $n = 3$, the phonons being generated in the quasiparticle RPA. The QPM has been adopted systematically in heavy spherical and deformed nuclei. Among the most recent investigations, we may mention the study of double giant resonances [9] and, at low energy, the investigation of the mixed symmetry states in spherical nuclei [10, 11]. In deformed nuclei, the QPM has been successfully adopted to study the 0^+ states discovered in large abundance in the rare earth region and the actinides [12] and to investigate the controversial nature of the 4^+ states in transitional nuclei [13].

Recently, we have formulated an equation of motion phonon method (EMPM) [14], that moves along the lines drawn within the multistep shell model [15, 16]. The method generates iteratively a basis of multiphonon states built of phonons obtained in the Tamm-Dancoff approximation (TDA). Such a basis greatly simplifies the structure of the Hamiltonian matrix, which can be easily brought to diagonal form yielding exact eigenvalues and eigenvectors. A numerical implementation was carried out on ^{16}O. A space spanned by all states up to three phonons and $3\hbar\Omega$ was adopted to compute spectra and nuclear responses [17]. The most meaningful results of this calculation will be discussed here.

2. The method

We consider [14] a two-body Hamiltonian H of general form

$$H = \sum_i \epsilon_i a_i^\dagger a_i + \frac{1}{4}\sum_{ijkl} V_{ijkl}\, a_i^\dagger a_j^\dagger a_l a_k \,, \tag{1}$$

where ϵ_i are the single-particle energies, V_{ijkl} the antisymmetrized matrix elements of the nucleon-nucleon interaction, $a_i^\dagger$ (a_i) the creation (annihilation) particle operators with respect to the physical vacuum.

We intend to generate a basis of n-phonon states $|n;\, \beta >$, having the structure

$$|n;\, \beta >= \sum_{\alpha\, ph} C_{\alpha\beta}^{(n)}(ph)\, b_{ph}^\dagger |n-1;\, \alpha >, \tag{2}$$

where $b_{ph}^\dagger = a_p^\dagger a_h$ creates a ph state out of the unperturbed Hartree-Fock vacuum and $|n-1;\, \alpha >$ are states with (n-1) phonons.

In order to construct such a basis we extend a well established equation of motion method. Indeed, under the request that the Hamiltonian be diagonal within each n-phonon subspace, spanned by the states $|n; \alpha >$, the following equations of motion hold

$$< n; \beta|\, [H, b_{ph}^\dagger]\, |n-1; \alpha >= (E_\beta^{(n)} - E_\alpha^{(n-1)}) X_{\alpha\beta}^{(n)}(ph), \tag{3}$$

where

$$X_{\alpha\beta}^{(n)}(ph) \equiv < n; \beta | b_{ph}^\dagger | n-1; \alpha > . \tag{4}$$

Upon expansion and linearization of the commutator $[H, b_{ph}^\dagger]$, one obtains for the n-phonon subspace the eigenvalue equations

$$\sum_{\gamma p'h'} A_{\alpha\gamma}^{(n)}(ph\ ;\ p'h')\ X_{\gamma\beta}^{(n)}(p'h') = E_\beta^{(n)}\ X_{\alpha\beta}^{(n)}(ph). \tag{5}$$

The matrix $A^{(n)}$ has the simple structure

$$\begin{aligned} A_{\alpha\gamma}^{(n)}(ph; p'h') \;&=\; \delta_{\alpha\gamma}^{(n)} \left[\delta_{hh'}\delta_{pp'}\left(\epsilon_p - \epsilon_h + E_\alpha^{(n-1)}\right) + V_{ph'hp'}\right] \\ &+\; \delta_{hh'} W_{\alpha\gamma}^{(n-1)}(pp') - \delta_{pp'} W_{\alpha\gamma}^{(n-1)}(hh'), \end{aligned} \tag{6}$$

Journal of Physics: Conference Series **205** (2010) 012008 doi:10.1088/1742-6596/205/1/012008

where

$$\begin{aligned} W^{(n)}_{\alpha\gamma}(pp') &= \sum_{h_1h_2} V_{p'h_1ph_2}\rho^{(n-1)}_{\alpha\gamma}(h_1h_2) + \frac{1}{2}\sum_{p_1p_2} V_{p'p_1pp_2}\rho^{(n-1)}_{\alpha\gamma}(p_1p_2) \\ W^{(n)}_{\alpha\gamma}(hh') &= \sum_{p_1p_2} V_{hp_1h'p_2}\rho^{(n-1)}_{\alpha\gamma}(p_1p_2) + \frac{1}{2}\sum_{h_1h_2} V_{hh_1h'h_2}\rho^{(n-1)}_{\alpha\gamma}(h_1h_2). \end{aligned} \tag{7}$$

Here

$$\rho^{(n)}_{\alpha\gamma}(kl) = < n;\, \gamma|a^{\dagger}_k a_l|n;\, \alpha > \tag{8}$$

is the density matrix.

We now insert the expansion (2) of $|n;\, \beta>$ into Eq. (4) obtaining

$$X^{(n)} = \mathcal{D}^{(n)}C^{(n)}, \tag{9}$$

where $\mathcal{D}^{(n)}$ is the overlap or metric matrix. In virtue of Eq. (9), the equation of motion (5) becomes the eigenvalue equation of general form

$$A^{(n)}\,\mathcal{D}^{(n)}C^{(n)} = E^{(n)}\mathcal{D}^{(n)}C^{(n)}, \tag{10}$$

where $A^{(n)}$ is still given by Eq. (6) and the overlap matrix $\mathcal{D}^{(n)}$ has the expression

$$\mathcal{D}^{(n)}_{\alpha\beta}(ph;p'h') = \sum_{\gamma}\left[\delta_{pp'}\delta_{\gamma\beta} - \rho^{(n-1)}_{\gamma\beta}(pp')\right]\rho^{(n-1)}_{\alpha\gamma}(hh'). \tag{11}$$

The density matrices can be computed by the recursive relations

$$\begin{aligned} \rho^{(n)}_{\alpha\beta}(p_1p_2) &= \sum_{ph\gamma\delta} C^{(n)}_{\alpha\gamma}(ph)\,X^{(n)}_{\delta\beta}(p_1h)\left[\delta_{pp_2}\delta_{\gamma\delta} - \rho^{(n-1)}_{\gamma\delta}(pp_2)\right], \\ \rho^{(n)}_{\alpha\beta}(h_1h_2) &= \sum_{ph\gamma\delta} C^{(n)}_{\alpha\gamma}(ph)X^{(n)}_{\delta\beta}(ph_2)\left[\delta_{hh_1}\delta_{\gamma\delta} - \rho^{(n-1)}_{\gamma\delta}(h_1h)\right]. \end{aligned} \tag{12}$$

Once the redundancy of the basis states is eliminated by the the Choleski decomposition method, the recursive formulas of $A^{(n)}$, $\mathcal{D}^{(n)}$ and $\rho^{(n)}$ allow to solve the generalized eigenvalue problem iteratively starting from $n = 0$, namely the $p - h$ vacuum, up to a n-phonon subspace. A basis of linearly independent multiphonon states, $\{| >, |1, \lambda >, \ldots |n, \alpha >\}$, is thereby generated.

This basis greatly simplifies the structure of the Hamiltonian matrix and makes its diagonalization much easier. The eigenvectors so obtained are exact and have the structure

$$| \Psi_\nu\rangle = \sum_{n,\beta} C^{(\nu)}_{\beta} \mid n, \beta\rangle. \tag{13}$$

Using these wave functions, we obtain for the transition amplitudes of the one-body operator

$$\mathcal{M} = \sum_{kl} \mathcal{M}_{kl}a^{\dagger}_k a_l \tag{14}$$

the formula

$$\mathcal{M}(i \to f) = \langle \Psi_{\nu_f} \mid \mathcal{M} \mid \Psi_{\nu_i}\rangle = \sum_{kln_in_f\alpha_i\beta_f} \mathcal{M}_{kl}C^{(\nu_i)}_{n_i\alpha_i}C^{*(\nu_f)}_{n_f\beta_f}\rho^{(n_in_f)}_{\alpha_i\beta_f}(kl). \tag{15}$$

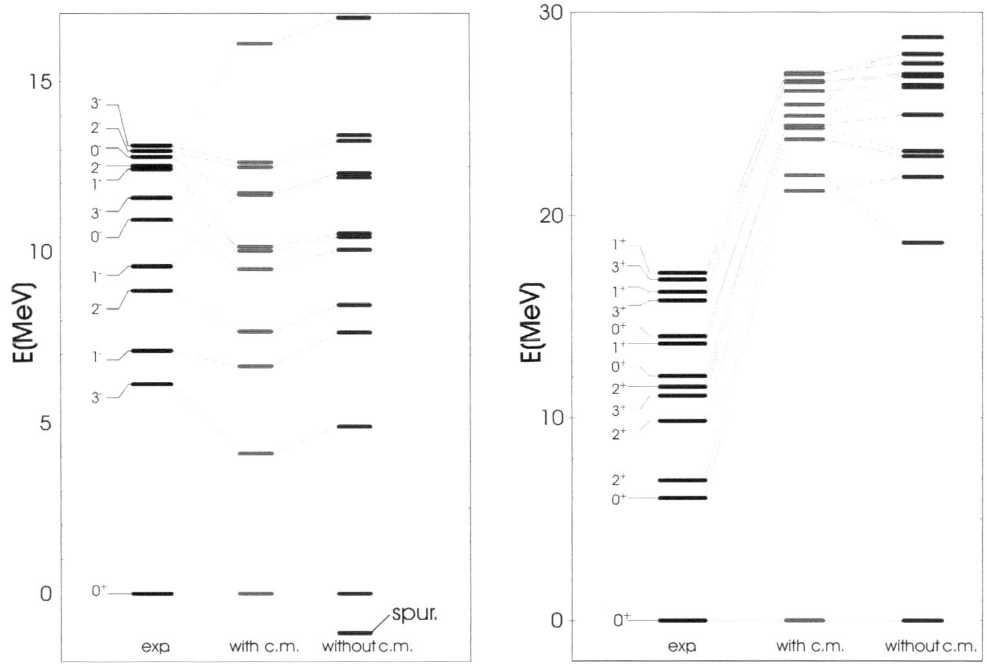

Figure 1. Positive and negative parity energy spectra computed in the EMPM with and without c.m. corrections

These amplitudes involve the density matrix, which can be computed through Eqs. (12) when $n_i = n_f$. When $n_i \neq n_f$, $\rho^{(n_i n_f)}_{\alpha_i \beta_f}(kl)$ is nothing but $X^{(n_i n_f)}_{\alpha_i \beta_f}(ph)$ (Eq. 4) or its complex conjugate.

The composite structure of the transition amplitudes deserves few words of comment. Let us look, for instance, at the wave functions considered in the present calculation. The ground state $|\Psi_0>$ is a linear combination of the ph vacuum $|>$, a very small one-phonon piece and a set of two-phonon states, while the excited states $|\Psi_\nu>$ contain up to three phonons. It follows that the transition amplitude can be written schematically

$$\begin{aligned} \mathcal{M}(0 \to \nu) \sim p_{01} < 1|\mathcal{M}| > + p_{11} < 1|\mathcal{M}|1 > + p_{12} < 2|\mathcal{M}|1 > \\ + p_{21} < 1|\mathcal{M}|2 > + p_{22} < 2|\mathcal{M}|2 > + p_{23} < 3|\mathcal{M}|2 >, \end{aligned} \tag{16}$$

where $i = 1, 2, 3$ indicates the number of phonons and $|p_{ij}| < 1$ the weights. The above equation makes clear that, within the EMPM, the transition amplitudes contain the Tam-Dancoff piece (one-phonon), quenched by a factor $|p_{01}| < 1$, plus many other corrective terms coming from the coupling, promoted by the operator $\mathcal{M}$, between states with $i = j$ or $i = j \pm 1$.

3. A numerical test ground: ^{16}O

We have chosen ^{16}O as bench mark for our method. Its low-energy positive parity spectrum, well known for its complex structure, was studied in large scale shell model calculations which included up to $4p - 4h$ and $4\hbar\omega$ configurations [18] and, more recently, up to $6\hbar\omega$ [19].

For our illustrative purposes, we have included all $p - h$ configurations up to $n = 3$ and $3\hbar\omega$ and used a Hamiltonian composed of a Nilsson unperturbed piece plus a bare G-matrix deduced from the Bonn-A potential.

In order to get a complete separation of the intrinsic from the center of mass motion, we have

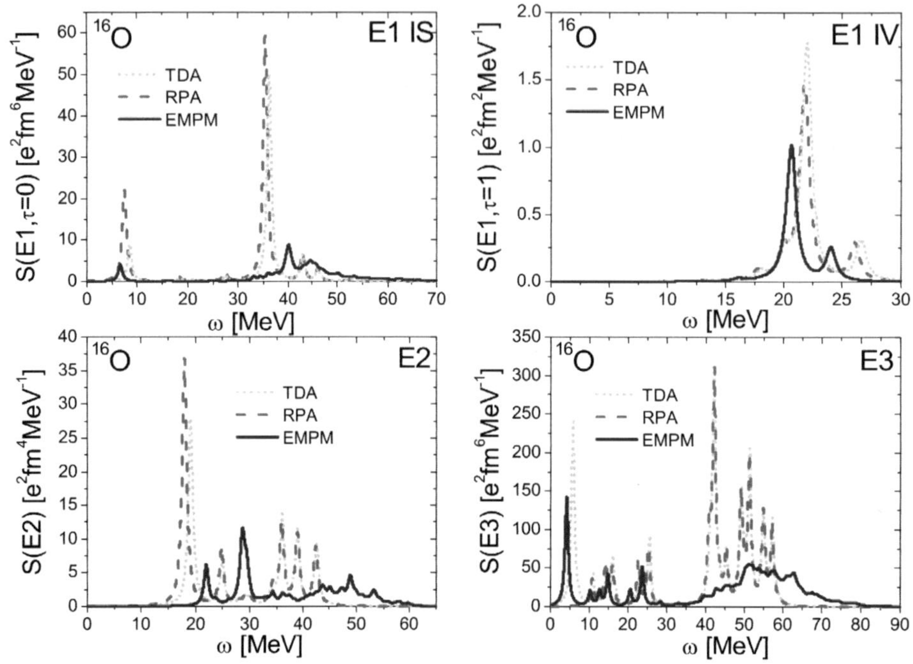

Figure 2. $E\lambda$ strength distributions in ^{16}O, computed in TDA, RPA and EMPM.

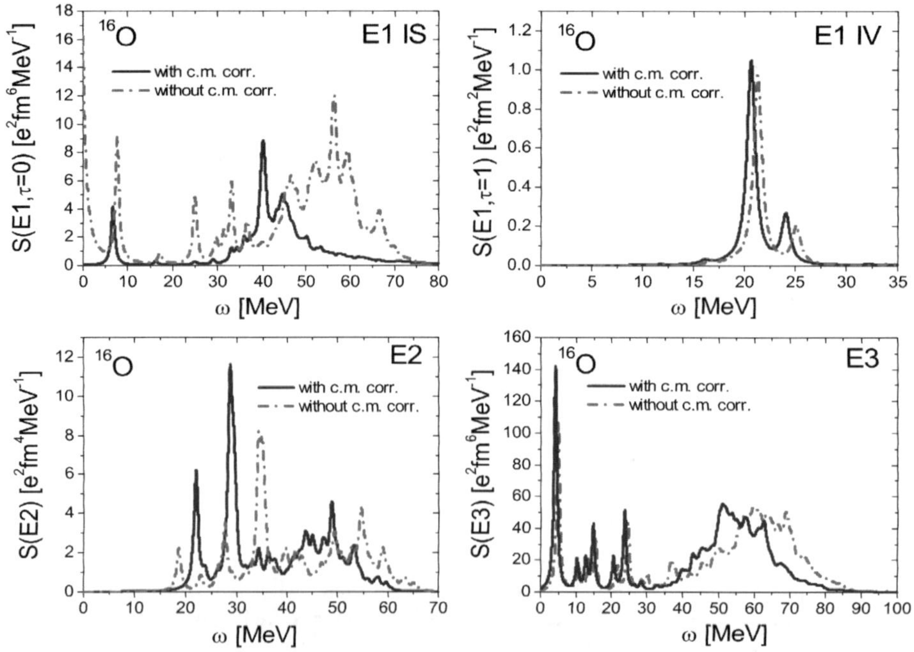

Figure 3. $E\lambda$ strength distributions computed for ^{16}O in the EMPM with and without c.m. spurious admixtures

adopted the method of Palumbo [20, 21], which separates the center of mass spurious mode from the intrinsic states and pushes such a spurious state high up in energy.

Being our space confined to 3 $\hbar\omega$, the ground state contains correlations up to 2-phonons only. These account for about 20% of the state, while the remaining 80% pertains to the $p-h$ vacuum. A no-core shell model calculation [19] yields about 60% for the $0p-0h$, 20% for $2p-2h$ and 20% for the other more complex configurations, excluded from our restricted space.

As shown in Fig. 1, the deviations of the computed spectra from the experimental ones, while reasonably small for negative parity levels, are dramatic for positive parity. This was largely expected, since one has to include four-phonon states, at least, in order to hope to approach the positive parity experimental levels [18, 19]. It might be worth pointing out the importance of removing the CM motion. If this is not removed, the spurious 1^- becomes the ground state!

As shown in Fig. 2, the multiphonon configurations have a very strong impact on the nuclear response. Only the isovector giant dipole resonance is not strongly altered, an indication of the basically harmonic character of this mode. The $E1$ spectrum is slightly shifted downward and keeps its shape while the prominent peak is considerably quenched with respect to the TDA or RPA. The strength distribution is compatible with the experimental dipole cross section [22, 23]. It is concentrated in the domain of the experimental cross section and has the main peaks in close correspondence with the observed ones. The nature of those peaks changes little with respect to TDA. They are described by states with dominant $J^\pi = 1^-, T = 1$ one-phonon components.

Strong is the impact on the $E3$ response. Because of the the strong coupling between one and three-phonon configurations, the dominantly one-phonon peaks are pushed down in energy, while the states contributing to the high energy broad hump acquire large two and, especially, three-phonon components, which become dominant in several instances.

Dramatic are the changes brought into the other two $E\lambda$ spectra. The $E2$ peaks are severely damped and the full strength strongly fragmented. Such a large effect is to be ascribed to the nature of the 2^+ states. These are dominated by low-energy two-phonon components like $|[3^- \otimes 3^-]^{2^+} >$.

Similarly, the strength of the isoscalar $E1$ response gets spread and its peaks severely reduced in magnitude as the multiphonon components come into play. The sensitivity of the isoscalar giant dipole resonance to multiphonon configurations was largely expected. Indeed, once the c.m. spurious mode is removed, most of the $p-h$ excitations entering the isoscalar 1^- states have an excitation energy of $3\hbar\omega$. For a consistent treatment of the mode, one has to include the huge number of $2p-2h$ and $3p-3h$ configurations of the same energy. In the EMPM, all these configurations are treated on the same footing of the $p-h$ states through the two- and three-phonon components.

In all cases, the multiphonon coupling has the effect of severely damping and strongly fragmenting the $E\lambda$ strengths. Once again, it is worth to point out the dramatic effect of removing the CM motion (Fig. 3)

4. Conclusions

The comparative analysis outlined in the previous section shows that, at least in ^{16}O, the multiphonon (or $np-nh$) configurations are important ingredients of the negative parity states and crucial constituents of the positive parity ones. They affect strongly the nuclear response for most $E\lambda$ transitions by inducing a severe damping and fragmentation of the strengths.

In computing the $E\lambda$ transitions, it becomes more and more compelling to remove the c.m. motion as the number of phonons increases. Otherwise, the c.m. admixtures would completely alter the strength distribution, at least for some multipolarities $E\lambda$.

Our calculation, though performed with no approximations, was carried out using a multiphonon basis composed of states with a maximum number of three TDA phonons. Because of these restrictions, the present calculation is far from reproducing the observed spectroscopic properties of this nucleus, especially those concerning the positive parity states. We need to include the $(n = 4)$-phonon subspace at least.

This extension is made possible by a new formulation of the method. This consists in replacing the basis states $a^\dagger_p a_h|n-1\alpha>$ with the more correlated ones $O^\dagger_\lambda|n-1\alpha>$, where $O^\dagger_\lambda$ is the TDA phonon operator. By so doing, a large number of $p-h$ states, with no restrictions on their energy, can be incorporated into the correlated TDA phonons. Moreover, it is possible to select the phonon states that are collective and discard the others thereby achieving a substantial truncation of each multiphonon subspace. Such a truncation should enable us to enlarge the phonon space by reaching subspaces with a large number of phonons $(n \geq 4)$ and, on the other hand, to apply the method to heavy nuclei. The upgrade of the method is being completed and its numerical implementation in progress.

References

[1] For a review Kneissl U, Pietralla N and Zilges A 2006 *J. Phys. G: Nucl. Part. Phys.* **32** R217
[2] For a review Kneissl U, Pitz H H and Zilges A 1996 *Prog. Part. Nucl. Phys.* **37** 439
[3] For a review Frascaria N 1988 *Nucl. Phys.* **A482** 245c
[4] Aumann T, Bortignon P F, Emling H 1998 *Annu. Rev. Nucl. Part. Sci.* **48** 351
[5] Bortignon P F, Bracco A and Broglia R A 1998 *Giant Resonances: Nuclear Structure at Finite Temperature* (Contemporary Concepts in Physics, V. 10) (the Netherland: Harwood Academic Publishers)
[6] Bortignon P F, Broglia R A, Bes D R and Liotta R J 1997 *Phys. Rep.* **30** 305
[7] Kamerdzhiev S, Speth S J, Tertychny G 2004 *Phys. Rep.* **393** 1
[8] Soloviev V G 1992 *Theory of Atomic Nuclei: Quasiparticles and Phonons* (Bristol: Institute of Physics)
[9] Ponomarev V Yu, Bortignon P F, Broglia R A, and Voronov V V 2000 *Phys. Rev. Lett.* **85** 1400
[10] Lo Iudice N and Stoyanov Ch 2002 *Phys. Rev.* C **65** 064304
[11] Lo Iudice N, Stoyanov Ch and Pietralla N 2009 *Phys. Rev.* C **80** 024312
[12] Lo Iudice N, Sushkov A V and Shirikova N Yu 2005 *Phys Rev* C **72** 034303
[13] Lo Iudice N and Sushkov A V 2008 *Phys. Rev.* C **78** 054304
[14] Andreozzi F, Knapp F, Lo Iudice N, Porrino A, and Kvasil J 2007 *Phys. Rev.* C **75** 044312
[15] Pomar C, Blomqvist J, Liotta R J, and Insolia A 1990 *Nucl. Phys.* **A515** 381
[16] Grinberg M, Piepenbring R, Protasov K V, Silvestre-Brac B 1996 *Nucl. Phys.* **A 597** 355
[17] Andreozzi F, Knapp F, Lo Iudice N, Porrino A, and Kvasil J 2008 *Phys. Rev.* C **78** 054308
[18] Haxton W C and Johnson C W 1990 *Phys. Rev. Lett.* **65** 1325
[19] Dytrych T, Sviratcheva K D, Bahri C, Draayer J P, Vary J P 2007 *Phys. Rev. Lett.* **98** 162503
[20] Palumbo F 1967 *Nuc. Phys.* **99** 100
[21] Glockner D H and Lawson R D 1974 *Phys. Lett.* **B 53** 313
[22] Berman B L and Fultz S C 1975 *Rev. Mod. Phys.* **47** 713
[23] Andreozzi F, Knapp F, Lo Iudice N, Porrino A and Kvasil J 2009 *Int. J. Mod. Phys.* E **10** 944

Journal of Physics: Conference Series **205** (2010) 012009 doi:10.1088/1742-6596/205/1/012009

Parity mixing in the single particle states of quadrupole-octupole deformed nuclei

N Minkov[1], S Drenska[1], M Strecker[2] and W Scheid[2]

[1] Institute of Nuclear Research and Nuclear Energy, Bulgarian Academy of Sciences, Tzarigrad Road 72, BG–1784 Sofia, Bulgaria

[2] Institut für Theoretische Physik der Justus-Liebig-Universität, Heinrich-Buff-Ring 16, D–35392 Giessen, Germany

E-mail: nminkov@inrne.bas.bg

Abstract. The effect of parity mixing in the single particle (s.p.) states of odd-mass nuclei with quadrupole-octupole deformations is examined through a reflection-asymmetric deformed shell model. A strong coupling scheme between the parity mixed s.p. state and a coherent quadrupole-octupole vibration mode in the core is considered. The Coriolis decoupling factor is obtained in a projected form corresponding to the good total parity of the system. The average parity of the s.p. state and the decoupling factor are evaluated in several nuclei as functions of the quadrupole and octupole deformation parameters β_2 and β_3. It is found that the average s.p. parity obtains various dominant (+ or −) values in the (β_2, β_3)-plane, while the s.p. wave function is strongly fragmented into components with different parities. It is shown that by comparing the behaviour of the decoupling factor in the (β_2, β_3)-plane to values obtained in a collective quadrupole-octupole model one can determine physically reasonable regions for the deformation parameters.

1. Introduction

In odd-mass atomic nuclei with quadrupole-octupole shapes the coupling of the collective rotations and vibrations to the single-particle (s.p.) motion of the unpaired nucleon provides a split parity-doublet structure of the spectrum [1, 2]. It is known that the octupole deformation of the core can induce respective deformation in the s.p. potential [3]. In this situation, known as a strong coupling, the state of the odd nucleon appears without good parity. On its turn the parity-mixed s.p. state couples to the core state in a way that the total state of the nucleus remains with a good parity. The relevant wave function of the total state can be constructed if a particular model assumption for the behaviour of the core with respect to the reflection asymmetric (octupole) degree of freedom is made.

In the present work a coupling of the particle to a core performing coherent quadrupole-octupole vibrations and rotations [4] is considered. It will be seen that such a coupling scheme provides a reasonable physical framework to examine the effect of the parity mixing in the s.p. state on the total behaviour of the system as well as on the interaction between the collective and s.p. motions of the nucleus. The magnitude of the parity mixing is characterized by the expectation value of the parity operator in the s.p. state. For the s.p. states with z- projection of the angular momentum $K = 1/2$ the effect of the Coriolis interaction between the odd particle and the rotating core has to be considered. Below it will be seen that the way in which this should be done essentially depends on the parity mixing.

The purpose of the present work is to examine the behaviour of the parity mixing and its effect on the Coriolis interaction in some odd-mass nuclei in dependence on the quadrupole and octupole deformations. The numerical evaluation of the Coriolis decoupling factor and the average parity is realized within a deformed shell model with reflection asymmetry in the basis of the axially deformed harmonic oscillator (ADHO) [5].

In sec. 2 the general Hamiltonian of the quadrupole-octupole vibrating and rotating core plus particle is considered. In sec. 3 the particle-core coupling scheme is considered. In sec. 4 the decoupling factor is determined. Numerical results and discussions are given in sec. 5.

2. Hamiltonian of quadrupole–octupole vibrating and rotating core plus particle

The Hamiltonian of an odd-mass nucleus with quadrupole–octupole degrees of freedom can be written in the form

$$H = H_{\text{qoc}}^{\text{coll}} + H_{\text{sp}} + H_{\text{coriol}}, \tag{1}$$

where $H_{\text{qoc}}^{\text{coll}}$ corresponds to the collective quadrupole-octupole vibrations and rotation of the system, H_{sp} describes the motion of the odd particle (neutron or proton) in the field of the even-even core, while H_{coriol} represents the Coriolis interaction between the core and the particle. The quadrupole and octupole vibration modes in $H_{\text{qoc}}^{\text{coll}}$ can be mixed through a centrifugal term if a dependence of the moment of inertia on the respective deformation variables β_2 and β_3 is assumed. The Coriolis part H_{coriol} can be superposed to the centrifugal term yielding the Hamiltonian [4]

$$\begin{aligned} H' &= H_{\text{qoc}}^{\text{coll}} + H_{\text{coriol}} = -\frac{\hbar^2}{2B_2}\frac{\partial^2}{\partial\beta_2^2} - \frac{\hbar^2}{2B_3}\frac{\partial^2}{\partial\beta_3^2} + \frac{1}{2}C_2{\beta_2}^2 + \frac{1}{2}C_3{\beta_3}^2 \\ &+ \frac{\hat{I}^2 - \hat{I}_z^{\,2} - \hat{I}_+\hat{\jmath}_- - \hat{I}_-\hat{\jmath}_+}{2(d_2\beta_2^2 + d_3\beta_3^2)}, \end{aligned} \tag{2}$$

where B_2 and B_3 are effective quadrupole and octupole mass parameters and C_2 and C_3 are stiffness parameters for the respective oscillation modes. Here $\hat{I}_\pm = \hat{I}_x \pm \imath \hat{I}_y$ are the total angular momentum operators, while $\hat{\jmath}_\pm = \hat{\jmath}_x \pm \imath\hat{\jmath}_y$ are the spherical components of the total intrinsic particle angular momentum $\hat{\jmath}$. In [4] the s.p. Hamiltonian H_{sp} was not considered explicitly and the contribution of the Coriolis part in (2) was treated phenomenologically by taking the decoupling factor as a free adjustable parameter. The Schrödinger equation for the Hamiltonian (2) was solved by assuming certain relations between the quadrupole and octupole stiffness and mass parameters which provide a coherent interplay between the both degrees of freedom. The obtained energy spectrum has a split parity-doublet structure with the wave function of the quadrupole-octupole oscillating core $\Phi_{\text{vib}} \equiv \Phi_{\text{core}}^{\pm}$ determining core vibrating states with given parity $\pi_c = (\pm)$ [4].

In the present work the single particle Hamiltonian H_{sp} is considered explicitly which allows one first, to construct an appropriate total wave function carrying detailed information for the motion of the odd nucleon in the reflection asymmetric s.p. potential, and second, to implement respectively a fully microscopic treatment of the Coriolis interaction by directly calculating the decoupling factor. For this purpose H_{sp} is taken in a form including a Woods-Saxon potential with axial quadrupole and octupole deformations [5]

$$H_{\text{sp}} = T + V_{\text{ws}} + V_{\text{s.o.}} + V_{\text{c}}, \tag{3}$$

where $V_{\mathrm{ws}}(\mathbf{r},\hat{\beta}) = V_0\left[1+\exp\left(\frac{\mathrm{dist}_\Sigma(\mathbf{r},\hat{\beta})}{a}\right)\right]^{-1}$ is the Woods-Saxon potential with $\hat{\beta} \equiv (\beta_2,\beta_3,\beta_4,\beta_5,\beta_6)$ and $\mathrm{dist}_\Sigma(\mathbf{r},\hat{\beta})$ being the distance between the point $\mathbf{r}$ and the nuclear surface represented by $R(\theta,\hat{\beta}) = c(\hat{\beta})R_0\left[1+\sum_{\lambda=2,3,\dots}\beta_\lambda Y_{\lambda 0}(\cos\theta)\right]$ ($c(\hat{\beta})$ is a volume conserving factor). $V_{\mathrm{s.o.}}$ and V_{c} are the spin-orbit and Coulomb terms whose analytic form is given in [5]. Hamiltonian (3) is diagonalized in the ADHO basis $|Nn_z\Lambda\Omega\rangle$. Due to the axial symmetry, Ω is equal to the third projection K of the total angular momentum I. The wave function for the odd particle is obtained as an expansion in the ADHO basis functions

$$\mathcal{F}_\Omega = \sum_{Nn_z\Lambda} C^{\Omega}_{Nn_z\Lambda}|Nn_z\Lambda\Omega\rangle. \tag{4}$$

The Hamiltonian (3) can be applied in the two cases of reflection-symmetric ($\beta_3 = 0$) and reflection-asymmetric ($\beta_3 \neq 0$) deformations of the s.p. potential. The respective s.p. wave functions, $\mathcal{F}_\Omega$, correspond to states with a good parity $\pi_{\mathrm{sp}} = (+)$ or $(-)$ ($\beta_3 = 0$) and to mixed-parity states without determined parity ($\beta_3 \neq 0$). Since the parity of the basis states $|Nn_z\Lambda\Omega\rangle$ is given by $(-1)^N$, in the first case the expansion (4) contains only terms with N =even ($\pi_{\mathrm{sp}} = +$), or N =odd ($\pi_{\mathrm{sp}} = -$). In the second case both terms with even and odd N are present in (4).

Thus in the case of a mixed s.p. state the expansion (4) contains components with both parities $\pi_{\mathrm{sp}} = (-1)^N = +1$ and -1. Then the wave function $\mathcal{F}_\Omega$ can be written as

$$\mathcal{F}_\Omega = \sum_{\pi_{\mathrm{sp}}=\pm 1} \mathcal{F}_\Omega^{(\pi_{\mathrm{sp}})} = \mathcal{F}_\Omega^{(+)} + \mathcal{F}_\Omega^{(-)}, \tag{5}$$

where $\mathcal{F}_\Omega^{(+)}$ contains only the positive parity components, while $\mathcal{F}_\Omega^{(-)}$ contains the negative ones with

$$\hat{\pi}_{\mathrm{sp}}\mathcal{F}_\Omega^{(\pm)} = \pm\mathcal{F}_\Omega^{(\pm)}. \tag{6}$$

Then the action of the s.p. parity operator $\hat{\pi}_{\mathrm{sp}}$ on the function $\mathcal{F}_\Omega$ gives

$$\hat{\pi}_{\mathrm{sp}}\mathcal{F}_\Omega = \mathcal{F}_\Omega^{(+)} - \mathcal{F}_\Omega^{(-)}. \tag{7}$$

The parity mixing in a given s.p. state is characterized by the expectation (average) value of the parity operator in this state

$$\langle\hat{\pi}_{\mathrm{sp}}\rangle = \langle\mathcal{F}_\Omega|\hat{\pi}_{\mathrm{sp}}|\mathcal{F}_\Omega\rangle = \sum_{Nn_z\Lambda}(-1)^N|C^{\Omega}_{Nn_z\Lambda}|^2.$$

3. Total particle-core wave function

The total wave function for the Hamiltonian (1) with the ansatz (2) and (3) can be given in a symmetrized form providing a good total parity π and $\mathcal{R}_1$ invariance of the system

$$\Psi^{\pi}_{IMK} = \frac{1}{2}\mathcal{N}(1+\mathcal{R}_1)D^I_{M\,K}(\boldsymbol{\theta})(1+\pi\hat{P})\Phi^{\pi_c}_{\mathrm{core}}\mathcal{F}_K. \tag{8}$$

Here, $D^I_{M\,K}(\boldsymbol{\theta})$ is the rotation function and $\mathcal{N}$ is a normalization factor. The operator $\hat{P} = \hat{\pi}_c\cdot\hat{\pi}_{\mathrm{sp}}$ is a composition of the core and s.p. parity operators $\hat{\pi}_c$ and $\hat{\pi}_{\mathrm{sp}}$, respectively. The operator $\mathcal{R}_1$ represents a rotation by an angle π about an axis perpendicular to the intrinsic z-axis and acts on the rotation, core and s.p. parts of the wave function, respectively, as follows

$$\mathcal{R}_1 D^I_{M\,K} = (-1)^{I-K}D^I_{M\,-K}, \qquad \mathcal{R}_1\Phi^{\pi_c}_{\mathrm{core}} = \hat{\pi}_c\Phi^{\pi_c}_{\mathrm{core}} = \pi_c\Phi^{\pi_c}_{\mathrm{core}}, \qquad \mathcal{R}_1\mathcal{F}_K = \mathcal{F}_{-K}. \tag{9}$$

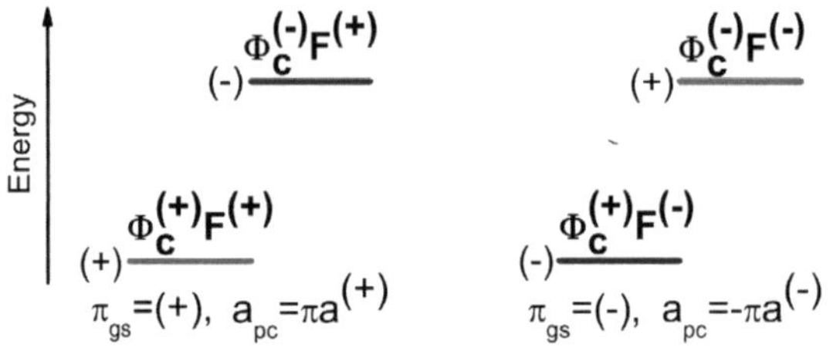

Figure 1. Particle-core states coupling for split parity-doublets with positive (left part) and negative (right side) ground-state parity. The respective expressions for the decoupling factor $a_{\rm pc} = \pm\pi a^{(\pm)}$ are also given.

The core parity π_c is fixed so as $\pi_c = (+)$ for the split doublet counterparts containing the ground state (states shifted down), and $\pi_c = (-)$ for the states shifted up. According to the model of coherent quadrupole–octupole motion [4] the lower $\pi_c = (+)$ oscillating-core state is characterized by the function $\varphi^+(\phi) = \sqrt{2/\pi}\cos(\phi)$, while the next upper $\pi_c = (-)$ state is presented by $\varphi^-(\phi) = \sqrt{2/\pi}\sin(2\phi)$, where ϕ is the quadrupole–octupole angular variable [see Eqs. (15) and (16) in [4]]. The full core wave function has the form $\Phi^{\pi_c}_{\rm core}(\eta,\phi) = \psi(\eta)\varphi^{\pi_c}(\phi)$, where $\psi(\eta)$ is the "radial" part obtained in terms of Laguerre polynomials for the effective quadrupole-octupole deformation η [see Eq. (14) in [4]].

The s.p. wave function $\mathcal{F}_K$ represents a mixture of two parts $\mathcal{F}_K^{(\pm)}$ with opposite parities. Under the above condition the operator $1+\pi\hat{P}$ projects out the component $\mathcal{F}_K^{(+)}$ or $\mathcal{F}_K^{(-)}$ from $\mathcal{F}_K$ providing a good total parity of the states in the split parity-doublet spectrum. Thus, if the lowest (ground) state (gs) of the doublet has a positive total parity, $\pi_{\rm gs} = (+)$, the projected component is $\mathcal{F}_K^{(+)}$, while for a negative-parity of the ground state, $\pi_{\rm gs} = (-)$, the projected component is $\mathcal{F}_K^{(-)}$. These two situations are shown schematically in Fig. 1. In the limiting case of $\beta_3 = 0$, the good parity of the s.p. state $\pi_{\rm sp} = (+)$, or $(-)$ is restored. Then only the component $\mathcal{F}_K^{(+)}$ or $\mathcal{F}_K^{(-)}$ remains non-zero and the scheme reduces to a s.p. wave function $\mathcal{F}_K^{\pi_{\rm sp}} \equiv \mathcal{F}_K^{(+)}$, or $\mathcal{F}_K^{(-)}$ with a good parity.

4. Decoupling factor for a parity-mixed s.p. state

When the system possesses reflection asymmetry the standard definition [1] for the Coriolis decoupling factor, $a = \langle\mathcal{F}_{1/2}|\hat{j}_+|\mathcal{F}_{-1/2}\rangle$, needs to be modified in dependence on the considered core+particle coupling scheme. The modified decoupling factor $a_{\rm pc}$ ('pc' means particle-core) is determined through the matrix element

$$-\langle\Psi^{\pi}_{IM\frac{1}{2}}|\hat{I}_-\hat{j}_+|\Psi^{\pi}_{IM\frac{1}{2}}\rangle = \mathcal{N}^2(-1)^{I+\frac{1}{2}}(I+\frac{1}{2})\cdot a_{\rm pc}. \tag{10}$$

Thus for the total wave function (8) with the transformation properties (9) and (7) the factor $a_{\rm pc}$ is obtained in the form

$$\begin{aligned} a_{\rm pc} &= \frac{1}{2}\pi_c\left(\left\langle\mathcal{F}_{\frac{1}{2}}|\hat{j}_+|\mathcal{F}_{-\frac{1}{2}}\right\rangle + \pi\pi_c\left\langle\hat{\pi}_{\rm sp}\mathcal{F}_{\frac{1}{2}}|\hat{j}_+|\mathcal{F}_{-\frac{1}{2}}\right\rangle\right) \\ &= \frac{1}{2}\pi_c\left[(1+\pi\pi_c)a^{(+)} + (1-\pi\pi_c)a^{(-)}\right], \end{aligned} \tag{11}$$

where

$$a^{(+)} = \langle\mathcal{F}^{(+)}_{1/2}|\hat{j}_+|\mathcal{F}^{(+)}_{-1/2}\rangle, \qquad a^{(-)} = \langle\mathcal{F}^{(-)}_{1/2}|\hat{j}_+|\mathcal{F}^{(-)}_{-1/2}\rangle, \tag{12}$$

Table 1. Particular forms of the decoupling factor $a_{\rm pc}$ in dependence on the ground-state parity $\pi_{\rm gs}$ and the parity π of a given state.

	$\pi = (+)$	$\pi = (-)$
$\pi_{\rm gs} = (+)$	$a_{\rm pc} = a^{(+)}$	$a_{\rm pc} = -a^{(+)}$
$\pi_{\rm gs} = (-)$	$a_{\rm pc} = -a^{(-)}$	$a_{\rm pc} = a^{(-)}$

with $\langle\mathcal{F}^{(+)}_{1/2}|\hat{j}_+|\mathcal{F}^{(-)}_{-1/2}\rangle = \langle\mathcal{F}^{(-)}_{1/2}|\hat{j}_+|\mathcal{F}^{(+)}_{-1/2}\rangle = 0$. The factors $a^{(+)}$ and $a^{(-)}$ represent projected matrix elements of the operator $\hat{j}_+$ in the subspaces of positive and negative parity components of the s.p. wave function, respectively. By using the expansion (4) they are obtained in the form

$$\begin{aligned} a^{\pm} &= \sum_{Nn_z\Lambda}\sum_{N'n'_z\Lambda'} C^{\frac{1}{2}*}_{N'n'_z\Lambda'} C^{\frac{1}{2}}_{Nn_z\Lambda} \\ &\times \left\langle N'n'_z\Lambda'\frac{1}{2}|\hat{j}_+|Nn_z-\Lambda-\frac{1}{2}\right\rangle , \end{aligned} \tag{13}$$

where the matrix element of $\hat{j}_+$ between ADHO basis states (with N even or odd) can be calculated through an available analytic expression [6] (see also Table 1 in Ref. [7]).

It can be straightforward seen, that according to Eq. (11), the decoupling factor $a_{\rm pc}$ is reduced to $\pi a^{(+)} = \pm a^{(+)}$ when the downwards shifted sequence (containing the ground state) has positive total parity $\pi_{\rm gs} = (+)$, and to $-\pi a^{(-)} = \mp a^{(-)}$ when this sequence has negative total parity $\pi_{\rm gs} = (-)$. This is illustrated in Table 1 (see also Fig. 1). (Note that $\pi_{\rm gs}$ is the *total parity* in the ground state, while π_c is the *core parity* considered to be always (+) in the ground state.) From Table 1 it is seen that for a given state I^π in the parity-doublet one can shortly write

$$a_{\rm pc} = \pi\pi_{\rm gs}a^{(\pi{\rm gs})} = \pm\pi a^{(\pm)}. \tag{14}$$

Now one can easily obtain a correspondence between the values of the decoupling factor $a_{\rm qoc}$ fitted in the model of coherent quadrupole–octupole motion [4] and the values of the projected decoupling factors $a^{(\pm)}$ in the present scheme. One has

$$a_{\rm coll} = \pi a_{\rm qoc} \leftrightarrow \pi\pi_{\rm gs}a^{(\pi{\rm gs})} = \pm\pi a^{(\pm)} = a_{\rm pc},$$

which gives

$$a_{\rm qoc} \leftrightarrow \pi_{\rm gs}a^{(\pi{\rm gs})} = \pm a^{(\pm)}. \tag{15}$$

In the limiting case $\beta_3 = 0$ of a good s.p. parity the quantity $a_{\rm pc}$ reduces to the standard Coriolis decoupling factor.

5. Numerical results and discussion

The numerical behaviour of the average s.p. model parity $\langle\hat{\pi}_{\rm sp}\rangle$ and of the parity-projected decoupling factor $a_{\rm pc}$ was examined as a function of the quadrupole and octupole deformations in several odd-mass nuclei. Numerical calculations were performed on a net in the two-dimensional space of the parameters β_2 and β_3.

In Fig. 2 the behaviour of the average s.p. model parity in the nuclei ^{237}U, and ^{249}Cm is shown (upper plots) as a function of the octupole deformation β_3 at β_2 corresponding to the

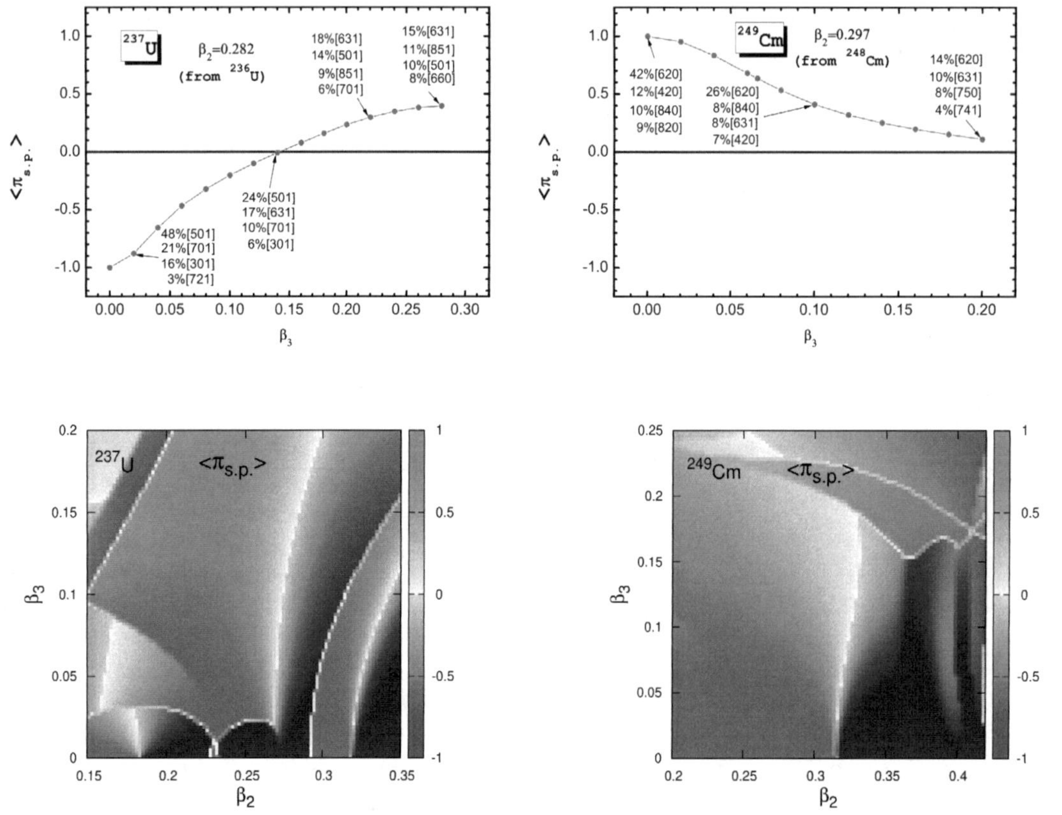

Figure 2. Average s.p. model parity in ^{237}U and ^{249}Cm as a function of β_3 at fixed β_2-values (upper plots) and as a two-dimensional function (lower plots) in the plane (β_2, β_3).

experimentally estimated quadrupole deformations of the respective even-even core nuclei and (lower plots) as a two-dimensional function (contour plot) in the plane (β_2, β_3). From the upper plots it is seen that $\langle\hat{\pi}_{\rm sp}\rangle$ smoothly changes with the increasing of β_3. In ^{237}U the average parity starting by -1 at $\beta_3 = 0$ goes through the zero value at $\beta_3 \sim 0.15$ and further increases to $\langle\hat{\pi}_{\rm sp}\rangle = 0.3 - 0.4$. This is an interesting observation showing that a presence of a dominant parity is possible at large β_3 deformations. In ^{249}Cm $\langle\hat{\pi}_{\rm sp}\rangle$ smoothly approaches the zero value for $\beta_3 \sim 0.2$ indicating the presence of a deformation region with strong parity mixing. Actually, the two-dimensional contour plots show that in the most regions in the (β_2, β_3)- plane the s.p. state is characterized with some dominant parity instead of becoming totally mixed ($\langle\hat{\pi}_{\rm sp}\rangle \sim 0$). Another important result is that in the considered nuclei the s.p. wave functions appear strongly fragmented into various Nilsson orbitals with different parities. This is indicated in Fig. 2 (upper plots) where the contributions (in %) of the largest $[N, n_z, \Lambda]$ components of the wave function are given for certain β_3 deformations.

The behaviour of the quantity $a^{(+)}$ (which determines the decoupling factor $a_{\rm pc}$ in the cases of a ground state with $K = 1/2^+$) as a function of β_2 and β_3 in the nucleus ^{239}Pu is illustrated in Fig. 3. The left plot shows a two-dimensional surface containing the values of $a^{(+)}$ in the (β_2, β_3)-regions with $K = 1/2$. The right plot contains a contour which corresponds to the intersection of the two-dimensional surface with the plane determined by the decoupling parameter value $a_{\rm coll} = -0.34$ fitted in the collective model [4]. This contour outlines deformation regions for

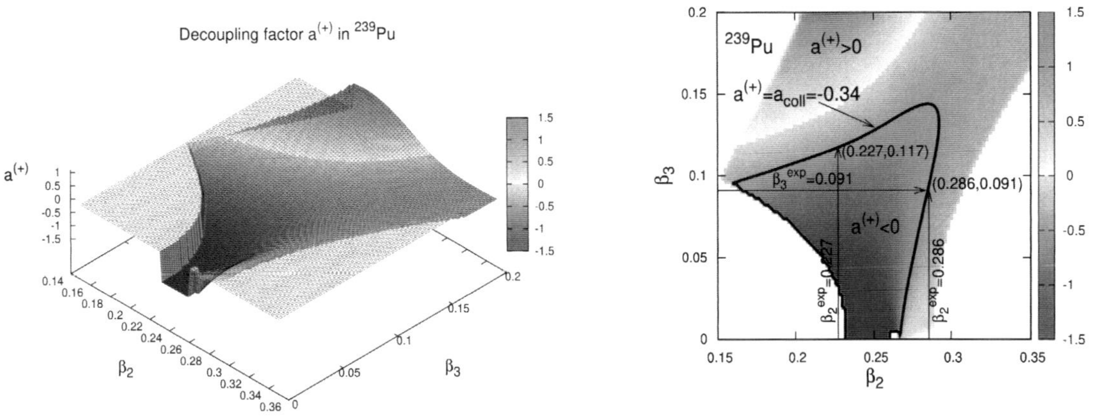

Figure 3. The decoupling factor $a_{\rm pc} = a^{(+)}$ in ^{239}Pu as a function of β_2 and β_3. The flat area at $a^{(+)} = 0$ in the left plot and the white area in the contour plot (right) correspond to $K \neq 1/2$.

which the microscopically calculated decoupling factor $a_{\rm pc}$ provides a reasonable description of the split parity-doublet structure, perturbed by the Coriolis interaction, of the experimental spectrum in ^{239}Pu. Two experimental guesses for the quadrupole deformation, $\beta_2 = 0.227$ (from an experimental data base for ^{239}Pu) and 0.286 (from the core nucleus ^{238}Pu) and one for the octupole deformation, $\beta_3 = 0.091$ (also from ^{238}Pu), are given in the right plot of Fig. 3. It is seen that the couple of ^{238}Pu core-deformations ($\beta_2 = 0.286, \beta_3 = 0.091$) lies exactly on the contour, while the value $\beta_2 = 0.227$ suggests a slightly larger octupole deformation $\beta_3 = 0.117$ for the odd-mass nucleus ^{239}Pu. This result shows that by comparing the behaviour of the microscopically obtained decoupling factor $a_{\rm pc}$ in the (β_2, β_3)-plane to a value obtained in the collective model one can determine physically reasonable regions of quadrupole and octupole deformations in the considered nucleus. However, more detailed analysis of the possible deformation values and the corresponding dominant s.p. parity $\langle \hat{\pi}_{\rm sp} \rangle$ should be done in order to draw a definite conclusion.

Finally, the obtained results indicate the possibility for a consistent collective and microscopic model description of the split parity-doublet spectra in odd-mass nuclei. This is the subject of further work.

Acknowledgements

This work is supported by DFG and by the Bulgarian Scientific Fund under contract F-1502/05.

References

[1] Bohr A and Mottelson B R 1975 *Nuclear Structure* vol 2 (New York: Benjamin)
[2] Butler P A and Nazarewicz W 1996 *Rev. Mod. Phys.* **68** 349
[3] Leander G A and Sheline R K 1984 *Nucl. Phys. A* **413** 375
[4] Minkov N, Drenska S, Yotov P, Lalkovski S, Bonatsos D and Scheid W 2007 *Phys. Rev. C* **76** 034324
[5] Cwiok S, Dudek J, Nazarevicz W, Skalski J and Werner T 1987 *Comp. Phys. Comm.* **46** 379
[6] Bonneau L 2008 *Eléments de matrice de $\hat{j}_+$ dans la base asymptotique* CENBG Bordeaux Preprint
[7] Boisson J P and Piepenbring R 1971 *Nucl. Phys. A* **168** 385

XVIII International School on Nuclear Physics, Neutron Physics and Applications IOP Publishing
Journal of Physics: Conference Series **205** (2010) 012010 doi:10.1088/1742-6596/205/1/012010

Covariant density functional theory beyond mean field and applications for nuclei far from stability

P Ring

Physics Department, Technical University of Munich, 85748 Garching, Germany

E-mail: ring@ph.tum.de

Abstract. Density functional theory provides a very powerful tool for a unified microscopic description of nuclei all over the periodic table. It is not only successful in reproducing bulk properties of nuclear ground states such as binding energies, radii, or deformation parameters, but it also allows the investigation of collective phenomena, such as giant resonances and rotational excitations. However, it is based on the mean field concept and therefore it has its limits. We discuss here two methods based based on covariant density functional theory going beyond the mean field concept, (i) models with an energy dependent self energy allowing the coupling to complex configurations and a quantitative description of the width of giant resonances and (ii) methods of configuration mixing between Slater determinants with different deformation and orientation providing are very successful description of transitional nuclei and quantum phase transitions.

1. Introduction

The structure of nuclei far from stability with extreme isospin is one the most exciting challenges of present nuclear physics. New experimental facilities with radioactive nuclear beams make it possible to investigate the nuclear chart to the very limits of nuclear binding. A wealth of structure phenomena in exotic nuclei have been reported and the next generation of radioactive-beam facilities will present new exciting opportunities to study not only the ground states but also excitations and spectra of these strongly interacting many-body systems. This situation has stimulated considerable new efforts on the theoretical side to understand the dynamics of the nuclear many-body problem by microscopic methods.

Density functional theory (DFT) provides a universal tool to provide a quantitative microscopic description for a large majority of nuclei. Considerable progress has been reported recently in constructing such functionals [1]. A very successful example is covariant density functional theory (CDFT) in the form of the Relativistic Hartree-Bogoliubov (RHB) model [2, 3, 4]. It combines a density dependence with pairing correlations based on an effective interaction of finite range.

In principle density functional theory can be used for the description of all properties depending on the single-particle density. It is therefore not only limited to the description of the ground state properties. The same density functionals have also been applied for a very successful description of excited states, such as rotational bands in normal and super-deformed nuclei [5, 6] and collective vibrations [7, 8, 9]. Relativistic RPA and QRPA provides a natural framework to investigate collective and non-collective excitations of *ph*- (or 2*qp*) character. It is successful in particular for the understanding of the position of giant resonances. Recently it has been also used for a theoretical interpretation of low lying E1-strengths observed in neutron rich isotopes (pygmy modes) [9] and for low-lying collective quadrupole excitations [10].

However, density functional theory in nuclei is based on intrinsic densities and on the mean field approach. Therefore it cannot provide an exact treatment of the full nuclear dynamics. It breaks down in several respects. We discuss here two examples of phenomena in nuclear structure, which cannot be described properly within this concept and we discuss methods based on CDFT to overcome these shortcomings: (i) the fact, that the self energy does not depend on energy, leads to pure single particle states, which are not coupled to more complex configurations and not fragmented as it is observed in experiment. In particular this theory does not provide the proper level density at the Fermi surface and the collective vibrations are only of *ph*-structure. The coupling the complex configurations is absent and therefore the width of giant resonances is not reproduced properly, (ii) conventional density functional theory breaks breaks also down in transitional nuclei, where the intrinsic density is not well defined and where one has to include correlations going beyond the mean field approximation by treating quantum fluctuations through a superposition of several mean field solutions, as for instance in the Generator Coordinate Method (GCM) [11].

2. Covariant density functional theory with complex configurations

The fact that DFT usually provides a poor approximation for the single-particle spectra, particularly for those of ideal shell model nuclei such as ^{208}Pb, is connected with the relatively small effective mass in such models. In finite nuclei the effective mass is influenced by the coupling to low-lying surface phonons. It leads to an energy dependent self energy. The single particle levels are then fragmented and shifted. These effects are not included in conventional DFT.

Already before density functional theory has been introduced in the sixties for the description of quantum mechanical many-body problems by Kohn and Sham [12] Landau has developed in the fifties his Fermi Liquid Theory [13] for infinite systems. It has been extended to the Theory of Finite Fermi Systems (TFFS) [14] by Migdal. This theory provides another very successful method for the description of low-lying nuclear excitations [15]. It has several general properties in common with density functional theory. First, both theories are know to be exact, at least in principle, but in practice, in nuclear physics, the parameters entering these theories have to be determined in a phenomenological way by adjustment to experimental data. Second, both theories are based on a single-particle concept. DFT uses the mean field concept with Slater determinants in an effective single-particle potential as a vehicle to introduce shell-effects in the exact density functional introduced by Hohenberg and Kohn [16]. Fermi liquid theory is based on the concept of quasi-particles obeying a Dyson equation, which are defined as the basic excitations of the neighboring system with odd particle number. Third, in practical applications both theories describe in the simplest versions the nuclear excitations in the RPA approximation, i.e. by a linear combination of *ph*-configurations in an average nuclear potential.

However, there are also essential differences between these two concepts. First, in contrast to DFT, TFFS does not attempt to calculate the ground state properties of the many-body system, but it describes the nuclear excitations in terms of Landau quasi-particles and their interaction. Therefore the experimental data used to fix the phenomenological parameters of the theories are bulk properties of the ground state in the case of density functional theory, and properties of single-particle excitations and of the collective excitations in the case of finite Fermi systems theory. Second, in DFT the mean field is determined in a self-consistent way and therefore the RPA spectrum contains Goldstone modes at zero energy. This is usually not the case in TFFS calculations, which are based on a non-relativistic shell-model potential, whose parameters are adjusted to the experimental single-particle spectra. Therefore, there is no self-consistency in the RPA calculations of TFFS and the Goldstone modes do not separate from the

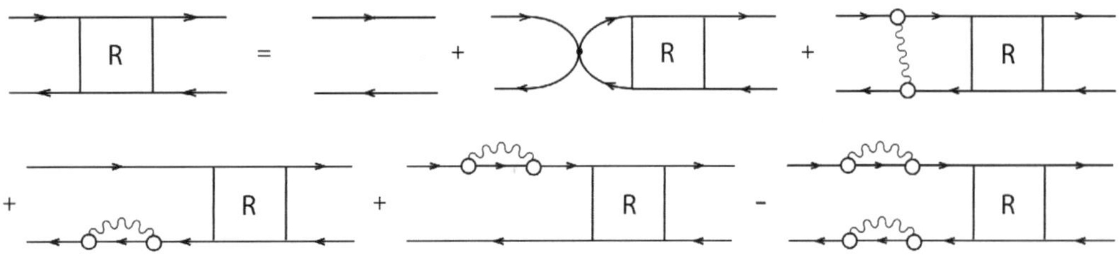

Figure 1. Bethe-Salpeter equation for the *ph*-response function R in graphical representation. Details are given in Fig. 1 and the small black circle means the static part of the residual *ph*-interaction V.

other modes. They are distributed among the low-lying excitations and have to be separated afterwards by special techniques. Third, modern versions of TFFS go much beyond the mean field approximation. The coupling between the particles and the phonons is investigated with Green's function techniques. Based on the phonons calculated in the framework of the RPA one has included particle-phonon coupling vertices and an energy dependence of the self-energy in the Dyson equation [17]. This leads also to an induced interaction in the Bethe-Salpeter equation caused by the exchange of phonons which also depends on the energy. The coupling of particles and phonons has also been derived from Nuclear Field Theory (NFT) and its extensions [18, 19].

Recently it has been attempted [20, 21, 22, 23] to find a combination of the basic ideas of covariant density functional theory and Landau-Migdal theory. The starting point is a covariant density functional $E[\rho]$ widely used in the literature, e.g. NL3 [24]. Without any additional parameter, it provides the necessary input of finite Fermi systems theory, such as the mean field and the single-particle spectrum as well as an effective interaction between the *ph*-configurations in terms of the second derivative of the same energy $E[\rho]$ with respect to the density. Thus the phenomenological input of Landau-Migdal theory is replaced by the results of density functional theory. The same interaction is used to calculate the vertices for particle-vibration coupling [20]. In a second step techniques of Landau-Migdal theory and its modern extensions are used to describe the coupling of one- and two-quasiparticle configurations. The main assumption of the quasiparticle-phonon coupling model is that two types of elementary excitations – two-quasiparticle and vibrational modes – are coupled in such a way that configurations of $1p1h \otimes phonon$ type with low-lying phonons strongly compete with simple $1p1h$ configurations close in energy or, in other words, that quasiparticles can emit and absorb phonons with rather high probabilities. In this way a fully consistent description of the many-body dynamics is obtained. As a result an induced additional interaction between single-particle and vibrational excitations provides a strong fragmentation of the pure RQRPA states causing the spreading width of giant resonances and the redistribution of the pygmy strength.

Nuclear dynamics of an even-even nucleus in a weak external field is described by the response function $R(14, 23)$, where This response function is the solution of the Bethe–Salpeter equation (BSE) in the *ph*-channel:

$$R(14, 23) = R^0(14, 23) - i \sum_{5678} R^0(16, 25)W(58, 67)R(74, 83), \quad (1)$$

where R^0 is the free response and W is the effective *ph*-interaction obtained as the derivative of the self energy with respect to the density. Neglecting the energy dependent term in the self energy we obtain the linearized Bethe-Salpeter equation with the static interaction V as the second derivative of the energy density functional with respect to the density. This is conventional linear response or RPA theory. In addition to the static interaction V the

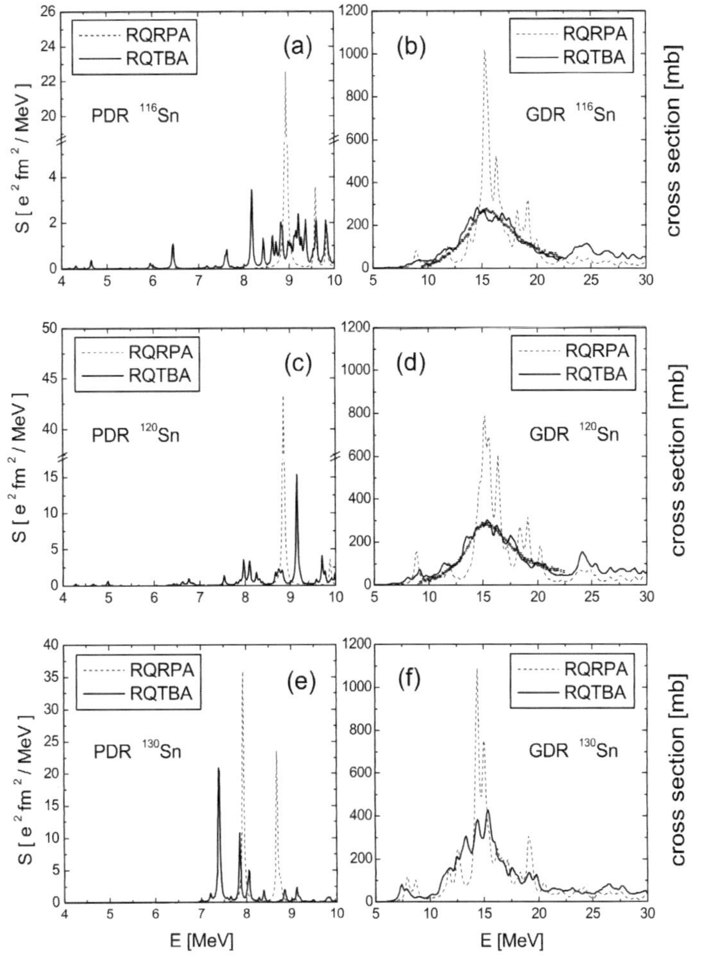

Figure 2. The calculated dipole spectra for the heavier tin isotopes ^{116}Sn, ^{120}Sn, ^{130}Sn, compared to data of Ref. [25] for ^{116,120}Sn. Right panels (b, d, f): photo absorption cross sections computed with the artificial width 200 keV. Left panels (a, c, e): the low-lying portions of the corresponding spectra in terms of the strength function, calculated with 20 keV smearing. Calculations within the RQRPA are shown by the dashed curves, and the RQTBA - by the solid curves.

effective interaction W contains terms derived from the energy dependent part of the self-energy. They contain diagrams with energy-dependent self-energies and an energy-dependent *induced interaction*, where a phonon is exchanged between the particle and the hole as shown in Fig. 1

To describe the observed spectrum of the excited nucleus in a weak external field D, as for instance a dipole field, one needs to calculate the strength function

$$S_D(E) = -\frac{1}{\pi}\mathrm{Im}\left(D^{\dagger}R((E))D\right).,\tag{2}$$

In Fig. 2 we show dipole spectra for the tin isotopes ^{116}Sn, ^{120}Sn, ^{130}Sn. The right panels show the photoabsorption cross section derived from the dipole strength function S_D. The left panels show the low-lying parts of the corresponding spectrum in terms of the strength function. A small imaginary part of 20 keV is used for the energy variable, in order to see the fine structure of the spectrum and sometimes individual levels in this region. RQRPA calculations are shown by dashed curves and the RQTBA by the solid curves. Experimental data are taken from the EXFOR database [25].

These figures clearly demonstrate how the two-quasiparticle states, which are responsible for the spectrum of the RQRPA excitations, are fragmented through the coupling to the collective vibrational states. The effect of the particle-vibration coupling on the low-lying dipole strength below and around the neutron threshold within the presented approach is shown in the left panels of the Fig. 2. Such calculations give us an example how the low-lying strength develops with the increase of the neutron excess. It is also found that the presence of pairing correlations causes a noticeably stronger fragmentation of both the GDR and the PDR modes as compared to the case of a normal system. This effect has the two reasons. First, pairing correlations lead to a diffuseness of the Fermi surface and, thus, increase the number of possible $2qp \otimes phonon$ configurations, and second, pairing correlations cause a considerable lowering of the energies and increased transition probabilities of the lowest 2^+ states. In spherical open-shell medium mass nuclei the highly collective first 2^+ states appear at energies around 1 MeV (and they are usually well reproduced in RQRPA calculations [26]) whereas in magic nuclei and often in nuclei near the shell closures they appear much higher, at about 3-4 MeV and have considerably reduced transition probabilities. This causes a strong configuration mixing in the case of presence of very low-lying vibrational states.

A systematic analysis of the transition densities of the RQRPA and the RQTBA states shows that the $2qp$ transition densities in the broad low-lying energy region dominated by the fragmentation of the RQRPA pygmy mode have a very similar behavior as the initial RQRPA state: proton and neutron components oscillate in phase in the nuclear interior and neutron components dominate on the surface in nuclei with noticeable neutron excess.

3. Shape fluctuations in transitional nuclei

An important area, where conventional density functional theory based on the mean field concept breaks down is the region of transitional nuclei. In this case it is not possible to define an intrinsic deformation of the system. In fact, in such cases the potential energy surfaces (PES) as a function of the deformation parameters are rather flat and many configurations with different deformations have nearly the same energy. One has to take into account fluctuations in deformation space by forming a superposition of wave functions with different deformation.

In the following we study as an example the effects of quantum fluctuation for triaxial shapes in the nucleus ^{16}C. First we generate a set of deformed intrinsic wave functions $|q\rangle$ with the quadrupole deformations $q = (\beta, \gamma)$ generated by constrained RMF+BCS calculations. Of course, for all the non-spherical points with $(\beta \neq 0)$ the intrinsic wave functions $|\beta \neq 0, \gamma\rangle$ have a certain orientation and rotational symmetry is broken. To restore this symmetry, a three-dimensional angular momentum projection (3DAMP) [27] is carried out. The left panel of Fig. 3 shows the unprojected (MF) and the J-projected ($J^\pi = 0^+$) energy surfaces in the β-γ plane.

As found already in the eighties [30] we observe in the J-projected surface a triaxial minimum This minimum is still very soft along the γ direction. One has, therefore, to allow for superpositions of wave functions with different deformations and to perform a GCM calculation in the full β-γ plane

$$|\Psi^J_M\rangle = \sum_K \int d^2q f_K(q) P^J_{MK}|q\rangle \tag{3}$$

including correlations due to restoration of broken symmetries and fluctuations of the deformation coordinates. The weight function $f(q)$ is determined by the solution of the Hill

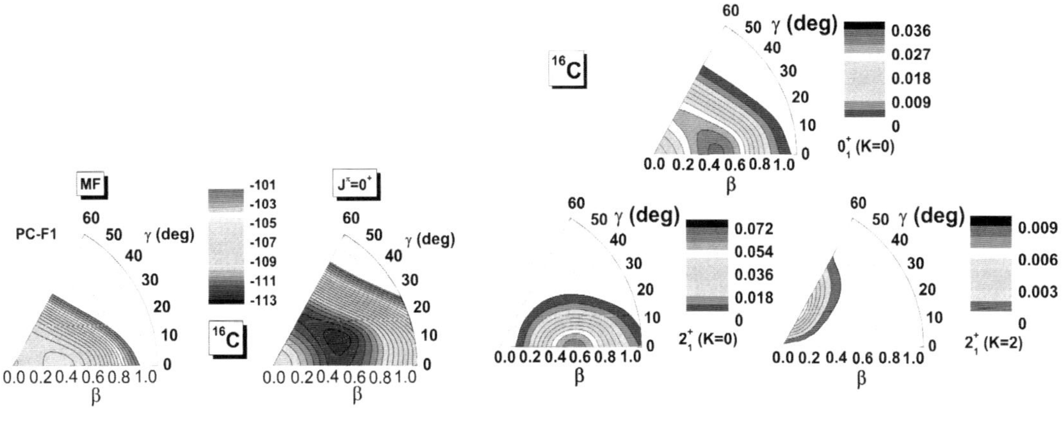

Figure 3. Left panel: Potential energy surfaces in β-γ plane for the nucleus ^{16}C obtained by triaxial relativistic mean-field calculations (MF) and with projection onto angular momentum after the variation ($J^{\pi} = 0^+$). Right panel: Contour plots of the probability distributions in β and γ from 3DAMP+GCM calculation for the ground state (0^+_1) and the first excited state (2^+_1)

Wheeler equation [11].

The right panel of Fig. 3 shows contour plots of the probability distributions derived from $f(q)$ in the β-γ plane resulting from 3DAMP+GCM calculations for both the ground state (0^+_1) and the first excited state (2^+_1). The ground state has a triaxial structure with almost uniform probability along the γ-direction which indicates an obvious quantum shape fluctuation in γ direction. For $J = 2$ we have K-mixing and we therefore show the two distributions for $K = 0$ and $K = 2$ separately. Both are concentrated along the axially symmetric configurations with $K = 0$ on the prolate and $K = 2$ on the oblate side. However, the $K = 0$ part exhausts 93.1% of the norm so that the nucleus has a strong prolate deformation in the (2^+_1) state with $\beta \simeq 0.6$.

To show the importance of shape fluctuations in γ for spectroscopic properties, we plot in

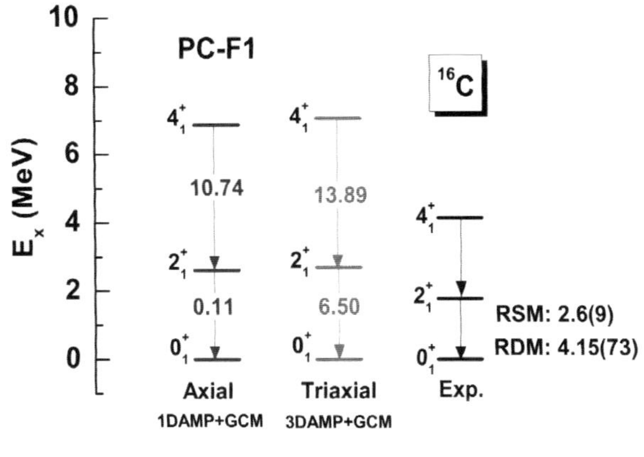

Figure 4. (Color online) The lowest energy levels of angular momentum $J = 0, 2, 4$ in ^{16}C. The $B(E2)$ values are in units of e^{2}fm^4. The experimental data are taken from Refs. [28, 29].

Fig. 4 the lowest energy levels with $J = 0, 2, 4$ obtained from the 1DAMP+GCM, where only axial symmetric configuratons $\gamma = 0$ are superimposed for various β-values and 3DAMP+GCM calculations and compare them with experiment [29]. The inclusion of triaxial states does not change the energy spectrum too much, but it improves the electric quadrupole transitions significantly. Only with a proper treatment of the shape fluctuations in γ, are the calculated $B(E2 : 2_1^+ \to 0_1^+) = 6.50\ \mathrm{e}^2\mathrm{fm}^4$ in agreement with recent data.

The predicted ratio $R_{4/2} \equiv E_4/E_2 = 2.63$ is a little larger than the experimental value $R_{4/2} = 2.35$. Both are close to the $R_{4/2} = 2.50$, a typical value for γ-soft nuclei. The predicted deviations of $R_{4/2}$ from experiment go in the direction of a rotor ($R_{4/2} = 3$). As observed in many calculations using projection after variation [31, 32, 33, 34] the calculated spectrum is systematically stretched. It would be far more compressed by the inclusion of a cranking term in the mean-field calculation [5].

Acknowledgments

I am very grateful to all my collaborators participating in the work presented here, in particular to E. Litvinova, J. Meng, V. Tselyaev, and J.M. Yao. This work was supported in part by the DFG cluster of excellence "Origin and Structure of the Universe" (www.universe-cluster.de).

References

[1] Bender M, Heenen P H and Reinhard P G 2003 *Rev. Mod. Phys.* **75** 121
[2] Gonzales-Llarena T, Egido J L, Lalazissis G A and Ring P 1996 *Phys. Lett.* **379** 13
[3] Ring P 1996 *Prog. Part. Nucl. Phys.* **37** 193
[4] Vretenar D, Afanasjev A V, Lalazissis G A and Ring P 2005 *Phys. Rep.* **409** 101
[5] Afanasjev A V, König J, Ring P, Egido J L and Robledo L M 2000 *Phys. Rev.* C **62** 054306
[6] Afanasjev A V, Ring P and König J 2000 *Nucl. Phys.* A **676** 196
[7] Ring P, Ma Z Y, Giai N V, Vretenar D, Wandelt A and Cao L G 2001 *Nucl. Phys.* A **694** 249
[8] Ma Z Y, Wandelt A, Giai N V, Vretenar D, Ring P and Cao L 2002 *Nucl. Phys.* A **703** 222
[9] Paar N, Ring P, Nikšić T and Vretenar D 2003 *Phys. Rev.* C **67** 034312
[10] Ansari A 2005 *Phys. Lett.* B **623** 37
[11] Ring P and Schuck P 1980 *The Nuclear Many-Body Problem* (Berlin: Springer-Verlag)
[12] Kohn W and Sham L J 1965 *Phys. Rev.* **137** A1697
[13] Landau L D 1959 *Sov. Phys. JETP* **8** 70
[14] Migdal A B 1967 *Theory of Finite Fermi Systems: Applications to Atomic Nuclei* (New York: Wilson)
[15] Ring P and Speth J 1974 *Nucl. Phys.* A **235** 315
[16] Hohenberg P and Kohn W 1964 *Phys. Rev.* **136** B864
[17] Ring P and Werner E 1973 *Nucl. Phys.* A **211** 198
[18] Bohr A and Mottelson B 1975 *Nuclear Structure* vol II (Reading,Mass.: Benjamin)
[19] Bertsch G F, Bortignon P F and Broglia R A 1983 *Rev. Mod. Phys.* **55** 287
[20] Litvinova E and Ring P 2006 *Phys. Rev.* C **73** 044328
[21] Litvinova E, Ring P and Tselyaev V I 2007 *Phys. Rev.* C **75** 064308
[22] Litvinova E, Ring P and Tselyaev V I 2008 *Phys. Rev.* C **78** 014312
[23] Litvinova E, Ring P and Tselyaev V I 2009 *Phys. Rev.* C **79** 054312
[24] Lalazissis G A, König J and Ring P 1997 *Phys. Rev.* C **55** 540
[25] Experimental nuclear reaction data URL `http://www-nds.iaea.or.at/exfor/exfor00.htm`
[26] Ansari A and Ring P 2006 *Phys. Rev.* C **74** 054313
[27] Yao J M, Meng J, Ring P and Peña Arteaga D 2009 *Phys. Rev.* C **79** 044312
[28] Ong H J *et al.* 2008 *Phys. Rev.* C **78** 014308
[29] Wiedeking M *et al.* 2008 *Phys. Rev. Lett.* **100** 152501
[30] Hayashi A, Hara K and Ring P 1984 *Phys. Rev. Lett.* **53** 337
[31] Valor P and Bonche P 2000 *Nucl. Phys.* A **671** 145
[32] Bender M, Heenen P H and Reinhard P G 2003 *Rev. Mod. Phys.* **75** 121
[33] Rodríguez-Guzmán R, Egido J L and Robledo L M 2002 *Nucl. Phys.* A **709** 201
[34] Nikšić T, Vretenar D and Ring P 2006 *Phys. Rev.* C **73** 034308

Journal of Physics: Conference Series **205** (2010) 012011 doi:10.1088/1742-6596/205/1/012011

Clustering in N≠Z nuclei.

E D Johnson[1], G V Rogachev[1], V Z Goldberg[2], S Brown[3], D Robson[1], A M Crisp[1], P D Cottle[1], C Fu[2], J Giles[1], B W Green[1], K W Kemper[1], K Lee[1], B T Roeder[1], and R E Tribble[2]

[1]Department of Physics, Florida State University, 32306 FL
[2]Cyclotron Institute, Texas A&M University, College Station, TX 77843
[3]The University of Surrey, Guildford, Surrey, UK

E-mail: edj04@fsu.edu

Abstract. The structure of the ^{18}O nucleus at excitation energies above the α decay threshold was studied using ^{14}C+α resonance elastic scattering and (^{7}Li,t) α-transfer reactions. A number of states with large α reduced widths have been observed, indicating that the α-cluster degree of freedom plays an important role in this N≠Z nucleus. A 0^+ state with an α reduced width exceeding the single particle limit was identified at an excitation energy of 9.9±0.3 MeV. We discuss evidence that states of this kind are common in light nuclei and give possible explanations of this feature. Also, the astrophysical implications of the cluster structure of ^{18}O is discussed.

1. Introduction

α clustering manifests itself in light 4N nuclei such as ^{8}Be, ^{12}C, ^{16}O, and ^{20}Ne through the existence of twin quasi-rotational bands of states with alternating parities and α-particle reduced widths which are close to the single particle limit (see recent review by M. Freer [1] and references therein).

There are different approaches that attempt to describe the shell model and the α-cluster structure of nuclei on an equal footing in order to shed light on the interplay between the α-cluster and single particle degrees of freedom [1]. Data on the nucleon decay of α-cluster states would be instrumental for such efforts. However, this data is practically absent due to the much higher nucleon decay thresholds in comparison with the thresholds for the α decay in the 4N nuclei. It is more promising to observe nucleon decay from the α-cluster states in N≠Z nuclei where the nucleon and α-particle thresholds are close to each other. The study of non-self-conjugate nuclei has an advantage in that one can investigate α-cluster states in mirror systems and use the Coulomb shift to extract information on the relationship between the cluster and single particle degrees of freedom. Unfortunately, the data on the α-cluster structure of N≠Z nuclei is generally very limited.

Current interest in the α-particle interaction with N≠Z nuclei is also strongly motivated by astrophysics [2]. Even if astrophysical reactions involving helium do not proceed through the strong α-cluster states (because of high excitation energy), these states can provide α width to the states that are closer to the region of astrophysical interest through configuration mixing [3]. In order to better understand this process we investigated the α-clusterization of the states in ^{18}O ranging from just below the α-decay threshold to ~4.5 MeV above the threshold. The properties of the higher lying states were studied through ^{14}C+α resonance elastic scattering, while the states nearest to the threshold were determined using sub-Coulomb (^{6,7}Li,d/t) α-transfer reactions along with the Asymptotic Normalization Coefficient (ANC) technique.

2. A Study of the ^{18}O Cluster Structure Through Resonance Elastic Scattering

We report here new measurements of α+^{14}C resonance elastic scattering made using the Thick Target Inverse Kinematics (TTIK) method [4]. This experimental approach provided good statistics and excitation functions which were remarkably free of any background. The data provided a basis for a successful analysis of the excitation functions using a complete multi-level, multi-channel **R**-matrix approach [5]. As a result, we found that in ^{18}O there is a system of strong α-cluster states. Both α and nucleon reduced widths of these α-cluster states were determined from the multi-channel **R**-matrix fit.

The experiment was carried out at the Florida State University John D. Fox Superconducting Accelerator Laboratory. The 25 MeV ^{14}C beam was projected into the scattering chamber, which consisted of two compartments. The first compartment was under vacuum, the second was filled with 99.9% pure helium gas (^{4}He). The two compartments were separated with a 1.27 μm Havar foil. The helium gas pressure in the second compartment was adjusted to stop the incoming beam before an array of Silicon detectors located at a distance of $\sim$40 cm from the Havar foil. The accuracy of the absolute normalization of the cross section is 15%. Details of the method are given in [6].

The analysis of the excitation functions was performed using a multi-level, multi-channel **R**-matrix approach [5]. The ^{14}C+α excitation functions measured in this experiment are continuous which excludes the possibility of missing a narrow resonance. The sensitivity of the data is demonstrated by the fact that even the 2^+ state at 8.213 MeV, which has a width of $\sim$1 keV, is still clearly visible in the 180° excitation function (see inset in Fig. 1). In addition, data from previous measurements of the ^{14}C(α,α) [7] and ^{14}C(α,n) [8] excitation functions were used. The **R**-matrix fit to the ^{14}C(α,α) data is shown in Fig. 1. As can be expected most of the states also have substantial neutron widths, which are obtained through the (α,α) fit. We verified that the ^{14}C(α,n) total cross section from Ref. [8] is reproduced rather well by **R**-matrix calculations performed using the parameters from the ^{14}C(α,α) fit. One can notice that the **R**-matrix fit underestimates the experimental cross section at the lowest energies for angles far from 180°. This is understood to be an effect of the finite dimensions of the beam spot.

The fourth panel in Fig. 1 is 90° data from [7]. The spectrum at 90° in c.m. is only influenced by states with even angular momentum and positive parity. This is an important simplification and makes the spectrum at 90° very valuable for the **R**-matrix analysis. Clearly, our data contains this information. However, for the purpose of a more clear representation of the data we used the spectrum of [7].

Twenty-four resonances were used to fit the data, some of them were previously known. Levels with large α-cluster reduced widths ($\theta_\alpha^2 > 0.1$) are given in Table 1.

Five levels with dimensionless α reduced width greater than 0.1 have been observed in ^{18}O in the narrow excitation energy range between 9.1 and 9.9 MeV. Three of them have been suggested in previous publications [9–11]. The strong α-cluster state at 9.0 ± 0.2 MeV was first suggested in [9], where the 1^- spin parity assignment was made on the basis of population of this state in ^{18}N β decay. A more recent ^{18}N β decay experiment [11] confirmed the 1^- state at 9.16 MeV with a width of $\Gamma = 420 \pm 200$ keV. Our **R**-matrix fit requires a 1^- state at an excitation energy of 9.17$\pm$0.03 MeV in good agreement with [11]. The width of this state is lower in the present work (230$\pm$50 keV) but still within the error bars of [11]. Another 1^- state observed at 9.85$\pm$0.5 MeV in [11] with a width of 560$\pm$200 keV is in very good agreement with the results of this work. The 3^- at 9.39 MeV was previously suggested in [10] and has excellent agreement with the present results. The strong α-cluster 2^+ state at 9.77 MeV is identified for the first time in this work.

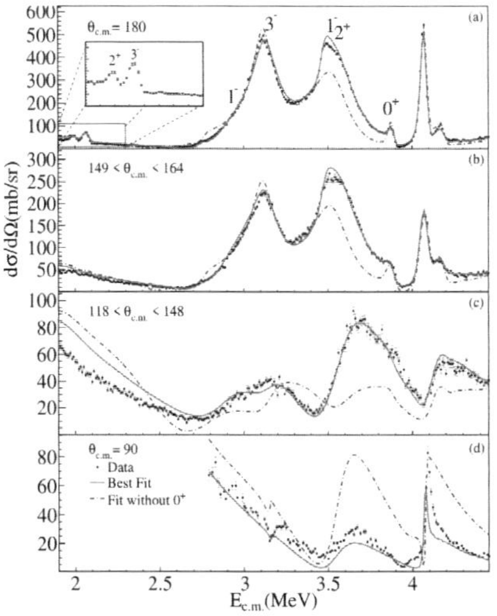

Figure 1. (Color online) The excitation functions for α+^{14}C elastic scattering measured at various angles, demonstrating the influence of the broad $J^\pi = 0^+$ state reported here. The red curves show the best **R**-matrix fit with the broad $J^\pi = 0^+$ state while the blue dash-dotted curve is the best fit without this state.

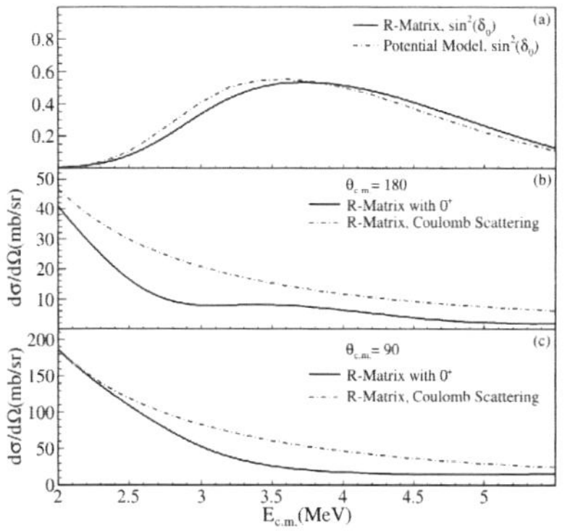

Figure 2. The top panel is the $\sin^2 \delta_{\ell=0}$ from the **R**-matrix fit (solid curve) compared to the potential model prediction (dash-dotted curve). The middle and the bottom panel demonstrate influence of the broad 0^+ state on the cross section at. The dash-dotted curve corresponds to Coulomb scattering and the solid curve shows the interference of the broad 0^+ with Coulomb scattering.

The most surprising finding of this work was the observation of a very broad 0^+ state at 9.90±0.3 MeV. Generally it is not easy to identify very broad resonances because they are disguised by interference with the sharp ones. This is especially true for the 0^+ resonances. Nevertheless, existence of the broad α-cluster 0^+ state in the spectrum of ^{18}O is certain. The effect of this state on the cross section is demonstrated in Fig. 2. The cross section is lowered dramatically due to the destructive interference of the 0^+ state with Coulomb scattering. The Rutherford cross section and the Rutherford with the broad 0^+ state at 9.90 MeV (3.7 MeV c.m.) are shown at two c.m. angles, 90° and 180°, in Fig. 2. Without this destructive interference it is not possible to reproduce the experimental data. As seen in the bottom panel of Fig. 1, the cross section calculated without the broad 0^+ state is significantly larger than the experimental one. This is a clear indication of a 0^+ resonance. All other resonances would not produce the right interference with the Rutherford scattering and could not be broad enough. One can follow the effect of the 0^+ level at different angles in Fig. 1 and reach the same conclusion. We also considered the possibility of two narrower 0^+ states instead of one. The decisive factor against it was that the characteristic interference pattern between two nearby 0^+ states was not observed experimentally. The α particle reduced width amplitude of the 0^+ state with channel radii of 5.2 and 6.5 fm are 1.38 and 0.66 MeV$^{1/2}$, respectively. Formally, both values exceed the single-particle limit. Using a classical approach one can interpret this as evidence that the α particle resides at a large distance from the ^{14}C core.

A more detailed description of the observed 0^+ state in ^{18}O can be given using the potential model approach. The parameters of this model were extracted starting from the potential model

Table 1. Levels with large α-reduced width in ^{18}O. Γ_{tot}, Γ_α and Γ_n are the total and partial α and neutron widths, respectively. $\theta^2_\alpha = \gamma^2_\alpha/\gamma^2_{SP}$ is the dimensionless reduced width for the α channel, where $\gamma^2_{SP} = \hbar^2/\mu R^2$ is the single particle limit, calculated at channel radius 5.2 fm.

E_{exc} (MeV)	J^π	Γ_{tot} (keV)	Γ_α (keV)	θ^2_α	Γ_n (keV)
9.17(3)	1^-	229(50)	200(50)	0.24	29(7)
9.39(3)	3^-	151(50)	100(48)	0.45	51(15)
9.75(4)	1^-	628(100)	585(100)	0.43	43(10)
9.77(3)	2^+	251(50)	172(45)	0.20	79(20)
9.9(3)	0^+	2100(500)	2100(500)	2.6	N/A

for ^{8}Be given in [12]. First, this potential model was used to investigate the ^{16}O+α interaction by assuming the α-cluster model [13] for the ground state of ^{16}O, properly modified [14] to include antisymmetry and the strong repulsion between nucleons. The strong interaction between the incoming α and the target "alphas" is obtained by folding the Buck interaction [12] with the target α-density in the ground state. The result is to produce a bound state for the ground state of ^{20}Ne and a broad 0^+ state at 4 MeV in c.m. as observed in [15] with the same 0^+ phase shift behavior. The corresponding density distribution for this 0^+ resonance has an inner peak at 2 fm and an outer peak at 5 fm. The outer peak is at a separation distance beyond the sum of the charge radii of ^{4}He and ^{16}O and it contains most of the probability indicating that the broad 0^+ resonance in ^{20}Ne appears to be a state with extreme α-clustering.

The ^{18}O system was then investigated using the ^{20}Ne potential as a starting point but small changes in the strengths of the folded and repulsive potentials were used to reproduce the ground state of ^{18}O. This potential produces a broad 0^+ state at 3.5 MeV in c.m. The experimental s-wave resonance phase shift from the **R**-matrix fit (solid black curve in Fig. 2a) is reproduced rather well by such a potential (dash-dotted blue curve in Fig. 2a). The density distribution for the broad 0^+ state in ^{18}O is similar to the one in ^{20}Ne. The outer peak is at 5.5 fm which is considerably larger than the sum of the charge radii for ^{4}He and ^{14}C. Both systems appear to show well separated α-cluster configurations that correspond to extreme α-clustering.

Due to the large distance between the α cluster and the core, one can speculate that levels of this kind should be a general feature of the interaction between an α particle and a core nucleus which is independent of the specific structure of the core. If this is the case then resonances of this kind should be present in all nearby nuclei at excitation energies on the order of 10 MeV above the first s-wave cluster state. Indeed, a broad (3 MeV) 0^+ level at an excitation energy of 10.3 MeV appears in ^{12}C [16]. The most recent parameters, as given in [17], are an excitation energy of 10.73 MeV and a width of 1.72 MeV. However, there are still significant uncertainties in these values. Also, a broad 0^+ level ($\Gamma > 800$ keV) at an excitation energy of $\sim$8.5 MeV, was observed in ^{20}Ne [15].

In order to understand the possible influence that these broad low spin resonances may have on astrophysically important reaction rates one must first determine the contribution of the near threshold states to the reaction rates. The contributions of the near threshold states in ^{18}O to the reaction ^{14}C(α,γ) are discussed in the following section.

Journal of Physics: Conference Series **205** (2010) 012011 doi:10.1088/1742-6596/205/1/012011

3. Influence of Near Threshold states in ^{18}O to the ^{14}C(α,γ) Reaction Rate

The near threshold states in ^{18}O are known to play a significant role in several astrophysical processes through the ^{14}C(α,γ) reaction. It was suggested in [18] that the ^{14}N(e$^-$,ν)^{14}C(α,γ)^{18}O reaction (NCO reaction) may trigger the helium flash in the core of low mass stars earlier than the 3α reaction. Also, it was suggested by Nomoto and Sugimoto [19], and later by Hashimoto, et al. [20], that the NCO reaction may trigger the helium flash in accreting helium white dwarfs at a lower temperature and density than the 3α reaction. However, this conclusion is rather sensitive to the actual value of the ^{14}C(α,γ) reaction rate.

The ^{14}C(α,γ) reaction is also important for production of ^{19}F in asymptotic giant branch stars. Observations by Jorissen, et al., [21] show enhanced fluorine abundance in the atmosphere of K, M, MS, S, SC, and C asymptotic giant branch (AGB) stars. A detailed investigation of the ^{19}F production in AGB stars has been performed recently by Lugaro, et al., [22] and it was determined that the major uncertainties in the production of ^{19}F are associated with the uncertainties in the ^{14}C(α,γ)^{18}O and ^{19}F(α,p)^{22}Ne reaction rates.

The relevant temperature range for accreting helium white dwarfs is between 0.03 and 0.1 GK and for ^{19}F nucleosynthesis in AGB stars it is ~0.1 GK. At these energies(temperatures) the cross section for the α capture reaction on ^{14}C is too low to be measured directly and has to be extrapolated. However, near threshold resonances in ^{18}O may have a very strong influence on this extrapolation. Three states with excitation energies close to the α decay threshold at 6.227 MeV are known in ^{18}O. These are the 1^- at 6.198 MeV, the 2^- at 6.351 MeV and the 3^- at 6.404 MeV. The 2^- state at 6.351 MeV is an unnatural parity state and cannot contribute to the ^{14}C(α,γ) reaction at any significant level.

In order to extract the cluster properties of these states the sub-Coulomb α-transfer reactions ^{6}Li(^{14}C,d) and ^{7}Li(^{14}C,t) were performed at the John D. Fox Superconducting Accelerator Laboratory at Florida State University. The use of inverse kinematics, ^{14}C beam and ^{6,7}Li target, eliminates the background associated with (^{6,7}Li,d/t) reactions on the unavoidable ^{12}C admixture to the ^{14}C target and provides a boost to the light recoils making them easier to detect.

Using the relation between the measured cross section and the DWBA calculated cross section we were able to extrat the 3^- partial α width and the 1^- ANC. The partial α width of the 3^- state at 6.404 MeV in ^{18}O is $\Gamma_\alpha = (7.8 \pm 2.7) \times 10^{-14}$ eV and the squared Coulomb modified ANC of the 1^- state at 6.198 MeV in ^{18}O is 2.8$\pm$0.7 fm^{-1}. See [6] for the details of this method.

With the Γ_α for the 3^- state at 6.404 MeV and ANC for the 1^- at 6.198 MeV in ^{18}O, the ^{14}C(α,γ) reaction rate can be reliably extrapolated down to very low temperatures. At temperatures below 1 GK four major contributors to the ^{14}C(α,γ) reaction rate can be identified. These are the direct capture (DC), resonance capture through the 4^+ and 3^- states at 7.12 and 6.404 MeV and the sub-threshold resonance capture through the 1^- state at 6.198 MeV.

The reaction rates due to resonant and non-resonant capture are shown in Figure 3. At T>0.3 GK the ^{14}C(α,γ) reaction rate is dominated by the 4^+ state at 7.12 MeV. The 4^+ resonance strength was measured directly [23; 24] and the uncertainty of the ^{14}C(α,γ) reaction rate in this region is 17%. In the temperature range between 0.03 and 0.3 GK the 3^- state at 6.404 MeV dominates. This temperature range is of particular interest for helium accreting white dwarfs and AGB stars. The 3^- resonance strength was determined in this work with an uncertainty of 35%, which determines the new uncertainty for the ^{14}C(α,γ) reaction rate in the temperature range of 0.03-0.3 GK. Below 0.03 GK, direct capture and capture due to the sub-threshold 1^- state at 6.198 MeV dominate. While direct capture was extrapolated from direct measurements in [23] and the S-factor due to the 1^- sub-threshold resonance was determined in this work with an uncertainty of 30% the reaction rate at this low energy region is still uncertain by two orders of magnitude due to the unknown interference between the direct and the 1^- sub-threshold resonance capture amplitudes.

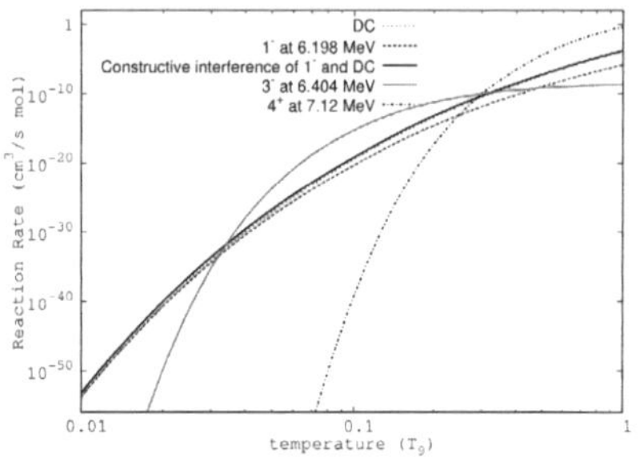

Figure 3. (Color online) The ^{14}C(α,γ) reaction rate due to resonant and non-resonant capture.

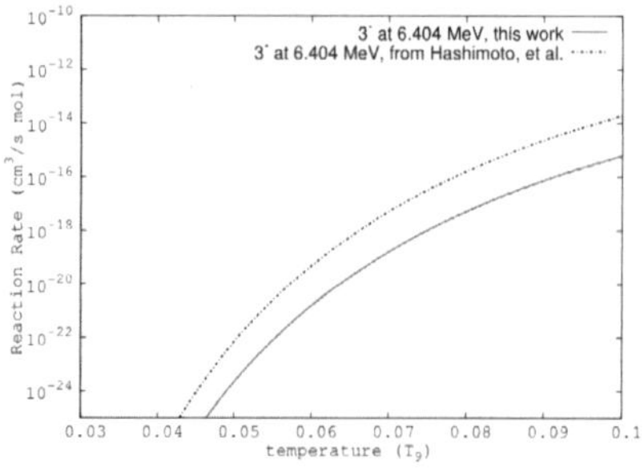

Figure 4. (Color online) Comparison of the new ^{14}C(α,γ) reaction rate in the temperature range most relevant for the helium accreting white dwarfs to the rate used by Hashimoto, et al., [20].

The new ^{14}C(α,γ) reaction rate may have the most profound effect on the evolution of accreting helium white dwarfs. It was suggested by [20] that under certain conditions the NCO reaction dominates over the 3α reaction and triggers a helium flash. However, this suggestion is sensitive to the actual ^{14}C(α,γ) reaction rate. According to [20] ^{14}C burning is ignited within helium white dwarfs when the central temperature reaches 0.066 GK while the 3α reaction does not begin until 0.080 GK. This prediction was based on a hypothetical ^{14}C(α,γ) reaction rate which turned out to be a factor of 30 higher than the rate determined from our measurements. Comparison of the reaction rate in the temperature range of interest for helium accreting white dwarfs used by Hashimoto and the one determined from our experimental data is shown in Figure 4. The new, much lower reaction rate would change the NCO ignition temperature to higher values. Detailed calculations with the new ^{14}C(α,γ) reaction rate are necessary to know whether the NCO reaction has an effect on the evolution of the helium accreting white dwarf.

It was found in [22] that uncertainties in the ^{14}C(α,γ) and the ^{19}F(α,p) reaction rates are the main contributing factors to the uncertainty of the production of ^{19}F in AGB stars. The authors of [22] cite 5 orders of magnitude uncertainty in the ^{14}C(α,γ) reaction rate due to the unknown spectroscopic factor of the 3^- state at 6.404 MeV. This uncertainty is eliminated now. Surprisingly, the "recommended" ^{14}C(α,γ) reaction rate used in [22] is very similiar to the reaction rate determined from our measurements. This fortuitous coincidence means that the yields of ^{19}F calculated in [22] with the "recommended" ^{14}C(α,γ) reaction rate are accurate and no new calculations are necessary.

4. Summary

In summary, we are only in the beginning phases of studying the α-cluster structure of light N$\neq$Z nuclei. Here we report a measurement of the ^{14}C+α elastic scattering excitation functions including a successful analysis using the complete **R**-matrix framework. We identified that the α-cluster states in ^{18}O have many surprising properties, foremost of which is the discovery of a broad, ℓ=0 state which we suggest may be present in other nuclei in this mass range. This conclusion is strengthened through comparison with previous results [15–17; 25–27]. There is much work still ahead to prove this point. The discovery of the broad states in question within odd-even nuclei is an especially important and difficult task. Nevertheless, the possible influence that these broad low spin resonances may have on astrophysically important reaction rates along with the insight they can give into α clustering justifies these efforts.

We also studied the influence of the cluster structure of near threshold states in ^{18}O to the ^{14}C(α,γ) reaction rate. The asymptotic normalization coefficients (ANCs) for the 1^- state at 6.198 MeV and 3^- at 6.404 MeV in ^{18}O were measured in the α-transfer reactions (^{7}Li,t) and (^{6}Li,d) on ^{14}C in order to determine their contribution to the reaction rate. The new ^{14}C(α,γ) reaction rate is a factor of 30 lower than the one used in [20]. Therefore, the importance of the NCO chain as a trigger for helium flashes in helium accreting white dwarfs suggested in [20] is reduced, if not eliminated all together. The "recommended" ^{14}C(α,γ) reaction rate used in [22] for ^{19}F nucleosynthesis calculation in AGB stars fortuitously coincides with that determined from our experimental data and the uncertainty of the ^{19}F production in AGB stars associated with the ^{14}C(α,γ) reaction rate is now eliminated. We hope that the new information on the ^{14}C(α,γ) reaction rate will help to resolve the question of whether or not the NCO reaction chain can trigger helium flashes in cores of low-mass stars before the 3α reaction does.

This work was supported by the NSF under grant number PHY-456463 and the DOE under grant number DE-FG02-93ER40773.

References

[1] Freer M 2007 *Rep. Prog. Phys.* **70** 2149–2210
[2] Aprahamian A, Langanke K and Wiescher M 2005 *Prog. Part. Nucl. Phys.* **54** 535
[3] Fu C, Goldberg V, Rogachev G *et al.* 2008 *Phys. Rev.* C **77** 064314
[4] Artemov K *et al.* 1990 *Sov. J. Nucl. Phys.* **52** 408
[5] Lane A and Thomas R 1958 *Rev. Mod. Phys.* **30** 257
[6] Johnson E 2008 *The Cluster Structure of Oxygen Isotopes* Ph.D. thesis Florida State University, http://etd.lib.fsu.edu/theses/available/etd-11102008-102305/unrestricted/JohnsonEDissertation.pdf
[7] Morgan G *et al.* 1970 *Nuc. Phys.* A **148** 480
[8] Bair J *et al.* 1966 *Phys. Rev.* **144** 799
[9] Zhao Z *et al.* 1989 *Phys. Rev.* C **39** 1985
[10] Goldberg V *et al.* 2004 *Phys. of Atomic Nuclei* **68** 1123
[11] Buchmann L *et al.* 2007 *Phys. Rev.* C **75** 012804(R)
[12] Buck B 1984 *International Conference on Clustering Aspects of Nuclear Structure and Nuclear Reactions* (Dordrecht, Holland: D. Reidel Publishing Company) p 71
[13] Wheeler J 1937 *Phys. Rev.* **52** 1083
[14] Robson D 1979 *Phys. Rev. Lett.* **42** 876
[15] McDermott L *et al.* 1960 *Phys. Rev.* **118** 175
[16] Tilley D *et al.* 1998 *Nucl. Phys.* A **636** 247
[17] Diget C *et al.* 2005 *Nucl. Phys.* A **760** 3–18
[18] Kaminisi K, Arai K and Yoshinaga K 1975 *Prog. of Theor. Phys.* **53** 1855
[19] Nomoto K and Sugimoto D 1977 *Pub. of the Astr. Soc. Japan* **29** 765
[20] Hashimoto M, Nomoto K, Arai K and Kaminisi K 1986 *Astrophys. J.* **307** 687
[21] Jorissen A, Smith V and Lambert D 1992 *Astron. and Astrophys.* **261** 164
[22] Lugaro M, Ugalde C, Karakas A, Görres J, Wiescher M, Lattanzio J and Cannon R 2004 *Astrophys. J.* **615** 934
[23] Görres J, Graff S, Wiescher M, Azuma R, Barnes C and Wang T 1992 *Nucl. Phys.* A **548** 414

[24] Gai M, Reddy R, Bromley D, Olness J and Warburton E 1987 *Phys. Rev.* C **36** 1256

[25] Fynbo H *et al.* 2005 *Nature* **433** 136–139 URL `http://dx.doi.org/10.1038/nature03219`

[26] Ajzenberg-Selove F 1988 *Nucl. Phys.*
/ A **490** 1

[27] Lui Y W, Clark H and Youngblood D 2001 *Phys. Rev. C* **64** 064308

Journal of Physics: Conference Series **205** (2010) 012012 doi:10.1088/1742-6596/205/1/012012

Parton promenade into the nucleon

Eric Voutier

Laboratoire de Physique Subatomique et de Cosmologie, CNRS/IN2P3, Université Joseph Fourier, INP, 53 rue des Martyrs, 38026 Grenoble cedex, France

E-mail: voutier@lpsc.in2p3.fr

Abstract. Generalized parton distributions (GPDs) offer a comprehensive picture of the nucleon struture and dynamics and provide a link between microscopic and macroscopic properties of the nucleon. These quantities, which can be interpreted as the transverse distribution of partons carrying a certain longitudinal momentum fraction of the nucleon, can be accessed in deep exclusive processes. This lecture reviews the main features of the nucleon structure as obtained from elastic and inelastic lepton scatterings and unified in the context of the GPDs framework. Particular emphasis is put on the experimental methods to access these distributions and the today experimental status.

1. Introduction

The nucleon, this most singular object of nuclear physics whose mass is quite larger than the mass of its constituents, did resist since several decades to theoretical and experimental investigations. It is only recently that a comprehensive picture of the nucleon structure started to develop within the framework of the generalized parton distributions (GPDs) [1, 2]. These distributions parametrize the partonic structure of the nucleon in terms of correlations between quarks, anti-quarks and gluons, and therefore contain information about the dynamics of this system. The power of this framework for the problem of the nucleon structure is certainly within the Mellin moments of the GPDs [3] which provide a natural link between microscopic and macroscopic properties of the nucleon.

GPDs can be accessed in the Bjorken regime of deep exclusive processes [4, 5], that is when the resolution power of the probe is large enough to resolve partons and when the momentum transfer to the nucleon is small enough to insure the separation of perturbative and non-perturbative scales. Pioneer measurements at HERMES [6] and CLAS [7], and recent experiments at the Jefferson Laboratory [8, 9] have established the relevance of the deeply virtual Compton scattering (DVCS) process for this studies. The real and imaginary parts of the DVCS amplitude are the quantities of interest that can be separated via beam charge, or beam polarization, or target polarization observables.

This lecture reviews the main steps of the story of the nucleon structure and its modern expression within the GPDs framework. The experimental methods to access these distributions and the today experimental status are also discussed.

2. From nucleon to partons...

From an experimental point of view, the story of the nucleon structure started in the fifties when deviations from the Mott cross section were observed in elastic electron scattering [10], meaning that the nucleon is not a pointlike object. The size of the nucleon is expressed by the so-called electromagnetic form factors which characterize the nucleon shape with respect to the electromagnetic interaction. This shape depends not-only on the resolution of the probe, controlled by the momentum transfer Q^2 to the nucleon, but also on the nature of the probe, this last feature being an early indication of the complexity of the nucleon structure. Within a non-relativistic framework, the electric (magnetic) form factor is the Fourier transform of the charge(current) density. Relativistic corrections to this picture modify this simple relationship but preserve this intimate link [11]. As today, there exists a significant enough knowledge of these form factors to question the experimental disagreement between Rosenbluth and recoil polarization measurements [12], as well as the role and magnitude of the two photons exchange mechanism [13].

At the end of the sixties, the nucleon revealed an unexpected behaviour in deep inelastic electron scattering (DIS). For excitation energies well beyond the resonance region, the cross section was found to weakly depend on the momentum transfer as compared to elastic scattering, indicating an interaction off a structureless object [14]. This behaviour was later identified as the first evidence of the existence of partons. The cross section for these experiments depends on the variable x_B, the fraction of the nucleon longitudinal momentum carried by the partons, and can be expressed in terms of the probability to find in the nucleon a parton carrying a given longitudinal momentum. This feature led to extensive measurements of the momentum distributions of partons into nucleons, the so-called parton distributions whose statisfactory knowledge has now been obtained after thirty years of experimental efforts. The nucleon appears as a system of three valence quarks globally equally sharing the nucleon momentum, and lying in a sea of quark-antiquark pairs and gluons.

Soon after this discovery, it was realized that deep inelastic scattering of polarized leptons off polarized targets allows to experimentally access the spin of the nucleon. Helicity conservation at the $\gamma^\star$-quark interaction vertex requires the $\gamma^\star$ coupling with opposite helicity quarks. Therefore, changing the polarization of the nucleon allows to measure alternatively the population of quarks with spin parallel or anti-parallel to the nucleon. Within the naive parton model, the difference between these populations is a measurement of the nucleon spin. This picture turns out to be too rough when it was found that quarks carry about 20-30 % of the total spin of the nucleon [15, 16]. In a more realistic partonic picture, the nucleon spin gets orbital and spin contributions of each constituent in a scale dependent manner, and the dynamics of the system expresses differently in the longitudinal (helicity) and transverse (transversity) directions yielding two spin sum rules [17]. If a fair knowledge of quark helicity distributions has now been obtained, investigations of the gluon contribution and transversity distributions are only starting, and the orbital momentum contributions are essentially unknown. Overall, the spin of the nucleon remains by many respects a terra incognita.

GPDs procure the cement of the nucleon spin puzzle and more generally of the hadron structure.

3. ... And conversely

GPDs are universal non-perturbative objects entering the description of hard scattering processes and correspond to the amplitude for removing a parton of longitudinal momentum fraction $x+\xi$ and restoring it with momentum fraction $x-\xi$ (fig. 1). In this process, the nucleon receives a four-momentum transfer $t = \Delta^2$ whose transverse component $\Delta_\perp$ is Fourier conjugate to the transverse position of partons. Consequently, GPDs can be interpreted as a distribution in the transverse plane of partons carrying a certain longitudinal momentum [18, 19, 20, 21], constituting a femto-tomography of the nucleon.

At the leading twist, the partonic structure of the nucleon [3, 22] is described by four quark helicity conserving and chiral even GPDs ($H^q, \widetilde{H}^q, E^q, \widetilde{E}^q$) and four quark helicity flipping and chiral odd GPDs ($H_T^q, \widetilde{H}_T^q, E_T^q, \widetilde{E}_T^q$), together with eight similar gluon GPDs. In the forward limit ($t \to 0, \xi \to 0$), the optical theorem links the H GPDs to the usual density, helicity, and tranversity distributions measured in DIS. However, the E GPDs, which involve a flip of the nucleon spin, do not have any DIS equivalent and then constitute a new piece of information about the nucleon structure. The first Mellin moments relate chiral even GPDs to form factors, as E^q with the Pauli electromagnetic form factor

$$\int_{-1}^{+1} dx\; E^q(Q^2, x, \xi, t) = F_2^q(t)\,, \tag{1}$$

and the second Mellin moments relate GPDs to the nucleon dynamics, particularly the total angular momentum carried by the partons, following Ji's sum rule [23]

$$J^q = \frac{1}{2}\int_{-1}^{+1} dx\; x\left[H^q(Q^2, x, \xi, 0) + E^q(Q^2, x, \xi, 0)\right]\,. \tag{2}$$

Here, E is of particular interest since it is not constrained by any DIS limit and remains essentially unknown. Similar relations have been proposed which relate chiral odd GPDs to the transverse spin-flavor dipole moment and the correlation between quark spin and angular momentum in an unpolarized nucleon [24].

4. Deep exclusive processes

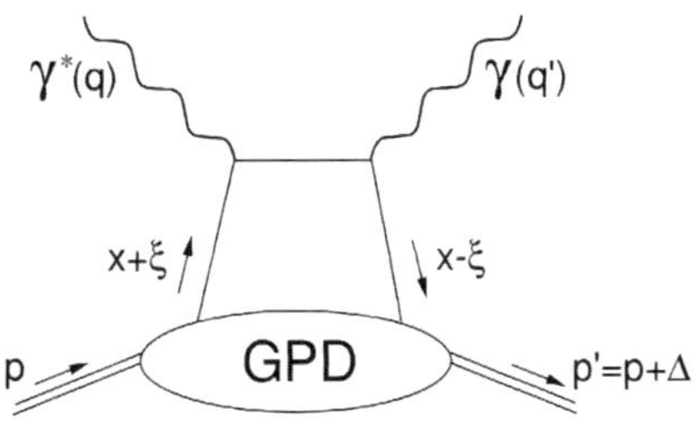

Figure 1. Lowest order (QCD) amplitude for the virtual Compton process. The momentum four-vectors of the incident and scattered photon are q and q', respectively. The momentum four-vectors of the initial and final proton are p and p', with Δ=$(p' - p)$=$(q - q')$. The DIS scaling variable is $x_{\rm B}$=$Q^2/(2p \cdot q)$ and the DVCS scaling variable is ξ=$x_{\rm B}/(2 - x_{\rm B})$.

DVCS, corresponding to the absorption of a virtual photon by a quark followed quasi-instantaneously by the emission of a real photon, is the simplest reaction to access GPDs. In the Bjorken regime, $-t \ll Q^2$ and Q^2 much larger than the quark confinement scale, the leading contribution to the reaction amplitude is represented by the so-called handbag diagram (fig. 1) which figures the convolution of a known $\gamma^* q \to \gamma q$ hard scattering kernel with an unknown soft matrix element describing the partonic structure of the nucleon parametrized by GPDs. Consequently, GPDs enter the reaction cross section through a Compton form factor which involves an integral over the intermediate quark propagator and leads to a complex DVCS amplitude.

In addition to the DVCS amplitude, the cross section for electroproduction of photons gets contributions from the Bethe-Heitler (BH) process where the real photon is emitted by the initial or final lepton, leading to [25]

$$\sigma(ep \to ep\gamma) = \sigma_{BH} + \sigma_{DVCS} + P_l\, \widetilde{\sigma}_{DVCS} + e_l\, \sigma_{INT} + P_l e_l\, \widetilde{\sigma}_{INT} \tag{3}$$

where the $\sigma(\widetilde{\sigma})$'s are even(odd) function of the out-of-plane angle between the leptonic and hadronic planes; P_l and e_l are the lepton polarization and charge, respectively. Though undistinguishable from DVCS, the BH cross section is known and exactly calculable from the electromagnetic form factors. The pure DVCS and interference contributions contain the information of interest, particularly $\sigma_{INT}(\ \widetilde{\sigma}_{INT})$ is proportionnal to the real(imaginary) part of the DVCS amplitude. From the above relation, it is obvious that four different measurements combining beam charge and/or polarization allow to extract the four unknown contributions involving different Compton form factor combinations. Polarized target observables add other combinations with different sensitivity to the GPDs. For instance, E can be accessed from the difference between polarized cross section off the neutron for opposite lepton helicities, and from the target spin asymmetry for a transversely polarized proton.

Deeply Virtual Meson Production (DVMP), where the real photon is replaced by a meson, is another channel to access GPDs additionnally providing an elegant flavor decomposition. In this case, the factorization of the cross section applies only to longitudinal virtual photons and the GPDs entering the Compton form factors are further convoluted with a meson distribution amplitude. The measurement of the angular distribution of the decay products of the vector mesons allows to extract the longitudinal part of the cross section and the longitudinal polarization of the vector mesons, assuming s-channel helicity conservation. Other reaction mechanisms, like the 2-gluon exchange from a $q\bar{q}$ fluctuation of the virtual photon, may contribute to the production process. Similarly to DVCS, polarization observables help to single-out the pure handbag contributions, particularly the production of longitudinal ρ^0 off a transversely polarized proton target is an observable very sensitive to E [26].

5. Experimental status

The pioneering studies of the electro-production of photons at DESY [6, 27] and JLab [7] did prove the existence of the DVCS mechanism by measuring sizeable beam spin asymmetries (BSA) in the valence region and significant cross sections in the gluon sector. Other limited studies showed the importance of the beam charge [28] and the target polarization observables [29]. A recent remarkable JLab result is the strong indication for scaling in the valence region at Q^2 as low as 2 GeV^2, and the observation of an unexpected DVCS amplitude magnitude at JLab energies [8]. This early scaling is also supported by the φ angular dependence of the BSA measured at JLab with an unprecedented accuracy over a wide kinematic range [9]. In general, GPD based calculations provide a reasonable but unsatisfactory agreement with these data which turn out to be fairly reproduced by a more conventional Regge approach [30]. The significance of this duality has not yet been resolved.

In the meson sector, the experimental status with respect to GPDs remains controversial. Sizeable BSAs have been reported for exclusive π^0 electro-production at JLab energies [31], which suggest that both longitudinal and transverse amplitudes contribute to the process. On the one hand, this forbids a direct GPD based interpretation of the data, and on the other hand, a Regge based approach fails to reproduce them. The longitudinal cross section for the electro-production of longitudinally polarized ρ^0 was recently measured at JLab [32]. Standard GPD calculations, particularly successful at high energies, do not reproduce data in the valence region, while calculations based on hadronic degrees of freedom are in very good agreement over the complete energy range scanned by the world data. The observation that strongly modified GPDs allow for a partonic interpretation of these measurements raises the question whether current GPD parametrizations must be revisited or the existence of strong higher twist corrections percludes a GPD wise interpretation in the valence region.

Concerning the quest for the angular momentum, DVCS measurements off the neutron [33] and off a transversely polarized proton [34], as well as longitudinal ρ^0 production off a transversely polarized proton [35] were reported. The interpretation of these data in terms of GPDs can lead to a constraint on the angular momentum of the u and d quarks [36]. This constraint is however highly model dependent. It is the goal of the future experimental programs to provide enough data to map out the GPDs and allow for a model independent determination of the angular momentum of the partons.

6. Conclusion

Since a decade, the study of the nucleon structure entered a new era with the advent of GPDs which unify within the same framework form factors, parton distributions, helicity and transversity distributions. DVCS appears as the golden chanel to access GPDs, showing an early scaling behaviour supported by recent JLab experimental results. The lack of experimental data does not yet allow for a precise and model-independent extraction of these quantities from experimental observables. The future experimental programs at JLab 12 GeV and CERN are expected to overcome this situation, and hopefully provide a model independent determination of the contribution of the quark angular momentum to the spin of the nucleon.

Acknowledgments

This work was supported in part by the U.S. Department of Energy (DOE) contract DOE-AC05-06OR23177 under which the Jefferson Science Associates, LLC, operates the Thomas Jefferson National Accelerator Facility, the National Science Foundation, the French Atomic Energy Commission and National Center of Scientific Research, and the GDR n°3034 Physique du Nucléon.

References

[1] Müller D, Robaschick D, Geyer B, Dittes F M , Hořejši J. 1994 *Fortschr. Phys.* **42** 101
[2] Radyushkin A V 1997 *Phys. Rev.* D **56** 5524
[3] Diehl M 2003 *Phys. Rep.* **388** 41
[4] Ji X, Osborne J 1998 *Phys. Rev.* D **58** 094018
[5] Collins J C, Freund A 1999 *Phys. Rev.* D **59** 074009
[6] Airapetian A et al. 2001 *Phys. Rev. Lett.* **87** 182001
[7] Stepanyan S et al. 2001 *Phys. Rev. Lett.* **87** 182002
[8] Muñoz Camacho C, Camsonne A, Mazouz M, Ferdi C, Gavalian G, Kuchina E et al. 2006 *Phys. Rev. Lett.* **97** 262002
[9] Girod F-X, Niyazov R A et al. 2008 *Phys. Rev. Lett.* **100** 162002
[10] Hofstadter P, McAllister R W 1995 *Phys. Rev.* **98** 217
[11] Kelly J J 2002 *Phys. Rev.* C **66** 065203
[12] Perdrisat C F, Punjabi V, Vanderhaeghen M 2007 *Prog. Part. Nucl. Phys.* **59** 694
[13] Guichon P A M, Vanderhaeghen M 2003 *Phys. Rev. Lett.* **91** 142303
[14] Breidenbach M, Friedman J I, Kendall H W, Bloom E D, Coward D H, DeStaebler H, Drees J, Mo L W, Taylor R W 1969 *Phys. Rev. Lett* **23** 935
[15] Ashman J et al. 1988 *Phys. Lett.* B **206** 364
[16] Ashman J et al. 1989 *Nucl. Phys.* B **328** 1
[17] Bakker B L, Leader E, Trueman T L 2004 *Phys. Rev.* D **70** 114001
[18] Burkardt M 2000 *Phys. Rev.* D **62** 071503
[19] Ralston J P, Pire B 2002 *Phys. Rev.* D **66** 111501
[20] Diehl M 2002 *Eur. Phys. Jour.* C **25** 223
[21] Belitsky A V, Müller D 2002 *Nucl. Phys.* A **711** 118c
[22] Belitsky A V, Radyushkin A V 2005 *Phys. Rep.* **418** 1
[23] Ji X 1997 *Phys. Rev. Lett.* **78** 610
[24] Burkardt M 2005 *Phys. Rev.* D **72** 094020
[25] Diehl M 2009 *CLAS12 European Workshop* (Genova, Italy); *http://www.ge.infn.it/~clas12/talks/thursday_session6/diehl-genova.pdf*
[26] Goeke K, Polyakov M V, Vanderhaeghen M 2001 *Prog. Part. Nucl. Phys.* **47** 401
[27] Adloff C et al. 2001 *Phys. Lett.* B **517** 47
[28] Airapetian A et al. 2007 *Phys. Rev.* D **75** 011103
[29] Chen S.et al. 2006 *Phys. Rev. Lett.* **97** 072002

[30] Laget J-M et al. 2007 *Phys. Rev.* C **76** 052201
[31] De Masi R, Garçon M, Zhao B et al. 2008 *Phys. Rev.* C **77** 042201(R)
[32] Morrow S A, Guidal M, Garçon M, Laget J-M, Smith E S et al. 2008 *Preprint* hep-ex/0807.3834
[33] Mazouz M, Camsonne A, Muñoz Camacho C, Ferdi C, Gavalian G, Kuchina E et al. 2007 *Phys. Rev. Lett.* **99** 242501
[34] Airapetian A et al. 2008 *J. High Energy Phys.* JHEP06(2008)066
[35] Airapetian A et al. 2009 *Preprint* gr-qc/0906.5160
[36] Ellinghaus F, Nowak W-D, Vinnikov A V, Ye Z 2006 *Eur. Phys. J.* C **46** 729

Journal of Physics: Conference Series **205** (2010) 012013 doi:10.1088/1742-6596/205/1/012013

Relativistic mean field formulation of clustering in heavy nuclei

S.M. Wyngaardt[1], H. W. Groenewald[1], T.T Ibrahim[1,3,4], J. Ndayishimye[1], S.M. Perez[2,3]

[1]University of Stellenbosch, Stellenbosch, 7600, South Africa
[2]Department of Physics, University of Cape Town, Rondebosch, 7700, South Africa
[3]iThemba LABS, PO Box 722, Somerset West 7129, South Africa
[4]Department of Physics, University of Ilorin, PMB1515, Ilorin, Nigeria

E-mail: shaunmw@sun.ac.za

Abstract. Very little is known about clustering in heavy nuclei and in particular the interaction between the correlated cluster nucleons and remaining core nucleons. Currently the phenomenological Saxon-Woods plus cubic Saxon-Woods core-cluster potential successfully predicts the alpha decay half-life and energy band spectra of a number of heavy nuclei. This model, however, lacks a microscopic understanding of clustering phenomenon in these heavy nuclear systems. A fully relativistic microscopic formalism is presented, which generates the core-cluster potential by means of the McNeil, Ray and Wallace based double folding procedure. The core and cluster baryon densities are calculated by using a relativistic mean field approach. The Lorentz covariant IA1 representation of the nucleon-nucleon interaction is folded with the core and cluster densities. Theoretical predictions of the ground-state decay half-life and positive parity energy band of ^{212}Po are obtained with the relativistic mean field formalism and which are compared to the results from the phenomenological Saxon-Woods plus cubic Saxon-Wood core-cluster potential and microscopic M3Y interaction.

1. Introduction

In nuclear dynamics, as seen in light stable nuclei, clustering phenomenon is one of the essential features which has been studied in great detail [1]. In the physics of unstable nuclei, clustering is one of the central areas of study. The cluster-core interaction lies central to the identification of clustering in the nuclear matter and the description of clustering phenomenon in various nuclei. During the last decade the modified phenomenological Saxon-Woods plus Cubic Saxon-Woods cluster potential has successfully described various phenomenon related to alpha clustering in light as well as even-even heavy nuclei. In order to fully describe clustering in nuclear systems one would have to develop a microscopic model of the phenomenon at the nucleon-nucleon scale.

At a more microscopic level the core-cluster interaction may be constructed from a nucleon-nucleon interaction. Prior to the development of the Saxon-Wood plus Saxon-Wood cubed potential form, such a microscopic interaction had been employed in various forms to describe α cluster bound states in light nuclei [2] and the exotic decays in heavy nuclei [3]. In recent years the microscopic M3Y-type potential model has been extended to describe the alpha decay half-lives and the structure of heavy nuclei [4], [5], and [6]. An application of the interaction to ^{94}Mo and ^{212}Po in particular suggests a good amount of α clustering in these nuclei [7].

Relativistic mean field theory (RMFT) [8] has proven to be very successful in describing various properties of nuclear structure [10]. In this work a RMFT description of clustering is presented and a comparison is made between the experimental ground-state decay half-lives and band energy spectral of ^{212}Po and cluster model predictions of these quantities which were obtained from the Saxon-Woods + cubic Saxon-Woods potential, double folded M3Y nucleon-nucleon interaction [9] and the microscopic RMFT based core-cluster interaction.

2. Formulation of Binary cluster model

This model is based on the preformed binary cluster model for which the decay half life is given by

$$T_{1/2} = \hbar\frac{\ln 2}{\Gamma}, \tag{1}$$

where Γ represents the cluster decay width. For the breakup of a nucleus into the core and cluster the decay width is defined by the relationship

$$\Gamma = P\frac{\hbar^2}{2\mu}\frac{exp(-2\int_{r_2}^{r_3} k(r)dr)}{\int_{r_1}^{r_2}[k^{-1}(r)]dr} \tag{2}$$

with (P) being the core-cluster preformation probability in the parent nucleus, μ is reduced mass of the core-cluster system and $k(r)$ is cluster wavenumber. The wavenumber depends on both the decay energy (E) and the core-cluster potential $V(r)$, and is given by

$$k(r) = \sqrt{\frac{2\mu}{\hbar^2}|E - V(r)|}. \tag{3}$$

The energy band structure of the quasi-boundstates can be obtained from a combination of the Bohr-Sommerfeld (BS) quantization integral

$$\int_{r_1}^{r2}\sqrt{\frac{2\mu}{\hbar^2}\left[E_l - V(r)\right]}\,dr = (2n+1)\frac{\pi}{2} \tag{4}$$

and the Wildermuth condition $G = 2n + l$, where n is the number of nodes of the radial wavefunction and l is the orbital angular momentum of the cluster state. G defines the global quantum number of the core-cluster relative motion. The interaction between the core and cluster, $V(r)$, is described by the sum of the attractive nuclear cluster-core potential $U(r)$, the Coulomb potential between the two charged centres, and the centrifugal potential.

3. Core cluster potentials

3.1. Phenomenological core-cluster interaction

The recently developed modified Saxon-Wood with an additional cubic Saxon-Woods core-cluster phenomenological potential

$$U(r) = U_0 \left[\frac{x}{1 + exp\left(\frac{r-R}{a}\right)} + \frac{1-x}{1 + exp\left(\frac{r-R}{3a}\right)^3} \right] \tag{5}$$

is found to consistently reproduce not only the alpha and exotic decay half-lives, but also correctly predict the level properties of nuclei in the rare earth and the actinide region. This potential is parameterized in terms of the potential depth (U_0), nuclear radius (R), diffuseness (a), and x is a mixing parameter. Despite its success this potential model tells us very little about the microscopic nature of clustering in closed shell nuclei.

3.2. Relativistic mean field construction of the cluster-core potential

In the IA1 representation of the nucleon-nucleon scattering amplitude [12]

$$F = F^S I^a I_b + F^V \gamma_a^\mu \gamma_{\mu b} + F^{PS} \gamma_a^5 \gamma_{5b} + F^T \sigma_a^{\mu\nu} \sigma_{\mu\nu b} + F^A \gamma_a^5 \gamma_a^\mu \gamma_b^5 \gamma_{\mu b}, \tag{6}$$

Lorentz covariance, parity conservation, isospin invariance, and the constraint that the free nucleons are on the mass shell imply that the invariant NN scattering operator F be written in terms of the five complex functions for pp and five for pn scattering. The quantities $\lambda_i^L = (I, \gamma^\mu, \gamma^5, \sigma^{\mu\nu}, \gamma^5\gamma^\mu)$ represent the five Dirac gamma matrices [13], and the index ($i = a, b$) labels the two interacting nucleons. The index L labels the scalar, vector, pseudo-scalar, tensor and axial terms.

Out of the Lorentz covariant McNeil, Ray and Wallace (MRW) construction of the optical potential for nucleon-nucleus scattering [11], arises the double folded MRW form which describes the cluster-core potential

$$U^L(r,\epsilon) = -\frac{4\pi i p}{Mc^2} \int \frac{d^3q}{(2\pi)^3} e^{i\mathbf{q}\cdot\mathbf{r}} F^L(q,\epsilon) \int d^3r' e^{-i\mathbf{q}\cdot\mathbf{r}'} \rho_1^L(r') \int d^3r'' e^{-i\mathbf{q}\cdot\mathbf{r}''} \rho_2^L(r''), \tag{7}$$

where r represents the separation distance between the cluster (1)and core (2)center, and ϵ is the laboratory energy of the nucleons in the cluster. The momentum of the nucleons in the nucleon-nucleon (NN)center of mass system is given by p while M represents the nucleon mass. Equation (7) contains the Lorentz covariant nucleon-nucleon scattering amplitudes $F^L(q,\epsilon)$, which are functions of the NN centre of mass momentum transfer ($\vec{q}$) and nucleon laboratory energy (ϵ), as well as the respective cluster and core densities ρ_1^L and ρ_2^L.

The Walecka model is based on a relativistic mean field theory with an effective Lagrangian which describes the NN interaction via the electromagnetic interaction and the effective meson fields [8]. The dynamical equation which results from the Lagrangian is given by

$$\hat{H}\psi(\mathbf{r}) = \left(i\alpha\cdot\nabla - g_v\gamma^0 V^0(r) + \beta[M - g_s\phi(r)]\right)\psi(\mathbf{r}) = E\psi(\mathbf{r}) \tag{8}$$

with the Dirac Hamiltonian operator ($\hat{H} = i\alpha\cdot\nabla - g_v\gamma^0V^0(r) + \beta[M - g_s\phi(r)]$), vector and scalar fields g_v and g_s respectively, as well as the zeroth component vector field (V_0) and scalar field (ϕ). Equation (8) has both positive and negative solutions $U(\mathbf{r})$ and $V(\mathbf{r})$, and thus the field operator can be expanded as

$$\hat{\psi}(\mathbf{r}) = \sum_{\Lambda} \left[A_\Lambda U_\Lambda(\mathbf{r}) + B_\Lambda^\dagger V_\Lambda(\mathbf{r}) \right]. \tag{9}$$

The baryon and antibaryon creation operators $A_\Lambda^\dagger$ and $B_\Lambda^\dagger$ satisfy the standard anticommutator relationships and the index Λ specifies the full set of single-particle quantum numbers, which for a spherically symmetric and parity conserving system, are the usual angular momentum and parity quantum numbers, as given by reference [13]. The positive-energy spinor can be written as

$$U_\Lambda \equiv U_{njlmt}(\mathbf{r}) = \begin{pmatrix} i\,[G_{njlt}(r)/r]\,\Phi_{jlm} \\ [F_{njlt}(r)/r]\,\Phi_{jl+1m} \end{pmatrix} \varsigma_t, \tag{10}$$

where Φ_{jlm} is the angular momentum and spin dependent part of the solution and n and ς_t represents the principal quantum number and two component isospinor which is labeled by the isospin projection t. The functions $G(r)$ and $F(r)$ represent the radial wave functions for the upper and lower components of the positive energy spinor U_Λ
Neglecting the negative-energy spinors the local baryon (ρ_B) and scalar (ρ_s) densities can be derived from the positive-energy solutions

$$\left.\begin{matrix} \rho_B(\mathbf{r}) \\ \rho_s(\mathbf{r}) \end{matrix}\right\} = \sum_{\Lambda} \bar{U}_\Lambda(\mathbf{r}) \begin{pmatrix} \gamma^0 \\ I \end{pmatrix} U_\Lambda(\mathbf{r}). \tag{11}$$

4. Model predictions and conclusions

For the BMP phenomenological form of the cluster-core potential prediction of the positive parity alpha band energy structure of ^{212}Po the parameters $U_0 = 208$ MeV, $a = 0.66$ fm, $x = 0.30$ and $R = 6.784$ were used with $G = 18$ [15].
The Walecka based RMFT prediction uses the experimental masses $M = 939$ MeV, $m_v = m_\omega = 738$ MeV, $m_\rho = 770$ MeV, $m_s = 520$ MeV, and $\alpha = e^2/4\pi = 1/137.36$ are used. The coupling constants for the scalar, vector, and ρ-meson are $g_s^2 = 109.6$, $g_v^2 = 190.4$, and $g_\rho^2 = 65.23$ respectively. We apply the Dirac-Hartree code Timora [14] to calulate the scalar and vector densities for both the protons and neutrons. The densities of the core and cluster systems are inturn used to calculate the core-cluster potential by means of the double folded MRW method. The results of the calculated α-decay half-life of the ground-state as predicted by the phenomenological BMP, the microscopic M3Y with phenomenological core and cluster baryon densities [15], and MRW double folded relativistic mean field nucleon densities with Lorentz covariant NN scattering amplitudes are compared with experimental data [16] in Table I. Table II compares the predicted band energy structure from the BMP, M3Y and RMFT model calculations with available experimental data [16].
From the results in Table (1) and (2) one see that the Saxon-Woods plus cubic Saxon-Woods potential gives a reasonable prediction of the half-life of the 0^+, while the microscopic M3Y and RMFT based models underpredict the ground-state alpha decay half-life of ^{212}Po by a factor of approximately 2. Furturemore the energy spectra of the excited α states are predicted reasonably well by the Saxon-Woods plus cubic Saxon Woods core-cluster potential where as both the microscopic M3Y and RMFT model potentials show a clear inversion of the energy spectra. The energy bands structure as predited by the RMFT model is wider spread at large l values as compared to the M3Y predictions.

Journal of Physics: Conference Series **205** (2010) 012013 doi:10.1088/1742-6596/205/1/012013

Table 1. The experimental ground state decay of ^{212}Po and the corresponding values obtained with the BMP phenomenological potential given by equation (5), double folded M3Y potential with phenomenological Saxon-Woods and Gaussian baryon density distributions (M3Y), and the double folded RMFT potential (RMFT).

$T_{1/2}$(Exp) ns	$T_{1/2}$(BMP) (ns)	$T_{1/2}$(M3Y) (ns)	$T_{1/2}$(RMFT) (ns)
300	348.0	157.4	153.8

Table 2. The experimental energy level scheme of ^{212}Po and the calculated spectra obtained with the BMP phenomenological potential given by equation (5) and the double folded RMFT potential.

J^{π}	E_{exp} (MeV)	E_{BMP} MeV	E_{M3Y}	E_{RMFT} (MeV)
0^+	0.000	(0.495)	−0.004	−0.007
2^+	0.727	0.659	−0.067	−0.136
4^+	1.132	0.948	−0.229	−0.575
6^+	1.355	1.318	−0.508	−1.584
8^+	1.476	1.730	−0.930	−3.346
10^+	1.834	2.145	−1.538	−5.917
12^+	2.702	2.519	−2.358	−9.407
14^+	2.885	2.805	−3.437	−14.226
16^+	–	2.941	−4.800	−21.000
18^+	2.921	2.841	−6.477	−30.040

References

[1] M.Freer and A.C.Merchant, J.Phys.G23,261(1997).
[2] B. Buck, C. B. Dover and J. P. Vary, Phys. Rev. C11, (5), 1803 (1975).
[3] Buck, B. Buck, A. C. Merchant, Phys. Rev. C39, (5) 2097 (1989).
[4] F. Hoyler, P. Mohr and G. Staudt, Phys. Rev. C50, (5) 2631 (1994).
[5] C. Xu and Z. Ren, Nucl. Phys. A753, 174 (2005); Nucl. Phys. A760, 303 (2005).
[6] D. N. Basu, J. Phys. G: Nucl. Part. Phys. 29, 2079 (2003):Phys. Lett. B566 90 (2003).
[7] S. Ohkubo, Phys. Rev. Lett. 74, (12) 2176 (1995).
[8] J. D. Walecka, Ann. Phys. (N.Y) (83) (1974) 491.
[9] G. R. Satchler and W. G. Love, Phys. Rep, 55, 183 (1979).
[10] P. Ring, A. V. Afanasjev and J. Meng, Lecture notes in Physic, vol. 482, (Springer Berlin Heidelberg, 2007).
[11] J. A. McNeil, L. Ray and S. J. Wallace, Phys. Rev. C27, 2123 (1983).
[12] J. A. McNeil, J. Shepard, and S. J. Wallace, Phys. Rev. Lett. 50, 1439 (1983); J. Shepard, J. A. McNeil, and S. J. Wallace *ibid* 50, 1443 (1983).
[13] J. D. Bjorken and S. D. Drell, *Relativistic Quantum Mechanics*, 2nd edition (McGraw-Hill,New York, 1964).
[14] K. Langanke, J. A. Maruhn, S. E. Koonin, *Computational Nuclear Physics*, vol. 1, Springer Berlin Heidelberg)
[15] T. T. Ibrahim, PhD Thesis, unpublished (2007).
[16] A. B. Garnsworthy et al, J. Phys. G: Nucl. Part. Phys. 31, 1851 (2005).

Journal of Physics: Conference Series **205** (2010) 012014 doi:10.1088/1742-6596/205/1/012014

Deuteron breakup effects on activation cross sections at low and medium energies

M. Avrigeanu and V. Avrigeanu

"Horia Hulubei" National Institute for Physics and Nuclear Engineering, P.O. Box MG-6, 76900 Bucharest, Romania

E-mail: mavrig@ifin.nipne.ro; vavrig@ifin.nipne.ro

Abstract. An extended analysis of the reaction mechanisms involved in the deuterons interaction with light and medium nuclei, ^{27}Al and ^{63,65}Cu, at incident energies from 3 to 60 MeV starting with elastic scattering until the evaporation from fully equilibrated compound system is presented. An increased attention is devoted to the breakup mechanism, all its components, the elastic, inelastic (fusion), and total breakup, being carefully analysed. Moreover, the corresponding stripping mechanism contribution to the (d,p) and (d,n) reaction cross sections through population of the discrete levels of the residual nuclei, was calculated. Finally, following the decay path, the pre-equilibrium and evaporation cross sections, corrected for the breakup and stripping decrease of the total reaction cross section, have completed the deuteron activation cross sections analysis. The overall agreement between the measured data and model calculations proves the correctness of nuclear mechanism description taken into account for the deuteron-nucleus interaction.

1. Introduction

Recent evaluation of the deuteron-induced activation in the International Fusion Materials Irradiation Facility (IFMIF), e.g. [1], based on preliminary deuteron part of the EAF library [2], shows that the deuterons are much more important than the neutrons by a factor of about 70 for the ratio of the deuteron– and neutron–induced activity [3]. However, despite the high technologies like the fusion reactor projects request increased accuracy deuteron nuclear data, the corresponding evaluations are less extensive and mature than in the case of neutrons, e.g. Refs. [3, 4]. Therefore, the analysis of deuteron-induced reactions at low and medium incident energies needs improved model calculations and further measurements if the deuteron data libraries are to approach the standard of the established neutron-data bases.

The difficulties to interpret the deuteron-induced reaction data in terms of the usual reaction mechanism models are the result of the high enhancement of a variety of reactions, even at low bombarding energies, induced by the neutrons and protons coming from the deuteron breakup. Then, the weak binding of the deuteron (2.22 MeV) results in significant contributions of the breakup channel which hampered so far a comprehensive analysis of deuteron interactions with nuclei.

The accurate description of the deuteron interactions appears at low incident energies, below e.g. 20 MeV, is additionally difficult since there is no global optical model potential (OMP) which describes the scattering data sufficiently well over a wide range of nuclei and energies

[5, 6]. Because the OMP is a basic ingredient of almost all nuclear model analyses, there is a significant uncertainty of the calculated deuteron-induced reaction cross sections.

The present lecture, concerning the analysis of the nuclear reaction mechanisms involved in the deuteron interaction with nuclei, has paid larger attention to the deuteron breakup consideration as well as the OMP parameters which are able to describe the bulk of deuteron elastic scattering data. The ultimate results of the corresponding assumptions and forms on the calculated deuteron-induced reaction cross sections are also pointed out.

2. OMP analysis

In order to increase the accuracy of the optical potential parameterization, the OMP analysis may take the advantage of a two-steps approach [7, 8, 9, 10]. In the first step, a semi-microscopic optical potential analysis provides the imaginary and spin-orbit parameterization, while within a second step a full phenomenological OMP can be obtained.

2.1. Semi-microscopic optical potential analysis

A semi-microscopic OMP may consist of the real Double Folding (DF) part, U_{DF}, and phenomenological imaginary volume and surface derivative Woods-Saxon potentials, as well as spin-orbit potential of the Thomas form[11]:

$$\begin{aligned} U(r) &= V_C(r) + U_{DF}(r) + iW_V\, f(r, R_V, a_V) + iW_D\, g(r, R_D, a_D) \\ &\quad + V_{so}(r)\, (\hbar/m_\pi c)\, \mathbf{LS}\, 1/r\, d/dr[f(r, R_{so}, a_{so})] \end{aligned} \tag{1}$$

where $V_C(r)$ is the Coulomb potential of a uniformly charged sphere of radius R_C=$r_C A^{1/3}$, and r_C= 1.30 fm, $f(r, r_i, a_i)$=(1+exp[(r-$r_i A^{1/3}$)/a_i])$^{-1}$, $g(r, r_i, a_i)$=-4a_id/dr[$f(r, r_i, a_i)$], and $(\hbar/m_\pi c)^2$ is the square of the pion Compton wavelength.

The basic formulas for calculations of the DF real part of the OMP [12] U_{DF} as well as the basic model assumptions have been discussed elsewhere [7]. The direct and exchange components of the real microscopic optical potential U_{DF} are given in terms of the projectile and target nuclear densities, which are folded with the Paris M3Y effective NN interaction [13]. The knock-on exchange term of the folded potential has been calculated by using the approximation of Campi and Bouyssy [14], which preserves the first term of the expansion given by Negele-Vautherin [15] for the realistic density-matrix while the average relative momenta of nucleons are obtained from the modified Thomas-Fermi approximation of Krivine and Treiner [16] for the kinetic-energy density.

The deuteron density distribution has been obtained from the experimental charge form factors measured by Abbott et al. [17], and the nuclear density distribution of the target nuclei has been described by means of a two-parameter Fermi-type function with Negele's [18] parameters chosen to reproduce the electron scattering data.

Similarly to the previous studies [7, 8], within our two-step OMP approach the semi-microscopic analysis of the available experimental angular distributions of elastic–scattered deuterons provided the energy-dependent depths of the phenomenological imaginary surface W_D and volume W_V optical potential components. We have made use in this respect also of the corresponding dispersive correction $\Delta U(r, E)$ [19], to the microscopic "parameter free" DF real potential. Concerning the spin-orbit potential we found the parameterization of Ref. [20] quite suitable and have preserved it within present analysis.

It should be emphasized that no adjustable parameter or normalization constant has been involved within this analysis for the real part of Eq. (1) in order to determine the phenomenological parameterization, so that the predictive power of this semi–microscopic potential is preserved. This can be seen from the comparison between the experimental [21]

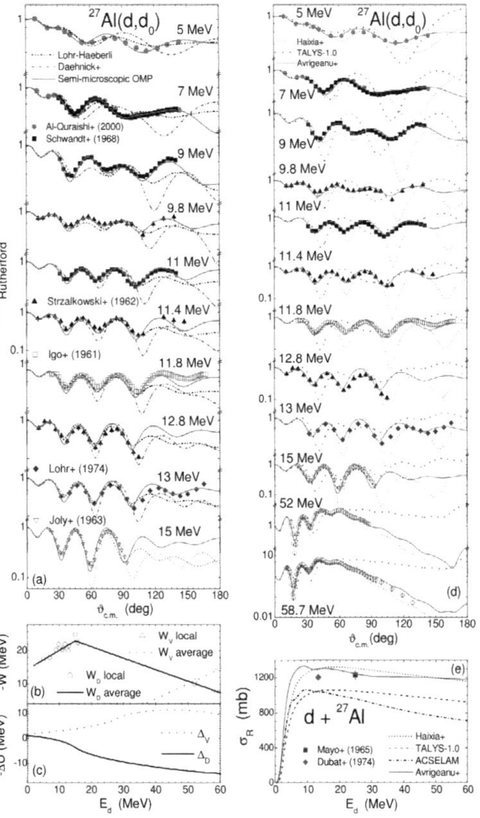

Figure 1. Optical model potential analysis of the deuterons interaction with ^{27}Al (see text for more details).

and semi-microscopical angular distributions for the ^{27}Al target nucleus and incident energies from 5 up to 15 MeV [21] shown in Fig. 1 (a). Angular distributions calculated with Daehnick *et al.* [20] and Lohr-Haeberli [22] global OMPs are also shown. The average energy dependence of the local imaginary volume and surface potential depths of the semi-microscopic OMP are shown in Fig. 1 (b), and the corresponding dispersive corrections to the real microscopic potential are shown in Fig. 1(c). The differential elastic scattering cross sections have been calculated by a modified version of the code SCAT2[23] which includes the DF model potential as an option for the OMP real part.

2.2. Phenomenological OMP

In the second step of our OMP analysis [5, 7, 8] the microscopic real potential is replaced by a phenomenological Woods-Saxon component. A full phenomenological OMP was then obtained from the analysis of the same experimental data, but keeping frozen the imaginary and spin-orbit potentials already determined. The advantage of having well settled already more than half of the usual OMP parameters, increases obviously the accuracy of fitting the data. The phenomenological elastic–scattering angular distributions for ^{27}Al target nucleus, are shown in Fig. 1(d) and compared to predictions of the recent global OMP [24] and the default OMP of the computer code TALYS [25] based on Watanabe [26] folding approach. Neither the previous global OMPs nor the default OMP within the code TALYS are able to describe the experimental angular distributions of the deuteron elastic scattering at the incident energies from 5 to 20 MeV, while a good agreement has been obtained with our energy–dependent phenomenological OMP parameter set, Table II of Ref. [10].

Moreover, the measured total reaction cross sections [21] of deuterons have also been compared with the predictions of the present OMP, the global OMPs [20, 24], the default OMP of TALYS code, as well as the evaluated data of the ACSELAM library [27]. The results shown in Fig. 1(e) for ^{27}Al point out differences of at least 20%, up to the incident energies of 60 MeV, between the more recent OMPs and the former ones. Nevertheless, they support once more the present phenomenological OMP.

3. Direct Reactions (DR)

3.1. Deuteron breakup

Following the deuteron elastic scattering analysis, an increased attention has been devoted to the direct interactions. They are dominant for deuteron energies below and around the Coulomb barrier, while other reaction mechanisms like pre–equilibrium emission (PE) or evaporation from the fully equilibrated compound nucleus (CN) become also important increasing the incident energy. The deuteron breakup mechanism responsible for the enhancement of a large variety of reactions is important along the whole incident energy range and its contribution to the reaction cross sections has to be taken explicitly into account [10].

The physical picture of the deuteron breakup in the Coulomb and nuclear fields of the target nucleus considers two distinct chains, namely (i) the elastic breakup (EB) in which the target nucleus remains in its ground state and none of the deuteron constituents interacts with it, and (ii) the inelastic breakup or breakup fusion (BF) where one of these deuteron constituents interacts with the target nucleus while the remaining one is detected. An empirical parametrization of the total proton-emission breakup fraction, σ^p_B/σ_R, and elastic breakup fraction, σ_{EB}/σ_R, have been obtained [9] from the analysis of the experimental systematics [28, 29, 30, 31, 32] of the proton–emission spectra and angular distributions from deuteron–induced reactions on nuclei from Al to Pb at incident energies from 15 to 80 MeV. Their dependence of the charge (Z), the atomic number of the target nucleus (A), and the deuteron incident energy (E) is [9]:

$$f_B^{(p)} = 0.087 - 0.0066Z + 0.00163ZA^{1/3} + 0.0017A^{1/3}E - 0.000002ZE^2, \tag{2}$$

$$f_{EB} = 0.031 - 0.0028Z + 0.00051ZA^{1/3} + 0.0005A^{1/3}E - 0.000001ZE^2. \tag{3}$$

Comparing with total proton- and neutron- emission breakup cross sections parametrization of Kalbach [33]:

$$\sigma_B^{p(n)} = K_{d,p(n)} \frac{(A^{1/3} + 0.8)^2}{1 + exp\frac{(13-E)}{6}}, \quad K_{d,p} = 21, \ \ K_{d,n} = 18. \tag{4}$$

the former parametrization [9] considers equal breakup fractions for proton and neutron emission, but supplementary gives all breakup total, elastic, and inelastic components.

In Fig. 2(a,b) it is shown that both parameterizations of Refs. [9, 33] are very close for ^{63,65}Cu target nuclei, for deuteron incident energies above ~7 MeV, while at lowest incident energies the total breakup cross sections given by Kalbach parametrization [33] exceed even the deuteron total reaction cross sections.

On the whole, the deuteron total breakup reaction cross section reduces the amount of the total reaction cross section that would be shared among different outgoing channels. On the other hand, contributions to different reaction channels, enhancing the secondary chance emission of particles from the original deuteron-nucleus interaction are brought by the inelastic

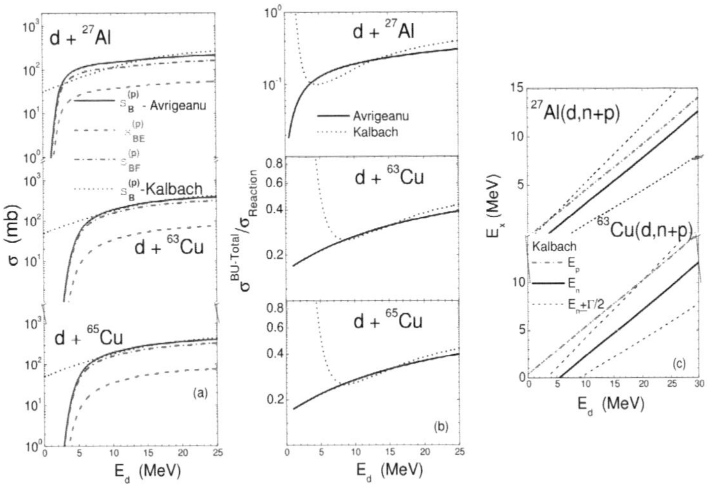

Figure 2. (a) The energy dependence of proton breakup cross sections components given by Ref.[9] and of the total proton-emission breakup cross sections given by Ref. [33] for the deuterons interactions with ^{27}Al and ^{63,65}Cu target nuclei. (b) The energy dependence of the total deuteron breakup fraction given by Refs. [9] and [33] on the same target nuclei. (c) The centroid E_x of assumed Gaussian line shape [33] for deuteron breakup-peak energies of emitted neutrons (solid) and protons (dash-dotted), and the corresponding $E_n \pm \Delta$ values (dashed) calculated for deuterons interacting with ^{27}Al and ^{63}Cu.

breakup process where one of deuteron constituents interacts with the target forming a secondary composite nucleus. Thus, the interactions of the neutron following the breakup proton emission contributes mainly to the enhancement of the $(d,2p)$ and $(d,p\alpha)$ reaction cross sections. In order to calculate this enhancement, the inelastic breakup proton-emission cross section $\sigma_{BF}^{(p)}$ has been multiplied by the fraction of the neutron total cross section σ_{tot} leading to the above-mentioned reactions, i.e. $\sigma_{(n,i)}/\sigma_{tot}$, where i stands for the γ, p, or α outgoing channels [10]. The cross section $\sigma_{BF}^{(p)}$ has been expressed as a function of the deuteron incident energy while the Kalbach [33] formula has been used for the centroid energy of the breakup neutron responsible for the specific reaction enhancements [10]. The quite large widths of the assumed Gaussian line shape of these peaks as showed in Fig.2 (c) emphasize however the broad approximation of this method. Therefore, the related contributions to the deuteron activation reaction cross sections have been additionally smoothed by using an average energy width of 3 MeV, with the results shown in Fig. 3 (d,e,f) by the dot-dashed curves.

A similar procedure has been followed in order to obtain the contribution to the $(d,2n)$ reaction cross section due to the protons which, following the breakup neutron emission, are absorbed in further interactions with the target nucleus. The inelastic breakup contributions $\sigma_{BF}^{(n)}\sigma_{(p,i)}/\sigma_R$, where σ_R is the proton total reaction cross section, and i=n, p or α, have been considered in this respect.

The inelastic breakup component still remains difficult to be handled in the reaction calculations by using the available codes. Thus, Kalbach's [33] absolute total neutron and proton breakup cross sections are considered in the TALYS-1.0 code for the first–chance neutron and respectively proton emission spectra while the consideration of a particular contribution following the absorption of one BF proton or neutron by the target remains an open question.

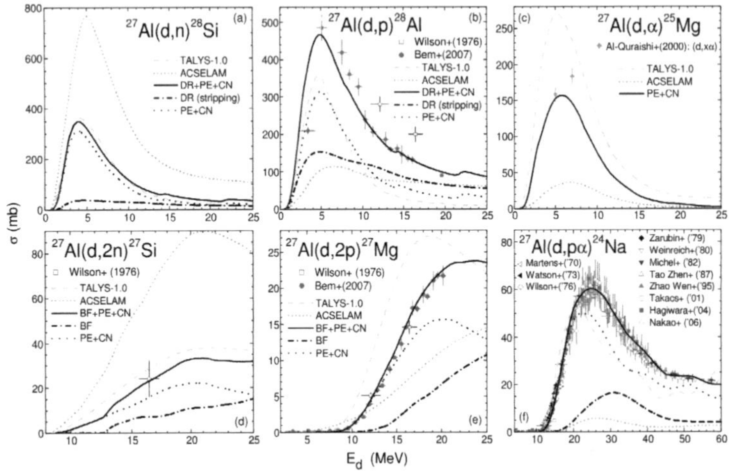

Figure 3. Comparison of measured ([21]) and calculated reaction cross sections provided by the code TALYS (dashed curves), the library ACSELAM (short-dotted), and local analysis, [41] (solid) using the DR results (dash-dotted) for the (d,p) and (d,n) reactions and deuteron breakup for the $(d,2p)$, $(d,2n)$, and $(d,p\alpha)$ reactions, and the PE+CN components (dotted).

3.2. Single-nucleon stripping

The direct-reaction, stripping mechanism contributions to the (d,p) and (d,n) reaction cross sections through population of the low-lying discrete levels of the residual nuclei were calculated by means of the code FRESCO [34]. The post form distorted wave transition amplitude with finite-range interaction has been chosen in this respect. Apart from the incident channel, for which the OMP was subject of the unitary deuteron interaction analysis [9, 10, 35], for the neutron and proton outgoing channels the Koning and Delaroche [36] optical model potentials have been chosen. The neutron–proton interaction V_{np} was assumed to have a Gaussian shape

$$V(r) = -V_0 e^{-(r/r_0)^2}, \tag{5}$$

where V_0=72.15 MeV and r_0=1.484 fm [37], determined from the fit of the deuteron binding energy. The nucleon bound states were generated in a Woods–Saxon real potential with global reduced radius of 1.25 fm and diffuseness of 0.65 fm, while its depth has been adjusted to the nucleon binding energies in the residual nuclei.

For the ^{27}Al(d,p) reaction cross section calculations a number of 35 final states of the odd-odd residual nucleus ^{28}Al extending up to 5.135 MeV have been considered, using the corresponding spectroscopic factors obtained experimentally from proton angular distribution measurements [38]. Similarly, the ^{27}Al(d,n) reaction cross section calculation has taken into account 24 final states of the even–even residual nucleus ^{28}Si, up to 11.4 MeV, with the spectroscopic factors given by analysis of neutron angular–distributions [39]. Finally the energy dependence of the stripping reaction cross section contributions to the activation cross sections ^{27}Al(d,p)^{28}Al and ^{27}Al(d,n)^{28}Si are shown by dot-dashed curves in Fig. 3(a,b).

The single neutron stripping, ^{63,65}Cu(d,p), reaction cross section calculations involved transitions to 63 final states of the odd-odd residual nucleus ^{64}Cu and 52 final states of the residual nucleus ^{66}Cu extending up to ~3 MeV for which spectroscopic factors exist [40].

4. Pre-equilibrium and statistical emission

Following the decay path, the pre-equilibrium (PE) and evaporation cross sections (CN), corrected for the breakup and stripping decrease of the total reaction cross section, completed the deuteron activation cross sections calculations by using the code STAPRE-H [41]. The main assumptions and parameters involved in Geometry Depndent Hybrid Model and Hauser-Feschbach formalism for PE and respectively CN calculations have recently been described elsewhere [42], while the specific parameters for the mass range A<30 are given in Ref. [10].

The comparison of the measured and calculated activation reaction cross sections corrsponding to deuteron interaction with ^{27}Al is presented in Fig. 3 (a-f), including the global (TALYS code [25]) and local analysis results described here, as well as the ACSELAM library [27] results. The common statement for all activation cross sections shown in Fig. 3 is that of the need to update the ACSELAM library.

The results of the local analysis of the final (d, n) and (d, p) activation cross sections of ^{27}Al shown in Figs. 3(a) and 3(b) include all three mechanisms components i.e. the DR cross sections provided by the code FRESCO and the PE+CN contributions supplied by STAPRE-H. The latter is alone rather close to the TALYS predictions while the complete local approach has led to much better agreement with the most recent (d, p) reaction data [10] especially due to the DR stripping contribution.

Concerning the complete description of the $(d, 2n)$,$(d, 2p)$, and $(d, p\alpha)$ reaction cross sections, the inelastic breakup mechanism may contribute up to 50% of the activation cross sections for deuteron incident energies of ~25 MeV, as it is shown by the dot-dashed curves in Fig. 3(d), 3(e), and 3(f) respectively. A simpler analysis concerns the (d, α) reaction, Fig. 3(c) which has no other breakup corrections except the corresponding decrease of the total reaction cross section. The DR effects have also been overlooked, due to the symmetrical experimental (d, α) reaction angular distributions [21] which are already reproduced by the PE+CN contributions. Thus the PE+CN components provide alone the corresponding activation cross sections [10].

5. Conclusions

The main applied–oriented aim of this lecture has been to present an unitary analysis of the nuclear reaction mechanisms responsible for the deuteron interactions with nuclei. An extended analysis of elastic-scattering, breakup and direct-reaction of deuterons followed by the pre-equilibrium and statistical emissions has been considered.

The overall agreement between the measured data and model calculations proves the correctness of nuclear mechanism description taken into account for the deuteron-nucleus interaction. Also, the simultaneous analysis of the deuteron elastic scattering and induced activation appears essential for a consistent input into the nuclear reaction model studies. Finally the comparison of the present calculations with predictions of the TALYS code stress out the importance of the appropriate consideration of the deuteron breakup mechanism contribution to the activation cross-section calculations.

Acknowledgments

Work carried out within the framework of the European Fusion Development Agreement and CNCSIS–Bucharest PN-II Contract No. 149/2007.

References

[1] A. Ibarra, A. Möslang, R. Lässer, R. Ferdinand, R. Andreani, E. Surrey, B. Riccardi, V. Heinzel, H. Klein, U. Fischer, R. Forrest and M. Gasparotto 2007 Fusion Eng. Design **82** 2422
[2] R.A. Forrest, J. Kopecky and J.-Ch. Sublet 2004 The European Activation File: EAF-2005 cross section library, UKAEA FUS 515
[3] R.A. Forrest and I. Cook 2007 Fusion Eng. Design **82** 2478

[4] S.P. Simakov, P. Bém, V. Bujan, U. Fischer, Mötz, M. Honusek, V. Kroha, J. Novák and E. Šimečková 2007 Fusion Eng. Design **82** 2510
[5] M. Avrigeanu, W. von Oertzen, U. Fischer and V. Avrigeanu 2005 Nucl. Phys. A **759** 327
[6] M. Avrigeanu, H. Leeb, W. von Oertzen, F. L. Roman and V. Avrigeanu 2008 *Proc. Int. Conf. on Nuclear Data for Science and Technology - ND2007* (Nice) Eds. O. Bersillon *et al.* (EDP Sciences), p 219
[7] M. Avrigeanu, W. von Oertzen, A.J.M. Plompen and V. Avrigeanu 2003 Nucl. Phys. **A723** 104; M. Avrigeanu and V. Avrigeanu 2006 Phys. Rev. **C 73** 038801
[8] M. Avrigeanu, A.C. Obreja, F.L. Roman, V. Avrigeanu and W. von Oertzen 2009 At. Data Nucl. Data Tables **95** 501
[9] M. Avrigeanu, W. von Oertzen, R. A. Forrest, A.C. Obreja, F. L. Roman and V. Avrigeanu 2009 Fusion Eng. Design **84** 418
[10] P. Bém, E. Šimečková, M. Honusek, U. Fischer, S. P. Simakov, R. A. Forrest, M. Avrigeanu, A. C. Obreja, F. L. Roman and V. Avrigeanu 2009 Phys. Rev. C **79** 044610
[11] M. Avrigeanu, W. von Oertzen, H. Leeb, F.L. Roman and V. Avrigeanu 2006 *Proc. Int. Conf. on Nuclear Reaction Mechanisms* (Varenna) ed E. Gadioli (Milan: Ricerca Scientifica ed Educazione Permanente) p. 193
[12] Dao T. Khoa and W. von Oertzen, Phys. Lett. 1993 **B304** 8 ; 1995 **B342** 6; Dao T. Khoa, W. von Oertzen and H.G. Bohlen 1994 Phys. Rev. C **49** 1652
[13] N.Anantaraman, H. Toki and G. Bertsch 1983 Nucl. Phys. A **398** 279
[14] X. Campi and A. Bouyssy 1978 Phys. Lett. B **73** 263
[15] J.W. Negele and D. Vautherin 1972 Phys. Rev. C **5** 1472
[16] H. Krivine and J. Treiner 1979 Phys. Lett. B **88** 212
[17] D. Abbott, A. Ahmidouch, H.Anklin, J. Arvieux, et. al. 2000 Phys. Rev. Lett. **84** 5053
[18] J.W. Negele 1970 Phys. Rev. C **1** 1260
[19] C. Mahaux, H. Ngo and G.R. Satchler 1986 Nucl. Phys. A **449** 354
[20] W. W. Daehnick, J. D. Childs and Z. Vrcelj 1980 Phys. Rev. C **21** 2253
[21] *Experimental Nuclear Reaction Data (EXFOR)*; www-nds.iaea.or.at/exfor.
[22] J.M. Lohr and W. Haeberli 1974 Nucl. Phys. A **232** 381
[23] O. Bersillon 1992 Code SCAT2, Note CEA-N-2227
[24] Haixia An and Chonghai Cai 2006 Phys. Rev. C **73** 054605
[25] A.J. Koning, S. Hilaire and M.C. Duijvestijn 2008 Ref. [6], p 211
[26] S. Watanabe 1958 Nucl. Phys. **8** 484
[27] N. Yamano, Table of Isotope Production Cross Sections (ACSSELAM Library), http://wwwndc.jaea.go.jp/ftpnd/sae/acl.html
[28] J. Pampus, J. Bisplinghoff, J. Ernst, T. Mayer-Kuckuk, J. Rama Rao, G. Baur, F. Rosel and D. Trautmann 1978 Nucl. Phys. A **311** 141
[29] J.R. Wu, C.C. Chang and H.D. Holmgren 1979 Phys. Rev. C **19** 370
[30] J. Kleinfeller, J. Bisplinghoff, J. Ernst, T. Mayer-Kuckuk, G. Baur, B. Hoffmann, R. Shyam, F. Rosel and D. Trautmann 1981 Nucl. Phys. A **370** 205
[31] N. Matsuoka, M. Kondo, A. Shimizu, T. Saito, S. Nagamachi, H. Sakaguchi, A. Goto and F. Ohtani 1980 Nucl. Phys. A **345** 1
[32] M.G. Mustafa, T. Tamura and T. Udagawa 1987 Phys. Rev. C **35** 2077
[33] C. Kalbach 2002-2003 TUNL Prog. Rep. **XLII** pp. 80-83
[34] I.J. Thompson 1988 Comput. Phys. Rep. **7** 167; Version FRES 2.3, May 2007.
[35] M. Avrigeanu and V. Avrigeanu 2009 *Second Int. Workshop CNR*09, 05-08 October 2009* Bordeaux
[36] A.J. Koning and J.P. Delaroche 2003 Nucl. Phys. A **713** 231
[37] M. Kamimura, M. Yahiro, Y. Iseri, Y. Sakuragi, H. Kameyama and M. Kawai 1986 Prog. Theor. Phys. Suppl. No. **89** 1
[38] S. Chen, J. Rapaport, H. Enge and W.W. Buechner 1972 Nucl. Phys. A **197** 97; T.P.G. Carola and J.G. Van Der Baan 1971 Nucl. Phys. A **173** 414; J.V. Maher, H.T. Fortune, G.C. Morrison and B. Zeidman 1972 Phys. Rev. C **5** 1313
[39] W. Bohne, H. Fuchs, K. Grabisch, M. Hagen, H. Homeyer, U. Janetzki, H. Lettau, K.H. Maier, H. Morgenstern, P. Pietrzyk, G. Roschert and J.A. Scheer 1969 Nucl. Phys. A **131** 273
[40] Y.S. Park and W.W. Daehnick 1969 Phys. Rev. **181** 1082; W.W. Daehnick and Y.S. Park 1969 Phys. Rev. **180** 1062
[41] M. Avrigeanu and V. Avrigeanu 1995 IPNE Report NP-**86**-1995, Bucharest, 1995, and Refs. therein; News NEA Data Bank **17** 22
[42] M. Avrigeanu , S. Chuvaev, A. A. Filatenkov, R. A. Forrest, M. Herman, A. J. Koning, A. J. M. Plompen, F. L. Roman and V. Avrigeanu 2008 Nucl. Phys. **A806** 15

Journal of Physics: Conference Series **205** (2010) 012015 doi:10.1088/1742-6596/205/1/012015

Nuclear effects in neutrino-nucleus interactions

Maria B. Barbaro

University of Torino and INFN, Via Giuria 1, I-10125 Torino, Italy

E-mail: `Barbaro@to.infn.it`

Abstract. An accurate description of the nuclear response functions for neutrino scattering in the Gev region is essential for the interpretation of present and future neutrino oscillation experiments. Due to the close similarity of electromagnetic and weak scattering processes, we will review the status of the scaling approach and of relativistic modeling for the inclusive electron scattering response functions in the quasielastic and Δ-resonance regions. In particular, recent studies have been focused on scaling violations and the degree to which these imply modifications of existing predictions for neutrino reactions. We will discuss sources and magnitude of such violations, emphasizing similarities and differences between electron and neutrino reactions.

1. Introduction

The motivation of the recent interest in neutrino-nucleus interactions is twofold. One one hand ongoing and future high precision neutrino oscillation experiments increasingly rely on our knowledge of neutrino-nucleus interactions in both Charged Current (CC) processes of the type $\nu_l + A \to l + N + (A-1)$, where a virtual W boson is exchanged, and Neutral Current (NC) reactions $\nu_l + A \to \nu_l' + N + (A-1)$, where a Z boson is exchanged. Typical neutrino energies involved in these experiments are of a few GeV. In this region a large contribution to the cross section comes from quasielastic and resonance processes, where the nuclear dynamics is known to play an important role. Since many electron scattering data exist at these kinematics (see Ref. [1] for a recent review), we should be able to use them to predict neutrino cross sections. As we will show, the superscaling properties of inclusive (e, e') data represent a bridge between electron and neutrino scattering reactions off nuclei.

A second motivation consists in the fact that neutrinos can reveal informations on the nuclear structure complementary to electrons and they can be used to probe the internal structure of the nucleon (in particular the axial mass and the strange form factors), providing the nuclear content of the problem is under control.

In this contribution we will first review the phenomenon of superscaling in inclusive electron-nucleus scattering [2, 3, 4, 5] and then show some predictions for ν-A cross sections based on the SuperScaling Approximation ("SuSA") [6, 7]. We will then discuss the microscopic origin of the superscaling function, with particular focus on two models, the Relativistic Mean Field (RMF) [8] and a BCS-like correlated Fermi gas [9], which seem to correctly reproduce the shape of the (e, e') experimental data, at variance with other relativistic models. Finally we will address the issue of scaling violations, in particular the ones connected to meson-exchange currents.

Journal of Physics: Conference Series **205** (2010) 012015 doi:10.1088/1742-6596/205/1/012015

2. Superscaling

The experimental quasielastic (e, e') cross sections display the so-called "first kind scaling" [5] behaviour: at high momentum transfer (q larger than about 0.5 GeV/c) the reduced quasielastic cross section

$$F = \frac{d\sigma}{d\Omega d\omega} / \left(Z\sigma_{ep} + N\sigma_{en}\right) \tag{1}$$

does not depend on two variables (q, ω) but only on one combination of them, called the scaling variable and indicated by $y(q, \omega)$ in a non-relativistic scheme and $\psi(q, \omega)$ in a relativistic context. These two scaling variables represent respectively the minimum momentum ($-y$) or kinetic energy ($m_N\psi$) that a nucleon inside the target nucleus must have in order to participate to the reaction. The conditions $y = 0$ and $\psi = 0$ correspond, in the non-relativistic and relativistic case respectively, the quasielastic peak (QEP). In Ref. [4] it was shown that if the reduced function (1) corresponding to different kinematical conditions is plotted as a function of y it fulfills with good accuracy first kind scaling in the so-called scaling region $y < 0$, whereas at $y > 0$ scaling is broken.

Second kind scaling consists instead in the observation that for different target nuclei the reduced cross section scales as the inverse of the Fermi momentum k_F, so that the so-called *superscaling function* $f = k_F \times F$ is independent of the specific nucleus. This has been illustrated in Ref. [5], where it has been shown that scaling of second kind is excellent in the scaling region and slighthly violated at the right of the QEP.

If first and second kind scaling are simultaneously respected the cross sections are said to "superscale". A closer inspection of the Rosenbluth-separated longitudinal and transverse response functions shows that the scaling violations observed at $\psi > 0$ mainly reside in the transverse channel, essentially due to the Δ resonance contribution.

Before proceeding, let us briefly remind the basic formalism for inclusive electron scattering, focussing on the longitudinal channel where superscaling is seen to work better. In plane wave impulse approximation the longitudinal quasielastic response function can be expressed as

$$R_L(q, \omega) = \frac{2\pi m_N^2}{q} \int\int_{\Sigma} dp\, d\mathcal{E}\, S(p, \mathcal{E}) \mathcal{R}_L(q, \omega; p, \mathcal{E}), \tag{2}$$

where $\Sigma(q, \omega)$ is the kinematically allowed region in the $(\mathcal{E}, p)$ plane, $\mathcal{E}$ being the excitation energy of the residual nucleus and p its momentum, $S(p, \mathcal{E})$ is the nuclear spectral function and $\mathcal{R}_L$ the single nucleon response, which depends on the electric and magnetic nucleon's form factors.

Assuming that the single nucleon response $\mathcal{R}_L$ is smoothly varying in the $(\mathcal{E}, p)$-plane and can be factored out of the integral one gets

$$R_L(q, \omega) = \mathcal{R}_L(q, \omega; p = p_{\min}, \mathcal{E} = 0) \times F_L(q, \omega) \tag{3}$$

where

$$F_L(q, \omega) = \frac{2\pi m_N^2}{q} \int\int_{\Sigma(q,\omega)} dp\, d\mathcal{E} S(p, \mathcal{E}) \tag{4}$$

is the longitudinal scaling function. Note that the above factorization can be only approximately true in a relativistic context and it is violated to some degree due to off-shell effects and breaking of independent particle models.

The simplest model where one can perform exact relativistic calculations is the Relativistic Fermi Gas (RFG), namely an independent-particle model in which on-shell nucleons are described by Dirac spinors. In spite of its simplicity the RFG model retains some fundamental properties: it is Lorentz covariant and gauge invariant and is therefore a good starting point

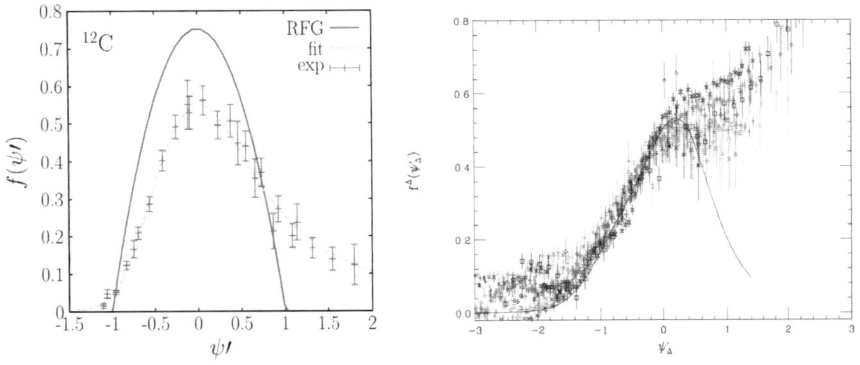

Figure 1. The experimental superscaling function in the quasielastic (left panel) [10] and Δ resonace (right panel) regions.

for more sophisticated calculations. The RFG spectral function is $S^{RFG}(p,\mathcal{E}) = \theta(k_F - p)\delta(\mathcal{E} - T_F + T_{\mathbf{p}})$ and the corresponding scaling function is

$$F_L(q,\omega) = \left(\frac{m_N}{q}\right)\left(\frac{1}{k_F}\right)\frac{3}{4}\left(1-\psi^2\right)\theta\left(1-\psi^2\right) \tag{5}$$

where the parabola $\frac{3}{4}\left(1-\psi^2\right)\theta\left(1-\psi^2\right) \equiv f(\psi)$ is the superscaling function and $\psi \equiv \psi(q,\omega)$ is the RFG scaling variable. The RFG model exactly superscales, as the data do in the so-called scaling region. However the comparison with the averaged data, illustrated in the left panel of Fig. 2, shows that the corresponding superscaling function is very different from the experimental one and in particular it fails the reproduce the long tail displayed by the data at large positive ψ. However a simple phenomenological approach can be taken for describing the QE region: a fit of the experimental superscaling function can be extracted from the data and plugged into the RFG formulas in place of the parabola (5). Note that the fit involves only four parameters for describing *all* kinematics and *all* nuclei. Moreover, it represents a strong constraint on nuclear models, since any reasonable model used to describe the quasielastic region must reproduce this shape.

The scaling analysis has been recently extended to the Δ resonance region [6]. In order to isolate the contributions in this region, the impulsive contributions arising from quasielastic eN scattering, calculated assuming superscaling is valid, has been removed from the experimental total cross section and the result has been divided by the elementary $N \to \Delta$ cross section, thus leading to a new superscaling function

$$f_\Delta(q,\psi_\Delta) \equiv k_F \frac{\left[d^2\sigma/d\Omega_e d\omega\right]_\Delta}{\sigma_M\left[v_L G_L^\Delta + v_T G_T^\Delta\right]}\,. \tag{6}$$

When the latter is plotted against the appropriate scaling variable $\psi_\Delta \equiv \psi(q\rho,\omega\rho)$, which accounts for the different kinematics through the inelasticty paramater $\rho = 1 + (m_\Delta^2 - m_N^2)/(4|Q^2|)$, the result shown in the right panel of Fig. 2 is obtained. Obviously this approach can work only at $\psi'_\Delta < 0$, since at $\psi'_\Delta > 0$ other resonances and the tail of DIS contribute.

The two above superscaling functions can then be used to reconstruct the (e,e') cross sections in the QE and Δ regions, giving a satisfactory description of the data, as shown in Ref. [6].

3. Predictions for neutrino scattering: the SuSA approach

We shall now illustrate how the superscaling properties of inclusive electron scattering can be exploited to predict neutrino cross sections off nuclei.

Charged current neutrino-nucleus scattering can be easily linked to inclusive (e, e') reactions by replacing the virtual photon by a wirtual W-boson. The main difference is that in the neutrino case, beyond the longitudinal and transverse response functions, a third response is involved, arising from the axial component of the weak current. In the RFG model the three response functions turn out to be

$$\widetilde{R}_i = \widetilde{G}_i \times f(\psi) , \tag{7}$$

where the functions $\widetilde{G}_i$ depend on the weak nucleon form factors and the information about the nuclear dynamics is carried by the superscaling function $f(\psi)$. The basic assumption of the SuSA approximation is that the nuclear dynamics governing the neutrino-nucleus process is the same entering the electron scattering reactions and therefore the corresponding cross sections in the QE and Δ regions can be obtained by simply replacing in (7) the RFG superscaling function with the corresponding phenomenological fits: $f_{RFG} \Rightarrow f_{QE,\Delta}$.

In the left panel of Fig. 3 the double differential cross section in the QE and Δ regions is shown for (see Refs. [6] for more results): it clearly appears that in both regions the SuSA result lies significantly lower and extends over a wider range in k' than the RFG. In the right panel the SuSA fully integrated quasielastic cross section [11] on ^{12}C is shown as a function of the neutrino energy and compared not only with the RFG but also with other relativistic models: the relativistic mean field (RMF) model, the relativistic plane wave impulse approximation (RPWIA) and a semi-relativistic shell model without (SRWS) or with (SRWS-tot) inclusion of discrete states of ^{12}N. Note that the RMF yields results which are very similar to the SuSA prediction, whereas the other models are closer to the RFG. The comparison of these results with the recent MiniBooNE data [12], which lie above the RFG and seem to point to a very high value of the axial mass ($M_A c^2$ = 1.35 GeV in place of the usual value 1.03, which we are using in our calculations), is somehow puzzling and still needs to be carefully explored.

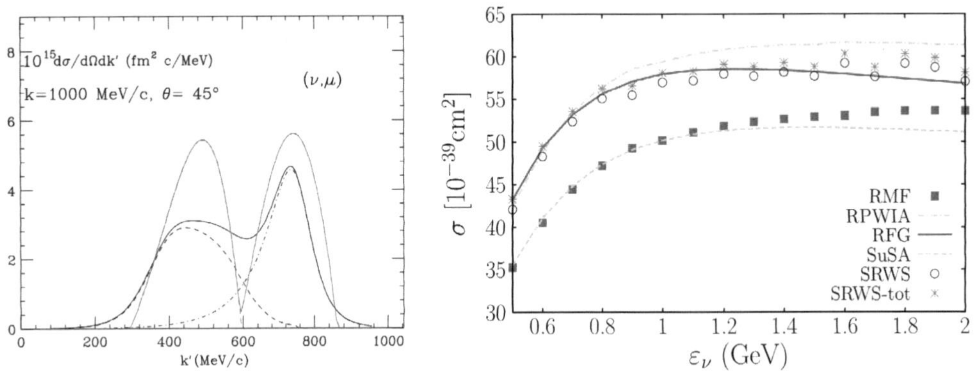

Figure 2. Left: CC neutrino reaction cross sections on ^{12}C versus the outgoing muon momentum k', showing a comparison of the SuSA results (black online) with the RFG (red onleine); the separate QE and Δ contributions are shown with dashed lines. Right: integrated QE cross section versus the neutrino energy evaluated in different models.

Concerning neutral current neutrino reactions, it has been proved in Ref. [7] that, although the kinematics is not the same of (e, e') reactions because the outgoing detected particle is the knocked nucleon instead of the lepton, the SuSA approach can still be used in the QE channel to predict cross sections. The result is qualitatively similar to the CC case, namely the SuSA predictions are lower the RFG and the response estends to a wider region. As anticipated an interesting feature of NC reactions is their sensitivity to the strangeness content of the nucleon, which allows in principle to extract informations on the strange hadronic form factors by different

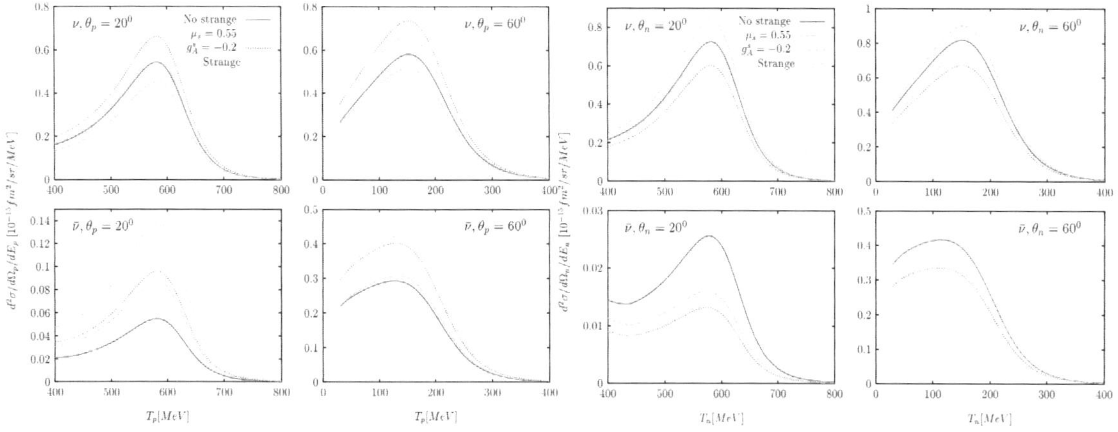

Figure 3. NC neutrino (upper panels) and antineutrino (lower panels) cross sections for proton (left) and neutron (right) knockout. The effect of strangeness is shown in the various curves.

combinations of neutrino and antineutrino scattering off protons and neutrons [13, 14, 15]. This is illustrated in the examples of Figs. 3, where it is seen that for proton knockout (left panels) the magnetic strangeness decreases (increases) the ν ($\overline{\nu}$) cross section, while the axial strangeness increases both ν and $\overline{\nu}$ cross sections. For neutron knockout (right panels) and ν scattering the situation is reversed with respect to p-knockout, while for $\overline{\nu}$ scattering the effect of magnetic strangeness is very small due to cancellation between the vector and axial-vector responses.

4. Microscopic origin of the superscaling function

In this section we will briefly illustrate the results of two relativistic models which, although sensibly different, seem to reproduce the basic properties of the experimental quasielastic superscaling function, namely a scaling behaviour of both first and second kind and a long high energy tail.

The first model is the Relativistic Mean Field model, based on a Lagrangian containing strong scalar and vector potentials (corresponding to σ, ω and ρ mesons) and describing the bound nucleon states as self-consistent Dirac-Hartree solutions. The final state interactions (FSI) are described in distorted wave impulse approximation using the same relativistic potentials [8]. As can be seen in Fig. 4 (left panel) the RMF model provides the correct amount of asymmetry, unlike other relativistic models, also shown in the same plot. These correspond to the absence of FSI (RPWIA) and to a treatment of FSI through a real optical potential (rROP). It clearly emerges from the results that not only relativity, but also a consistent treatment of the initial and final states are required in order to get the observed asymmetry of the scaling function. Morevoer, as shown in Ref. [8], the model fulfills superscaling with good accuracy.

A completely different model, which we will refer to as "BCS-like", has been shown to give very similar results [9], as illustrated in Fig. 4 (right panel). It is a simple covariant extension of RFG which accounts for NN correlations through the following wave functions

$$|\Phi>=\prod_k(u_k+v_k a^{\dagger}_{k\uparrow}a^{\dagger}_{-k\downarrow})|0> \quad \text{and} \quad |D(p)>=\frac{1}{|v_p|}a_{p\uparrow}\prod_k[u_k+v_k a^{\dagger}_{k\uparrow}a^{\dagger}_{-k\downarrow}]|0> \tag{8}$$

for the initial and final states, respectively. The v_k-coefficients are taken phenomenologically as Fermi-Dirac distributions with one free parameter β, controlling the tail of the momentum

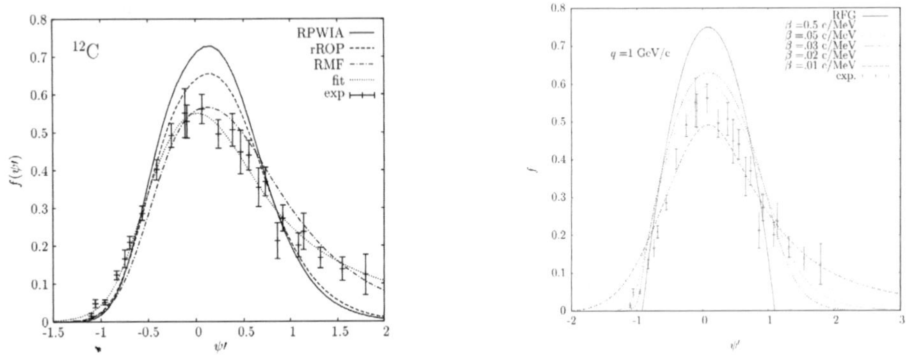

Figure 4. Left: QE superscaling function f evaluated in the RMF model compared with other models (see text); right: f in the BCS-like model for different values of the parameter β.

distribution v_k^2. The associated superscaling function f turns out to be asymmetric in ψ, the asymmetry increasing with the value of β, and to reproduce the data for values of $\beta \simeq 0.02$ c/MeV. The superscaling properties of the BCS-like model have been explored in Ref. [9], where it was shown that both kinds of scaling are good at the QEP and broken to some degree at the right and left of it.

5. Scaling Violations

At low momentum and energy transfer (which can be defined approximately by $q < 400$ MeV/c and $\omega < 50$ MeV) the scaling approach is bound to fail because collective effects, like giant resonances, become important. However even at higher energies scaling violations are observed and it is very interesting to understand their origin.

Among other possible contributions, like *e.g.* short range correlations, meson-exchange currents (MEC) can certainly play a significant role in breaking superscaling in the transverse channel. The MEC, being two-body currents, contribute to both the 1p-1h and the 2p-2h sectors of the response and their calculation is far from trivial in a relativistic context. Moreover in order to preserve gauge invariance, in parallel with the pure MEC diagrams, where the virtual photon attaches to the exchanged meson, it is necessary to take into account the corresponding correlation diagrams, where the photon attaches to the nucleon. Up to now an exact gauge invariant relativistic calculation has been performed only in the RFG system and in the 1p-1h channel [16, 17, 18]. Here the diagrams are the ones shown in Fig. 5, where thick lines in (d)-(g) represent either a nucleon or a Δ. It turns out that the diagrams involving the Δ give the dominat contribution and they break both kinds of scaling, as illustrated in the right panel of Fig. 5. Moreover a recent calculation [19] performed in a semi-relativistic shell model essentially equivalent to the RMF with FSI has shown that the MEC contribution to the high energy tail of the response can become very large due to the dynamical nature of the pion propagator, an effect which is emphasized by the presence of FSI.

In the 2p-2h sector the pure MEC have been calculated on the RFG basis [20] and shown to give a substantial contribution at high negative and positive values of ψ, but a full gauge invariant calculation is still missing.

6. Conclusions

We have shown how the superscaling properties of (e, e') data can be exploited to predict charged and neutral current neutrino scattering cross sections and we have proposed two microscopic

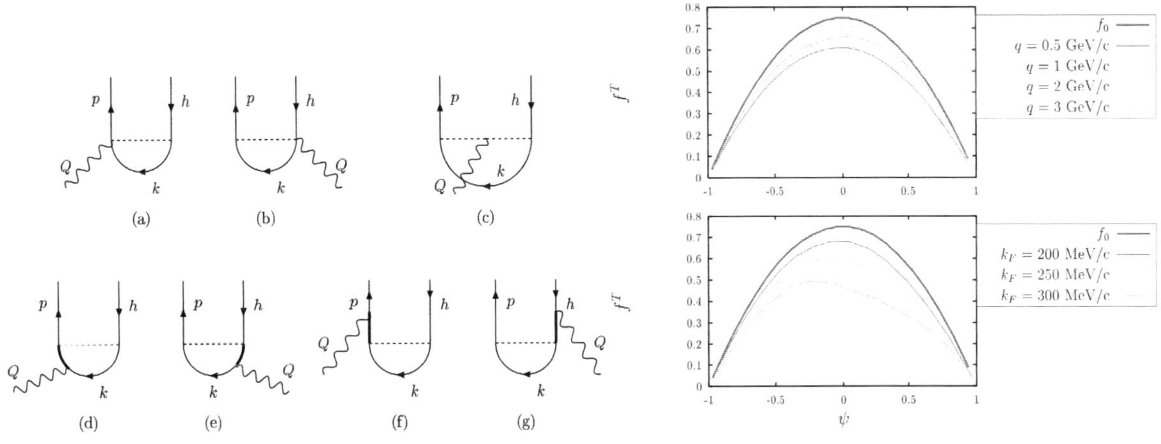

Figure 5. 1p-1h MEC diagrams and results for f in the RFG (f_0) and including the Δ-MEC diagram for different values of q (upper panel) and k_F (lower panel).

relativistic models, the RMF model and the BCS-like correlated Fermi gas, which are capable of reproducing the experimental superscaling function, in particular its long high energy tail.

However, in order to assess the applicability of the SuSA approach and eventually improve the method we need not only to justify the microscopic origin of superscaling but also to understand its violations, in particular the ones associated with meson exchange currents. This has been partially accomplished in recent work [21] and more effort will be done in this direction in order to provide a consistent description of the nuclear dynamics in the quasielastic, resonance and deep inelastic regions.

Acknowledgments

The work presented here is the result of collaborations with J.E. Amaro, J.A. Caballero, R. Cenni, T.W. Donnelly, C. Maieron, A. Molinari, I. Sick and J.M. Udias.

References

[1] O. Benhar, D. Day and I. Sick, arXiv:nucl-ex/0603032.
[2] G. B. West, Phys. Rept. **18** (1975) 263.
[3] W. M. Alberico *et al.*, Phys. Rev. C **38** (1988) 1801.
[4] D. B. Day, J. S. McCarthy, T. W. Donnelly and I. Sick, Ann. Rev. Nucl. Part. Sci. **40** (1990) 357.
[5] T. W. Donnelly and I. Sick, Phys. Rev. Lett. **82** (1999) 3212.
[6] J. E. Amaro *et al.*, Phys. Rev. C **71** (2005) 015501.
[7] J. E. Amaro, M. B. Barbaro, J. A. Caballero and T. W. Donnelly, Phys. Rev. C **73** (2006) 035503.
[8] J. A. Caballero *et al.*, Phys. Rev. Lett. **95** (2005) 252502.
[9] M. B. Barbaro, R. Cenni, T. W. Donnelly and A. Molinari, Phys. Rev. C **78** (2008) 024602.
[10] J. Jourdan, Nucl. Phys. A **603** (1996) 117.
[11] J. E. Amaro, M. B. Barbaro, J. A. Caballero and T. W. Donnelly, Phys. Rev. Lett. **98** (2007) 242501.
[12] T. Katori [MiniBooNE Collaboration], Proceedings of NUINT09, arXiv:0909.1996 [hep-ex].
[13] M. B. Barbaro *et al.*, Phys. Rev. C **54** (1996) 1954.
[14] W. M. Alberico *et al.*, Phys. Lett. B **438** (1998) 9.
[15] A. Meucci, C. Giusti and F. D. Pacati, Nucl. Phys. A **773** (2006) 250.
[16] J. E. Amaro *et al.*, Nucl. Phys. A **643** (1998) 349.
[17] J. E. Amaro *et al.*, Phys. Rept. **368** (2002) 317.
[18] J. E. Amaro *et al.*, Nucl. Phys. A **723** (2003) 181.
[19] J. E. Amaro *et al.*, arXiv:0906.5598 [nucl-th].
[20] A. De Pace *et al.*, Nucl. Phys. A **726** (2003) 303.
[21] C. Maieron *et al.*, Phys. Rev. C **80** (2009) 035504.

Journal of Physics: Conference Series **205** (2010) 012016 doi:10.1088/1742-6596/205/1/012016

Theory of few body reaction frameworks: Application to Halo nuclei

R Crespo

Departamento de Física, Instituto Superior Técnico, Taguspark,
Av. Prof. Cavaco e Silva, Taguspark, 2780-990 Porto Salvo, Oeiras, Portugal
Centro de Física Nuclear, Av Prof. Gama Pinto, 2, 1699, Portugal

E-mail: raquel.crespo@ist.utl.pt

Abstract. The theory of Faddev/AGS few body reaction frameworks is presented. A comparison with other reaction formalisms (DWBA, CDCC) is made. Sucesses and shortcomings of the scattering approaches for breakup are analysed. Some set of calculated reaction observales for resonant and nonresonant breakup are presented to obtain insight into the physics incorporated on the scattering approaches.

1. Introduction

The use of beams of radioactive nuclei, allowing the exploration of the nuclear chart away from the stability line, has open a new era for the field of Nuclear Physics. From the fundamental point of view the study of extreme states of matter near the drip lines will allow a deeper understanding of the nuclear matter.

Away from the stability line the many-body concept may not be useful and cluster structure models are frequently used especially for light systems [1, 2, 3]. From the point of view of scattering, the study of these exotic nuclei is both exciting and challenging. The need to take into account the halo few-body nature ab initio into the reaction formalism must be addressed. This calls for novel developments in the reaction theory, in order to handle in a convenient way at least four body problems. In addition, a very tight control on the reaction theory must be performed on the reaction formalisms, in order to extract reliable information from increasingly high precision data.

In this contribution we make a detailed review of the available theories of fewbody reaction frameworks and its application for breakup.

2. The Faddeev/AGS multiple scattering framework

The Faddeev/Alt, Grassberger, and Sandhas(AGS) formalism [4, 5, 6] is a three-body multiple scattering reaction framework that treats all open channels (elastic, breakup and transfer in equal footing). In addition all resonant and nonresonant states of each pair are automatically included. Let us consider 3 particles (1,2,3) interacting by means of two-body potentials. We shall be using in this section the odd man out notation (appropriate for 3-body problems) which means for example that the interaction between the pair (2,3) is denoted as v_1. We assume that the system is non-relativistic and we write its total Hamiltonian as

$$H = H_0 + \sum_\gamma v_\gamma \, , \tag{1}$$

with H_0 the kinetic energy operator of the system and v_γ the interaction for the pair γ. The Hamiltonian can be rewritten as

$$H = H_\alpha + V^\alpha \, , \tag{2}$$

where H_α is the hamiltonian for channel α

$$H_\alpha = H_0 + v_\alpha \, , \tag{3}$$

and V^α represents the sum of interactions external to partition α

$$V^\alpha = \sum_{\gamma \neq \alpha} v_\gamma \, . \tag{4}$$

The total transition amplitude which describes the scattering from the initial state α to the final state β is given in the post form by (see Refs. [6, 7] for a derivation)

$$T_+^{\beta\alpha}(z) = V^\beta + \sum_{\gamma \neq \alpha} T_+^{\beta\gamma}(z) G_0 t_\gamma \, . \tag{5}$$

with the transition amplitude

$$t_\gamma = v_\gamma + v_\gamma G_0 t_\gamma \tag{6}$$

and the free resolvent

$$G_0(z) = (z - H_0)^{-1}. \tag{7}$$

The transition amplitude t_γ, althought containing the two-body potential v_γ, is a many body operator due to its dependence on the kinetic energy of the three particles. Equivalently in the prior form we get

$$T_-^{\beta\alpha}(z) = V^\alpha + \sum_{\gamma \neq \beta} t_\gamma G_0 T_-^{\gamma\alpha}(z) \, . \tag{8}$$

One then defines the operators $U^{\beta\alpha}$ which are equivalent to the transition amplitudes on the energy shell, such that

$$\begin{aligned} U^{\beta\alpha} &= \bar{\delta}_{\beta\alpha} G_\alpha^{-1} + T_+^{\beta\alpha} \\ &= \bar{\delta}_{\beta\alpha} G_\beta^{-1} + T_-^{\beta\alpha} \, , \end{aligned} \tag{9}$$

where $\bar{\delta}_{\beta\alpha} = 1 - \delta_{\beta\alpha}$. From Eqs. (5-8)

$$U^{\beta\alpha} = \bar{\delta}_{\beta\alpha} G_0^{-1} + \sum_\gamma U^{\beta\gamma} G_0 t_\gamma \bar{\delta}_{\gamma\alpha} \, , \tag{10}$$

or

$$U^{\beta\alpha} = \bar{\delta}_{\beta\alpha} G_0^{-1} + \sum_\gamma \bar{\delta}_{\beta\gamma} t_\gamma G_0 U^{\gamma\alpha} \, . \tag{11}$$

The solution of the Faddeev/AGS equations can be found by iteration

$$\begin{aligned} U^{\beta\alpha} &= \bar{\delta}_{\beta\alpha}G_0^{-1} + \sum_{\gamma} \bar{\delta}_{\beta\gamma} t_\gamma \bar{\delta}_{\gamma\alpha} \\ &+ \sum_{\gamma} \bar{\delta}_{\beta\gamma} t_\gamma \sum_{\xi} G_0 \bar{\delta}_{\gamma\xi} t_\xi \bar{\delta}_{\xi\alpha} \\ &+ \sum_{\gamma} \bar{\delta}_{\beta\gamma} t_\gamma \sum_{\xi} G_0 \bar{\delta}_{\gamma\xi} t_\xi \sum_{\eta} G_0 \bar{\delta}_{\xi\eta} t_\eta \bar{\delta}_{\eta\alpha} \\ &+ \cdots . \end{aligned} \quad (12)$$

The successive terms of this series can be considered as terms of zero order (which contribute only for rearrangment transitions), first order (single scattering), second order (double scattering) and so on in the transition operators t_α.

The solution of the Faddeev/AGS equations are in standard approches solved in momentum space and discretization of all momentum variables. The Coulomb interaction can be included as described in [9, 10].

We consider here the breakup series which is represented diagramatically up to third order in Figs. 1-2 where the upper particle is taken as particle one scattering from the bound state.

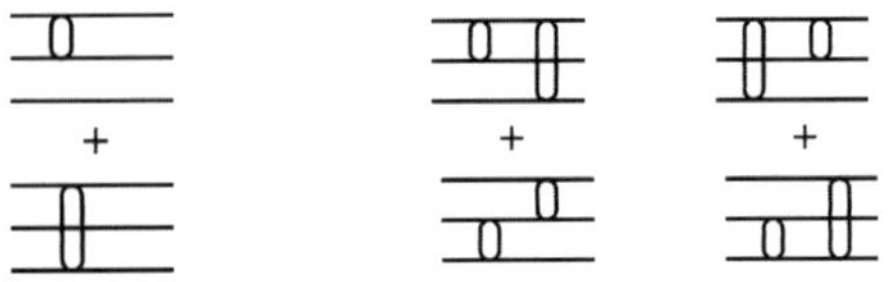

Figure 1. Multiple scattering diagrams (first to second order) for breakup in the Faddeev multiple scattering framework.

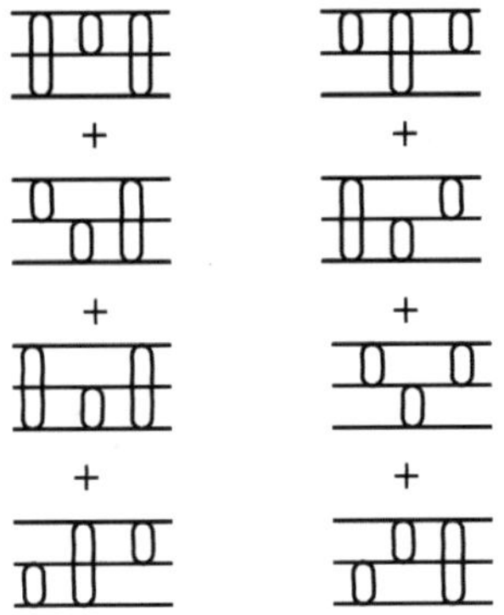

Figure 2. Multiple scattering diagrams (third order) for breakup in the Faddeev multiple scattering framework.

In order to solve the Faddeev/AGS equations we need to provide all the pair interactions. In the case examples we are presenting here envolving the scattering of a halo nucleus (well described by a core C and a valence particle n) on a proton target, p, we need then to provide the three pair interactions: $p - n$, $p - C$ and $n - C$ as described for example in the work of [11] for the case of p-^{11}Be scattering.

3. The DWIA/PWIA

Let us consider the reaction $A(a, ab)B$ where an incident particle a knocks out a nucleon or a bound cluster b in the target nucleus A resulting in three particles (a, b, B) in the final state.

This reaction has been analysed within the approximated reaction formalism, the distorted-wave impulse-approximation (DWIA) as described for example in the work of Chant and Roos [12]. This reaction framework relies mainly in two basic assumptions: (i) that the projectile strucks the ejected particle freely and (ii) that the 3-body wave function $\eta^{(-)}_{Bab}$ can be written as a factorized product of two wave functions that describe the $a + B$ and $b + B$ scattering. The transition amplitude is then written as

$$T_{AB} = \langle \eta^{(-)}_{aB} \eta^{(-)}_{bB} | t_{ab} | \phi_{Bb} \eta^{(+)}_{aA} \rangle, \tag{13}$$

where $\eta^{(-)}_{aB}$, $\eta^{(-)}_{bB}$ and $\eta^{(+)}_{aA}$ describe the relative motion of the particles in the entrance and exit channels: for the entrance channel one has

$$(T_{aA} + V_{aA} - V_{ab} - \epsilon_{aA}) \eta^{(+)}_{aA} = 0, \tag{14}$$

where ϵ_{aA} is the relative kinetic energy and T_{aA} the relative kinetic energy operator. For the exit channel, the relative wave functions satisfy:

$$(T_{aB} + V_{aB} - \epsilon_{aB}) \eta^{(+)}_{aB} = 0, \tag{15}$$

and

$$(T_{bB} + V_{bB} - \epsilon_{bB}) \eta^{(+)}_{bB} = 0. \tag{16}$$

The potentials V_{aB} and V_{bB} are taken to be the optical potentials which describe the $a + B$ and $b + B$ scattering at energies ϵ_{aB} and ϵ_{bB} respectively. Introducing plane waves in the exit channel and assuming

$$\eta^{(+)}_{aA} \sim (1 + G_0 t_{aB}) \chi^{(+)}_{aA}, \tag{17}$$

we then obtain

$$T_{AB} \sim \langle \chi^{(-)}_{aB} \chi^{(-)}_{bB} | T^{\text{exit}} + T^{\text{entrance}} | \phi_{Bb} \chi^{(+)}_{aA} \rangle, \tag{18}$$

where T^{exit} is given by

$$T^{\text{exit}} = t_{ab} + t_{aB} G_0 t_{ab} + t_{bB} G_0 t_{ab} + t_{bB} G_0 t_{aB} G_0 t_{ab}. \tag{19}$$

The first order term is the single scattering and contains the scattering from the struck particle only. This is the only term retained in plane-wave impulse approximation calculations. The second and third order terms in Eq. (19) represented in Fig. 4 are due to distortion in the exit channel. The component T^{entrance} is

$$T^{\text{entrance}} = t_{ab} G_0 t_{aB} + t_{aB} G_0 t_{ab} G_0 t_{aB} + t_{bB} G_0 t_{ab} G_0 t_{aB} + \mathcal{O}(t^4), \tag{20}$$

and the corresponding terms are represented in Fig. 3.

We refer as DWIA-full the *truncated* multiple scattering expansion $T^{\text{exit}} + T^{\text{entrance}}$ given respectively by the terms in Eqs. (19) and (20). We note that further approximations are usually made in standard applications of the DWIA when evaluating the transition amplitude, namely (i) The factorization approximation, which is only exact in PWIA (ii) On-shell approximation of the transition amplitude. We refer to this as DWIA-standard.

The Faddeev/AGS calculations do not make use of such approximations in the summation of the multiple scattering expansion.

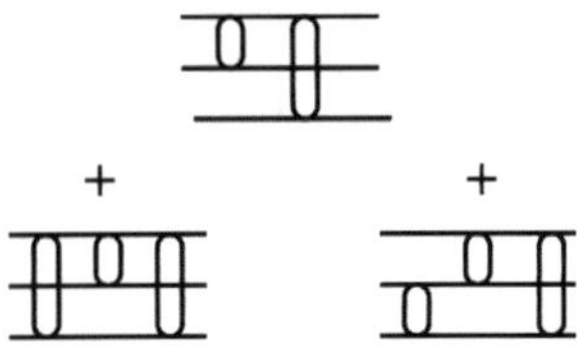

Figure 3. Diagrams due to the distortion in the income channel in the distorted-wave impulse approximation scattering framework.

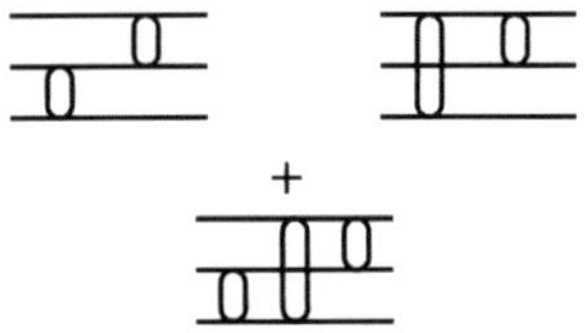

Figure 4. Diagrams due to the distortion in the exit channel in the distorted-wave impulse approximation scattering framework.

4. The CDCC scattering formalism

The continuum discretized coupled chanels (CDCC) [13, 14] reaction framework consists in solving the Schrödinger equation in a model space in which the three-body wavefunction is expanded in the internal states (bound and continuum resonant and nonresonant states) of the two-body projectile. The exact three-body wave function satisfies the Schrödinger equation

$$(H - E)\Psi^{+}_{\mathbf{K}_0}(\mathbf{R}, \mathbf{r}) = 0 \ , \tag{21}$$

where H is the three-body Hamiltonian, E is the total c.m. energy of the system, $\mathbf{R}$ the relative distance between the c.m. of the projectile and the target and $\mathbf{r}$ the relative distance between the valence particle and the core. Finally $\mathbf{K}_0$ is the incident wave number of the projectile in the c.m. frame.

The Hamiltonian H for this system can be written as a sum of three terms

$$H = T_R + U_{pt}(\mathbf{R}, \mathbf{r}) + H_p \ , \tag{22}$$

where T_R is the c.m. kinetic energy, U_{pt} the potential between the projectile constituents and the target which for a two-body target described by a core C and a valence particle v, is given by

$$U_{pt}(\mathbf{R}, \mathbf{r}) = \sum_{j=C,v} V_{jt}(\mathbf{R}, \mathbf{r}) \ , \tag{23}$$

and H_p internal Hamiltonian of the projectile. In the CDCC approach the exact wave function is expressed as an expansion in terms of the states (bound and bin) $\phi_\alpha(\mathbf{r})$ of the two-body Hamiltonian H_p as following

$$\Psi^{\mathrm{CDCC}}_{\mathbf{K}_0}(\mathbf{R}, \mathbf{r}) = \sum_{\alpha=0}^{N} \phi_\alpha(\mathbf{r})\omega_\alpha(\mathbf{R}) \tag{24}$$

where α=0 refers to the projectile ground state. The bin states includes both the resonant and the nonresonant part of the continuum. The wave functions $\omega_\alpha(\mathbf{R})$ of the projecticle-target relative motion are solutions of the coupled-channels equations

$$\left(E_\alpha - T_R - V_{\alpha\alpha}(\mathbf{R})\right)\omega_\alpha(\mathbf{R}) = \sum_{\beta\neq\alpha} V_{\alpha\beta}(\mathbf{R})\omega_\beta(\mathbf{R}) \ , \tag{25}$$

where $E_\alpha = E - \varepsilon_\alpha$ and the coupling potentials are

$$V_{\alpha\beta}(\mathbf{R}) = \langle\phi_\alpha|U_{pt}(\mathbf{R}, \mathbf{r})|\phi_\beta\rangle \ . \tag{26}$$

5. Bridging scattering approaches for breakup

In this section we show several calculated observables for breakup scattering making use of the previously discussed fewbody reaction frameworks. As an working example we have considered the case of the scattering of a ^{11}Be on a proton target at intermediate energies. We aim to pin down the relevant phyisics content of each fewbody scattering framework and to explore the validity of standard approximations in the calculated scattering observables.

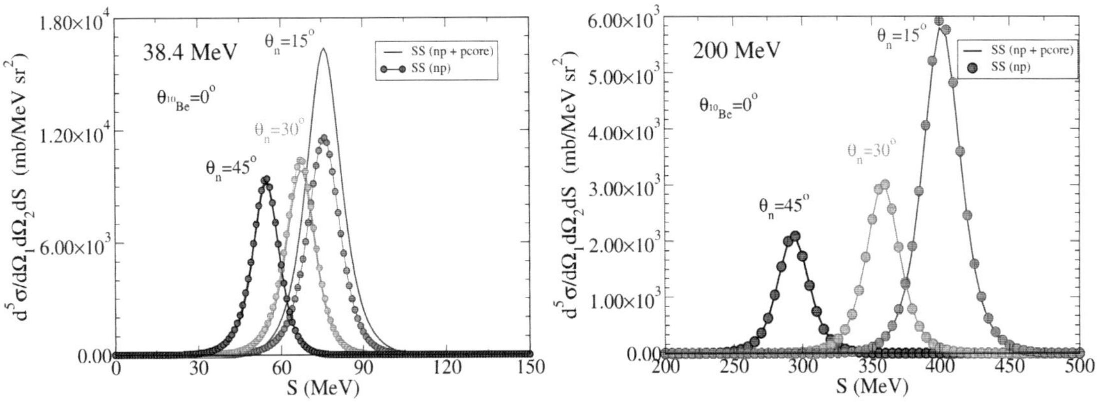

Figure 5. (Color online) Cross section for the breakup ^{11}Be(p,pn) at 38.4 MeV and 200 MeV including only the single scattering terms.

In Figs. 5 and 6 it is shown the Faddeev-type calculations performed for ^{11}Be breakup on proton target at 38.4 and 200 MeV/u incident energies at quasi free scattering conditions as described in the work of [15].

The curves represented in the figures, correspond to the calculated fivefold differential breakup cross section $d^5\sigma/d\Omega_1 d\Omega_2 dS$ versus the arclength S taking into account the single scattering term only. The kinematical configurations are characterized by the polar and azimuthal angles (θ_i, ϕ_i) of the two detected particles labeled 1 and 2 in Fig.10 of [15] . We assume those particles to be the ^{10}Be core (1) and the neutron (2). We take three quasi free scattering configurations: (conf 1) with $(\theta_1, \phi_1) = (0°, 0°)$ and $(\theta_2, \phi_2) = (45°, 180°)$; (conf 2) with $(\theta_1, \phi_1) = (0°, 0°)$ and $(\theta_2, \phi_2) = (30°, 180°)$; (conf 3) with $(\theta_1, \phi_1) = (0°, 0°)$ and $(\theta_2, \phi_2) = (15°, 180°)$.

Fig. 5 shows the breakup cross section calculated taking into account only the single scattering contribution of the Faddeev/AGS series. The case where both the proton-neutron and proton-core contributions is represented by the solid line. The circles correspond to the PWIA, i.e., without the proton-core scattering contribution. At 38.4 MeV/u this contribution is very small in the configurations 1 and 2, but is sizable when the neutron is emmited in the forward region. Therefore, even at the single scattering level, the DWIA calculations become innacurate in some configurations at low energies. At high energies the proton-core scattering contribution is negligible in all configurations.

Fig. 6 shows the Faddeev/AGS results taking multiple scattering terms up to second order (solid line) (shown in Fig 1) since the series is converged to this order.

The DWIA-full assumes that the contribution of the scattering between the proton and the neutron is dominant and therefore the single scattering contribution due to the scattering of the proton from the ^{10}Be core is neglected. The circles include the first and second order diagrams originated from the distortions in the exit channel represented in Fig. 4, and the dashed curve includes, in addition, the second order diagrams originated from the distortion in the incoming channel represented in Fig. 3. The DWIA-full results to second order (dashed line) provide a poor approximation to the Faddeev/AGS at 38.4 MeV in any configuration. As the energy increases

Journal of Physics: Conference Series **205** (2010) 012016 doi:10.1088/1742-6596/205/1/012016

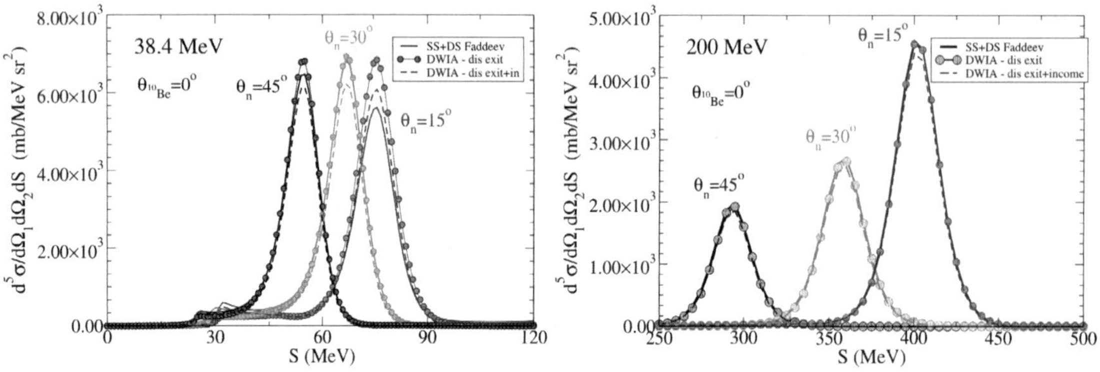

Figure 6. (Color online) Cross section for the breakup ^{11}Be(p,pn) at 38.4 MeV and 200 MeV including single and double scattering term.

DWIA-full approaches the exact result. However one should keep in mind that DWIA-standard calculations assume further approximations

As a final example Fig. 7 shows the breakup angular distribution for ^{11}Be breakup on a proton target @ 63.7 MeV/nucleon. The figure shows the inclusive angular distribution $d\sigma/d\Omega_{\mathrm{c.m.}}$. Very small angles $\theta_{\mathrm{c.m.}} < 5°$ (where there is no data) were not included because the convergence of the results with respect to the Coulomb screening radius is slow.

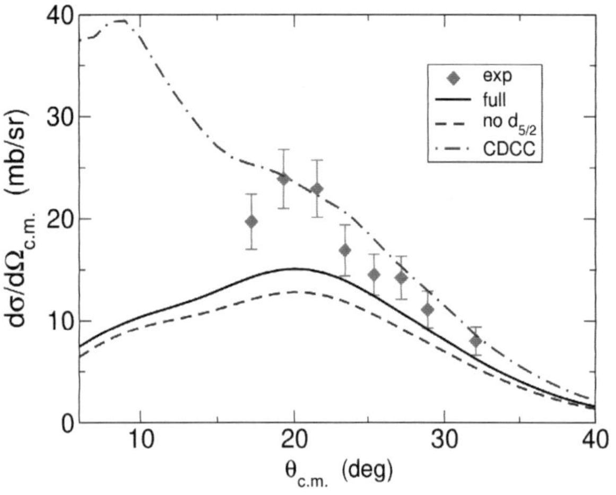

Figure 7. (Color online) Angular distribution for the breakup p(^{11}Be,p)^{10}Be n at 63.7 MeV/u integrated over the energy range $E_{\mathrm{rel}} = 0 - 2.5$ MeV. The solid line was obtained by solving the Faddeev/AGS equations with all partial waves. The dashed line represents the calculated observable excluding the $d_{\frac{5}{2}}$ partial wave contribution.

Due to the energy resolution of the experimental setup, the relative core-neutron energy was integrated around the resonance $E_{\mathrm{r}} = 1.275$ MeV in the energy range $E_{\mathrm{rel}} = 0 - 2.5$ MeV. The solid curve which includes both the resonant and nonresonant contributions reproduces fairly well the experimental data except for $\theta_{\mathrm{c.m.}} \leq 20°$. The dashed line represents the results excluding the $d_{\frac{5}{2}}$ partial wave. The dashed-dotted lines were obtained using the CDCC scattering framework with similar inputs as those used in the work of ref. [16]. A large disagreement between the two calculations for $\theta_{\mathrm{c.m.}} \leq 20°$ is evident and the source of discrepency needs to be investigated.

6. Conclusions

We have reviewed the theory of the Faddeev/AGS few body reaction framework along with other standard formalisms (DWIA and CDCC). In the Faddeev/AGS framework the sum of the series can be included. Practical applications of the DWIA approach envolves a truncated multiple scattering terms which may lead to significant contributions. A strong disagreement between calculated breakup observables was found between the Faddeev and the CDCC framework that needs to be understood.

Acknowledgements:

The author would like to thank A. Moro for calculating the CDCC resonant and nonresonant breakup p(^{11}Be,p)^{10}Be n at 63.7 MeV/u shown in the text. This work was supported in part by the FCT under the Grant POCTI/ISFL/2/275 and PTDC/FIS/65736/2006.

References

[1] Thompson I J, Danilin B V, Efros V D, Vaagen J S, Bang J M and Zhukov M V 2000 *Phys. Rev.* **C61** 1.
[2] Cobis A, Fedorov D V and Jensen A S 1998 *Phys. Rev* **C58** 1403.
[3] Danilin B *et al* 1997 *Phys. Rev. C* **55** R577.
[4] Faddeev L D 1960 *Zh. Eksp. Theor. Fiz.* **39** 1459 [1961 *Sov. Phys. JETP* **12** 1014].
[5] Alt E O, Grassberger P and Sandhas W 1967 *Nucl. Phys.* **B2** 167.
[6] Glöckle W, *The Quantum Mechanical Few-Body Problem* 1983 (Springer-Verlag, Berlin/Heidelberg).
[7] Crespo R, Cravo E, Deltuva A, Rodríguez-Gallardo M, and Fonseca A C 2007 *Phys. Rev.* **C76** 014620.
[8] Baker G A and Gammel J L eds. 1970 *The Padé Approximation in Theoretical Physics* (Academic, New York,).
[9] Deltuva A 2006 *Phys. Rev.* **C74** 064001.
[10] Deltuva A, Moro A M, Cravo E, Nunes F M, and Fonseca A C, 2007 *Phys. Rev.* **C76** 014611.
[11] Cravo E, Crespo R, Deltuva A, Fonseca A C, 2009 *Phys. Rev.* **C79** 064610.
[12] Chant N S and Roos P G 1997 . Rev. **C 15** 57.
[13] Autern N, Iseri Y, Kamimura M, Kawai M, Rawitscher G and Yahiro M 1987 *Phys. Rep.* **154** 125.
[14] Thompson I J 1988 *Comp. Phys. Rep.* **7** 167.
[15] Crespo R, Deltuva A, Cravo E, Rodríguez-Gallardo M, and Fonseca A C 2008 *Phys. Rev.* **C77**, 024601.
[16] Summers N C and Nunes F M 2007 *Phys. Rev.* **C76** 014611.

Journal of Physics: Conference Series **205** (2010) 012017 doi:10.1088/1742-6596/205/1/012017

Few-body nuclear reactions

A Deltuva

Centro de Física Nuclear, University of Lisbon, P-1649-003 Lisbon, Portugal

E-mail: deltuva@cii.fc.ul.pt

Abstract. Three- and four-body scattering is described in the framework of exact Faddeev-type integral equations. They are solved numerically using momentum-space partial-wave basis. The Coulomb interaction between charged particles is included using a novel implementation of the screening and renormalization method. The technique is applied to three- and four-nucleon scattering. Furthermore, the method has been extended successfully to elastic, transfer, and breakup reactions in three-body-like nuclear systems. Examples are deuteron scattering on stable nuclei ranging from ^{4}He to ^{58}Ni and proton scattering on weakly bound two-body system such as ^{11}Be, ^{15}C, and ^{17}O. These calculations allow to evaluate the accuracy of traditional approximate nuclear reaction approaches like the continuum-discretized coupled-channels (CDCC) method but also to test novel dynamical models such as nonlocal optical potentials.

1. Introduction

The application of exact Faddeev three-body theory to the description of nuclear reactions that are dominated by three-body degrees of freedom has been precluded in the past by the difficulty in dealing with the long-range Coulomb interaction between charged particles. Instead, approximate theoretical methods like the distorted-wave Born approximation (DWBA) or the continuum-discretized coupled-channels (CDCC) method [1] were used to analyze data resulting from the three-body-like nuclear reactions. However, recently also the exact Faddeev-type description of few-body nuclear reactions has become possible due to the novel implementation [2, 3] of the screening and renormalization method [4, 5, 6, 7] for including the Coulomb interaction between charged particles. This latter framework, though being technically and computationally most complicated and expensive, has an advantage that, once numerically well-converged results are obtained, all discrepancies with the experimental data can be attributed solely to the shortcomings of the used interaction models or to the inadequacy of the few-body model. The present paper recalls some recent achievements in the Faddeev-type description of few-body nuclear reactions.

Section 2 gives three-body scattering equations. Section 3 describes the inclusion of the Coulomb interaction. Section 4 presents selected results for three-body-like nuclear reactions. Four-nucleon scattering equations and example results are given in section 5. Section 6 contains summary.

2. Three-body scattering equations

An exact treatment of the quantum three-body scattering problem is provided by both Faddeev [8] and Alt, Grassberger, and Sandhas (AGS) equations [9] that are equivalent to the Schrödinger equation but are more suitable for the numerical solution due to the connectedness of the kernel. The Faddeev equations are formulated for the components of the wave function while the AGS equations,

$$U_{\beta\alpha} = \bar{\delta}_{\beta\alpha} G_0^{-1} + \sum_{\gamma=1}^{3} \bar{\delta}_{\beta\gamma} T_\gamma G_0 U_{\gamma\alpha}, \tag{1}$$

are a system of coupled integral equations for the transition operators $U_{\beta\alpha}$ whose on-shell matrix elements $\langle \psi_\beta | U_{\beta\alpha} | \psi_\alpha \rangle$ are scattering amplitudes and therefore lead directly to the observables. In Eq. (1) $\bar{\delta}_{\beta\alpha} = 1 - \delta_{\beta\alpha}$ and $G_0 = (E + i0 - H_0)^{-1}$ is the free resolvent, E being the available three-particle energy in the center of mass (c.m.) system and H_0 the free Hamiltonian. The two-particle transition matrix is a solution of the Lippmann-Schwinger equation

$$T_\gamma = v_\gamma + v_\gamma G_0 T_\gamma \tag{2}$$

where v_γ is the potential for the pair γ; we use the odd-man-out notation. The channel states $|\psi_\gamma\rangle$ for $\gamma = 1, 2, 3$ are the eigenstates of the corresponding channel Hamiltonian $H_\gamma = H_0 + v_\gamma$ with the energy eigenvalue E; thus, $|\psi_\gamma\rangle$ is a product of the bound state wave function for pair γ and a plane wave with fixed on-shell momentum corresponding to the relative motion of particle γ and pair γ in the initial or final state. The channel states $|\psi_0\rangle$ are the eigenstates of H_0 with the same eigenvalue E and describe the free motion of three particles. Observables of elastic scattering are calculated from the matrix elements with $\beta = \alpha$, those of breakup are given by $\beta = 0$ while $0 \neq \beta \neq \alpha$ correspond to transfer reactions.

We solve the AGS equations using momentum-space partial-wave basis where they become a system of integral equations with two continuous variables, the values of Jacobi momenta. The employed numerical techniques are described in great detail in Refs. [10, 11].

3. Inclusion of the Coulomb interaction

The AGS equations are applicable only to short-range potentials v_γ. The Coulomb interaction $w_{\gamma C}$, due to its long range, does not satisfy the mathematical properties required for the formulation of the standard scattering theory. However, since in nature the Coulomb potential is always screened, one could expect that the physical observables become insensitive to the screening provided it takes place at sufficiently large distances R and, therefore, the $R \to \infty$ limit should correspond to the proper Coulomb. This was proved by Taylor [4] in the context of the two-particle system: though the on-shell screened Coulomb transition matrix diverges in the $R \to \infty$ limit, after renormalization by (an equally) diverging phase factor it converges as a distribution to the well known proper Coulomb amplitude and therefore yields identical results for the physical observables. A similar renormalization relates screened and proper Coulomb wave functions [12]. Therefore, to include the long-range Coulomb force we follow the idea of screening and renormalization that was extended in Refs. [2, 6, 13, 14, 15] to three-particle scattering where only two particles are charged. The screened Coulomb potential

$$w_{\gamma R}(r) = w_{\gamma C}(r)\, e^{-(r/R)^n} \tag{3}$$

is added to the nuclear potential $v_{\gamma N}$, the standard scattering theory is applicable, and the AGS equations (1) are solved with the resulting potential $v_\gamma = v_{\gamma N} + w_{\gamma R}$ yielding the AGS transition operators $U^{(R)}_{\beta\alpha}$ that depend on the screening radius R. On-shell matrix elements $\langle \psi_\beta | U^{(R)}_{\beta\alpha} | \psi_\alpha \rangle$ do not have a $R \to \infty$ limit. However, as demonstrated in Refs. [2, 6, 13], the three-particle amplitudes can be decomposed into long-range and Coulomb-distorted short-range parts, where the quantities diverging in that limit are of two-body nature, i.e., the on-shell transition matrix $T^{\text{c.m.}}_{\alpha R}$ derived from the screened Coulomb potential between spectator α and the center of mass (c.m.) of the bound pair α, the corresponding wave function, and the screened Coulomb wave function for the relative motion of two charged particles in the final state. Those quantities, renormalized according to Ref. [4], in the $R \to \infty$ limit converge to the two-body Coulomb scattering amplitude $\langle \psi_\alpha | T^{\text{c.m.}}_{\alpha C} | \psi_\alpha \rangle$ (in general, as a distribution) and to the corresponding Coulomb wave functions, respectively, thereby yielding the three-particle scattering amplitudes in the proper Coulomb limit

$$\langle \psi_\beta | U^{(C)}_{\beta\alpha} | \psi_\alpha \rangle = \delta_{\beta\alpha} \langle \psi_\alpha | T^{\text{c.m.}}_{\alpha C} | \psi_\alpha \rangle + \lim_{R\to\infty} \{ Z^{-1/2}_{\beta R} \langle \psi_\beta | [U^{(R)}_{\beta\alpha} - \delta_{\beta\alpha} T^{\text{c.m.}}_{\alpha R}] | \psi_\alpha \rangle Z^{-1/2}_{\alpha R} \}, \quad (4)$$

$$\langle \psi_0 | U^{(C)}_{0\alpha} | \psi_\alpha \rangle = \lim_{R\to\infty} \{ z^{-1/2}_{R} \langle \psi_0 | U^{(R)}_{0\alpha} | \psi_\alpha \rangle Z^{-1/2}_{\alpha R} \}. \quad (5)$$

The renormalization factors $Z_{\alpha R}$ and z_R are diverging phase factors given in Refs. [2, 4, 5, 13]. The $R \to \infty$ limit in Eqs. (4) and (5) has to be calculated numerically, but due to the short-range nature of the corresponding operators it is reached with sufficient accuracy at finite R. Thus, for including the Coulomb interaction in three-body nuclear reactions with only two charged particles one only needs to solve the standard AGS equations (1) with short-range potentials. However, the screened Coulomb interaction, due to its longer range, compared to the nuclear interaction, brings additional difficulties: quasisingular nature of the potential and slow convergence of the partial-wave expansion. The right choice of the screening, i.e., the parameter n in Eq. (3), is essential in resolving those difficulties. For the fast convergence with R we have to ensure that the screened Coulomb potential $w_{\gamma R}(r)$ approximates well the proper Coulomb potential $w_{\gamma C}(r)$ for distances r smaller than the screening radius R and simultaneously vanishes smoothly but rapidly for $r > R$, providing a comparatively fast convergence of the partial-wave expansion. In Refs. [2, 13] we found the optimal values $3 \leq n \leq 6$.

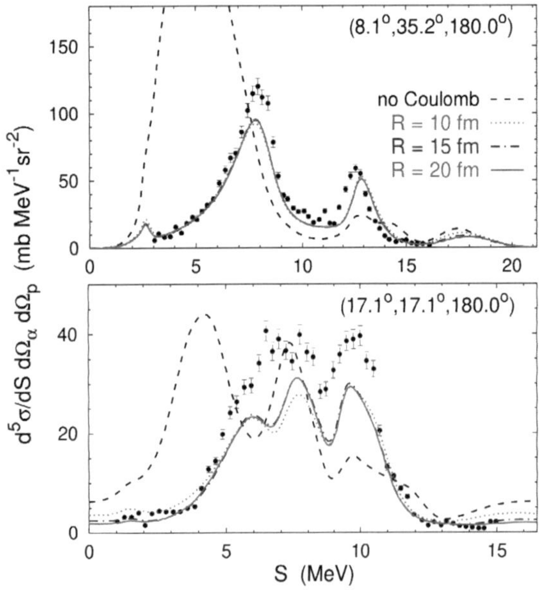

Figure 1. Convergence of the $\alpha + d \to \alpha + p + n$ breakup cross section with the screening radius R in selected kinematical configurations at 15 MeV α lab energy. Results obtained with screening radius R = 10 fm (dotted curves), 15 fm (dash-dotted curves), and 20 fm (solid curves) are compared. Results without Coulomb (dashed curves) are given as reference for the size of the Coulomb effect. The experimental data are from Ref. [16].

The most important criterion for the reliability of the screening and renormalization method is the convergence of the observables with the screening radius R used to calculate the Coulomb-distorted short-range part of the amplitudes. In Fig. 1 it is shown for $\alpha + d \to \alpha + p + n$ breakup reaction in two kinematical configurations; more examples can be found in Refs. [2, 13, 17]. In most cases the convergence is impressively fast and only becomes slower for the observables at very low energies. Furthermore, as demonstrated in Ref. [18], our results for proton-deuteron elastic scattering agree well over a wide energy range with those of Ref. [19] obtained from the variational solution of the three-nucleon Schrödinger equation in configuration space with the inclusion of an *unscreened* Coulomb potential and imposing the proper Coulomb boundary conditions explicitly.

4. Results for three-body nuclear reactions

The method of screening and renormalization in the framework of momentum-space AGS equations has been successfully applied in the three-nucleon system [2, 13, 20], i.e., for proton-deuteron elastic scattering, breakup, and radiative capture, and for photo- and electrodisintegration of ^{3}He. Furthermore, it was extended to three-body nuclear reactions [17, 21] like deuteron (d) scattering on a stable nucleus A or proton (p) scattering on a weakly bound two-body system (An) consisting of the core A and the neutron n; all elastic, transfer, charge-exchange, and breakup reactions, allowed by the chosen Hamiltonian, were calculated on the same footing. Those reactions have been calculated previously using approximate methods like DWBA or CDCC. The exact Faddeev-type calculations allow to evaluate the accuracy of those approaches. Recent comparison [22] of Faddeev/AGS and CDCC results revealed that CDCC is indeed a reliable method to calculate $d + A$ elastic and breakup cross sections but may lack accuracy for transfer reactions such as $p + {}^{11}\mathrm{Be} \to d + {}^{10}\mathrm{Be}$ and for breakup of one-neutron halo nuclei $p + {}^{11}\mathrm{Be} \to p + n + {}^{10}\mathrm{Be}$. Furthermore, the Faddeev/AGS momentum-space framework allows for an inclusion of a novel dynamic input like energy-dependent [23] or nonlocal [24, 25] optical potentials (OP). Especially important nonlocality effects were found in deuteron stripping and pickup reactions (d, p) and (p, d) involving both stable [26] and exotic nuclei [27]. Examples are given in Figs. 2 — 4.

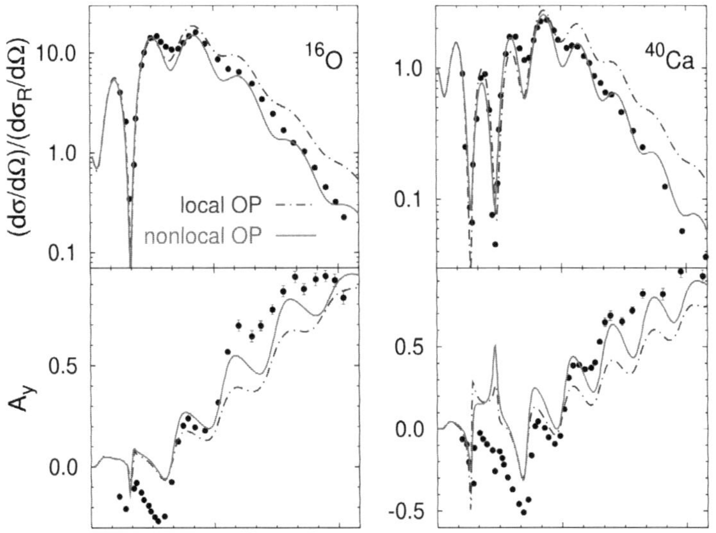

Figure 2. Differential cross section divided by Rutherford cross section and deuteron vector analyzing power for deuteron elastic scattering from ^{16}O and ^{40}Ca nuclei at $E_d = 56$ MeV. Predictions of nonlocal OP (solid curves) and equivalent local OP (dashed-dotted curves) are compared with the experimental data from Ref. [28].

5. Four-nucleon scattering

Exact description of the four-nucleon scattering is given by the Faddeev-Yakubovsky equations [32] for the wave-function components or by the equivalent AGS equations [33]. We use isospin formalism and solve the symmetrized AGS equations [34] for the transition operators $\mathcal{U}_{\beta\alpha}$. We consider only reactions with initial (final) two-cluster states, i.e., 1+3 and 2+2, that correspond

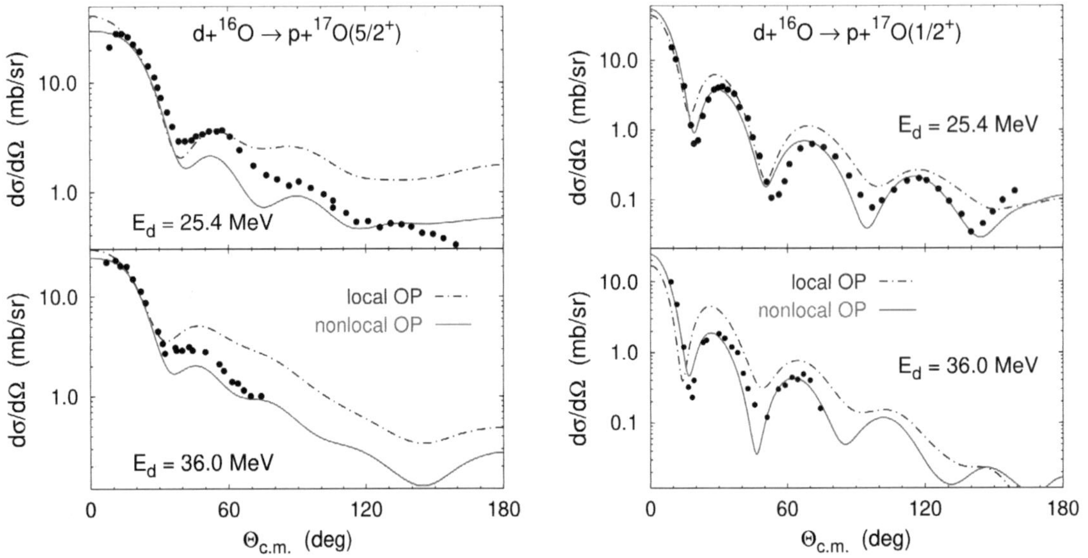

Figure 3. Differential cross section for $d + {}^{16}\mathrm{O} \to p + {}^{17}\mathrm{O}$ transfer to the ${}^{17}\mathrm{O}$ ground state $5/2^+$ (left side) and excited state $1/2^+$ (right side) at $E_d = 25.4$ and 36.0 MeV. Curves as in Fig. 2. The experimental data are from Ref. [29].

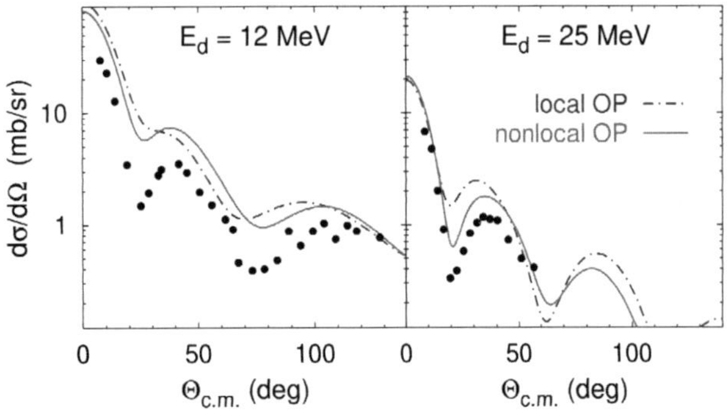

Figure 4. Differential cross section for $d + {}^{10}\mathrm{Be} \to p + {}^{11}\mathrm{Be}$ transfer to the ${}^{11}\mathrm{Be}$ ground state $1/2^+$ at $E_d = 12$ and 25 MeV as function of the c.m. scattering angle. Curves as in Fig. 2. The experimental data are from Refs. [30, 31].

to $\alpha(\beta) = 1$ and 2, respectively:

$$\begin{aligned}
\mathcal{U}_{11} &= -(G_0 T G_0)^{-1} P_{34} - P_{34} U_1 G_0 T G_0 \mathcal{U}_{11} + U_2 G_0 T G_0 \mathcal{U}_{21}, && (6)\\
\mathcal{U}_{21} &= (G_0 T G_0)^{-1} (1 - P_{34}) + (1 - P_{34}) U_1 G_0 T G_0 \mathcal{U}_{11}, && (7)\\
\mathcal{U}_{12} &= (G_0 T G_0)^{-1} - P_{34} U_1 G_0 T G_0 \mathcal{U}_{12} + U_2 G_0 T G_0 \mathcal{U}_{22}, && (8)\\
\mathcal{U}_{22} &= (1 - P_{34}) U_1 G_0 T G_0 \mathcal{U}_{12}. && (9)
\end{aligned}$$

Here G_0 is the free resolvent, P_{ij} is the permutation operator of particles i and j, and $T = v + vG_0T$ is the two-nucleon transition matrix, v being the potential. The subsystem $1+3$ and $2+2$ AGS transition operators are calculated as

$$\begin{aligned}
U_\alpha &= P_\alpha G_0^{-1} + P_\alpha T G_0 U_\alpha, && (10)\\
P_1 &= P_{12} P_{23} + P_{13} P_{23}, && (11)\\
P_2 &= P_{13} P_{24}. && (12)
\end{aligned}$$

The on-shell elements of the four-nucleon AGS transition operators $\langle \Psi_\beta | \mathcal{U}_{\beta\alpha} | \Psi_\alpha \rangle$ with initial (final) state Faddeev amplitudes

$$|\Psi_\alpha\rangle = G_0 T P_\alpha |\Psi_\alpha\rangle \tag{13}$$

are closely related to the scattering amplitudes [34].

In order to include the Coulomb interaction between protons we use the screening and renormalization approach. Long- and Coulomb-distorted short-range parts in the scattering amplitudes are separated [35, 36]. The former is of two-body nature and its $R \to \infty$ limit is known analytically. The Coulomb-distorted short-range part is calculated by solving symmetrized AGS equations numerically where the screened Coulomb potential is added to the nuclear proton-proton potential; the $R \to \infty$ limit is reached with sufficient accuracy at finite screening radii R as demonstrated in Ref. [35].

In Figs. 5 through 7 we show examples for elastic, charge-exchange, and transfer reactions in the four-nucleon system. The two-nucleon interactions we use are charge-dependent Bonn (CD Bonn) [37], AV18 [38], inside nonlocal outside Yukawa (INOY04) potential by Doleschall [39], and the one derived from chiral perturbation theory at next-to-next-to-next-to-leading order (N3LO) [40]. Although here we do not include a three-nucleon force, its presence is simulated by using the potential INOY04 that fits both ^{3}He and ^{3}H experimental binding energies (7.72 MeV and 8.48 MeV, respectively). The results for the two-baryon potential CD Bonn + Δ [41] allowing for a virtual excitation of a nucleon to a Δ-isobar and thereby yielding consistent effective three- and four-nucleon forces are qualitatively similar and can be found in Ref. [42]. Most of the experimental data are quite well described [35, 36, 43] at least by some of the used two-nucleon force models, but there exist also several discrepancies, e.g., for the neutron-^{3}H total cross section [44] and for the proton analyzing power in the proton-^{3}He elastic scattering [35, 45, 46] and in the $p + {}^3\text{H} \to n + {}^3\text{He}$ charge-exchange reaction [36].

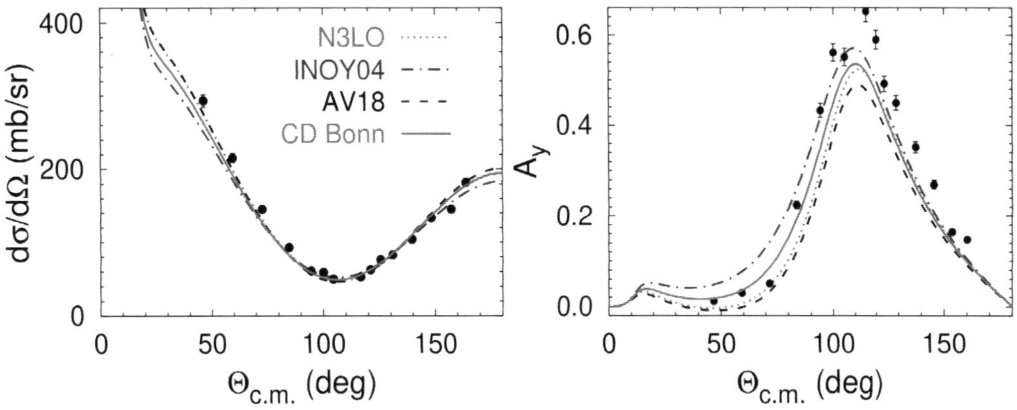

Figure 5. Differential cross section and proton analyzing power of elastic proton-^{3}H scattering at 4.15 MeV proton lab energy. Results obtained with realistic two-nucleon potentials CD Bonn (solid curves), AV18 (dashed curves), INOY04 (dashed-dotted curves), and N3LO (dotted curves) are compared with the experimental data from Ref. [47].

6. Summary

Exact Faddeev/AGS equations in the momentum-space framework were solved for the description of few-body nuclear reactions. The Coulomb interaction was included using the method of screening and renormalization. Fully converged results were obtained for three- and four-nucleon scattering and for three-body-like nuclear reactions.

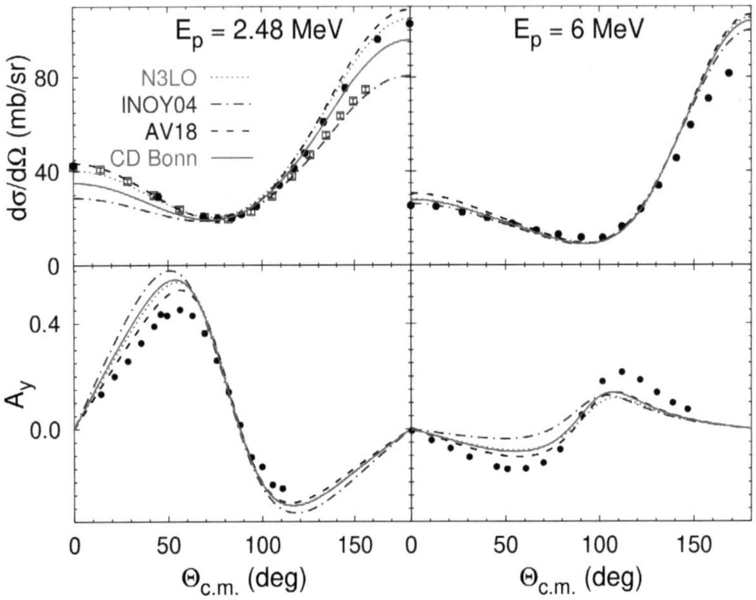

Figure 6. Differential cross section and proton analyzing power A_y of $p + {}^3\text{H} \rightarrow n + {}^3\text{He}$ charge-exchange reaction at 2.48 and 6 MeV proton lab energy. Curves as in Fig. 5. The cross section data are from Refs. [48] (circles) and [49] (squares) at 2.48 MeV, and from Ref. [50] at 6 MeV. A_y data are from Ref. [51] at 2.48 MeV and from Ref. [52] at 6 MeV.

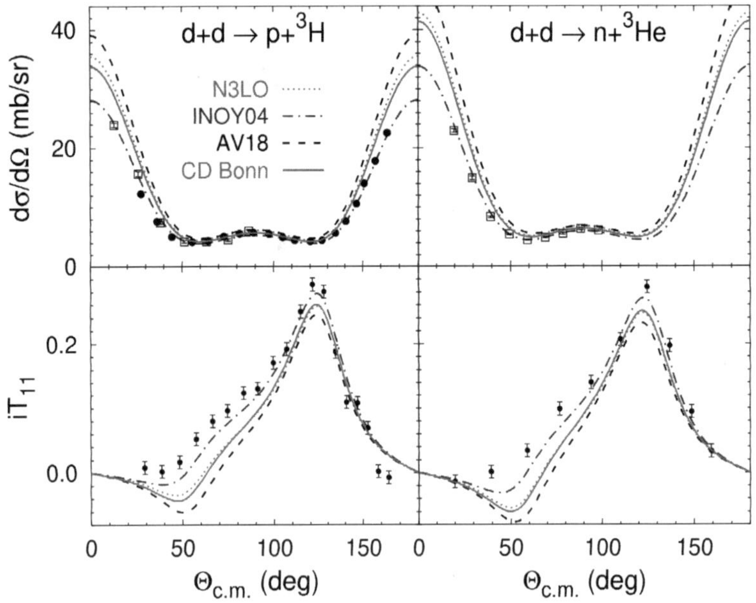

Figure 7. Differential cross section and deuteron analyzing power iT_{11} of $d + d \rightarrow p + {}^3\text{H}$ and $d + d \rightarrow n + {}^3\text{He}$ reactions at 3 MeV deuteron lab energy. Curves as in Fig. 5. The cross section data are from Refs [53] (squares) and [54] (circles). Analyzing power data are from Ref [54] for $d + d \rightarrow p + {}^3\text{H}$ and from Ref. [55] for $d + d \rightarrow n + {}^3\text{He}$.

References

[1] Austern N, Iseri Y, Kamimura M, Kawai M, Rawitscher G and Yahiro M 1987 *Phys. Rep.* **154** 125
[2] Deltuva A, Fonseca A C and Sauer P U 2005 *Phys. Rev. C* **71** 054005
[3] Deltuva A, Fonseca A C and Sauer P U 2005 *Phys. Rev. Lett.* **95** 092301
[4] Taylor J R 1974 *Nuovo Cimento B* **23** 313
[5] Alt E O, Sandhas W and Ziegelmann H 1978 *Phys. Rev. C* **17** 1981
[6] Alt E O and Sandhas W 1980 *Phys. Rev. C* **21** 1733
[7] Deltuva A, Fonseca A C and Sauer P U 2008 *Annu. Rev. Nucl. Part. Sci.* **58** 27
[8] Faddeev L D 1960 *Zh. Eksp. Teor. Fiz.* **39** 1459
[9] Alt E O, Grassberger P and Sandhas W 1967 *Nucl. Phys.* **B2** 167
[10] Chmielewski K, Deltuva A, Fonseca A C, Nemoto S and Sauer P U 2003 *Phys. Rev. C* **67** 014002
[11] Deltuva A, Chmielewski K and Sauer P U 2003 *Phys. Rev. C* **67** 034001
[12] Gorshkov V G 1961 *Sov. Phys.-JETP* **13** 1037
[13] Deltuva A, Fonseca A C and Sauer P U 2005 *Phys. Rev. C* **72** 054004
[14] Deltuva A, Fonseca A C and Sauer P U 2006 *Phys. Rev. C* **73** 057001
[15] Alt E O, Mukhamedzhanov A M, Nishonov M M and Sattarov A I 2002 *Phys. Rev. C* **65** 064613
[16] Koersner I, Glantz L, Johansson A, Sundqvist B, Nakamura H and Noya H 1977 *Nucl. Phys.* **A286** 431
[17] Deltuva A 2006 *Phys. Rev. C* **74** 064001
[18] Deltuva A, Fonseca A C, Kievsky A, Rosati S, Sauer P U and Viviani M 2005 *Phys. Rev. C* **71** 064003
[19] Kievsky A, Viviani M and Rosati S 2001 *Phys. Rev. C* **64** 024002
[20] Kistryn S *et al* 2006 *Phys. Lett. B* **641** 23
[21] Deltuva A 2009 *Nucl. Phys.* **A821** 72
[22] Deltuva A, Moro A M, Cravo E, Nunes F M and Fonseca A C 2007 *Phys. Rev. C* **76** 064602
[23] Deltuva A and Fonseca A C 2009 *Phys. Rev. C* **79** 014606
[24] Giannini M M and Ricco G 1976 *Ann. Phys. (NY)* **102** 458
[25] Giannini M M, Ricco G and Zucchiatti A 1980 *Ann. Phys. (NY)* **124** 208
[26] Deltuva A 2009 *Phys. Rev. C* **79** 021602
[27] Deltuva A 2009 *Phys. Rev. C* **79** 054603
[28] Matsuoka N *et al* 1986 *Nucl. Phys.* **A455** 413
[29] Cooper M D, Hornyak W F and Roos P G 1974 *Nucl. Phys.* **A218** 249
[30] Auton D L 1970 *Nucl. Phys.* **A157** 305
[31] Zwieglinski B, Benenson W, Robertson R and Coker W 1979 *Nucl. Phys.* **A315** 124
[32] Yakubovsky O A 1967 *Yad. Fiz.* **5** 1312
[33] Grassberger P and Sandhas W 1967 *Nucl. Phys.* **B2** 181
[34] Deltuva A and Fonseca A C 2007 *Phys. Rev. C* **75** 014005
[35] Deltuva A and Fonseca A C 2007 *Phys. Rev. Lett.* **98** 162502
[36] Deltuva A and Fonseca A C 2007 *Phys. Rev. C* **76** 021001
[37] Machleidt R 2001 *Phys. Rev. C* **63** 024001
[38] Wiringa R B, Stoks V G J and Schiavilla R 1995 *Phys. Rev. C* **51** 38
[39] Doleschall P 2004 *Phys. Rev. C* **69** 054001
[40] Entem D R and Machleidt R 2003 *Phys. Rev. C* **68** 041001
[41] Deltuva A, Machleidt R and Sauer P U 2003 *Phys. Rev. C* **68** 024005
[42] Deltuva A, Fonseca A C and Sauer P U 2008 *Phys. Lett. B* **660** 471
[43] Lazauskas R 2009 *Phys. Rev. C* **79** 054007
[44] Lazauskas R and Carbonell J 2004 *Phys. Rev. C* **70** 044002
[45] Viviani M, Kievsky A, Rosati S, George E A and Knutson L D 2001 *Phys. Rev. Lett.* **86** 3739
[46] Fisher B M, Brune C R, Karwowski H J, Leonard D S, Ludwig E J, Black T C, Viviani M, Kievsky A and Rosati S 2006 *Phys. Rev. C* **74** 034001
[47] Kankowsky R, Fritz J C, Kilian K, Neufert A and Fick D 1976 *Nucl. Phys.* **A263** 29
[48] Drosg M 1978 *Nucl. Sci. Eng.* **67** 190
[49] Jarvis G A 1956 *Los Alamos Scientific Lab Reports No.* **2014** 35
[50] Wilson W E, Walter R L and Fossan D B 1961 *Nucl. Phys.* **27** 421
[51] Doyle M A, Clark H W, Dries L J, Regner J L, Donoghue T R and Hale G M 1981 *Nucl. Phys.* **A371** 225
[52] Jarmer J J, Haight R C, Simmons J E, Martin J C and Donoghue T R 1974 *Phys. Rev. C* **9** 1292
[53] Blair J M, Freier G, Lampi E, Sleator W and Williams J H 1948 *Phys. Rev.* **74** 1599
[54] Grüebler W, König V, Schmelzbach P A, Risler R, White R E and Marmier P 1972 *Nucl. Phys.* **A193** 129
[55] Dries L J, Clark H W, Detomo R and Donoghue T R 1979 *Phys. Lett.* **80B** 176

Mixing of fusion-fission and quasifission products in reaction with massive nuclei

A K Nasirov[1,2], A I Muminov[2], G Giardina[3], G Mandaglio[3], M Manganaro[3]

[1]Joint Institute for Nuclear Research, Joliot-Curie 6, 141980, Dubna, Russia
[2]Institute of Nuclear Physics, Ulugbek, 100214, Tashkent, Uzbekistan
[3]INFN, Sezione di Catania, and Dipartimento di Fisica dell'Università di Messina, Salita Sperone 31, 98166 Messina, Italy

E-mail: nasirov@jinr.ru, akhtam@uzsci.net, giardina@nucleo.unime.it, gmandaglio@unime.it and manganarom@unime.it

Abstract. It is known the fusion cross section plays crucial role in the choice of the reaction to synthesize new superheavy elements. An estimation of the fusion cross section in the reactions with massive nuclei is difficult task when the mass (charge) and angular distributions of the quasifission and fusion-fission fragments strongly overlap. The measured yields of evaporation residues, fusion-fission and quasifission fragments in the ^{48}Ca+^{154}Sm reactions are analyzed in the framework of the method based on the dinuclear system concept and advanced statistical model. The experimental data of the fission fragments are decomposed into contributions coming from fusion-fission, quasifission, and fast fission. Our investigations showed the synthesis of the new element Z=120 (A=302) is more preferable in the ^{54}Cr+^{248}Cm reaction in comparison with the ^{58}Fe+^{244}Pu and ^{64}Ni+^{238}U reactions because the excitation function of the evaporation residues of the former reaction is some orders of magnitude larger than that for the last two reactions.

25.70.Jj, 25.70.-z

1. Introduction

The synthesis of the new superheavy elements is one of modern problems in nuclear physics. The correct estimation of the cross section of the compound nucleus formation in the reactions with massive nuclei is important to choose favorable reaction and optimal values of bombarding energy allowing us to observe an events corresponding to synthesis of the new expected element. The experimental methods used to estimate the fusion probability depend on the unambiguity in identifying of the complete fusion reaction products among the quasifission products. The difficulties arise when the mass (charge) and angular distributions of the quasifission and fusion-fission fragments strongly overlap depending on the reaction dynamics. As a result, the complete fusion cross sections may be overestimated. We know that quasifission fragments show anisotropic angular distributions [1, 2]. This is a way to separate them from the fusion-fission fragments which should have isotropic angular distributions. But fission fragments in reactions with heavy ions also show anisotropic angular distributions which is explained by the assumption that an equilibrium K-distribution is not reached (K is the projection of the total spin of the compound nucleus on its axial symmetry axis). At the same time the angular distribution of

Journal of Physics: Conference Series **205** (2010) 012018 doi:10.1088/1742-6596/205/1/012018

the quasifission fragments may be isotropic when the dinuclear system decays having a large angular momentum [3].

This paper is devoted to analyze reasons for the disappearance of the quasifission feature in the experimental data for the ^{48}Ca+^{154}Sm reactions studied in the paper [4] by Knyazheva *et al.* The same method of analysis is applied to study the problem of the synthesis of the new superheavy element Z=120. The three reactions ^{54}Cr+^{248}Cm, ^{58}Fe+^{244}Pu and ^{64}Ni+^{238}U are compared with the aim to answer the question which of these reactions is preferable to obtain Z=^{302}120.

2. Mixing the mass and angular distributions of the fusion-fission and quasifission products

The reaction channels with the full momentum transfer at low collision energies take place through the capture stage at which dinuclear system (DNS) is formed. The lifetime of DNS should be enough for its transformation into compound nucleus during its evolution. There is a hindrance at formation of the compound nucleus (CN) in reactions with massive nuclei: not all of DNS formed at capture of the projectile by the target-nucleus can be transformed into CN. DNS may break up into two parts after multinucleon transfer changing strongly its mass and charge asymmetry or with the masses and charges close ones of the projectile and target.The decay of the DNS into two fragments bypassing the stage of the CN formation is called a quasifission. The mass distribution of the quasifission products depends on the landscape of potential energy surface, on the initial mass and charge asymmetry, as well as on the beam energy and impact parameter of the collision. So due to quasifission of DNS the number of events corresponding to formation of CN decreases. Another process which decreases the probability of CN formation is the fast fission. It is the inevitable decay of the fast rotating mononucleus into two fragments without reaching the equilibrium compact shape of CN. Such a mononucleus is formed from DNS which survived against quasifission. At large values of the angular momentum $\ell > \ell_f$, where ℓ_f is a value of ℓ at which the fission barrier of the corresponding compound nucleus disappears, the mononucleus immediately decays into two fragments [8]. As distinct from fast fission, the quasifission can occur at all values of ℓ at which capture occurs. Both of processes can produce fragments which have overlap of their mass (or charge) and angular distributions with ones of fusion-fission reactions.

In Ref. [4] the authors established the fusion suppression and the presence of quasifission for the ^{48}Ca+^{154}Sm reaction at energies near and below the Coulomb barrier. They consider the mass and angular distribution of the fission-like products with masses in the range $55 < A < 145$. The authors did not analyze the products of the quasifission with masses outside of the range $55 < A < 145$ where the quasifission fragments yield presents (see Fig. 2). Authors accepted for quasifission products only yields corresponding to the "shoulders" peaked around the masses 65 and 140 in the mass distribution of fission fragments appearing as an "asymmetric fission mode" at $E^*_{\rm CN}$=49 and 57 MeV. Quasifission cross sections of this reaction have been extracted from the total fission-like events by the analysis of their mass and angular distributions. The anisotropy of the angular distribution of the products corresponding to the "asymmetric shoulders" points to the quasifission nature of this component. The contribution of the quasifission components to the total mass distribution of fission fragments increases, with respect to the one of the symmetric compound nucleus-fission, as the ^{48}Ca projectile energy decreases (see Fig. 3 in [4]). This phenomenon is seen in Fig. 1 where the theoretical results obtained in the framework of the DNS model [see Refs. [5, 6] are compared with the experimental results for the quasifission and fusion-fission excitation functions from Ref. [4]. The contribution of the fast fission channel increases by increasing the bombarding energy due to the increase in the angular momentum of the mononucleus.

We can discuss three reasons leading to the overestimation of the experimental quasifission

Journal of Physics: Conference Series **205** (2010) 012018 doi:10.1088/1742-6596/205/1/012018

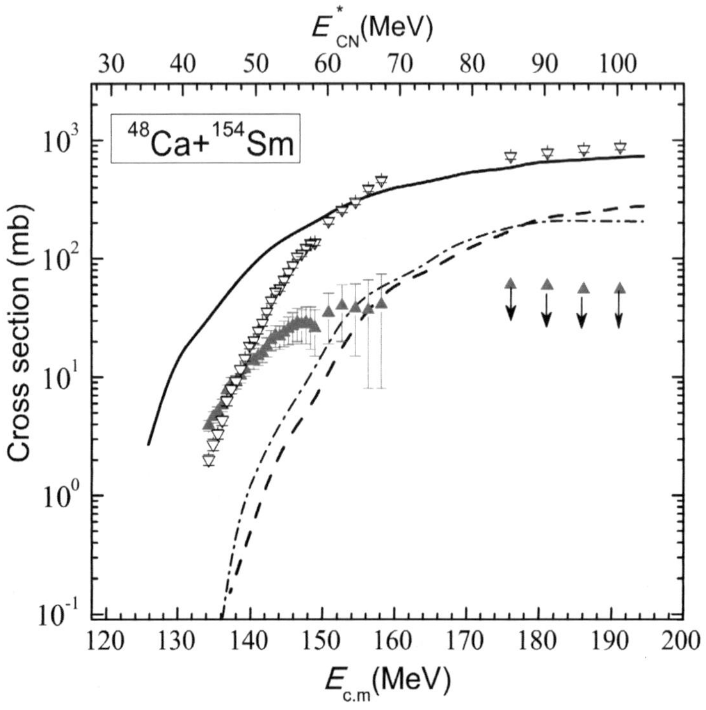

Figure 1. Comparison of the theoretical results obtained in the DNS model (curves) [5, 6] and experimental (symbols) [4] excitation functions for the quasifission (solid line and solid triangles), fusion-fission (dot-dashed line and inverted open triangles) and fast fission (dashed line) for the ^{48}Ca+^{154}Sm reaction. The theoretical values of the fusion-fission cross section calculated by the advanced statistical model [7]

(solid triangles) cross section by our calculation (solid line) in Fig. 1. The first reason is the excluding the reaction products having mass numbers outside the mass range $55 < A < 145$ from their analysis by the authors of Ref. [4]. Consequently they lost a part of the capture cross sections related to the contributions of the quasifission fragments with $A_{\rm qf} < 55$. Because our studies showed that capture events, *i.e.* events of the full momentum transfer, can lead to yields of fragments with masses $A_{\rm qf} < 55$. It is seen in the left panel of Fig. 2 where we present the evolution of mass distribution of the quasifission product yields 2 which were calculated by our method presented in Ref. [9]. The observed quasifission feature at low energies is connected with the peculiarities of the shell structure of the interacting nuclei.

The second reason is that at low energies the contribution of the fusion-fission to the yield of binary fragments is small in comparison with the quasifission contribution because the complete fusion is small in competition with quasifission. Third reason causing the smallness of the theoretical fusion-fission cross section is the large fission barrier ($B_{\rm f}$=12.33 MeV) for the ^{202}Pb nucleus according to the rotating finite range model by A. J. Sierk [10] and by the additional barrier $B_{\rm f}^{\rm (microscopic)} = -\delta W = -(\delta W_{\rm saddle-point} - \delta W_{\rm gs}) \cong 8.22$ MeV due to the nuclear shell structure.

At the large bombarding energies the contribution of fusion-fission (dash-double dotted line) and fast fission (dash-dotted line) components increases and become comparable with the components of the quasifission (short dashed line) in Fig. 1. The increase in the beam energy leads to a decrease of the shell effects and the yield of the quasifission fragments near the

Journal of Physics: Conference Series **205** (2010) 012018 doi:10.1088/1742-6596/205/1/012018

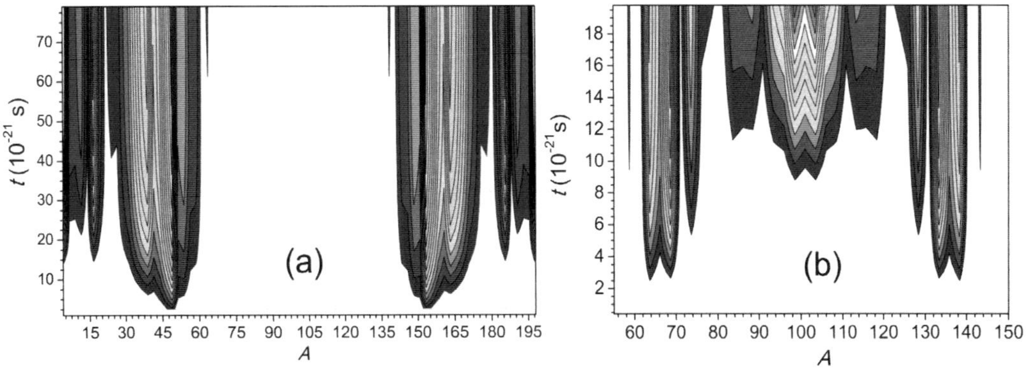

Figure 2. The mass distribution of the quasifission products yield in the ^{48}Ca+^{154}Sm reaction at $E_{\rm c.m.}$=140 MeV as a function of the lifetime of the dinuclear system formed at capture stage (a). The mass distribution of the quasifission product yields in the ^{48}Ca+^{154}Sm reaction at $E_{\rm c.m.}$=160 MeV as a function of the lifetime of the dinuclear system (b)

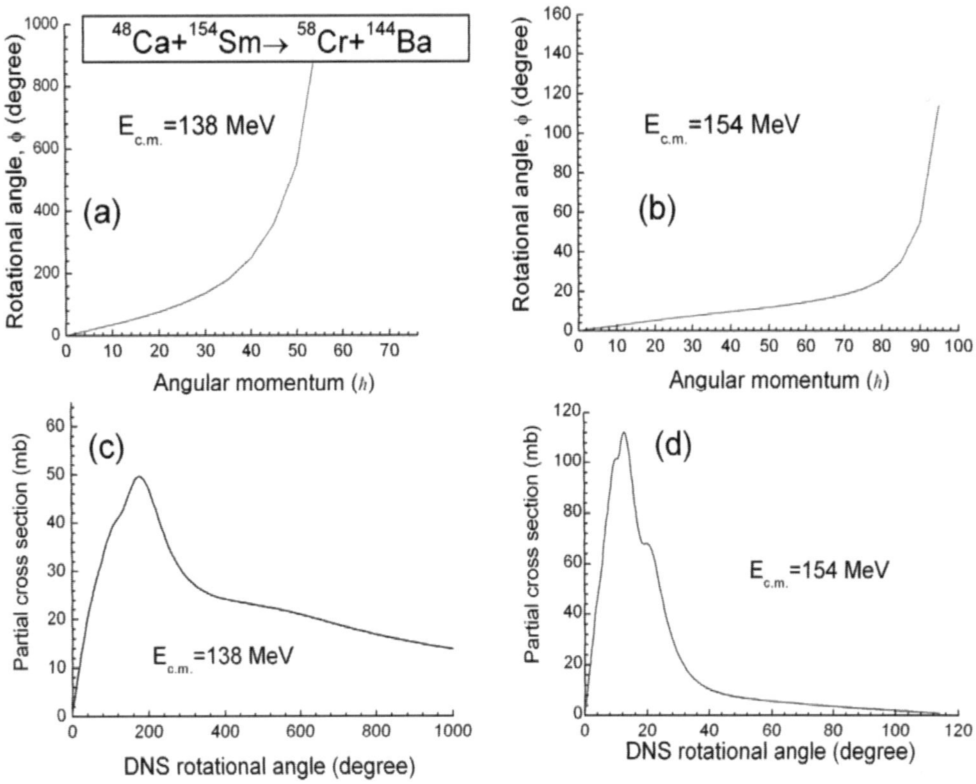

Figure 3. The rotational angle of the dinuclear system as a function of the orbital angular momentum (a) and (b), and angular distribution of the yield of quasifission fragments (c) and (d).

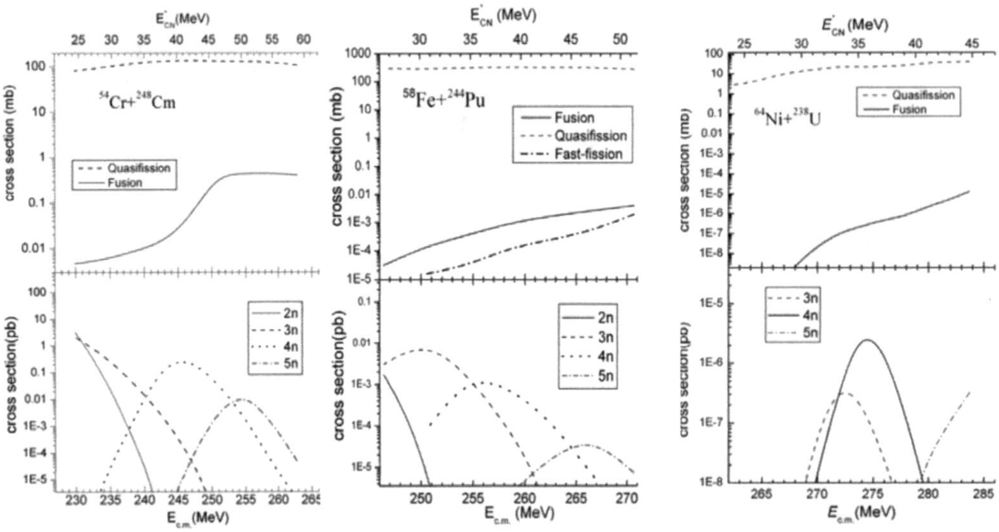

Figure 4. Calculated excitation functions for the capture, fusion and formation of the evaporation residues in the 2n, 3n, 4n, and 5n channels in the the ^{54}Cr+^{248}Cm (left panel), ^{58}Fe+^{244}Pu (middle panel), and ^{64}Ni+^{238}U (right panel) reactions.

asymmetric shoulders decreases. The main contribution to quasifission moves to the symmetric mass distribution (see the right panel of Fig. 2). The decay of the DNS formed with large angular momentum can lead to yield of products with isotropic angular distribution (see Fig. 3a). Therefore, these fragments are considered as fusion-fission fragments. This result was obtained by calculation of the lifetime of the rotating DNS as a function of its excitation function and depth of the potential well in the nucleus-nucleus potential [3]. This is one of mechanisms which are responsible for the disappearance of the "asymmetric shoulders" in the mass distribution of the fission fragments from the ^{48}Ca+^{154}Sm reactions at collision energies $E_{\rm c.m.} > 154$ ($E^*_{\rm CN} > 63$ MeV). The experimental data, which were identified as fusion-fission fragments by the authors of Ref. [4], increase strongly starting from the energies $E_{\rm c.m.} > 147$ ($E^*_{\rm CN} > 57$) MeV. According to our results, a large part of this increase belongs to the quasifission fragments (see Fig. 1). So we stress that, in the ^{48}Ca+^{154}Sm reaction, the quasifission (solid line in Fig.1) is the dominant channel in comparison with the fusion-fission (dot-dashed line) and fast fission (dashed line in Fig. 1) channels. The presented results in Fig. 1 are obtained by averaging over all orientation angles of the symmetry axis of ^{154}Sm which is a well deformed nucleus ($\beta_2 = 0.341$). The role of the target orientation angle relative to the beam direction during the formation of the fusion-fission and ER products in the ^{48}Ca+^{154}Sm reaction was analyzed in Ref. [11]. Indeed the fusion probability increases by increasing the beam energy due to the inclusion of the contributions from collisions with large orientation angles of the target-nucleus symmetry axis with respect to the beam direction. The favourableness of the large orientation angles for the formation of the compound nuclei was analyzed in Refs. [6, 11]. The fission probability increases because fission barrier $B_{\rm f}$ decreases at higher excitation energies and large values of angular momentum. We conclude that the experimental fusion-fission data obtained at the high energy collisions contain the contribution of quasifission fragments with masses $A > 83$ which show an isotropic distribution as presented in Ref. [4]. The products of quasifission should have less excitation energy than the ones of fusion-fission reaction because former products come from the decay of the fast rotating DNS. Therefore, the multiplicity of neutron emission accompanying quasifission products should be smaller than that accompanying fusion-fission products.

Calculation of angular distribution of the quasifission products as a function of the angular momentum and excitation energy of DNS shows that the angles corresponding to the peak of yields concentrate near forward and backward directions as shown in Figs. 3b and 3d. As a result sufficient part of them were not registered by detector placed at angles larger than 15°. This circumstance can be considered as the technical reason of disappearance of the yield of quasifission products at $E_{\rm c.m.} > 153$ MeV.

The experimental results confirming the large contribution of quasifission products into mass symmetric region appeared recently in Ref. [12, 13]. At the large energy $E_{\rm c.m.}$=154 MeV ($E^*_{\rm CN}$=63 MeV) the experimental values of the quasifission cross section are much lower than that of the fusion-fission cross section.

So, we have explained the large difference between the calculated and experimental capture cross sections at low collision energies and the decrease in this difference at high collision energies. The experimental data of fission-like fragments seem to include some part of the quasifission and fast fission fragments which overlaps with the mass and angular distributions of the fusion-fission fragments.

In the framework of our model [5, 6] we estimate the most preferable reaction for the synthesis of the superheavy element $Z = 120$. Among the three studied reactions, ^{54}Cr+^{248}Cm, ^{58}Fe+^{244}Pu, and ^{64}Ni+^{238}U, the first one is most preferable for the synthesis of the element Z=120. Because a more asymmetric reaction it has a smaller intrinsic fusion barrier and a larger quasifission barrier that lead to a larger fusion cross section (see [14]). The expected cross section for the synthesis of superheavy element Z=120 in the ^{54}Cr+^{248}Cm reaction is more than 1 pb for the 2n and 3n evaporation channels at $E_{\rm c.m.}$=233–245 MeV.

References

[1] Tõke J, Bock R, Dai G X, Gobbi A, Gralla S, Hildenbrand K D, Kuzminski J, Müller W F J, Olmi A and Stelzer H 1985 *Nucl. Phys.* A **440** 327
[2] Shen W Q *et al.* 1987 *Phys. Rev.* C **36** 115
[3] Nasirov A K, Muminov A I, Utamuratov R K, Fazio G, Giardina G, Hanappe F, Mandaglio G, Manganaro M, and Scheid W, 2007 *Eur. Phys. Jour.* A **34** 325
[4] Knyazheva G N *et al.* 2007 *Phys. Rev.* C **75**, 064602
[5] Fazio G, Giardina G, Lamberto A, Ruggeri R, Saccà C, Palamara R, Muminov A I, Nasirov A K, Yakhshiev U T, Hanappe F 2004 *Eur. Phys. Jour.* A **19** 89
[6] Nasirov A, Fukushima A, Toyoshima Y, Aritomo Y, Muminov A, Kalandarov Sh, Utamuratov R, 2005 *Nucl. Phys.* A **759** 342
[7] D'Arrigo A, Giardina G, Herman M, Ignatyuk A V, Taccone A 1994 *J. Phys.* G **20** 365
[8] Gregoire C, Ngo C, Tomasi, Remaud B, Scheuter F 1982 *Nucl.Phys.* A **387** 37.
[9] Fazio G, Giardina G, Mandaglio G, Hanappe F, Muminov A I, Nasirov A K, Scheid W, and Stuttgé: L 2005 *Mod. Phys. Lett.* A **20** 391
[10] Sierk A J 1988 *Phys. Rev.* C **33** 2039
[11] Fazio G, Giardina G, Hanappe F, Mandaglio G, Manganaro M, Muminov A I, Nasirov A K, Saccá C 2008 *J. Phys. Soc. Japan* **77** 124201
[12] Hinde D J *et al.* 2008 *Phys. Rev. Lett.* **101** 092701
[13] Knyazheva G N *et al.* 2008 *PEPAN Letters* **5** 40 (In Russian)
[14] Nasirov A K, Giardina G, Mandaglio G, Manganaro M, Hanappe F, Heinz S, Hofmann S, Muminov A I, Scheid W 2009 *Phys. Rev.* C **79** 024606

Journal of Physics: Conference Series **205** (2010) 012019 doi:10.1088/1742-6596/205/1/012019

Studies of the Nuclear Landscape and the Nuclear Equation of State using Peripheral Collisions near the Fermi Energy

George A. Souliotis

Laboratory of Physical Chemistry, Department of Chemistry,
National and Kapodistrian University of Athens, Athens, Greece, and
Cyclotron Institute, Texas A&M University, College Station, Texas, USA

E-mail: `soulioti@comp.tamu.edu`

Abstract.
Studies of peripheral (quasi-elastic and deep-inelastic) collisions near the Fermi energy are presented with a two-fold goal: first, to produce very neutron-rich nuclei towards the neutron drip line and, second, to obtain insight into the underlying dynamics and the nuclear equation of state (EOS). Towards the first goal, our studies indicated an enhanced production of neutron-rich nuclides suggesting that peripheral reactions at Fermi energies offer a novel way to access very neutron-rich rare isotopes in unexplored regions of the nuclear chart. Towards the second goal, we have made comparisons of our experimental data to detailed calculations using the semiclassical microscopic model CoMD (Constrained Molecular Dynamics). The CoMD code implements an effective interaction with a nuclear-matter compressibility of K=200 (soft EOS) or K=380 (stiff EOS) with several forms of the density dependence of the nucleon-nucleon symmetry potential and imposes a phase space constraint to restore the Pauli principle during the collision. Present results from these comparisons are consistent with a soft equation of state (K=200) with a rather stiff density dependence of the symmetry potential (symmetry energy).

1. Introduction

The exploration of the nuclear landscape, as well as the elucidation of the nuclear equation of state (EOS) and, especially, its isospin part (symmetry energy) have recently received an exceptionally strong attention by the nuclear physics community. The EOS, being intimately connected to the underlying nucleon-nucleon effective interaction, plays a pivotal role in a broad range of phenomena in both nuclear physics, e.g. the structure and dynamics of exotic/drip-line nuclei and in astrophysics, e.g. the structure and evolution of neutron stars and the dynamics of supernova explosions (e.g. [1, 2, 3] and references therein).

2. Neutron-rich rare-isotope production at Fermi energies

We have undertaken a systematic study of projectile-like fragment distributions from peripheral (quasi-elastic and deep-inelastic) collisions near the Fermi energy. The motivation of our studies was the understanding and optimization of the production of very neutron-rich rare isotopes in these collisions. The experimental procedures are described in detail in our recent articles [4, 5, 6, 7, 8].

The use of the high-resolution recoil separator MARS in combination with standard Bρ–ΔE–E (magnetic rigidity, energy-loss, residual-energy) and time-of-flight techniques provided high-resolution information on the atomic number Z, the ionic charge q, the mass number A and the velocity distributions of the projectile fragments. Summation over q provided the yield distributions with respect to Z, A and velocity from which production cross sections were extracted. The measurements were performed inside the grazing angles of the corresponding reactions at 25 MeV/nucleon and in a wide Bρ window that enabled efficient collection of heavy projectile residues produced in a broad range of energy damping, from quasielastic to deep-inelastic collisions.

As representative results, the mass distributions (cross sections) of germanium (Ge) isotopes from the reactions of 25 MeV/nucleon ^{86}Kr with ^{64}Ni, and ^{124}Sn are shown in Fig. 1 and compared with calculations. The simulations involve the deep-inelastic transfer (DIT) code of Tassan-Got [9] for the dynamic stage followed by the de-excitation code GEMINI [10, 11]. The calculations are shown by the thick solid lines. The dotted lines are predictions of the EPAX parameterization [12] of fragmentation cross sections. In high energy fragmentation (see, e.g., [13]) nucleon-pickup products are highly suppressed, compared to lower-energy peripheral collisions. As seen in Fig. 1, neutron-rich nuclides are produced with substantial yields. For near-projectile elements (e.g. Ge in Fig. 1), a large enhancement is observed that can not be described by the DIT/GEMINI simulations. We found that this enhancement can be described by an extension of DIT in which realistic density distributions of the nuclei are employed. Thus, we infer that the neutron skin of the neutron-rich targets plays an important role [5, 7].

The present data show the production of nuclides with up to 4–5 neutrons picked-up from the target, along with usual proton removal products [5, 6, 7] demonstrating the important role of the target in the production mechanism. This point can be clearly seen in Fig. 2 which provides an overall representation of the measured production cross sections of projectile fragments from the reaction ^{86}Kr (25 MeV/nucleon) + ^{64}Ni on the Z–N plane. In this figure, the cross section ranges are shown by open circles according to the key. The closed squares show the stable isotopes. The solid line indicates the astrophysical r-process path, whereas the dashed line denotes the location of the neutron drip-line according to [14]. In closing, we conclude that peripheral reactions around or below the Fermi energy offer a novel way to access very neutron-rich rare isotopes towards the r-process path and the neutron drip-line.

3. Studies related to the nuclear effective interaction and the nuclear EOS

In parallel to the rare isotope production studies, we employed the experimental results on residue distributions from peripheral reactions in efforts to extract information on the properties of the nuclear effective interaction as manifested in the mechanism of nucleon exchange taking place in these reactions and the course towards N/Z equilibration [15]. We focussed our interest on the possibility of extracting information on the isospin part of the nuclear equation of state by comparing our experimental data on heavy residues to detailed calculations using microscopic models of heavy-ion collisions at these energies [16]. After some initial efforts with one-body transport-type codes (BUU, BNV) [17], we turned our attention to the N-body quantum molecular dynamics approach (QMD) [18]. We have performed detailed calculations using the recent version of the Constrained Molecular Dynamics (CoMD) code of A. Bonasera and M. Papa [19, 20].

3.1. Brief description of the CoMD model

The CoMD code is especially designed for reactions near and below the Fermi energy. Following the general philosophy of the QMD approach [18], in the CoMD model, nucleons are described as localized Gaussian wave packets. The wave function of the N-body nuclear system is assumed

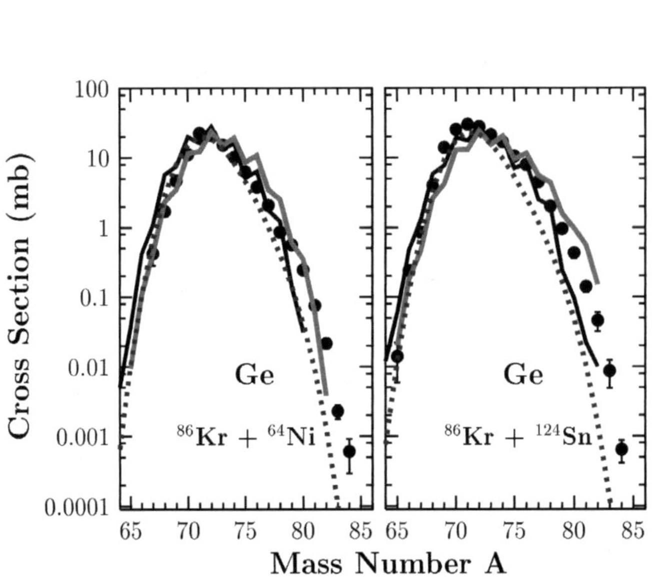

Figure 1. (Color online) Comparison of the experimental mass distributions of Ge (Z=32) from ^{86}Kr (25 MeV/nucleon) on ^{64}Ni [4] and ^{124}Sn [5] with DIT/GEMINI calculations without and with (lower and upper solid lines, respectively, on the neutron-rich side) realistic neutron and proton density distributions. Dotted line: EPAX prediction [12].

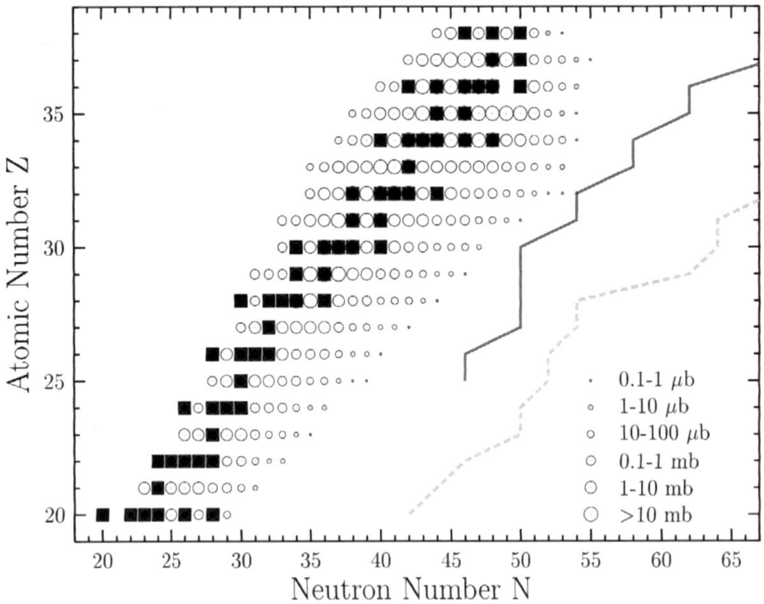

Figure 2. (Color online) Representation of the production cross sections of projectile fragments from the reaction ^{86}Kr (25 MeV/nucleon) + ^{64}Ni on the Z–N plane [4]. The cross section ranges are shown by open circles according to the key. The closed squares show the stable isotopes. The solid line indicates the astrophysical r-process path, whereas the dashed line denotes the neutron drip-line according to [14].

to be the product of these single-particle wave functions. With the Gaussian description, the N-body time-dependent Schöndinger equation leads to (classical) Hamilton's equations of motion

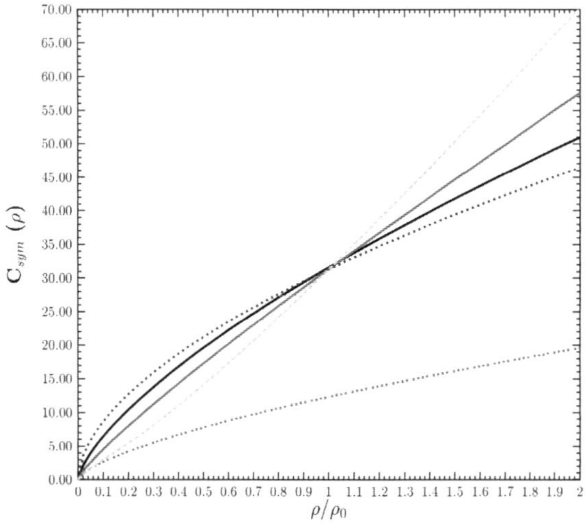

Figure 3. (Color Online) Density dependence of the nuclear symmetry energy $C_{sym}(\rho)$ corresponding to the forms of the nucleon-nucleon symmetry potential in the CoMD code. Blue (dotted): asy-soft, red (solid): asy-stiff, green (dashed): super asy-stiff and grey (lower dotted) line: no symmetry potential. The black (solid) line represents the form $31.6(\rho/\rho_0)^{0.69}$ consistent with the isoscaling analysis of fragments from central heavy-ion collisions [2].

for the centroids of the nucleon wavepackets. The potential part of the Hamiltonian consists of a Skyrme-like effective interaction. The isoscalar part of this effective interaction corresponds to a nuclear-matter compressibility of K=200 (soft EOS) or K=380 (stiff EOS). For the isovector part, several forms of the density dependence of the nucleon-nucleon symmetry potential are implemented. The corresponding symmetry energy functionals are shown in Fig. 3 and referred to as: "asy-soft": blue (dotted) line; "asy-stiff": red (solid) line; "super asy-stiff": green (dashed) line; and, "no symmetry potential": grey (dotted) line. The first three forms correspond to a dependence of the symmetry potential V_{sym} on the 1/2, 1 and 2 power of the density, respectively, whereas in the last case, the nucleon-nucleon symmetry potential is set to zero – thus, only the kinetic part of the symmetry energy plays a role in this case.

In the CoMD model, while not explicitly implementing antisymmetrization of the N-body wavefunction, a constraint in the phase space occupation for each nucleon is imposed, effectively restoring the Pauli principle at each time step of the (classical) evolution of the system. This constraint restores in an approximate way the fermionic nature of the nucleon motion in the interacting nuclei [19]. The short range (repulsive) nucleon-nucleon interactions are described as individual nucleon-nucleon collisions governed by the nucleon-nucleon scattering cross section, the available phase space and the Pauli principle, as usually implemented in transport codes [17]. The latest CoMD version (CoMD-II) fully preserves the total angular momentum (along with linear momentum and energy) [20], features which are critical for the accurate description of observables from heavy-ion collisions.

3.2. Comparisons of CoMD calculations to the experimental data

Results of the CoMD calculations (with compressibility K=200) and comparisons with our experimental data on heavy-residue distributions are shown in Figs. 4–7. Fig. 4 shows the calculated average quasiprojectile angle (upper panel) and excitation energy per nucleon

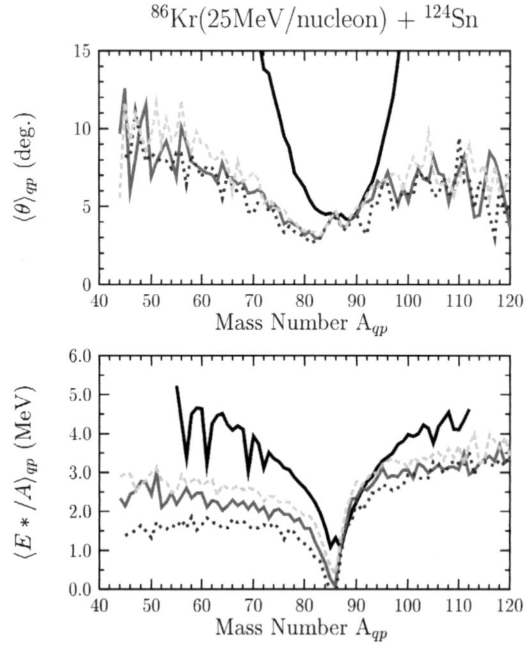

Figure 4. (Color Online) Mean quasiprojectile angle (upper panel) and excitation energy per nucleon (lower panel) as a function of quasiprojectile mass for the reaction ^{86}Kr(25MeV/nucleon) + ^{124}Sn. The black (solid) line represents DIT calculations. CoMD calculations are given by the following lines: blue (dotted): asy-soft, red (solid): asy-stiff, and green (dashed): super asy-stiff (see text).

(lower panel) as a function of the mass of the (primary) quasiprojectiles. The black (solid) line corresponds to the prediction of the deep-inelastic transfer (DIT) code of Tassan-Got that has been extensively used in our studies of quasiprojectile formation near the Fermi energy [9]. The remaining three lines are the results of CoMD calculations with symmetry potential options: "asy-soft": blue (dotted) line; "asy-stiff": red (solid) line; "super asy-stiff": green (dashed) line (see, also Fig. 3). The CoMD calculation was stopped at t=300 fm/c. We observe differences between the predictions of DIT and CoMD that we will try to further investigate and understand in the near future. Regarding CoMD, despite the observed fluctuations of the mean values, we may tentatively conclude that the mean quasiprojectile angle is not sensitive to the choice of the symmetry potential. However, the mean excitation energy shows some sensitivity in the choice that deserves further exploration.

In Fig. 5, the distributions of the mean angle, mean velocity and yield as a function of the mass of the (final) observable fragments are shown. The deexcitation of the primary fragments was done with the sequential decay code GEMINI [10]. The meaning of the curves is as before: black (solid) line: DIT, colored lines: CoMD. The top panel shows, along with the calculations, the angular acceptance of the MARS separator $\Delta\theta$=3°–6° for our measurements (dashed horizontal lines). In the middle and lower panels, the MARS data [5] are shown with solid symbols. The calculations in both cases are filtered with the angular acceptance of the separator. Additionally, in the lower panel, the thin lines show the calculations of the total residue yields. We observe an overall satisfactory agreement of the CoMD calculations with the data and again, in the CoMD calculations, an insensitivity to the choice of the symmetry

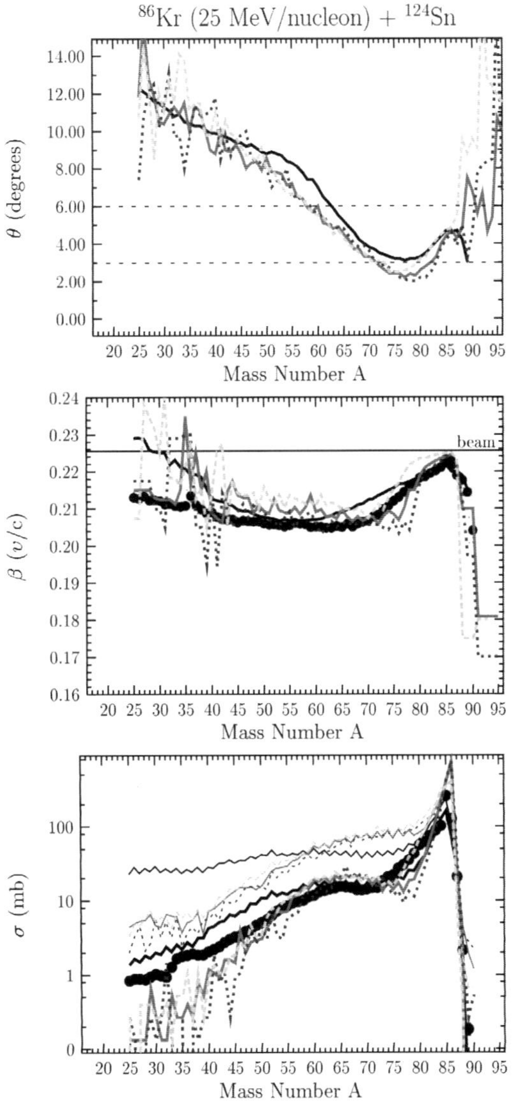

Figure 5. (Color Online) Mean angle (upper panel), mean velocity (middle panel) and yield (lower panel) as a function of projectile residue mass for the reaction ^{86}Kr (25MeV/nucleon) + ^{124}Sn. DIT calculations are given by the black (solid) line. CoMD calculations are given by the other three lines as in Fig. 4 (see text). Black points: MARS data [5].

potential. The situation is similar for the comparison of the mean Z/A (atomic number over mass number, or proton fraction) values of the observed residues with the CoMD calculations shown in Fig. 6.

In Fig. 7, upper panel, we show the calculated mean $(Z/A)^2$ of the primary quasiprojectiles as a function of the excitation energy per nucleon. The meaning of the curves is as in Fig. 4. The upper set of curves is for the ^{86}Kr (25 MeV/nucleon) + ^{112}Sn reaction and the lower set is for the ^{86}Kr (25MeV/nucleon) + ^{124}Sn reaction. The solid horizontal line corresponds to the $(Z/A)^2$ of the projectile, whereas the upper and lower dashed lines give the $(Z/A)^2$ of

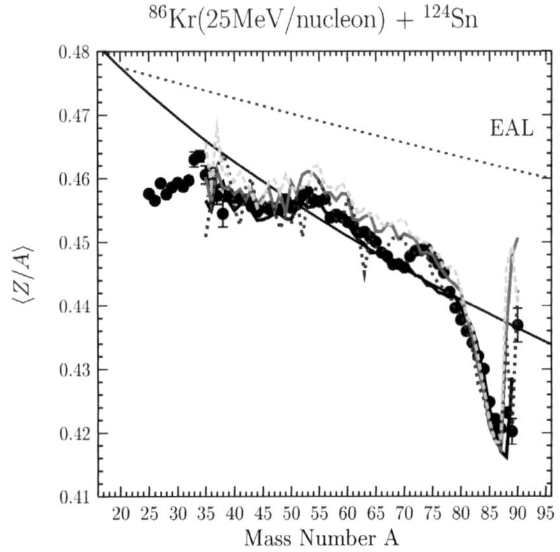

Figure 6. (Color Online) Mean projectile residue Z/A as a function of residue mass for the reaction ^{86}Kr(25MeV/nucleon)+^{124}Sn. DIT calculations are given by the black (solid) line that traces the data. CoMD calculations are given by the other three lines as in Fig. 4 (see text). Black points: MARS data [5]. Black near-diagonal (solid) line: stability line. EAL: evaporation attractor line [11].

the fully equilibrated systems in the two cases. In the lower panel of the figure we show the difference of the calculated mean $(Z/A)^2$ values, along with our data (solid and open points) from the heavy-residue isoscaling analysis of [15]. It is interesting to note that the CoMD calculations show sensitivity in the choice of the symmetry potential. However, this observation may be subject to the inherent uncertainty in the determination of the excitation energy of the quasiprojectiles. In the present calculations, the excitation energy has been determined from the difference of the binding energy of the (hot) quasiprojectiles as given by the CoMD code at t=300 fm/c and the corresponding binding energies of cold nuclei taken from mass tables. We have investigated the issues of the excitation energy determination and we believe that, except from very peripheral collisions (essentially corresponding to direct reactions) the CoMD code provides a reliable estimate of the excitation energies of the quasiprojectiles. The comparison presented in Fig. 7 suggests a rather stiff dependence of the symmetry energy on density (Fig. 3) in overall agreement with other studies, as presented in Ref. [2] (and references therein).

3.3. Calculation of ground state properties with CoMD

As part of our detailed studies with the CoMD model framework, we report in Fig. 8 the predicted neutron skin of the ^{86}Kr nucleus using the four options of the symmetry potential. The values of the skin show a small sensitivity to the density dependence of the symmetry potentials and are in agreement with expectations from microscopic SHF or Thomas-Fermi calculations. In the same vein, Fig. 9 presents the giant dipole resonance (GDR) spectrum of the ^{86}Kr nucleus obtained from the Fourier transform of the spatial oscillation of the neutron vs proton spheres within the CoMD model. The symmetry potentials employed seem to give reasonable values for the GDR energy centroids (although somewhat lower that the value 16.8 MeV expected from empirical systematics [21, 22]) and widths ~4 MeV in very good agreement

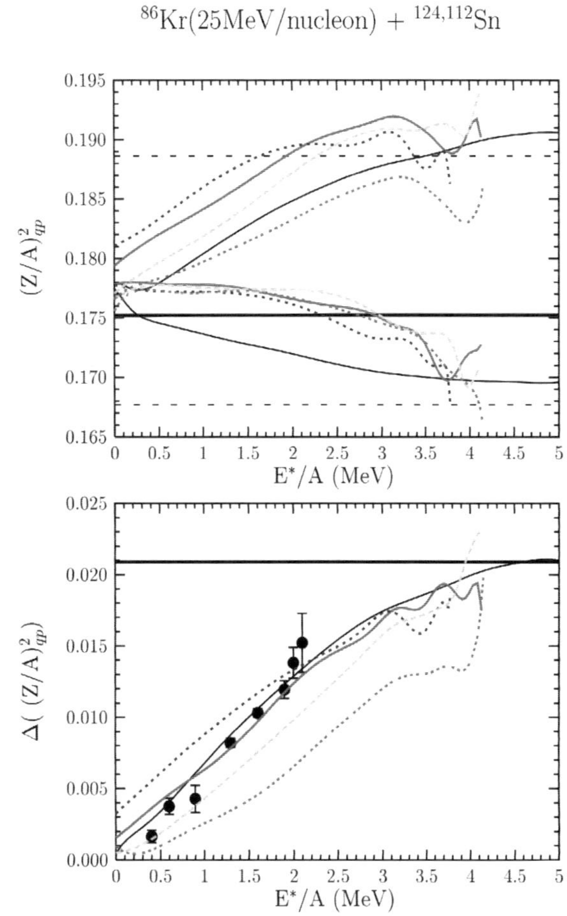

Figure 7. (Color Online) Upper panel: Mean $(Z/A)^2$ of quasiprojectiles as a function of excitation energy per nucleon for the 25 MeV/nucleon reactions: ^{86}Kr+ ^{112}Sn (upper set of curves) and ^{86}Kr+ ^{124}Sn (lower set of curves). Lower panel: Difference in quasiprojectile mean $(Z/A)^2$. Black (solid) lines represent DIT calculations. CoMD calculations are given by the following lines (as in the previous figures): blue (dotted): asy-soft, red (solid): asy-stiff, green (dashed): super asy-stiff and grey (dotted): no symmetry potential (see text). Black points: residue isoscaling data from [15].

with expectations for near ground-state nuclei [21].

4. Plans for future work

In regards to the production of rare isotopes and the studies of the nuclear EOS, in the near future we plan to extend our studies utilizing a broad set of recent experimental data with the MARS separator concerning the reactions of 15 MeV/nucleon ^{40}Ar and ^{86}Kr projectiles on ^{64,58}Ni and ^{124,112}Sn targets. The analysis of these data is approaching completion. Detailed calculations with the CoMD code will follow along the lines presented in this paper.

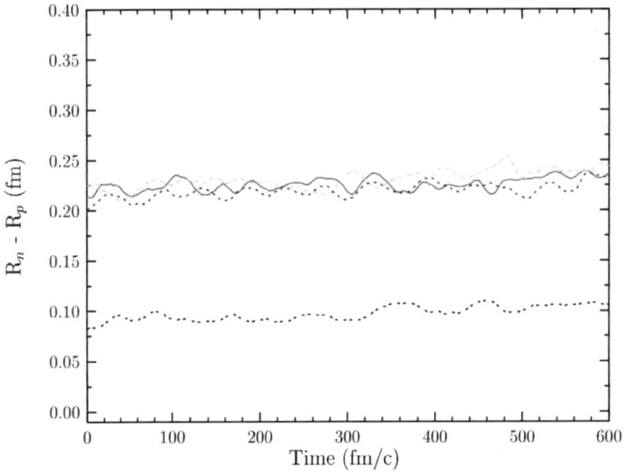

Figure 8. (Color Online) Time evolution of the neutron skin of an isolated ^{86}Kr nucleus predicted by CoMD. The various lines correspond to the choices of the nucleon symmetry potential as follows: blue (dotted): asy-soft, red (solid): asy-stiff, green (dashed): super asy-stiff and gray (lower dotted): no symmetry potential (see text).

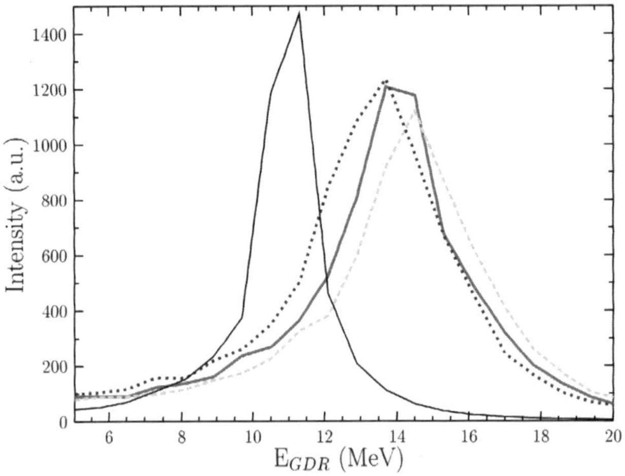

Figure 9. (Color Online) Giant dipole resonance (GDR) response of an isolated ^{86}Kr nucleus predicted by CoMD. The various lines correspond to the choices of the nucleon symmetry potential as follows: blue (dotted): asy-soft, red (solid): asy-stiff, green (dashed): super asy-stiff, and black (left solid): no symmetry potential. The expected value of the energy according to the empirical GRD systematics is 16.8 MeV [21, 22].

5. Summary and conclusions

Systematic studies of heavy residue distributions from quasi-elastic and deep-inelastic collisions near the Fermi energy were presented with a two-fold goal: first, to produce very neutron-rich nuclei towards the neutron drip line and, second, to obtain insight into the underlying dynamics and the nuclear equation of state (EOS). Towards the first goal, our studies indicated an enhanced production of neutron-rich nuclides in comparison with the traditional high-energy fragmentation mechanism which was attributed to the role of the N/Z and the nuclear periphery of the target. From a practical viewpoint, these reactions at Fermi energies offer a novel way to access very neutron-rich rare isotopes towards the r-process path and the neutron drip-line. Towards the second goal, we have made detailed comparisons of our experimental data to calculations with the semiclassical microscopic model CoMD (Constrained Molecular Dynamics). We believe that the CoMD model, being an N-body approach, offers an appropriate model framework to provide a unified description of a large body experimental data as the present heavy-residue data from peripheral collisions in the Fermi energy regime. The results from the present comparisons point to a soft equation of state (K=200) with a rather stiff density dependence of the symmetry energy part of the nucleon-nucleon effective interaction, in overall agreement with other heavy-ion studies (e.g. [2] and references therein).

6. Acknowledgement

The author is thankful to A. Bonasera and M. Papa for providing the CoMD code and for numerous fruitful discussions and support of the present efforts. The author also wishes to thank M. Veselsky, A. Keksis, B.C. Stein, S. Galanopoulos, M. Jandel, D.V. Shetty and S.J. Yennello for the help and support on the experimental work related to heavy-residue measurements at 15–25 MeV/nucleon at Texas A&M. The present work was supported in part by the Department of Energy through grant No. DE-FG03-93ER40773.

7. References

[1] Li B A, Chen L W and Ko C M 2008 *Phys. Rep.* **464** 113
[2] Shetty D V, Yennello S J and Souliotis G A 2007 *Phys. Rev.* C **76** 024606
[3] Baran V et al. 2005 *Phys. Rep.* **410** 335
[4] Souliotis G A et al. 2002 *Phys. Lett.* B **543** 163
[5] Souliotis G A et al. 2003 *Phys. Rev. Lett.* **91** 022701
[6] Souliotis G A et al. 2003 *Nucl. Instrum. Methods* B **204** 166
[7] Veselsky M and Souliotis G A 2006 *Nucl.Phys.* A **765** 252
[8] Souliotis G A et al. 2008 *Nucl. Instum. Methods* B **266** 4692
[9] Tassan-Got L and Stephan C 1991 *Nucl. Phys.* A **524** 121
[10] Charity R et al. 1988 *Nucl. Phys.* A **483** 391
[11] Charity R et al. 1998 *Phys. Rev.* C **58** 1073
[12] Summerer K and Blank B 2000 *Phys. Rev.* C **61** 034607
[13] Weber M et al. 1994 *Nucl. Phys.* A **578** 659
[14] Moller P, Nix J R and Kratz K L 1997 *At. Data Nucl. Data Tables* **66** 131
[15] Souliotis G A et al. 2004 *Phys. Lett.* B **588** 35
[16] Ono A and Randrup J 2006 *Eur. Phys. J.* A **30** 109
[17] Bonasera A, Gulminelli F, Molitoris J 1994 *Phys. Rep.* **243** 1
[18] Aichelin J 1991 *Phys. Rep.* **202** 233
[19] Papa M et al. 2001 *Phys. Rev.* C **64** 024612
[20] Papa M et al. 2005 *J. Comp. Phys.* **208** 403
[21] Santonocito D and Blumenfeld Y 2006 *Eur. Phys. J.* A **30** 183
[22] Trippa L, Colo G and Vigezzi E 2008 *Phys. Rev.* C **77** 061304

Journal of Physics: Conference Series **205** (2010) 012020 doi:10.1088/1742-6596/205/1/012020

A long sought result: Closed analytical solutions of the Bohr Hamiltonian with the Morse potential

Dennis Bonatsos[1,a], **I Boztosun**[2,b], **I Inci**[2,c]

[1]Institute of Nuclear Physics, N.C.S.R. "Demokritos", GR-15310 Aghia Paraskevi, Attiki, Greece

[2]Department of Physics, Erciyes University, Kayseri, Turkey

E-mail: [a]bonat@inp.demokritos.gr, [b]boztosun@erciyes.edu.tr, [c]inci@erciyes.edu.tr

Abstract. The determination of closed analytical solutions of the Morse potential for nonzero angular momenta has been an open problem for decades, solved recently by the Asymptotic Iteration Method (AIM) for solving differential equations. Closed analytical expressions have been obtained for the energy eigenvalues of the Bohr Hamiltonian in the γ-unstable case, as well as in an exactly separable rotational case with $\gamma \approx 0$, called the exactly separable Morse (ES-M) solution. All medium mass and heavy nuclei with known β_1 and γ_1 bandheads have been fitted by using the two-parameter γ-unstable solution for transitional nuclei and the three-parameter ES-M for rotational ones. It is shown that bandheads and energy spacings within the bands are well reproduced for more than 50 nuclei in each case. Comparisons to the fits provided by the Davidson potential, also soluble by the AIM, are made.

1. Introduction

The recent introduction of the critical point symmetries E(5) [1] and X(5) [2], which describe shape phase transitions between vibrational and γ-unstable/prolate deformed rotational nuclei respectively, has stirred much interest in special solutions of the Bohr Hamiltonian [3], describing collective nuclear properties in terms of the collective variables β and γ. Such solutions can describe nuclei in the whole region between different limiting symmetries, while critical point symmetries are appropriate for describing nuclei only at or near the critical point, in good agreement with experiment [4].

It has been known for a long time [5] that simple special solutions of the Bohr Hamiltonian, resulting from exact separation of variables in the relevant Schrödinger equation, can be obtained in the γ-unstable case, in which the potential depends only on β, as well as in the case in which the potential can be written in the separable form

$$u(\beta,\gamma) = u(\beta) + \frac{u(\gamma)}{\beta^2}, \tag{1}$$

in the special cases of $\gamma \approx 0$ or $\gamma \approx \pi/6$ [6]. An approximate separation of variables has also been attempted for potentials of the form

$$u(\beta,\gamma) = u(\beta) + u(\gamma) \tag{2}$$

in the cases of $\gamma \approx 0$ [2] or $\gamma \approx \pi/6$ [7].

Various solutions involving different β potentials (infinite well, harmonic oscillator, Coulomb, Kratzer, Davidson) have been worked out (see [8, 9] and references therein). These potentials are known to be exactly soluble for all values of angular momentum L. In the present work, we introduce special solutions for the Morse potential [10],

$$u(\beta) = e^{-2a(\beta-\beta_e)} - 2e^{-a(\beta-\beta_e)}, \tag{3}$$

which is known [11, 12] to be exactly soluble only for $L = 0$. The overall factor D of the Morse potential is set equal to unity, without affecting the method of solution, since it can be scaled out if ratios of energies are used, as in the present work. Analytical expressions for the spectra for any L are obtained by solving the relevant differential equation through the Asymptotic Iteration Method (AIM) [13, 14], after applying the Pekeris approximation [15]. Solutions for the γ-unstable case and the exactly separable rotational case with $\gamma \approx 0$ (to be called ES-M) are obtained.

It can be shown [16] that the AIM method can be applied to the Davidson and Kratzer potentials in the same cases (γ-unstable, exactly separable rotational with $\gamma \approx 0$), recovering the above-mentioned exact solutions which can be obtained in terms of special functions.

A few advantages of the present approach are mentioned here.

1) A well known problem of X(5) and related solutions is the overprediction of the energy spacings within the beta band by almost a factor of two [4]. It is known that this problem can be avoided by replacing the infinite-well potential of X(5) by a potential with sloped walls [17]. The present solution avoids this problem, since the right branch of the Morse potential imitates the sloped wall.

2) In X(5) and related models, using potentials of the form of Eq. (2), the ground state and beta bands depend only on the parameters of the β potential, while the gamma bands depend also on an additional parameter introduced by the γ potential [usually the stiffnes of the harmonic oscillator used as $u(\gamma)$]. When exactly separable potentials of the form of Eq. (1) are used, all bands (ground state, beta, gamma) depend on all parameters. Thus, all bands are treated on an equal footing.

2. Exactly separable solutions for $\gamma \approx 0$

The original collective Bohr Hamiltonian [3] is

$$H = -\frac{\hbar^2}{2B}\left[\frac{1}{\beta^4}\frac{\partial}{\partial\beta}\beta^4\frac{\partial}{\partial\beta} + \frac{1}{\beta^2\sin 3\gamma}\frac{\partial}{\partial\gamma}\sin 3\gamma\frac{\partial}{\partial\gamma} - \frac{1}{4\beta^2}\sum_{k=1,2,3}\frac{Q_k^2}{\sin^2\left(\gamma - \frac{2}{3}\pi k\right)}\right] + V(\beta,\gamma), \tag{4}$$

where β and γ are the usual collective coordinates which define the shape of the nuclear surface. Q_k (k=1, 2, 3) represents the angular momentum components in the intrinsic frame, and B is the mass parameter. Reduced energies and reduced potentials are defined as $\epsilon = 2BE/\hbar^2$, $v = 2BV/\hbar^2$ respectively [1]. If the potential has a minimum around $\gamma = 0$, exact separation of variables can be achieved [5, 8] for potentials of the form given in Eq. (1). We then assume wavefunctions of the form

$$\psi(\beta,\gamma,\theta_j) = \xi_L(\beta)\Gamma_K(\gamma)\mathcal{D}^L_{M,K}(\theta_j), \tag{5}$$

where θ_j ($j = 1, 2, 3$) are the Euler angles, $\mathcal{D}(\theta_j)$ represents Wigner functions of these angles, L stands for the eigenvalues of the angular momentum, while M and K are the eigenvalues of the projections of the angular momentum on the laboratory-fixed z-axis and the body-fixed z'-axis respectively. The Schrödinger equation is thus separated, as in Refs. [5, 8], into a "radial" part (depending on β) and a γ part.

Table 1. Comparison of theoretical predictions of the γ-unstable Bohr Hamiltonian with Morse potential to experimental data [18] of nuclei with $A \geq 100$, $R_{4/2} \leq 2.6$, and known 0_2^+ and 2_γ^+ states. The $R_{4/2} = E(4_1^+)/E(2_1^+)$ ratios, as well as the β and γ bandheads, normalized to the 2_1^+ state and labelled by $R_{0/2} = E(0_\beta^+)/E(2_1^+)$ and $R_{2/2} = E(2_\gamma^+)/E(2_1^+)$ respectively, are shown. The angular momenta of the highest levels of the ground state, β and γ bands included in the rms fit are labelled by L_g, L_β, and L_γ respectively, while n indicates the total number of levels involved in the fit and σ is the quality measure of Eq. (14).

nucleus	$R_{4/2}$ exp	$R_{4/2}$ th	$R_{0/2}$ exp	$R_{0/2}$ th	$R_{2/2}$ exp	$R_{2/2}$ th	β_0	a	L_g	L_β	L_γ	n	σ
^{98}Ru	2.14	2.25	2.0	2.4	2.2	2.3	3.84	0.44	18	0	4	12	0.659
^{100}Ru	2.27	2.29	2.1	2.8	2.5	2.3	4.43	0.36	28	0	4	17	0.291
^{102}Ru	2.33	2.25	2.0	2.3	2.3	2.2	3.78	0.42	16	0	5	12	0.341
^{104}Ru	2.48	2.33	2.8	2.9	2.5	2.3	7.57	0.10	8	2	8	12	0.433
^{102}Pd	2.29	2.28	2.9	2.6	2.8	2.3	4.30	0.34	26	4	4	18	0.219
^{104}Pd	2.38	2.28	2.4	2.6	2.4	2.3	4.15	0.41	18	2	4	13	0.300
^{106}Pd	2.40	2.26	2.2	2.4	2.2	2.3	3.93	0.43	16	4	5	14	0.343
^{108}Pd	2.42	2.28	2.4	2.5	2.1	2.3	4.36	0.30	14	4	4	12	0.313
^{110}Pd	2.46	2.29	2.5	2.0	2.2	2.3	4.01	0.26	12	10	4	14	0.338
^{112}Pd	2.53	2.33	2.6	2.5	2.1	2.3	4.11	0.60	6	0	3	5	0.485
^{114}Pd	2.56	2.31	2.6	2.9	2.1	2.3	5.12	0.24	16	0	11	18	0.727
^{116}Pd	2.58	2.34	3.3	3.5	2.2	2.3	7.44	0.13	16	0	9	16	0.626
^{106}Cd	2.36	2.37	2.8	2.9	2.7	2.4	4.45	0.62	12	0	2	7	0.196
^{108}Cd	2.38	2.26	2.7	2.5	2.5	2.3	3.97	0.43	18	0	5	13	0.688
^{110}Cd	2.35	2.24	2.2	2.2	2.2	2.2	3.66	0.47	16	6	5	15	0.269
^{112}Cd	2.29	2.23	2.0	2.1	2.1	2.2	3.55	0.50	12	8	11	20	0.542
^{114}Cd	2.30	2.21	2.0	2.0	2.2	2.2	3.43	0.51	14	4	3	11	0.359
^{116}Cd	2.38	2.29	2.5	2.7	2.4	2.3	4.10	0.47	14	2	3	10	0.408
^{118}Cd	2.39	2.29	2.6	2.7	2.6	2.3	4.11	0.47	14	0	3	9	0.313
^{120}Cd	2.38	2.28	2.7	2.6	2.6	2.3	4.09	0.44	16	0	2	9	0.379
^{118}Xe	2.40	2.31	2.5	2.7	2.8	2.3	5.40	0.19	16	4	10	19	0.343
^{120}Xe	2.47	2.35	2.8	3.8	2.7	2.4	7.05	0.16	26	4	9	23	0.652
^{122}Xe	2.50	2.43	3.5	3.5	2.5	2.4	5.03	0.66	10	0	9	13	0.501
^{124}Xe	2.48	2.35	3.6	3.8	2.4	2.4	6.88	0.17	20	2	11	21	0.562
^{126}Xe	2.42	2.33	3.4	3.2	2.3	2.3	6.12	0.18	12	4	9	16	0.576
^{128}Xe	2.33	2.33	3.6	3.3	2.2	2.3	4.74	0.39	10	2	7	12	0.522
^{130}Xe	2.25	2.27	3.3	2.6	2.1	2.3	4.05	0.44	14	0	5	11	0.476
^{132}Xe	2.16	2.17	2.8	1.4	1.9	2.2	3.16	0.66	6	0	5	7	0.731
^{134}Xe	2.04	2.11	1.9	1.0	1.9	2.1	2.91	0.74	6	0	5	7	0.753
^{130}Ba	2.52	2.46	3.3	3.3	2.5	2.5	5.58	0.77	12	0	6	11	0.416
^{132}Ba	2.43	2.29	3.2	2.7	2.2	2.3	4.63	0.29	14	0	8	14	0.609
^{134}Ba	2.32	2.26	2.9	2.4	1.9	2.3	3.82	0.50	8	0	4	7	0.483
^{136}Ba	2.28	2.18	1.9	1.7	1.9	2.2	3.24	0.60	6	0	2	4	0.454
^{142}Ba	2.32	2.44	4.3	4.3	4.0	2.4	5.45	0.60	14	0	2	8	0.605
^{134}Ce	2.56	2.37	3.7	4.1	2.4	2.4	6.00	0.27	34	2	8	25	0.502
^{136}Ce	2.38	2.24	1.9	2.2	2.0	2.2	3.66	0.47	16	0	3	10	0.618
^{138}Ce	2.32	2.23	1.9	2.1	1.9	2.2	3.55	0.50	14	0	2	8	1.308
^{140}Nd	2.33	2.19	1.8	1.7	1.9	2.2	3.27	0.60	6	0	2	4	0.265
^{148}Nd	2.49	2.32	3.0	2.9	4.1	2.3	6.40	0.14	12	8	4	13	0.810

Table 1. (continued)

nucleus	$R_{4/2}$ exp	$R_{4/2}$ th	$R_{0/2}$ exp	$R_{0/2}$ th	$R_{2/2}$ exp	$R_{2/2}$ th	β_0	a	L_g	L_β	L_γ	n	σ
^{140}Sm	2.35	2.38	1.9	1.9	2.7	2.4	4.20	0.77	8	0	2	5	0.153
^{142}Sm	2.33	2.20	1.9	1.7	2.2	2.2	3.33	0.61	8	0	2	5	0.173
^{142}Gd	2.35	2.28	2.7	2.7	1.9	2.3	4.17	0.42	16	0	2	9	0.188
^{144}Gd	2.35	2.36	2.5	2.5	2.5	2.4	4.24	0.65	6	0	2	4	0.102
^{152}Gd	2.19	2.26	1.8	2.4	3.2	2.3	3.93	0.40	16	10	7	19	0.436
^{154}Dy	2.23	2.28	2.0	2.7	3.1	2.3	4.22	0.38	26	10	7	24	0.371
^{156}Er	2.32	2.31	2.7	3.1	2.7	2.3	4.75	0.34	20	4	5	16	0.374
^{186}Pt	2.56	2.34	2.5	1.7	3.2	2.3	6.18	0.07	26	6	10	25	1.070
^{188}Pt	2.53	2.45	3.0	3.3	2.3	2.5	5.45	0.75	16	2	4	12	0.356
^{190}Pt	2.49	2.34	3.1	3.6	2.0	2.3	5.08	0.35	18	2	6	15	0.566
^{192}Pt	2.48	2.35	3.8	3.7	1.9	2.3	6.42	0.19	10	0	8	12	0.681
^{194}Pt	2.47	2.35	3.9	3.6	1.9	2.3	7.28	0.14	10	4	5	11	0.657
^{196}Pt	2.47	2.32	3.2	2.9	1.9	2.3	6.26	0.15	10	2	6	11	0.627
^{198}Pt	2.42	2.25	2.2	2.3	1.9	2.3	3.87	0.39	6	2	4	7	0.374
^{200}Pt	2.35	2.20	2.4	1.7	1.8	2.2	3.31	0.59	4	0	4	5	0.676

For the Morse potential [10], given in Eq. (3), using the Pekeris approximation [15] and solving the β equation through AIM (the details are given in Ref. [16]), we obtain the energy eigenvalues

$$\epsilon_{n,L} = \frac{\mu c_0}{\beta_e^2} - \left[\frac{\gamma_1^2}{2\beta_e\gamma_2} - \left(n+\frac{1}{2}\right)\frac{\alpha}{\beta_e}\right]^2, \tag{6}$$

where

$$c_0 = 1 - \frac{3}{\alpha} + \frac{3}{\alpha^2}, \qquad c_1 = \frac{4}{\alpha} - \frac{6}{\alpha^2}, \qquad c_2 = -\frac{1}{\alpha} + \frac{3}{\alpha^2}, \qquad \alpha = a\beta_e, \tag{7}$$

$$\gamma_1^2 = 2\beta_e^2 - \mu c_1, \qquad \gamma_2^2 = \beta_e^2 + \mu c_2, \qquad \mu = \frac{L(L+1)}{3} + 2 + \lambda. \tag{8}$$

λ in the last equation comes from the exact separation of variables and is determined from the γ equation. We use the γ potential $u(\gamma) = (3c)^2\gamma^2$, leading to [20]

$$\lambda = \epsilon_\gamma - \frac{K^2}{3}, \qquad \epsilon_\gamma = (3C)(n_\gamma + 1), \qquad C = 2c. \tag{9}$$

3. γ-unstable solutions

In this case, the reduced potential is assumed to be γ independent, $v(\beta,\gamma) = u(\beta)$. Then the wavefunction is assumed to be of the form [5]

$$\psi(\beta,\gamma,\theta_j) = R(\beta)\Phi(\gamma,\theta_j). \tag{10}$$

The equation which includes the Euler angles and γ has been solved by Bès [21]. In this equation, the eigenvalues of the second-order Casimir operator of SO(5) occur, having the form $\Lambda = \tau(\tau+3)$, where τ is the seniority quantum number, characterizing the irreducible representations of SO(5) and taking the values $\tau = 0, 1, 2, 3, \ldots$ [22].

The values of the angular momentum L are given by the algorithm

$$\tau = 3\nu_\Delta + \lambda, \qquad \nu_\Delta = 0, 1, 2, \ldots \qquad L = \lambda, \lambda+1, \ldots, 2\lambda-2, 2\lambda \tag{11}$$

Table 2. Comparison of theoretical predictions of the exactly separable Morse model [ES-M] to experimental data [18] of rare earth and actinides with $A \geq 150$, $R_{4/2} > 2.9$, and known 0_2^+ and 2_γ^+ states. Data for ^{228}Ra come from Ref. [19]. The $R_{4/2} = E(4_1^+)/E(2_1^+)$ ratios, as well as the β and γ bandheads, normalized to the 2_1^+ state and labelled by $R_{0/2} = E(0_\beta^+)/E(2_1^+)$ and $R_{2/2} = E(2_\gamma^+)/E(2_1^+)$ respectively, are shown. The angular momenta of the highest levels of the ground state, β and γ bands included in the rms fit are labelled by L_g, L_β, and L_γ respectively, while n indicates the total number of levels involved in the fit and σ is the quality measure of Eq. (14).

nucleus	$R_{4/2}$ exp	$R_{4/2}$ th	$R_{0/2}$ exp	$R_{0/2}$ th	$R_{2/2}$ exp	$R_{2/2}$ th	β_0	C	a	L_g	L_β	L_γ	n	σ
^{150}Nd	2.93	3.21	5.2	6.2	8.2	8.2	5.2	6.0	0.41	14	6	4	13	1.129
^{152}Sm	3.01	3.22	5.6	6.9	8.9	9.5	5.3	7.0	0.34	16	14	9	23	1.007
^{154}Sm	3.25	3.28	13.4	13.5	17.6	18.6	8.4	13.7	0.20	16	6	7	17	0.484
^{154}Gd	3.02	3.20	5.5	7.4	8.1	5.1	5.4	3.4	0.31	26	26	7	32	1.849
^{156}Gd	3.24	3.27	11.8	11.6	13.0	14.4	7.5	10.5	0.22	26	12	16	34	0.605
^{158}Gd	3.29	3.30	15.0	15.4	14.9	15.1	8.3	10.6	0.36	12	6	6	14	0.224
^{160}Gd	3.30	3.31	17.6	17.8	13.1	13.1	8.6	8.8	0.44	16	4	8	17	0.169
^{162}Gd	3.29	3.31	19.8	19.8	12.0	12.0	9.5	8.1	0.30	14	0	4	10	0.082
^{158}Dy	3.21	3.25	10.0	10.1	9.6	10.0	6.7	7.1	0.25	28	8	8	25	0.495
^{160}Dy	3.27	3.28	14.7	14.5	11.1	11.8	9.2	8.2	0.18	28	4	23	38	0.522
^{162}Dy	3.29	3.31	17.3	17.1	11.0	10.8	8.3	7.2	0.44	18	14	14	29	0.312
^{164}Dy	3.30	3.31	22.6	22.2	10.4	10.2	13.1	6.8	0.14	20	0	10	19	0.188
^{166}Dy	3.31	3.31	15.0	15.1	11.2	11.2	7.9	7.5	0.52	6	2	5	8	0.037
^{160}Er	3.10	3.20	7.1	7.9	6.8	3.9	5.6	2.5	0.30	26	2	5	18	1.790
^{162}Er	3.23	3.26	10.7	11.1	8.8	9.9	7.1	7.0	0.26	20	4	12	23	0.588
^{164}Er	3.28	3.27	13.6	12.9	9.4	9.6	9.3	6.6	0.14	22	10	18	33	0.827
^{166}Er	3.29	3.30	18.1	17.8	9.8	9.6	9.7	6.5	0.23	16	10	14	26	0.306
^{168}Er	3.31	3.32	15.3	15.5	10.3	10.2	8.1	6.7	0.59	18	6	8	19	0.176
^{170}Er	3.31	3.29	11.3	9.9	11.9	13.1	6.5	9.0	0.17	24	10	19	35	0.864
^{164}Yb	3.13	3.21	7.9	7.4	7.0	7.3	5.4	5.2	0.32	18	0	5	13	0.471
^{166}Yb	3.23	3.25	10.2	10.1	9.1	9.5	6.8	6.7	0.23	24	10	13	29	0.688
^{168}Yb	3.27	3.27	13.2	11.9	11.2	11.4	7.9	8.0	0.19	34	4	7	25	0.768
^{170}Yb	3.29	3.31	12.7	13.8	13.6	13.5	7.8	9.2	0.49	20	10	17	31	0.509
^{172}Yb	3.31	3.29	13.2	12.4	18.6	19.1	7.8	13.7	0.17	16	12	5	18	0.851
^{174}Yb	3.31	3.31	19.4	19.3	21.4	21.5	17.2	15.1	0.05	20	4	5	16	0.170
^{176}Yb	3.31	3.32	13.9	14.0	15.4	15.3	8.1	10.4	0.57	20	2	5	15	0.254
^{178}Yb	3.31	3.29	15.7	15.6	14.5	14.5	10.2	10.3	0.15	6	4	2	6	0.045
^{168}Hf	3.11	3.22	7.6	7.5	7.1	7.4	5.4	5.3	0.32	22	4	4	16	0.438
^{170}Hf	3.19	3.23	8.7	8.6	9.5	8.1	6.0	5.7	0.27	34	4	4	22	0.964
^{172}Hf	3.25	3.26	9.2	10.1	11.3	11.7	6.7	8.5	0.24	38	4	6	26	0.444
^{174}Hf	3.27	3.26	9.1	9.0	13.5	14.1	6.1	10.4	0.26	26	26	5	30	0.484
^{176}Hf	3.28	3.28	13.0	12.3	15.2	16.1	7.9	11.7	0.18	18	10	8	21	0.622
^{178}Hf	3.29	3.28	12.9	12.7	12.6	12.9	8.5	9.2	0.16	18	6	6	17	0.298
^{180}Hf	3.31	3.31	11.8	12.2	12.9	12.9	7.7	8.7	0.61	12	4	5	12	0.199
^{176}W	3.22	3.24	7.8	8.2	9.6	10.2	5.8	7.4	0.29	22	12	5	21	0.578
^{178}W	3.24	3.25	9.4	9.5	10.5	10.5	6.4	7.6	0.26	18	10	2	15	0.177
^{180}W	3.26	3.27	14.6	12.7	10.8	11.4	7.9	8.0	0.24	24	0	7	18	0.838
^{182}W	3.29	3.28	11.3	11.6	12.2	12.4	8.3	8.6	0.13	18	4	6	16	0.282
^{184}W	3.27	3.27	9.0	8.9	8.1	8.1	6.6	5.5	0.15	10	4	6	12	0.094
^{186}W	3.23	3.24	7.2	7.2	6.0	6.2	5.3	4.2	0.22	14	4	6	14	0.133

Table 2. (continued)

nucleus	$R_{4/2}$ exp	$R_{4/2}$ th	$R_{0/2}$ exp	$R_{0/2}$ th	$R_{2/2}$ exp	$R_{2/2}$ th	β_0	C	a	L_g	L_β	L_γ	n	σ
^{178}Os	3.02	3.19	4.9	5.8	6.6	7.1	4.8	5.1	0.39	16	6	5	15	0.724
^{180}Os	3.09	3.23	5.6	7.1	6.6	7.0	5.1	4.9	0.27	10	6	7	14	1.122
^{184}Os	3.20	3.24	8.7	9.4	7.9	8.5	6.3	6.0	0.28	22	0	6	16	0.600
^{186}Os	3.17	3.22	7.7	7.7	5.6	6.1	5.6	4.2	0.27	14	10	13	24	0.200
^{188}Os	3.08	3.20	7.0	7.3	4.1	4.3	5.3	2.8	0.34	12	2	7	13	0.213
^{228}Ra	3.21	3.27	11.3	11.3	13.3	13.0	7.3	9.5	0.24	22	4	3	15	0.387
^{228}Th	3.24	3.28	14.4	13.9	16.8	17.0	8.4	12.4	0.25	18	2	5	14	0.514
^{230}Th	3.27	3.28	11.9	11.7	14.7	14.8	7.6	10.7	0.19	24	4	4	17	0.276
^{232}Th	3.28	3.29	14.8	14.7	15.9	16.5	9.6	11.8	0.15	30	20	12	36	0.321
^{232}U	3.29	3.29	14.5	14.4	18.2	18.3	9.5	13.1	0.14	20	10	4	18	0.158
^{234}U	3.30	3.30	18.6	18.7	21.3	21.7	12.1	15.5	0.12	28	8	7	24	0.219
^{236}U	3.30	3.30	20.3	20.4	21.2	21.2	13.8	15.0	0.10	30	4	5	21	0.266
^{238}U	3.30	3.31	20.6	20.9	23.6	24.7	13.9	17.6	0.10	30	4	27	43	0.716
^{238}Pu	3.31	3.32	21.4	21.7	23.3	23.4	9.8	15.9	0.50	26	2	4	17	0.857
^{240}Pu	3.31	3.31	20.1	19.1	26.6	27.1	13.0	19.4	0.09	26	4	4	18	0.539
^{242}Pu	3.31	3.32	21.5	21.6	24.7	24.8	9.9	17.0	0.46	26	2	2	15	0.926
^{248}Cm	3.31	3.31	25.0	25.1	24.2	24.2	16.4	17.0	0.09	28	4	2	17	0.105
^{250}Cf	3.32	3.32	27.0	27.0	24.2	24.2	12.8	16.7	0.20	8	2	4	8	0.018

(with $2\lambda-1$ missing), where ν_Δ is the missing quantum number in the reduction $SO(5) \supset SO(3)$ [22]. The ground state band levels are determined by $L = 2\tau$ and $n = 0$.

Using the Pekeris approximation [15] and AIM (see Ref. [16] for the details), we obtain the energy eigenvalues

$$\epsilon_{n,\tau} = \frac{\nu c_0}{\beta_e^2} - \left[\frac{\gamma_1^2}{2\beta_e\gamma_2} - \left(n+\frac{1}{2}\right)\frac{\alpha}{\beta_e}\right]^2, \tag{12}$$

where

$$\gamma_1^2 = 2\beta_e^2 - \nu c_1, \qquad \gamma_2^2 = \beta_e^2 + \nu c_2, \qquad \nu = \tau(\tau+3)+2, \tag{13}$$

with the rest of the quantities given again by Eq. (7).

4. Numerical results

In order to test the applicability of the Morse potential in the description of nuclear spectra, we have fitted all nuclei with mass $A \geq 100$ and $R_{4/2} = E(4)/E(2) < 2.6$, for which at least the β_1 and γ_1 bandheads are known, using the γ-unstable solution of the Morse potential, which involves two free parameters (β_e, a). Results for 54 nuclei are shown in Table 1. The quality measure

$$\sigma = \sqrt{\frac{\sum_{i=1}^n (E_i(exp) - E_i(th))^2}{(n-1)E(2_1^+)^2}}, \tag{14}$$

used in the rms fits, remains below 1 in most cases.

The Morse potentials obtained for the $_{54}$Xe isotopes are shown in Fig. 1. The evolution of the parameters and the shapes of the potentials are clear. As one moves from $^{134}Xe_{80}$, which is just below the $N = 82$ magic number, to the mid-shell nucleus $^{120}Xe_{66}$, the β_e parameter (which is the position of the minimum of the potential) increases, while the parameter a, which corresponds to the steepness of the potential, decreases. As a result, one gradually obtains less

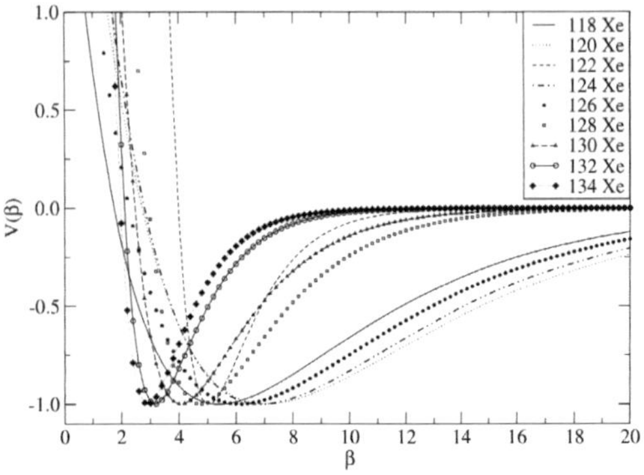

Figure 1. Evolution of Morse potential shapes for the $_{54}$Xe isotopes, with the parameters given in Table 1.

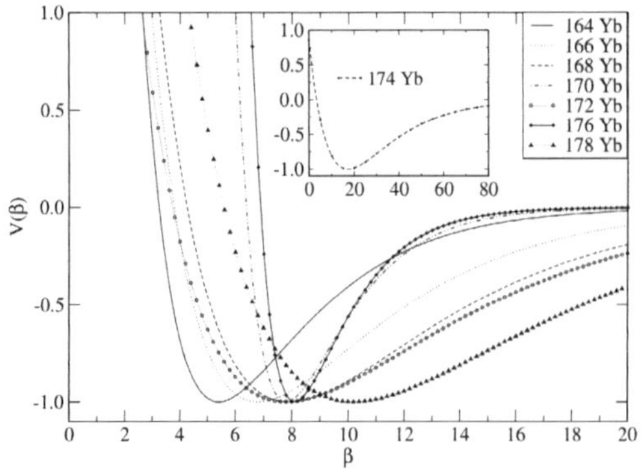

Figure 2. Evolution of Morse potential shapes for the $_{70}$Yb isotopes, with the parameters given in Table 2.

steep potentials with a minimum further away from the origin. The trends start to be reversed at ^{118}Xe$_{64}$, which is just below mid-shell.

We have also fitted all nuclei with mass $A \geq 150$ and $R_{4/2} = E(4)/E(2) > 2.9$ for which at least the β_1 and γ_1 bandheads are known, using the exactly separable rotational solution of the Morse potential with $\gamma \approx 0$ (ES-M), which involves three free parameters (the Morse parameters β_e and a, as well as the stiffness C of the γ potential). All bands are treated on an equal footing, depending on all three parameters. Results for 45 rare earths and 13 actinides are shown in Table 2. The quality measure σ of Eq. (14), used in the rms fits, remains below 1 in most cases.

The Morse potentials obtained for the $_{70}$Yb isotopes are shown in Fig. 2. The evolution of the parameters and the shapes of the potentials are again clear. As one moves from ^{164}Yb$_{94}$ to the mid-shell nucleus ^{174}Yb$_{104}$, the β_e parameter (which is the position of the minimum of the β-potential) again increases, while the parameter a, which corresponds to the steepness of the β-potential, again decreases. The C parameter, which is related to the stiffness of the γ-potential, increases. As a result, one gradually obtains less steep β-potentials with a minimum further away from the origin, while the γ-potentials get stiffer at the same time.

A notable exception occurs in the $N = 90$ isotones ^{150}Nd, ^{152}Sm, ^{154}Gd, which are known to be good examples of the X(5) critical point symmetry, along with ^{178}Os [4]. The relative failure of the Morse potential to describe critical nuclei is expected. The potential at the critical point is expected to be flat, as the infinite-well potential used in X(5), or to have a little bump in the middle [4]. Microscopic relativistic mean field calculations [23] of potential energy surfaces support these assumptions. Since the Morse potential cannot imitate a flat potential, with or without a bump in the middle, it is expected that it cannot describe these nuclei satisfactorily.

A comparison of the fits of Table 2 to the results provided by the Davidson potential in the exactly separable $\gamma \approx 0$ case [20] (ES-D), which contains two free parameters (β_0, c) instead of three (see Table 1 of Ref. [20]), shows that the extra parameter extends the region of applicability of the model in most nuclei to higher angular momenta, largely improving the quality of the fits.

A close examination of the spectra obtained with the Morse potential is revealing. For example, one can consider the spectra of ^{154}Dy (γ-unstable case) and ^{232}Th (exactly separable rotational case with $\gamma \approx 0$), given in Ref. [16]. The overall agreement between theory and experiment is very good in both cases. In the theoretical predictions for the gamma band of ^{154}Dy, the O(5) degeneracies are present, limiting the flexibility of the model to agree to experiment. Spacings within all bands of ^{232}Th, including the beta band (in which spacings in

X(5) are overpredicted by almost a factor of two [4]), are reproduced very accurately.

For the construction of complete level schemes, the calculation of $B(E2)$ transition rates is required, for which the wave functions are needed. Work in this direction is in progress.

5. Discussion

The Bohr Hamiltonian has been solved with the Morse potential for any angular momentum, both in the γ-unstable case and in the exactly separable rotational case with $\gamma \approx 0$ (in which a harmonic oscillator is used for the γ potential), labelled as ES-M. The solution has been achieved [16] through the Asymptotic Iteration Method (AIM) and has involved the Pekeris approximation.

Numerical results have been presented for both solutions, including all relevant medium mass and heavy nuclei for which at least the β_1 and γ_1 bandheads are known. The success of the present solutions in reproducing quite well both the bandeahds of and the spacings within the ground, β_1 and γ_1 bands indicate that a detailed study of γ_2 and other higher bands within this framework might be fruitful. The influence of the finite depth of the potential is also worth considering in further detail. From the findings of Ref. [24], where the E(5) case was solved for a finite well, the influence of the finite depth of the potential is expected to show up more clearly in the higher excited states. Work on the calculation of wave functions and $B(E2)$ transition rates is in progress.

References

[1] Iachello F 2000 *Phys. Rev. Lett.* **85** 3580
[2] Iachello F 2001 *Phys. Rev. Lett.* **87** 052502
[3] Bohr A 1952 *Fys. Medd. K. Dan. Vidensk. Selsk.* **26** 14
[4] Casten R F and McCutchan E A 2007 *J. Phys.* G **34** R285
[5] Wilets L and Jean M 1956 *Phys. Rev.* **102** 788
[6] Meyer-ter-Vehn J 1975 *Nucl. Phys.* A **249** 111
[7] Bonatsos D, Lenis D, Petrellis D and Terziev P A 2004 *Phys. Lett.* B **588** 172
[8] Fortunato L 2005 *Eur. Phys. J.* A **26** s01
[9] Bonatsos D, Lenis D and Petrellis D 2007 *Romanian Reports in Physics* **59** 273
[10] Morse P M 1929 *Phys. Rev.* **34** 57
[11] Flügge S 1974 *Practical Quantum Mechanics* (Berlin: Springer)
[12] Cooper F, Khare A and Sukhatme U 2001 *Supersymmetry in Quantum Mechanics* (Singapore: World Scientific)
[13] Ciftci H, Hall R L and Saad N 2003 *J. Phys.* A **36** 11807
[14] Ciftci H, Hall R L and Saad N 2005 *J. Phys.* A **38** 1147
[15] Pekeris C L 1934 *Phys. Rev.* **45** 98
[16] Boztosun I, Bonatsos D and Inci I 2008 *Phys. Rev.* C **77** 044302
[17] Caprio M A 2004 *Phys. Rev.* C **69** 044307
[18] as of December 2006 *Nuclear Data Sheets*
[19] Cocks J F C *et al.* 1999 *Nucl. Phys.* A **645** 61
[20] Bonatsos D, McCutchan E A, Minkov N, Casten R F, Yotov P, Lenis D, Petrellis D and Yigitoglu I 2007 *Phys. Rev.* C **76** 064312
[21] Bès D R 1959 *Nucl. Phys.* **10** 373
[22] Rakavy G 1957 *Nucl. Phys.* **4** 289
[23] Fossion R, Bonatsos D and Lalazissis G A 2006 *Phys. Rev.* C **73** 044310
[24] Caprio M A 2002 *Phys. Rev.* C **65** 031304

Phases of clusterized nuclei

J Cseh[1]

Institute for Nuclear Research, Hung. Acad. Sci., Debrecen, POB 51, Hungary-4001

E-mail: [1] cseh@atomki.hu

Abstract. After reviewing some basic features of the temperature-governed phase-transitions in macroscopic systems and in atomic nuclei we consider non-thermal phase-transitions of nuclear structure in the example of cluster states. The vibron model, as the algebraic model of the relative motion, is considered from this aspect. The extension by the coupling to the internal degrees of freedom of the clusters is discussed. Phenomenological and semimicroscopical algebraic cluster models with identical interactions are applied to binary cluster systems of closed and non-closed shell clusters. A phase-diagram for the binary cluster systems is proposed, as well as its relation to that of the shell-model. The role of the quasi-dynamical symmetry, as a possible signature of the phase of a finite quantum system, is supported.

1. Introduction

1.1. Phase-transitions in macroscopic systems

One of the best-known phase-transition is that of the water. E.g. at fixed pressure the system changes from the liquid phase to the gas phase by increasing the temperature. At the transition point the temperature (T), the pressure (p) and the chemical potential (μ) of the two phases are equal. The equilibrium of the system is determined by the second law of thermodynamics, via requiring the minimum of the free enthalpy (G), (or its portion for one particle ($\mu = G/N$), where (N) stands for the number of particles). Crossing the curve of phase-transition the chemical potential changes continuously, but its derivatives can be discontinuous. The order of phase-transition refers to the order of derivative, which is discontinuous. Below the critical point (at 647.1K, 22.06MPa) the transition is of first order, at the critical point it is of second order, and above that point it is analytical, or cross-over, i.e. each derivative is continuous.

1.2. Thermal phase-transitions in nuclei

In the ground-state region the atomic nucleus is usually considered to be a microscopic liquid drop. It is a long-standing question if there is a phase-transition to the gas phase with increasing temperature. After much experimental and theoretical investigations along this line the answer seems to be affirmative [1]. This is a first order phase-transition of the nucleonic matter.

At even higher temperature the nucleons dissolve and quarks and gluons move freely. As a consequence a drastic phase-transition may take place from the hadronic to the deconfined phase. At the RHIC, LHC and Early Universe energies (and densities) this transition is predicted to be analytical, but lower-order transitions are also expected (at smaller temperatures and larger densities). [2, 3].

These phase-transitions are governed by the temperature, thus we call them thermal transitions. The finite-size effects of the nucleus are very important, of course, they have to

Journal of Physics: Conference Series **205** (2010) 012021 doi:10.1088/1742-6596/205/1/012021

be taken into account and they modify the results in comparison with the infinite limit.

1.3. Non-thermal phase-transitions

In atomic nuclei other kind of phase-transitions can also be observed, which are not governed by the temperature, thus we call them non-thermal phase-transitions. Their appearance is even more closely related to the finite-size effects; in a sense we can say that they show up due to the finite size effects, since they have to do with the nuclear shape, or with the collective motion of the nucleus as a whole. These phenomena are sometimes called quantum phase-transitions, shape-phase transitions, $T = 0$ phase transitions, or ground-state phase-transitions.

In relation with the quadrupole collectivity of nuclei a large amount of work has been carried out along this line, and excellent review papers are available [4]. Here we discuss the basic concepts from the viewpoint of the dipole collectivity, which is the basic degrees of freedom of clusterization. Partly because the dipole collectivity is simpler than the quadrupole one, and also because it is less known, and it might be interesting to look at the problem from a different angle.

2. Algebraic cluster models

The clusterization of atomic nuclei is usually described by cluster models. The basic assumption of all the cluster models is that the relevant degrees of freedom of the nucleus are classified into two categories. Some of them account for the relative motion of the clusters, while others are related to the internal degrees of freedom of the clusters. The phenomenological and the semimicroscopical models apply cluster-cluster interactions, the difference between the two kinds is in the model space. In the semimicroscopical models the Pauli-forbidden states are excluded from the model space, while in the phenomenological model they are not, or not completely. (The fully microscopical models incorporate antisymmetrization, of course, and apply effective nucleon-nucleon forces, instead of cluster-cluster interactions.)

Algebraic models are especially suitable for the studies of phase-transitions, thus we apply here algebraic cluster models. The group-theoretical methods guide us in constructing the physical operators and calculating their matrix elements, but they do not decide about the inclusion of the Pauli-principle. Therefore, it is possible to build up phenomenological and semimicroscopical models with the same algebraic structure, including the same interactions, but having different model spaces.

We consider here binary cluster systems and describe them by phenomenological and semimicroscopical algebraic cluster models in such a way that the interactions of the two models are the same. We do so in order to check if the exclusion principle has an influence on the appearance of the phase-transition.

The relative motion is accounted for by the vibron model of $U(4)$ group-structure [5]. The building blocks of this model are the dipole and scalar bosons (as opposed to the interacting boson model of quadrupole collectivity [6], in which quadrupole and scalar bosons are taken). The vibron model has two dynamical symmetries, described by the (I) $U(4) \supset U(3) \supset O(3)$ and (II) $U(4) \supset O(4) \supset O(3)$ group-chains.

The internal structure of the clusters is described by the Elliott-model [7] of $U^{ST}(4) \otimes U(3)$ group structure, where $U^{ST}(4)$ is the spin-isospin group of Wigner [8], while $U(3)$ refers to the orbital part. The product wavefunctions of these models are contaminated by the Pauli-forbidden states, resulting in a phenomenological algebraic cluster model (PACM). When the Pauli-principle is appreciated, and the forbidden states are excluded from the model space we end up with the semimicroscopical algebraic cluster model (SACM)[9].

The interactions of the models are constructed from the group-generators, as usual in the algebraic description. When we are interested only in a single sector of the spin-isospin space, then Wigner's group is important only in the construction of the model space, but from the

Journal of Physics: Conference Series **205** (2010) 012021 doi:10.1088/1742-6596/205/1/012021

viewpoint of the interaction the group structure simplifies to: $U_{C_1} \otimes U_{C_2} \otimes U_R \supset U(3) \supset O(3)$, where C_i stands for the ith cluster, and R indicates relative motion. The Hamiltonian built up from the invariant operators of this chain has a $U(3)$ dynamical symmetry: H_{U3}.

The $U(3)$ limit of the algebraic cluster models corresponds to a soft vibrator in the collective model terms. (More precisely we should call it soft vibrational-rotational motion, but we follow here the simplified terminology.) In order to characterize the rotational-vibrational behaviour one can apply the parameter $R = (2E_{rot})/(E_{vib})$, where E_{rot} stands for the first rotational energy, while E_{vib} denotes the lowest-lying vibrational one. When $R \approx 1$ then the spectrum is considered to be a soft vibrator, and $R \ll 1$ corresponds to the rigid rotor [10]. From the microscopic viewpoint the $U(3)$ dynamical symmetry describes shell-model-like clusterization, when the cluster model state can be expressed in terms of a few (or a single) shell model basis states [11].

In the vibron model, applied for the description of the relative motion, there is also an $O(4)$ dynamical symmetry. We denote the $O(4)$ dynamically symmetric interaction by H_{O4}. This limit corresponds to a rigid rotor (in the language of the collective model), or to a rigid molecule-like cluster-configuration (from the microscopic viewpoint). This latter name indicates that the cluster wavefunction can be expressed only in a large number of shell-model basis states.

One can apply [12] a general Hamiltonian: $H = xH_{U3} + (1-x)H_{O4} + aH_{O3}$, thus $x = 0$ corresponds to the $O(4)$ limit, while $x = 1$ gives the $U(3)$ dynamical symmetry.

3. Phase-transitions in the Vibron Model

The vibron model, as an algebraic description of dipole motion, is extensively applied in molecular physics [10, 13], as well as in hadron spectroscopy [14], in addition to the studies of nuclear clusterization [15]. The characteristic features of the applications in different fields are discussed in [16]. The shape-phase-transition in the vibron model has been studied by different methods [17, 18].

First we consider the limiting case of large N, when the calculations can be carried out analytically. Here N is the number of particles, i.e. the total number of bosons. Its physical meaning is the largest possible number of dipole phonons (oscillator quanta) in the relative motion of the two clusters.

3.1. Large N case

The expectation value of the energy, which corresponds to the general Hamiltonian above can be calculated analytically with the coherent state technique [19]. The energy-functional depends on two parameters: r is the distance between the two clusters, and x is the control parameter, i.e. the relative weight of the interactions of different dynamical symmetries. The minimum with respect to the geometrical parameter determines the shape of the nucleus. We investigate the behaviour of this energy-minimum as a function of the control parameter. It varies continuously, as well as its first order derivative, but at a certain x_0 value the second-order derivative is discontinuous. Thus we speak about a second order phase-transition.

Let us summarize here the analogy between the macroscopic phase-transitions, and those of the cold finite quantum system, e.g. atomic nucleus. In both cases there is an energy-like physical quantity (e.g. the free enthalpy versus the ground-state energy), which determines the equilibrium of the system via its minimum (as a consequence of the second law of thermodynamics). There is a control-parameter, which governs the system (e.g. temperature versus relative weights of the interactions with different dynamical symmetries). The physical quantities changes continuously as a function of the control parameter, but some derivatives may show discontinuity at a critical point. Then we speak about a phase-transition of the order given by the discontinuous derivative. This similarity is the reason, why the language of phase-transitions is applied in relation with the cold nuclear structure problem.

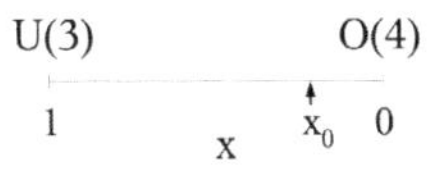

Figure 1. Phase-diagram of the vibron model.

3.2. Critical behaviour

The potential energies corresponding to the $U(3)$ and $O(4)$ dynamical symmetries have minima at different values of the geometrical parameter r. The competition of these two minima, as a function of the control parameter results in the phase-transition, as described in the previous subsection. For the parametrization, applied here, for large x values ($x \approx 1$) the minimum of the $U(3)$ dynamically symmetric parts dominates, and the minimum is at $r = 0$. With decreasing value of x the minimum gets less deep, and at the critical value of x_0 it is rather flat. One can try to approximate this flat potential by a square infinite well, in which case the description of the critical-point behaviour can be given analytically, in terms of the Bessel-equation. Then one speaks about the $E(3)$ critical-point symmetry [18], where the number 3 refers to the dimension of the problem, while E stands for the Euclidean group, related to the flat potential. (This symmetry is the 3-dimensional sister of the much studied $E(5)$ critical-point symmetry of the quadrupole model [20].)

3.3. Finite N case

The real physical systems represent the situation of finite boson number N. Then the eigenvalue-problem has to be solved by numerical diagonalization for different values of the control parameter. No analytic calculations can be carried out, which could show the appearance of the phase-transition, but numerical studies can show if a precursor of it can be observed as an abrupt but smooth change, due to the finite-size effects.

It turns out that both the energy-spectrum, and other physical quantities show this kind of behaviour, with sharper changes for larger N. In [18] the low-lying energy-spectrum, the overlap of the ground-state wave-function with that of the $U(3)$ limit, the expectation value of the dipole bosons over N (as the quantal order parameter), and some characteristic energy-ratios (E_{2_1}/E_{0_2} , E_{3_1}/E_{0_2}), as well as electromagnetic transition ratios ($\frac{B(E1:2_1\to 1_1)}{B(E1:1_1\to 0_1)}, \frac{B(E1:0_2\to 1_1)}{B(E1:1_1\to 0_1)}$) are found as signatures of the phase-transition.

4. Phase-transitions in the algebraic cluster models

4.1. Relative motion

The first straightforward question to investigate in the algebraic cluster model is if the phase-transition of the vibron model survives if we couple the internal cluster degrees of freedom. This problem was investigated in [12]. It turned out that the phase-transition between the $U(3)$ and $O(4)$ dynamical symmetries can be observed independent of the fact if the clusters have closed- or open-shell structure. Furthermore, both the phenomenological and the semimicroscopical description shows its appearance, i.e. the incorporation of the Pauli-principle does not change the situation qualitatively.

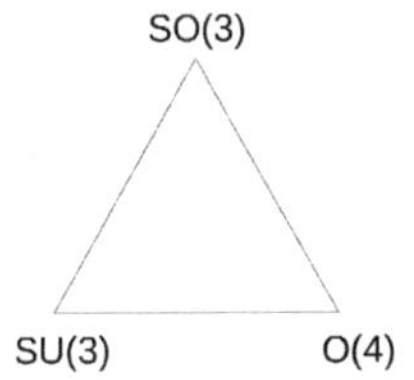

Figure 2. Phase-diagram of the binary-cluster model.

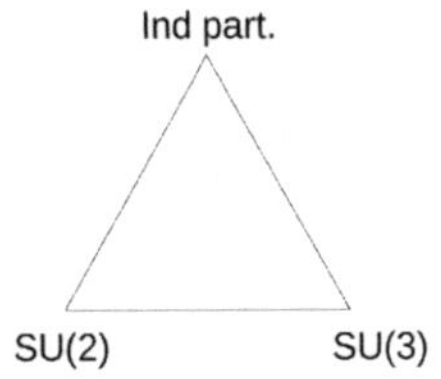

Figure 3. Phase-diagram of the shell-model. Ind. part. stands for independent particle.

4.2. Binary clusterization

In fact the algebraic model of a binary cluster system has three dynamical symmetries, not only two, like the vibron model. This is a consequence of the coupling to the internal cluster degrees of freedom, of course. In particular they are, as follows:

$$U_{C_1}(3) \otimes U_{C_2}(3) \otimes U_R(4) \supset U_C(3) \otimes U_R(3) \supset U(3) \supset SU(3) \supset SO(3) \quad (1)$$

$$U_{C_1}(3) \otimes U_{C_2}(3) \otimes U_R(4) \supset U_C(3) \otimes O_R(4) \supset SO_C(3) \otimes SO_R(3) \supset SO(3) \quad (2)$$

$$U_{C_1}(3) \otimes U_{C_2}(3) \otimes U_R(4) \supset U_C(3) \otimes U_R(3) \supset SO_C(3) \otimes SO_R(3) \supset SO(3). \quad (3)$$

The first two are sort of trivial extensions of the $U(3)$ and $O(4)$ symmetries of the vibron model, the third one is the so-called week-coupling limit, which contains the $U(3)$ group, but allows only angular momentum interactions (and not quadrupole interactions).

As a consequence of the three dynamical symmetries, the phase-diagram of the binary cluster model is two-dimensional, and can be represented as a triangle, shown in Figure 2.

4.3. Connection to the shell model

The interrelation between the shell-model and cluster model is a longstanding problem, which has been investigated from different aspects. From the viewpoint of the phase diagram the connection of these two approaches is also an interesting question.

The phase-diagram of the shell model is also a triangle [21], as shown in Figure 3. Here the $SU(2)$ limit corresponds to the jj-coupling, while the $SU(3)$ corresponds to the ls-coupling. This latter one is the common intersection with the cluster model. Therefore, the relation between

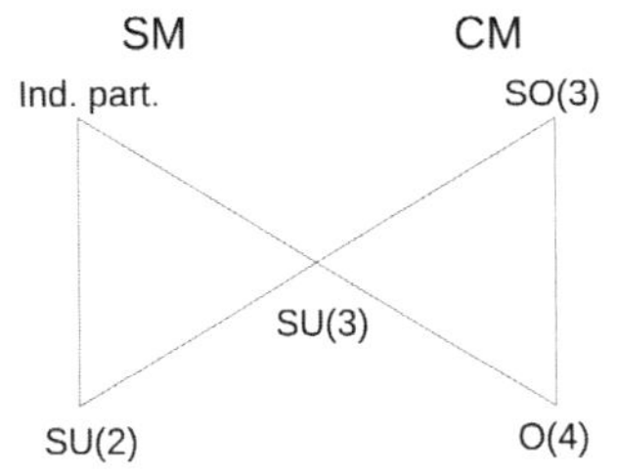

Figure 4. The connection of the phase diagrams of the cluster and shell models.

the phase diagrams of the shell model and cluster model can be illustrated, as shown in Figure 4.

Where the specific nuclei are located on this map is an important open problem, which remains to be investigated.

5. Signature of the phases

In the thermodynamics a phase is homogeneous and shows a common physical behaviour, e.g. it is described by a single (set of) equation of state. In case of the finite quantum systems, e.g. atomic nuclei it is not so trivial what the signature of the phase is (for a finite volume of the phase-space).

For reasons of simplicity let us come back to the one-dimensional phase-diagram of the vibron model (Figure 1) in studying this problem. The diagram is defined by the dynamical symmetries of the model, as ending points. Somewhere in between there is a phase-transition-like phenomenon. Thus the phase is determined by the end-point of the phase-diagram and the critical point of the transition. But the question of the common physical behaviour of a phase is not so simple. The unique and final answer to this question is not known yet, but there is an interesting suggestion put forward by D. Rowe [22], which is based on his numerical studies of the quadrupole collective model. In particular a quasi-dynamical symmetry seems to be present (practically) in the whole phase of the quantum system.

Our numerical studies [12] show that for the cluster-problems the quasi-dynamical $U(3)$ symmetry is the common physical content for the whole phase between the end-point of the real $U(3)$ symmetry at $x = 1$ and (very close to) the transitional point x_0.

The quasi-dynamical $U(3)$ symmetry is more general than the dynamical $U(3)$ symmetry[23]. In the latter case the energy-eigenvectors are $U(3)$ basis sates in spite of the fact the the interaction is not $U(3)$-invariant. In case of the quasi-dynamical symmetry, however, the eigenstates are linear combinations of basis states belonging to different irreducible representations of $U(3)$, but in such a special way that the matrix-elements of the generators between these states approximate those of the real symmetry. In a sense the quasi-dynamical symmetry is the most general symmetry-concept of the quantum mechanics, describing a situation when neither the operator, nor its eigenvectors are symmetric, yet the eigenvalue-equation is symmetric[24].

If this guess is correct, and the signature of the phase of a finite quantum system really proves to be the quasi-dynamical symmetry, then the situation resembles to that of Landau's theory of phase-transitions[25]. In particular, different phases are characterized by different symmetries, and the phase-transition involves the change of the symmetry.

6. Summary and conclusion

Today we know how the transition of the nucleonic matter takes place from the liquid phase to the gas phase. It is studied very intensively, how the phase transition from the hadronic matter to the deconfined matter happens. It is also an interesting question, what kind of phases the clusterized nuclear matter has, and what kind of transitions show up between them. Work along this line has also started, but it is at the very early stage. In this paper we tried to review the phenomenological aspects of the problem, from the angle of the algebraic models, which offer a convenient possibility for studying these problems quantitatively. Investigations from the side of the microscopic theories have also started [26].

The atomic nucleus is known to be a good prototype of mesoscopic systems, and several methods worked out in nuclear studies could be exported to other fields of research too. The better understanding of the phase-transitions of finite quantum systems might be fruitful for very exotic applications as well, like e.g. that of the protein-folding[27, 28].

Acknowledgments

This work was supported by the OTKA (Grant No. K72357) and by the exchange program of the Bulgarian and Hungarian Academies of Sciences. The author acknowledges collaboration on some parts of this work with H. Yepez-Martinez, P.O. Hess and J. Darai. Fruitful discussions with D. Rowe, N. Itagaki, M. Ploszajczak and A. Georgieva are also highly appreciated.

References

[1] Elliott J B et al. 2002 *Phys. Rev. Lett.* **88** 042701
[2] Aoki Y et al. 2006 *Nature* **443/12** 675
[3] www.gsi.de/fair/experiments/CBM
[4] Cejnar P, Jolie J 2009 Prog. Part. Nucl. Phys. **62** 210; Iachello F 2009 AIP Conf. Proc. **1165** 193
[5] Iachello F 1981 *Phys. Rev. C* **23** 2778
[6] Iachello F, Arima A 1987 The Interacting Boson Model, Cambridge University Press, Cambridge
[7] Elliott J P 1958 *Proc. Roy. Soc. A* **245** 128
[8] Wigner E P 1937 *Phys. Rev.* **51** 106
[9] Cseh J 1992 *Phys. Lett. B* **281** 173; Cseh J, Lévai G 1994 *Ann. Phys. (NY)* **230** 165
[10] Iachello F, Levine R D 1995 Algebraic Theory of Molecules, Oxford University Press, Oxford
[11] Cseh J, Darai J, Algora A, Yepez-Martinez H, Hess P O 2008 *Rev. Mex. Fis.* **54** S3 30
[12] Yepez-Martinez H, Cseh J, Hess P O 2006 *Phys. Rev. C* **74** 024319
[13] Frank A, Van Isacker P 1994 Algebraic Methods in molecular and Nuclear Structure Physics, Wiley, NY
[14] Iachello F Mukhopadhyay N C, Zhang L 1991 *Phys. Rev. D* **44** 898
[15] Lévai G, Cseh J, Scheid W 1992 *Phys. Rev. C* **46** 548; Cseh J, Lévai G, Scheid W 1993 *Phys. Rev. C* **48** 1724; Lévai G, Cseh J 1996 *Phys. Lett. B* **381** 1; Fulop Zs et al. 1996 *Nucl. Phys. A* **604** 286; Cseh J, Lévai G, Ventura A, Zuffi L 1998 *Phys. Rev. C* **58** 2144; Lepine-Szily A et al. 1999 *Phys. Rev. Lett.* **82** 3972; L. Hernández de la Peña et al. 2001 J. Phys. G **37**, 2019
[16] Iachello F, Cseh J, Lévai G 1995 *APH NS Heavy Ion Physics* **1** 91
[17] Van Roosmalen O S, Dieperink A E L 1982 *Ann. Phys. NY* **139** 198; Leviatan A, Kirson M 1988 *Ann. Phys. NY* **188** 142; Cejnar P, Iachello F 2007 *J. Phys. A* **40** 581
[18] Zhang Y et al. 2008 *Phys. Rev. C* bf 78 024314
[19] Hess P O, Lévai G, Cseh J 1996 *Phys. Rev. C* **54** 2345; and references therein.
[20] Iachello F 2000 *Phys. Rev. Lett.* **85** 3580
[21] Frank A, Jolie J, Van Isacker P 2009 Symmetries in Atomic Nuclei, Springer, NY
[22] Rowe D J 2004 Nucl. Phys. A **475** 47
[23] Jarrio M, Wood J L, Rowe D J 1991 *Nucl. Phys. A* **528** 409
[24] Cseh J 2006 *Proc. IV Int. Symp. on Quantum Theory and Symmetries (Varna)* (Sofia: Heron Press) p 918
[25] Landau L D, Lifsic E M 1976 Teoreticheskaya Fisica V, Nauka, Moskva
[26] Itagaki N, Masui H, Ito M, Aoyama S 2005 *Phys. Rev. C* **71** 064307
[27] Broglia N A, Tiana G, Provasi D 2004 *J. Phys: Cond. Matt.* **16** R111
[28] Yakubovich A V, Solov'yov I A, Solov'yov A V, Greiner W 2006 *Eur. Phys. J. D* **40** 363

Journal of Physics: Conference Series **205** (2010) 012022 doi:10.1088/1742-6596/205/1/012022

Exotic shapes and clusterization of atomic nuclei

J. Darai[1], J. Cseh[2], A. Lépine-Szily[3], A. Algora[4], P. O. Hess[5], N. V. Antonenko[6], R. V. Jolos[6], W. Scheid[7]

[1] Institute of Experimental Physics, University of Debrecen, Debrecen, Pf. 105, Hungary-4010
[2] Institute of Nuclear Research, Hung. Acad. Sci., Debrecen, Pf. 51, Hungary-4001
[3] Instituto de Fisica, Universidade de São Paulo, CP 66318, 05389-970, São Paulo, Brasil
[4] IFIC, CSIC-Universidad de Valencia, A. C. 22085, E 46071, Valencia, Spain
[5] Instituto de Sciencias Nucleares, UNAM, A. P. 70-543, 04510 Mexico D. F.
[6] Joint Insitute for Nuclear Research, 141980 Dubna, Russia
[7] Insitute of Theoretical Physics, Justus Liebig University, Giessen, Germany

E-mail: `darai@atomki.hu`

Abstract. Shape isomers, including superdeformed and hyperderformed states can be determined from the quasi-dynamical SU(3) symmetry based on the Nilsson-model. We investigate the possible binary clusterizations of these shape isomers. The allowed cluster-configurations give a hint for the favourable reaction channels for populating these states. Our semimicroscopic approach is largely based on symmetry considerations, combined with energetic preference calculations. As illustrative examples some results for the ^{36}Ar, ^{40}Ca and ^{56}Ni nuclei are shown.

1. Introduction

The exotic shapes of atomic nuclei attract much attention recently both from the experimental and from the theoretical sides. Theoretically they are usually obtained from shell-model or mean-field calculations as the minima of the energy-surface. Here we discuss an alternative possibility of determining these states from the Nilsson-model, via their quasi-dynamical (or effective) $SU(3)$ symmetry [1, 2]. The study of the possible clusterizations of these exotic states are important not only for the better understanding of their structure, but also for predicting the appropriate reaction channels for their formation.

Some superdeformed bands in $N = Z$ nuclei were observed experimentally during the last decade. As for their hyperdeformed states are concerned, practically they belong to a domain for theoretical studies.

A remarkable example is that of the ^{36}Ar nucleus. The superdeformed shape of the ^{36}Ar nucleus was detected in 2000 [3], and some theoretical predictions were done on the hyperdeformed state from alpha-cluster [4], and binary cluster studies [5]. A recent study on molecular resonances in heavy-ion reactions seems to justify these predictions [6] not only from the respect of the moment of inertia of the band, but also concerning its clusterization, i.e. the preferred reaction channels for its population. Furthermore, the above-mentioned shape-calculation from Nilsson-model + quasi-dynamical symmetry gives a hyperdeformed state completely in line with the cluster models and with the experimental finding [7]. The superdeformed band of the ^{40}Ca nucleus has recently been observed experimentally [8] and the same band has been detected in an independent measurement [9]. For the ^{56}Ni two

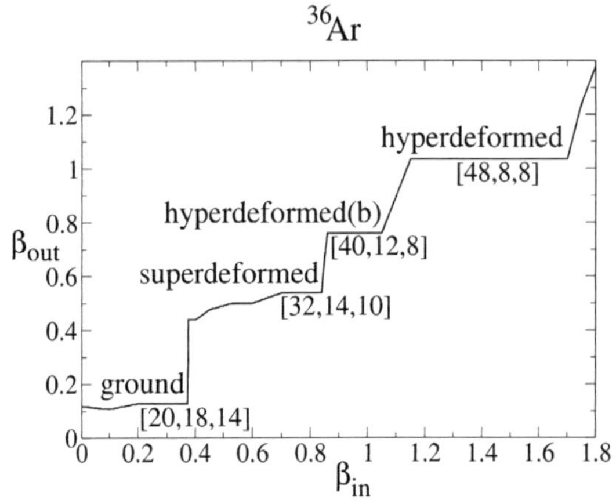

Figure 1. Quadrupole deformation of the ^{36}Ar nucleus from the Nilsson-model with the effective $U(3)$ quantum numbers at the plateaus.

deformed bands were observed in [10], and these bands were considered later on as examples of superdeformed bands [8].

2. Shape isomers

Nilsson-model calculations are very illuminating in searching for stable deformations and shape isomers [11]. Here we present the results of a new kind of calculation, based on the Nilsson-model. In particular, it consists in a self-consistency check with respect to the quadrupole deformation. It is done in terms of $U(3)$ symmetries in the following way. The $U(3)$ symmetry, which is an approximately good symmetry of light nuclei [12], is known to be uniquely related to the quadrupole shape [13]. Furthermore, when the real $U(3)$ symmetry breaks down, e.g. with increasing excitation energy (due to some symmetry breaking interactions, like spin-orbit, pairing, etc), a generalized version of it, called quasi-dynamical or effective $U(3)$ symmetry still survives [1]. The quasi-dynamical $U(3)$ quantum numbers can be obtained from Nilsson-calculations [14, 15]. Thus, the self-consistency calculation consists in the continuous variation of the quadrupole deformation, as an input for a Nilsson-calculation, and determination of the effective $U(3)$ quantum numbers or, from them, the corresponding β_{out} quadrupole deformation.The β_{out} as a function of the β_{in} is typically a "stair-like" function, as it can be seen in Figure 1. where the results for the input prolate deformations of the ^{36}Ar nucleus are shown. In this figure it is not the minima, rather the horizontal plateaus, which correspond to the stable shapes. They are stable in the sense, that they are insensitive to the small changes of the input parameter. Furthermore, at these deformations a good agreement between the input and output values can be observed. That is these deformations fulfill the self-consistency argument, to a reasonable approximation, between the input and output deformation-parameters.

For the ^{36}Ar nucleus this kind of calculation gives the ground state, of course, as the first stable shape. When proceeding towards larger deformation, the next one corresponds to the superdeformed shape, representing 4 nucleon excitation, being very much in line with the more recent Nilsson, as well as with the shell-model calculations [11, 3].

With increasing β-values two further stable plateaus appear, a less-pronounced one around 0.8 (β_{out}), and a very big one at cca 1.1. The first one coincides with the prolate state of [11]. The second one is in complete agreement with the prediction of the alpha-cluster model for the hyperdeformed state [4], as well as with the experimentally observed largely prolate band [6], as

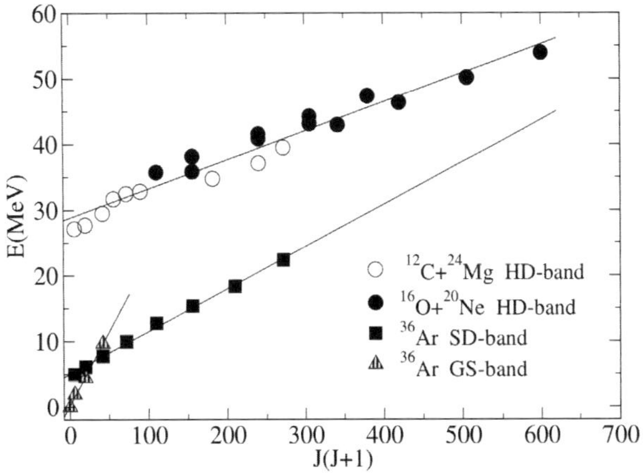

Figure 2. Excitation energy versus the angular momentum for the ground, superdeformed, and the recently observed hyperdeformed bands in the ^{36}Ar nucleus.

shown by Figure 2.

For ^{40}Ca and ^{56}Ni nuclei figures similar to Figure 1. have been obtained. For the ^{40}Ca the [20,20,20], [38,17,13] and [54,12,10] $U(3)$ quantum numbers have been found as characterizing the ground, superdeformed and hyperdeformed states, respectively. For the superdeformed state the $8\hbar\omega$ excitation is completely in line with the joint conclusion of the experimental and previous theoretical investigations [8, 16, 17]. For the ^{56}Ni the ground, superdeformed and hyperdeformed states have been identified as well. The superdeformed state is characterized by $4\hbar\omega$ excitation quanta similarly to other theoretical calculations [10] describing the experimental data.

For light nuclei, like ^{24}Mg and ^{28}Si, where detailed comparison could be made, the results of this self-consistency calculation are in very good agreement with that of the energy-minima calculations [18, 19].

3. Cluster structure

When studying the possible cluster structure of a specific nuclear state, like the ones of the previous section, one has to take into account the effects of two basic natural laws: the energy-minimum principle and the Pauli-exclusion principle. Fully microscopic cluster models are able to handle both of them in a completely satisfactory manner, however, their range of applicability is limited due to the large-scale calculations. We are interested in applying a semimicroscopic approach, which can be extended even to heavy nuclei and exotic clusterization [5, 20]. Symmetry-considerations can be helpful along this line, and exclude the Pauli-forbidden states without carrying out the antisymmetrization explicitly. For light nuclei even the real $SU(3)$ can lead us to useful results, but for the heavy ones we need to change to the more general quasi-dynamical symmetry. (This is applicable also for light nuclei, of course, as illustrated e.g. by the previous section. When the real $SU(3)$ is approximately valid, then it coincides with the quasi-dynamical one [15].)

There are two simple recipes, which are based on the microscopic picture, yet they are easy to apply systematically. These are the $U(3)$ selection rule [21], and Harvey's prescription [22]. In the case of a binary clusterization the $U(3)$ selection rule reads:

$$[n_1, n_2, n_3] = [n_1^{(1)}, n_2^{(1)}, n_3^{(1)}] \otimes [n_1^{(2)}, n_2^{(2)}, n_3^{(2)}] \otimes [n^{(R)}, 0, 0], \tag{1}$$

where $[n_1, n_2, n_3]$ is the set of $U(3)$ quantum numbers of the parent nucleus, the superscripts (i)

stand for the ith cluster, (R) indicates relative motion. A given clusterization is allowed if there is a matching between $[n_1, n_2, n_3]$ and the product representations. When a cluster configuration is forbidden, we can characterize its forbiddenness quantitatively [23]. The Harvey-prescription says that when two nuclei amalgamate, then only the number of quanta along the molecular axis can change, while numbers along the other two directions remain unchanged. Both of these rules applies the harmonic oscillator basis, thus there is a considerable similarities between them. However, they are not identical, rather, they are complementary to each other in a sense. Therefore, they should be applied in a combined way [19].

Since the $U(3)$ symmetry is related to the quadrupole deformation of the nucleus, the $U(3)$ selection rule can be interpreted as a consistency check of the quadrupole deformation, between the shell model (or collective model) states and the corresponding cluster states. It should also be mentioned that the $U(3)$ selection rule, which deals with the space-symmetry of the states, is always accompanied by a similar $U^{ST}(4)$ [24] selection rule for the spin-isospin degrees of freedom.

As for the energetic stability of the cluster-configuration is concerned, either binding-energy arguments [25], or double-folding calculation can be applied [26]. The criterium of maximal stability requires the largest value of the summed differences of the measured binding energies and the corresponding liquid drop values. In the dinuclear system model the potential energy U is minimized with respect to the mass asymmetry for each fixed charge asymmetry. Both methods indicate that the alpha-like (i.e. $N = Z =$ even) configurations are energetically preferred. Thus we concentrate here on alpha-like binary clusterizations.

The superdeformed state of the ^{36}Ar nucleus shows a little shape-uncertainty. This is a joined conclusion of several theoretical studies. In our method this uncertainty is reflected by the appearance of close-lying, but not completely identical $U(3)$ representations. Therefore, we have considered the shapes corresponding to the [32,14,10] effective quantum numbers, and to the [32,16,8], which correspond to a simple harmonic oscillator state. The latter one allows (^{4}He, ^{8}Be, ^{12}C, ^{16}O)+core clusterizations, while the former one allows only the last two configurations (though the other ones are not very strongly forbidden either).

The hyperdeformed state on the other hand seems to have a very well-defined symmetry, and it allows only ^{24}Mg+^{12}C and ^{20}Ne+^{16}O binary configurations. (In these considerations the clusters are supposed to be in their ground intrinsic states. E.g. a ^{28}Si+^{8}Be configuration is allowed, too with the prolate shape of the ^{28}Si, but it is thought to be not the dominant component of the ground-state wavefunction.)

It is also remarkable that the ^{24}Mg+^{12}C clusterization is possible both in the ground-sate, and in the superdeformed, as well as in the hyperdeformed state. The difference between these configurations is the relative orientation of the two deformed clusters with respect to the molecular axis. This observation could be made due to the facts that i) the Pauli-principle was taken into account, and ii) no oversimplifying model assumptions (e.g. spherically or cylindrically symmetric cluster shapes) were made.

Concerning the ^{40}Ca nucleus in its superdeformed state the allowed clusterizations are ^{8}Be+^{32}S, ^{12}C+^{28}Si and ^{20}Ne+^{20}Ne. In the hyperdeformed state of this nucleus the ^{16}O+^{24}Mg and the ^{20}Ne+^{20}Ne clusterizations are allowed.

We determined the possible binary alpha-like cluster-configurations for the super- and hyperdeformed states of ^{56}Ni as well. Similarly to the previous nuclei in its hyperdeformed state the more symmetric clusterizations are allowed, while in the superdeformed state asymmetric and symmetric clusterizations both appear.

4. Conclusions

With this contribution we wished to demonstrate that symmetry-considerations can be helpful in studying both the shape isomers and the possible clusterizations of atomic nuclei. They are able to shed some light also on the interrelations of these two phenomena. In this respect they can be helpful in predicting the favourable reactions for the population of superdeformed and hyperdeformed states. The hyperdeformed state of the ^{36}Ar nucleus serves as a very interesting example. Its existence was first foreseen from alpha-cluster model [4], then binary-cluster studies predicted their population in ^{24}Mg+^{12}C and ^{20}Ne+^{16}O reactions [5]. Recent reaction studies [6] seem to justify these predictions, and structure-calculations [7] show that the same state is obtained from the Nilsson-model via the quasi-dynamical $SU(3)$ symmetry.

Acknowledgments

This work was supported by the OTKA (Grant No. K72357), by the NKTH (project no.ES-26/2008), by the MTA-CONACyT exchange program, by the MTA-JINR (EAI-2009/001) and MTA-BAS bilateral projects. Fruitful discussions with Prof. A. Georgieva on this topic is kindly acknowledged.

References

[1] Rochford P, Rowe D J 1988 *Phys. Lett.* B **210**, 5; Rowe D J, Rochford P, Repka J 1988 *J. Math. Phys.* **29**, 572
[2] Cseh J 2006 *Proc. IV Int. Symp. on Quantum Theory and Symmetries (Varna)* (Sofia: Heron Press) p. 918
[3] Svensson C E, Machiavelli A O, Juodagalvis A, Poves A, Ragnarsson I, Aberg S, Appelbe D E, Austin R A E, Baktash C, Ball G C et al 2000 *Phys. Rev. Lett.* **85** 2693
[4] Rae W D M, Merchant A C 1992 *Phys. Lett.* **B279** 207
[5] Cseh J, Algora A, Darai J, Hess P O 2004 *Phys. Rev.* C **70** 034311
[6] Sciani W, Otani Y, Lepine-Szily A, Benjamin E A, Chamon L C, Lichtenthaler R, Darai J, Cseh J 2009 *Phys. Rev.* C **80** 034319
[7] Cseh J, Darai J, Sciani W, Otani Y, Lepine-Szily A, Benjamin E A, Chamon L C, Lichtenthaler R 2009 *Phys. Rev.* C **80** 034320
[8] Ideguchi E et al. 2001 *Phys. Rev. Lett.* **87** 222501; Chiara C J et al. 2003 *Phys. Rev.* C. **67** 041303
[9] Torilov S et al. 2004 *Eur. Phys. J.* A **19** 307
[10] Rudolph D et al 1999 *Phys. Rev. Lett.* **82** 3763
[11] Leander G, Larsson S E 1975 *Nucl. Phys.* A **239**, 93
[12] Elliott J P 1958 *Proc. Roy. Soc.* A **245** 128 and 562
[13] Rowe D J 1985 *Rep. Prog. Phys.* **48** 1419
[14] Jarrio M, Wood J L and Rowe D J 1991 *Nucl. Phys.* A **528** 409
[15] Hess P O, Algora A, Hunyadi M, Cseh J 2002 *Eur. Phys. J.* A **15** 449
[16] Gerace W J, Greene A M 1967 *Nucl. Phys.* A **93** 110; Gerace W J, Greene A M 1969 *Nucl. Phys.* A **123** 241; Zheng D C et al. 1990 *Phys. Rev.* C **87** 1004
[17] Kanada-Enyo Y, Kimura M, Horiuchi H 2003 *AIP Conf. Proc., Exotic clustering: 4th Catania Rel. Ion Stud. (Catania)* 644 p 188
[18] Cseh J, Darai J, Algora A, Yepez-Martinez H, Hess P O *Rev. Mex. Fis.* S. **54** (3) 30
[19] Cseh J, Darai J 2008 *AIP Conf. Proc., Fusion08 (Chicago)* 1098: Fusion08 p 225
[20] Darai J, Cseh J, Antonenko N V, Algora A, Hess P O, Jolos R V and Scheid W 2009 *AIP Conf. Proc., Nucl. Structure and Dynamics (Dubrovnik),* vol 1165, ed Milin M, Niksic T et al, pp 37-38
[21] Cseh J 1993 *J. Phys.* G **19** L97, and references therein
[22] Harvey M 1975 *Proc. 2nd Int. Conf. on Clustering Phenomena in Nuclei* (College Park) USDERA report ORO-4856-26 p 549
[23] Algora A, Cseh J 1996 *J. Phys.* G **22** L39
[24] Wigner E P 1937 *Phys. Rev.* **51** 106
[25] Buck B, Merchant A C, Perez S M 2000 Few-Body Systems **29** 53
[26] Cseh J, Darai J, Antonenko N V, Algora A, Hess P O, Jolos R V, Scheid W 2006 *Rev. Mex. Fis.* S. **52** (4) 11, and references therein

Journal of Physics: Conference Series **205** (2010) 012023 doi:10.1088/1742-6596/205/1/012023

The sextic oscillator and $E(5)$ type nuclei

Géza Lévai

Institute of Nuclear Research of the Hungarian Academy of Sciences (ATOMKI), Debrecen, Pf. 51, Hungary 4001

E-mail: levai@atomki.hu

Abstract. The sextic oscillator is applied as a γ-independent potential to describe the energy spectrum of nuclei in regions where the occurrence of $E(5)$ symmetry has been assumed. Experimental levels of Ru, Pd, Cd, Te, Xe and Ba isotopes have been assigned to model states characterized by the $\xi = 1,\ 2$ quantum numbers, and the two model parameters a and b have been determined from a least square fit procedure. The resulting potentials have a spherical or deformed minimum. Most of the typical $E(5)$ candidates, i.e. ^{104}Ru, ^{102}Pd, ^{106,108}Cd and ^{124}Te correspond to parameters lying close to the parabola separating the two domains in the $a - b$ plane.

1. Introduction

The collective motion of nuclei results in characteristic shapes near the equilibrium states of nuclei. The most well-known shapes, such as the spherical vibrator, the axially deformed rotor and the γ-unstable rotor are linked with the U(5), SU(3) and O(6) symmetries, respectively [1]. A rather effective method of the consistent quantum mechanical descriptiom of these collective features on the phenomenological level is the application of the Bohr Hamiltonian, which uses a potential picture to describe quadrupole collective excitations in terms of the intrinsic shape variables β, γ. The minima of the $V(\beta,\gamma)$ potential define the equilibrium shapes, while the potential itself depends on various parameters characterizing the given nucleus, e.g. the proton and neutron number, etc. Changing these parameters gradually (e.g. by proceeding along an isotope chain) the potential shape and its minima change and may give rise to transition from one equilibrium shape (and symmetry) to another one. This procedure can be interpreted as a shape phase transition of some order that goes through a critical point.

In the simplest case the potential is independent from the γ variable, in which case the Bohr Hamiltonian reduces to a Shrödinger-like radial equation. Such a situation can be used to describe the transition from the spherical (U(5)) to the γ-unstable (O(6)) shape phases. Iachello approximated this second-order phase transition with the infinite square well [2] and introduced the concept of the $E(5)$ critical point symmetry. This problem can be solved exactly in terms of Bessel functions, so parameter-free analytical results are available for the key spectroscopic properties (energy spectrum, electric quadrupole transition rates). Comparison of these quantities with the spectroscopic data of actual nuclei revealed that several nuclei (e.g. ^{134}Ba) are close to the E(5) critical point symmetry. Later the $X(5)$ symmetry was also introduced [3] to give account of the first-order phase transition between the the spherical and axially deformed shape phases. In this case the γ shape variable also has to be incorporated into the formalism through certain approximation schemes.

In certain cases the Bohr Hamiltonian (or the one-dimensional equation derived from it) can be solved exactly. These cases are typically adapted to the five-dimensional problem from solvable radial problems of three-dimensional potentials. In order to obtain a complete picture, potentials with centrifugal-like term have to be considered. The list of exactly solvable problems of this type is finite (see e.g. [4]). Besides the infinite square well only the Kratzer and Davidson potentials can be solved for arbitrary quantum numbers. Recently we introduced a rather flexible potential [5], which is quasi-exactly solvable meaning that the closed analytical solutions are available for a finite subset of states for certain potential parameters. However, this set of states contains the most important low-lying levels necessary to characterize the collective excitations of the nucleus within the scheme of the Bohr Hamiltonian. This potential, the sextic oscillator can describe situations in which a nucleus has spherical minimum (at $\beta = 0$), deformed minimum (at $\beta > 0$) or both. This feature is certainly advantageous when one wishes to describe transitions between different shape phases. In its first application the sextic oscillator was considered as a γ-independent potential describing the transition between the spherical and the γ-unstable shape phases. Here we employ this potential to describe the energy spectrum of a number of nuclei that have been proposed as candidates for $E(5)$ symmetry.

2. The sextic oscillator

The general Bohr Hamiltonian is written as [6]

$$H = -\frac{\hbar^2}{2B}\left(\frac{1}{\beta^4}\frac{\partial}{\partial\beta}\beta^4\frac{\partial}{\partial\beta} + \frac{1}{\beta^2\sin 3\gamma}\frac{\partial}{\partial\gamma}\sin 3\gamma\frac{\partial}{\partial\gamma} - \frac{1}{4\beta^2}\sum_k \frac{Q_k^2}{\sin^2(\gamma - \frac{2}{3}\pi k)}\right) + V(\beta,\gamma) \ . \quad (1)$$

If the potential is assumed to be independent of the γ variable, i.e. $V(\beta,\gamma) = U(\beta)$, then the β-dependent part can be separated by the substitution $\Psi(\beta,\gamma,\theta_i) = \beta^{-2}\phi(\beta)\Phi(\gamma,\theta_i)$, leading to a form similar to the usual radial Schrödinger equation

$$-\frac{\mathrm{d}^2\phi}{\mathrm{d}\beta^2} + \left(\frac{(\tau+1)(\tau+2)}{\beta^2} + u(\beta)\right)\phi = \epsilon\phi \ , \quad (2)$$

where $\epsilon = \frac{2B}{\hbar^2}E$ and $u(\beta) = \frac{2B}{\hbar^2}U(\beta)$. In (2) τ originates from the angular equation and essentially plays the rôle of the angular momentum in five spatial dimensions.

In Ref. [5] the $u(\beta)$ potential was chosen ast the sextic oscillator

$$u(\beta) = (b^2 - 4ac^\pi)\beta^2 + 2ab\beta^4 + a^2\beta^6 + u_0^\pi \ , \quad (3)$$

which is a quasi-exactly solvable potential [7] meaning that only a subset of its lowest-energy solutions can be obtained exactly for certain combinations of the potential parameters. $u(\beta)$ in (3) depends on three parameters, a, b and c^π, but c^π is related to the τ parameter via $2c^\pi = (\tau + 2M + \frac{7}{2})$ in order to account for the "centrifugal" term. Here M is a non-negative integer number that sets the number $M+1$ of solutions that can be obtained exactly. In typical applications of the sextic oscillator in the Bohr Hamiltonian it is sufficient to consider $M = 0$ and $M = 1$: with this the most characteristic levels are included in the description. The potential is slightly different for even and odd values of τ (or the $\tau + 2M$ combination): in the former case $c^+ = 11/4$, while in the latter one $c^- = 13/4$ corresponding to $\tau + 2M = 2$ and 3, respectively. This slight difference manifests itself near $\beta = 0$ in the coefficient of the β^2 term. In Ref. [5] this slight ambiguity was handled by selecting the constant terms u_0^+ and u_0^- such that the minima of the two potentials are at the same energy.

The solutions are written as

$$\phi_{n,\tau}(\beta) = N_{n,\tau}P_n(\beta^2)(\beta^2)^{\frac{\tau}{4}+1}\exp\left(-\frac{a}{4}\beta^4 - \frac{b}{2}\beta^2\right) \ , \quad (4)$$

Table 1. Explicit form of the lowest few energy eigenvalues and wavefunctions. Here $\lambda_\pm = 2b \pm 2(b^2 + 10a)^{1/2}$, $\tilde{\lambda}_\pm = 2b \pm 2(b^2 + 14a)^{1/2}$, and $u_0 = u_0(c = 11/4) - u_0(c = 13/4)$ hold.

ξ	τ	M	$E_{\xi,\tau}$	$E^*_{\xi,\tau} = E_{\xi,\tau} - E_{1,0}$	$\phi_{\xi,\tau}/\exp(-\frac{a\beta^4}{4} - \frac{b\beta^2}{2})$
1	0	1	$5b + \lambda_-$	0	$N_{10}(\beta^2 - \lambda_-/10)$
1	1	1	$7b + \tilde{\lambda}_- + u_0$	$2b + \tilde{\lambda}_- - \lambda_- + u_0$	$N_{11}\beta(\beta^2 - \tilde{\lambda}_-/14)$
1	2	0	$9b$	$4b - \lambda_-$	$N_{12}\beta^4$
1	3	0	$11b + u_0$	$6b - \lambda_- + u_0$	$N_{13}\beta^5$
2	0	1	$5b + \lambda_+$	$\lambda_+ - \lambda_-$	$N_{20}(\beta^2 - \lambda_+/10)$
2	1	1	$7b + \tilde{\lambda}_+ + u_0$	$2b + \tilde{\lambda}_+ - \lambda_- + u_0$	$N_{21}\beta(\beta^2 - \tilde{\lambda}_+/14)$

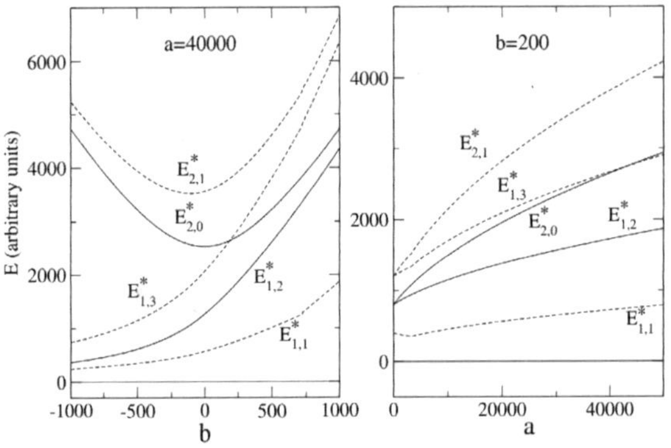

Figure 1. Excitation energies $E^*_{\xi,\tau} = E_{\xi,\tau} - E_{1,0}$ as with $a = 40000$ fixed as a function of b (left panel) and with $b = 200$ fixed as a function of a (right panel).

where $P_{n,\tau}(\beta^2)$ is a polynomial of the order $n = 0,\ 1$. The $N_{n,\tau}$ normalization constants and the matrix elements necessary to calculate B($E2$) values can be expressed exactly in terms of parabolic cylinder functions [5]. Note that normalizability requires $a \geq 0$ and that $a = 0$ corresponds to the harmonic oscillator limit. Table 1 displays the energy eigenvalues and the corresponding (unnormalized) eigenfunctions, while Figure 1 shows the trend of the energy eigenvalues for some fixed values of a and b.

Note that the a and b parameters can be fitted exactly to two energy eigenvalues as $a = E^*_{1,2}[E^*_{2,0} - E^*_{1,2}]/40$, $b = [2E^*_{1,2} - E^*_{2,0}]/4$. In practice this means that knowing the energies of the 0^+ state with $\xi = 2$ and that of the 2^+_2 or 4^+_1 state (with $\xi = 1$ and $\tau = 2$) immediately provides us with a first approximation of the model parameters.

Obviously, the extrema of the potential depend on the sign of b and $b^2 - 4ac^\pi$, i.e. on the coefficients of the β^2 and β^4 terms in (3). If $b^2 > 4ac^\pi$ and $b > 0$ hold then the potential has a minimum at $\beta = 0$ and it increases with β. When $b^2 < 4ac^\pi$ (irrespective of the sign of β) the minimum shifts to a finite value of β, while for $b^2 > 4ac^\pi$ and $b < 0$ (i.e. $b < -2(ac^\pi)^{1/2}$) there are two minima (one at $b = 0$ and one at $\beta > 0$) separated by a local maximum. The $b^2 = 4ac^\pi$ parabola thus separates the $a(b)$ plane into three domains: to the right of it the minimum is spherical ($\beta_{\text{min}} = 0$), above the parabola the minimum is deformed ($\beta_{\text{min}} > 0$), while to the left there are two minima such that ($\beta_{\text{min2}} > \beta_{\text{min1}} = 0$). Crossing the right leg of the parabola thus corresponds to the transition from the spherical to the deformed phase, i.e. it is expected to be close to the $E(5)$ critical point. The domain to the left of the parabola corresponds to potential shapes typical for the $X(5)$ symmetry, so the sextic oscillator may also be appropriate to describe the β-dependent part of the $V(\beta,\gamma)$ potential of such systems.

3. Application to some $E(5)$ candidates

The signatures of the $E(5)$ symmetry originate from the parameter-free calculations with the five-dimensional infinte well potential in the β shape variable [2] mimicking the situation in which a second-order phase transition occurs between the spherical and γ-unstable phases. The ratios of certain energy eigenvalues and electric quadrupole transition rates have been compared with experimental data and several nuclei have been proposed as candidates for the $E(5)$ symmetry. The first example was the ^{134}Ba nucleus [8]. Later the ^{104}Ru [9] and ^{102}Pd [10] nuclei have also been discussed in terms of this scheme. Later in a systematic search for $E(5)$ candidates the ^{106}Cd, ^{108}Cd, ^{124}Te and ^{128}Xe nuclei have also been proposed [11]. More recently the ^{124}Xe isotope has also been mentioned as a promising candidate [12].

From these nuclei the sextic oscillator has already been applied to ^{134}Ba, although not in a full scale fit, rather as an illustration of the model [5]. It turned out that the sextic oscillator was able to reproduce the main signatures of the $E(5)$ symmetry by applying model parameters corresponding to a potential with a deformed minimum. As another application the energy spectrum of the chain of Ru isotopes has been fitted and the resulting potentials have been analyzed [13, 14]. It was shown that starting from ^{98}Ru the spectrum is close to the $U(5)$ harmonic limit, then at ^{104}Ru a flat potential develops, which evolves further to a potential with a deformed minimum for ^{106}Ru and ^{108}Ru.

Here we extend these studies to a number of isotopes near those proposed recently as $E(5)$ candidates. For this we first associate experimental levels, notably 2_1^+; 4_1^+, 2_2^+; 0_2^+, 3_1^+, 4_2^+, 6_1^+ and 0_3^+ with model states corresponding to quantum numbers $(\xi,\tau)=(1,1)$, $(1,2)$, $(1,3)$ and (2,0). The identification of these levels is not too complicated, except for the second and third 0^+ levels. In identifying these levels the predictions of the $E(5)$ scheme on the $B(E2)$ rates can be used: the $0^+(2,0)$ state is known to decay strongly to the $2^+(1,1)$ level, while the $0^+(1,3)$ state prefers to decay to the $2^+(1,2)$ level. In ^{134}Ba, for example, the 0_3^+ experimental level is thought to be the one with $(\xi,\tau)=(2,0)$. According to the analysis of Ref. [11] the situation is similar in the case of ^{102}Pd and ^{128}Xe too, while in ^{106,108}Cd and ^{124}Te the 0_2^+ experimental state is associated with the $(\xi,\tau)=(2,0)$ level. The situation with the $B(E2)$ rates justifies the same treatment of ^{100,102,104}Ru too [17].

In our analysis we considered altogether 6 Ru, 3 Pd, 2 Cd, 3 Te, 6 Xe and 2 Ba isotopes for which the data set was reasonably complete. Their energy levels are associated with the model states in the upper half Tables 2, 3 and 4. The experimental data set [17] was sometimes incomplete: in some cases only the 0_2^+ level was know, and often there is no information on the $B(E2)$ rates. In identifying the 0^+ levels thus we followed the arguments discussed above for each member of the given isotope chain.

In the next step we performed a least square fit of the a and b parameters such that states with ambiguous J^π assignment were considered with weight 0.5. One further special feature of the $E(5)$ symmetry is that the 2_2^+ and 4_1^+ levels are degenerate, as are the 3_1^+, 4_2^+, 6_1^+ and 0_2^+ (or 0_3^+) levels. Obviously, this degeneracy is not fulfilled in the experimental spectrum, so the fit concerned an average value of these energy levels.

In the case of the light Ru isotopes a turned out to be a small negative value, which is clearly not allowed in the wave functions, so in these cases we chose $a=0$ corresponding to the harmonic oscillator limit. Indeed, the energy spectrum of these nuclei is close to a harmonic vibrator. The resulting parameters are displayed in the middle section of the tables, while the calculated energy eigenvalues are displayed in the bottom section. The last line in the Tables indicates whether the corresponding potentials have spherical or deformed minimum. As discussed before, this is clearly determined by the sign of $b^2-4ac^+=b^2-11c$.

Table 2. Experimental levels (in keV) of Ru and Pd nuclei corresponding to the $\phi_{\xi,\tau}$ model states, the a and b model parameters and the $E_{\xi,\tau}$ energy eigenvalues calculated from them. Experimental energies in parentheses indicate ambiguous J^π assignment. The last row displays the minimum of the corresponding $V(\beta)$ potential.

A	^{98}Ru	^{100}Ru	^{102}Ru	^{104}Ru	^{106}Ru	^{108}Ru	^{102}Pd	^{104}Pd	^{106}Pd
N	54	56	58	60	62	64	56	58	60
$E_{1,1}^{\text{Exp.}}(2^+)$	652	540	475	358	270	242	556	556	512
$E_{1,2}^{\text{Exp.}}(2^+)$	1414	1362	1103	893	792	(708)	1534	1341	1128
$E_{1,2}^{\text{Exp.}}(4^+)$	1398	1226	1106	888	(715)	665	1276	1323	1129
$E_{1,3}^{\text{Exp.}}(0^+)$		1741	1837				1593	1334	1134
$E_{1,3}^{\text{Exp.}}(3^+)$	1797		1522	1242	(1092)	(945)	2112	1821	1558
$E_{1,3}^{\text{Exp.}}(4^+)$	2267	(2075)	(1799)			(1183)	2138	2082	1932
$E_{1,3}^{\text{Exp.}}(6^+)$	2223	2076	1873	1556	(1296)	1240	2111	2249	2076
$E_{2,0}^{\text{Exp.}}(0^+)$	1322	1130	944	988	991	976	1658	1793	1706
a	0	0	0	1496	4190	5154	8918	15692	15812
b	347	318	283	216	143	114	283	191	134
$E_{1,1}^{\text{Th.}}$	695	635	566	409	308	297	528	511	461
$E_{1,2}^{\text{Th.}}$	1389	1270	1131	930	786	735	1388	1261	1108
$E_{1,3}^{\text{Th.}}$	2084	1905	1697	1362	1157	1108	1999	1906	1708
$E_{2,0}^{\text{Th.}}$	1389	1270	1131	994	999	1016	1645	1759	1679
$\beta_{\text{min.}}$	H.O.	H.O.	H.O.	0	>0	>0	>0	>0	>0

4. Discussion

Although the degree of completeness of the data sets is different for the members of the isotope chains, the parameters a and b change rather smoothly. The only exception is ^{122}Xe, which strays away from the rest of the chain occupying a rather straight line with small b and a increasing with increasing mass number A. This behavious may be explained by the fact that the 0_3^+ level is not known in ^{122}Xe, so the fit might be distorted. The other heavy isotopes, ^{134}Ba and ^{136}Ba are in the same region of the $a - b$ parameter space, although there a decreases with increasing A. It is also remarkable that the paramers chosen in Ref. [5] to generate the spectroscopic properties of ^{134}Ba ($a = 40000$, $b = 200$) are not too far from the fitted values.

The evolution of the Ru chain is also rather smooth: a and b move monotonously with increasing A. Furthermore, in the process the chain crosses the parabola separating the shape phases with spherical and deformed potential minimum. In fact, it is the ^{104}Ru nucleus that is closest to the parabola, which is in agreement with the findings of [9]. The Te chain also crosses this boundary, to which ^{124}Te (and also ^{122}Te) lies rather close. The former isotope was proposed as another example for $E(5)$ symmetry in [11]. The Pd chain also behaves in a similar way, and although it does not cross the parabola, the ^{102}Pd isotope, another candidate in [11] gets to it rather close from the deformed side. Both the ^{106}Cr and ^{108}Cr, the two remaining $E(5)$ candidates of [11] are also close to the parabola from the side of the spherical minimum.

Table 3. The same as Table 2 for Cd, Te and Ba nuclei.

A	^{106}Cd	^{108}Cd	^{122}Te	^{124}Te	^{126}Te	^{134}Ba	^{136}Ba
N	58	60	70	72	74	78	80
$E_{1,1}^{\text{Exp.}}(2^+)$	633	633	564	603	666	605	819
$E_{1,2}^{\text{Exp.}}(2^+)$	1717	1602	1257	1326	1420	1170	1551
$E_{1,2}^{\text{Exp.}}(4^+)$	1494	1508	1181	1249	1361	1401	1867
$E_{1,3}^{\text{Exp.}}(0^+)$	2144	1913	(1750)	1883		1761	1579
$E_{1,3}^{\text{Exp.}}(3^+)$	(2254)	2146		2483	2128	1643	2431
$E_{1,3}^{\text{Exp.}}(4^+)$	2104	2239	1909	1958	2013	1970	2054
$E_{1,3}^{\text{Exp.}}(6^+)$	2492	2541	1751	1747	1776	(2118)	2207
$E_{2,0}^{\text{Exp.}}(0^+)$	1795	1720	1357	1657	1873	(2160)	2141
a	7920	6989	2921	11917	17131	27308	20510
b	350	339	286	251	206	65	223
$E_{1,1}^{\text{Th.}}$	633	616	539	529	540	516	589
$E_{1,2}^{\text{Th.}}$	1600	1538	1240	1355	1336	1184	1455
$E_{1,3}^{\text{Th.}}$	2301	2217	1812	1989	2015	1890	2195
$E_{2,0}^{\text{Th.}}$	1798	1720	1334	1707	1849	2107	2019
$\beta_{\text{min.}}$	0	0	0	>0	>0	>0	>0

It is remarkable that the best $E(5)$ candidates are rather close to the boundary separating the shape phases corresponding to spherical and deformed minimum. The two exceptions are ^{128}Xe and ^{134}Ba, which differ from the remaining $E(5)$ candidates in that it is the 0_3^+ state that is assigned to $\xi = 2$ level, and the 0_2^+ level is substantially lower (400 keV). (Although 0_3^+ plays the same role in ^{102}Pd, the energy difference between the 0_2^+ and 0_3^+ states is much smaller.)

It has to be noted that we did not adjust the B inertia parameter in (1), although it is known to vary with the mass A approximately as $B \sim A^{5/3}$ [6]. Neglecting this effect does not influence the results significantly within an isotope chain. The effect is more significant within the whole mass range considered here, but it can be handled by a rescaling of a, b and c.

Due to the lack of space we did not consider the electric quadrupole transition rates, although they can be calculated in closed form [5]. This task will be completed elsewhere. Another possible task for the future is developing the formalism to treat nuclei with $X(5)$ symmetry. For this the γ degrees of freedom also have to be incorporated into the potential. There are several ways to approximate this effect [3, 15, 16], so the application of the sextic oscillator to $X(5)$ nuclei seems not only desirable but also possible. For this potential shapes with two minima (corresponding to $b < 0$ and $b^2 > 4ac^{\pi}$) seem suitable.

Acknowledgments
This work was supported by the OTKA grant No. T49646 (Hungary).

Table 4. The same as Table 2 for Xe nuclei.

A	^{120}Xe	^{122}Xe	^{124}Xe	^{126}Xe	^{128}Xe	^{130}Xe
N	66	68	70	72	74	76
$E_{1,1}^{\mathrm{Exp.}}(2^+)$	322	331	354	389	443	536
$E_{1,2}^{\mathrm{Exp.}}(2^+)$	876	(843)	846	880	969	1122
$E_{1,2}^{\mathrm{Exp.}}(4^+)$	796	828	879	942	1033	1205
$E_{1,3}^{\mathrm{Exp.}}(0^+)$	909	1149	1269	1314	1583	1793
$E_{1,3}^{\mathrm{Exp.}}(3^+)$	1271	1214	1248	1318	1430	1633
$E_{1,3}^{\mathrm{Exp.}}(4^+)$	1401	(1402)	1438	1488	1604	(1808)
$E_{1,3}^{\mathrm{Exp.}}(6^+)$	1397	1467	1549	1634	1737	1944
$E_{2,0}^{\mathrm{Exp.}}(0^+)$	1623		1690	1760	1877	2017
a	16148	9521	17760	19239	21875	24822
b	-17	95	2	5	25	65
$E_{1,1}^{\mathrm{Th.}}$	350	349	380	398	437	493
$E_{1,2}^{\mathrm{Th.}}$	770	835	846	888	986	1134
$E_{1,3}^{\mathrm{Th.}}$	1267	1293	1380	1447	1594	1809
$E_{2,0}^{\mathrm{Th.}}$	1609	1291	1686	1755	1873	2010
$\beta_{\mathrm{min.}}$	>0	>0	>0	>0	>0	>0

References

[1] Iachello F and Arima A 1987 *The Interacting Boson Model* (Cambridge: Cambridge University Press)
[2] Iachello F 2000 *Phys. Rev. Lett.* **85** 3580
[3] Iachello F 2001 *Phys. Rev. Lett.* **87** 052502
[4] Fortunato L 2005 *Eur. Phys. J. A* **26** 1
[5] Lévai G and Arias J M 2004 *Phys. Rev. C* **69** 014304
[6] Bohr A and Mottelson B 1975 *Nuclear structure Vol. II* (Reading, MA: Benjamin)
[7] Ushveridze A G 1994 *Quasi-exactly solvable models in quantum mechanics* (Bristol: IOP Publishing)
[8] Casten R F and Zamfir N V 2000 *Phys. Rev. Lett.* **85** 3584
[9] Frank A, Alonso C E and Arias J M 2001 *Phys. Rev. C* **65** 014301
[10] Zamfir N V, Caprio M A, Casten R F, Barton C J, Beausang C W, Berant Z, Brenner D S, Chou W T, Cooper J R, Hecht A A, Krücken R, Newman H, Novak J R, Pietralla N, Wolf A and Zyromski K E 2002 *Phys. Rev. C* **65** 044325
[11] Clark R M, Cromaz M, Deleplanque M, Diamond R M, Fallon P, Lee I Y, Macchiavelli A O, Mahmud H, Rodriguez-Vieitez E, Stephens F S and Ward D 2004 *Phys. Rev. C* **69** 064322
[12] Pietralla N 2009 private communication.
[13] Lévai G 2006 *Rev. Mex. Fis. S* **52** 75
[14] Lévai G. *AIP Conf. Proc.* **1165** 211.
[15] De Baerdemacker S, Fortunato L, Hellemans V and Heyde K 2006 *Nucl. Phys. A* **769** 16
[16] Fortunato L, De Baerdemacker S and Heyde K 2006 *Phys. Rev. C* **74** 014310
[17] *Isotope Explorer 3.0 b1* http://ie.lbl.gov/ensdf/

Journal of Physics: Conference Series **205** (2010) 012024 doi:10.1088/1742-6596/205/1/012024

Microscopic description of shape evolution in medium-mass nuclei

P Sarriguren[1], R R Rodríguez-Guzmán[2], L M Robledo[3]

[1] Instituto de Estructura de la Materia, CSIC, Serrano 123, E-28006 Madrid, Spain
[2] Departamento de Física Aplicada, Universidad de Huelva, 21071 Huelva, Spain
[3] Departamento de Física Teórica C-XI, Universidad Autónoma de Madrid, 28049-Madrid, Spain

E-mail: sarriguren@iem.cfmac.csic.es

Abstract. The evolution of the ground-state nuclear shapes with the number of nucleons is studied within a self-consistent Hartree-Fock-Bogoliubov formalism based on Gogny and Skyrme density-dependent interactions. Potential energy surfaces including triaxial degrees of freedom are studied in various isotopic chains. Signatures for the E(5) critical point symmetry are analyzed in Pd, Xe, and Ba isotopes, while those for the X(5) are studied in Nd, Sm, Gd, and Dy isotopes. Oblate to prolate shape transitions are also studied in Yb, Hf, W, Os, and Pt isotopic chains. The transitions are discussed in terms of the deformation dependence of the single-particle energies.

1. Introduction

The evolution of the atomic nuclei ground-state shapes as the number of their constituents changes is an active research field [1] that has been recently addressed both experimentally [2] and theoretically [3, 4, 5, 6, 7, 8, 9]. From the experimental side, low-lying spectroscopy is one of the most powerful sources of information about nuclear shapes and shape transitions, since one can establish signatures correlating the excitation energies with the deformation properties. From the theoretical side, the problem has been addressed from many different points of view. Geometrical and algebraic models, as well as microscopic models based on shell model or mean field approaches have been considered. The shape evolution is described differently depending on the approach adopted. Thus, in algebraic models like the Interacting Boson Model (IBM) [10], shape changes correspond to the breaking of the dynamic symmetries of the model and are considered as phase transitions. In the shell model, nuclear deformation and its evolution occur due to configuration mixing. From a mean field point of view, shape changes arise when the deformed single-particle levels are energetically favored to a different degree in open shell nuclei.

Iachello introduced some years ago, the E(5) and X(5) critical point symmetries, which provide parameter free (up to scale factors) predictions of excitation spectra and strengths for nuclei at the critical point of a phase shape transition [11]. These symmetries were obtained within the framework of the collective Bohr Hamiltonian under some simplifying approximations. In particular, the potential in the β degree of freedom was approximated by a simple square well potential, which is decoupled from the potential in the γ-variable. In the case of E(5), which corresponds to the transition from spherical vibrational U(5) to deformed γ-unstable O(6), the potential is constant in the γ-direction. In the case of X(5), related to the transition from U(5)

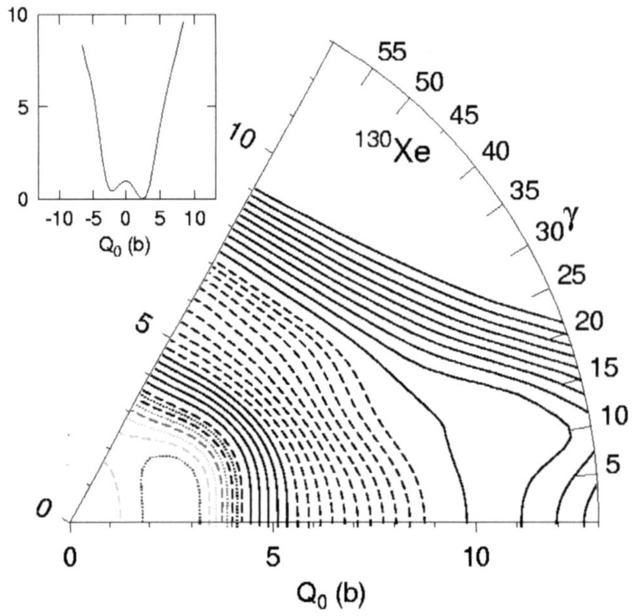

Figure 1. Contour plot of the PES for ^{130}Xe with the Gogny D1S force.

to axially symmetric prolate SU(3), a harmonic oscillator potential is used in the γ-direction. Empirical evidence of these transitional symmetries at the critical points has been observed in several isotopes of Ba, Pd, and Xe for E(5), and in ^{152}Sm, ^{150}Nd, ^{154}Gd, and ^{156}Dy for X(5).

In this work we investigate shape transitions within non-relativistic microscopic models, based on effective interactions between nucleons that provide a unified description of nuclear properties along the nuclear chart. In particular, we study whether the assumptions made on the β-γ potentials to construct the point symmetries are justified microscopically. Thus, we study various isotopic chains in which the occurrence of shape transitions has been predicted and we show results for the potential energy surfaces (PES) corresponding to constrained mean field calculations [4, 5, 6, 7]. We also focus our attention to the oblate-prolate shape transition in neutron-rich Yb, Hf, W, Os, and Pt isotopes. The theoretical framework is based on both the self-consistent Hartree-Fock-Bogoliubov approximation with the finite-range and density-dependent Gogny interaction [12] with the parametrization D1S [13], and the self-consistent Hartree-Fock + BCS with short-range density-dependent Skyrme interactions (SLy4) [14] and a zero-range density-dependent interaction in the pairing channel. In the latter case we use the code EV8 [15] that solves the HF equations in a coordinate space mesh. The role of triaxiality is also considered and discussed in those nuclei where this degree of freedom could be relevant.

2. Results

In Figure 1 we can see the contour plot of the PES for ^{130}Xe, obtained from Gogny D1S, as a function of the quadrupole moment Q_0 (b) and γ (deg) angle. This nucleus has been proposed as a candidate for the E(5) symmetry. Along with the triaxial plot, there is a cut corresponding to the axially symmetric shape ($\gamma = 0^o$ and $\gamma = 60^o$). We can see that the potential is rather flat in β (Q_0) around sphericity and that it is practically constant in the γ-direction at the β deformation corresponding to the minimum of the energy. It is also interesting to notice that the oblate axial minimum becomes a saddle point when triaxiality is considered. All of this

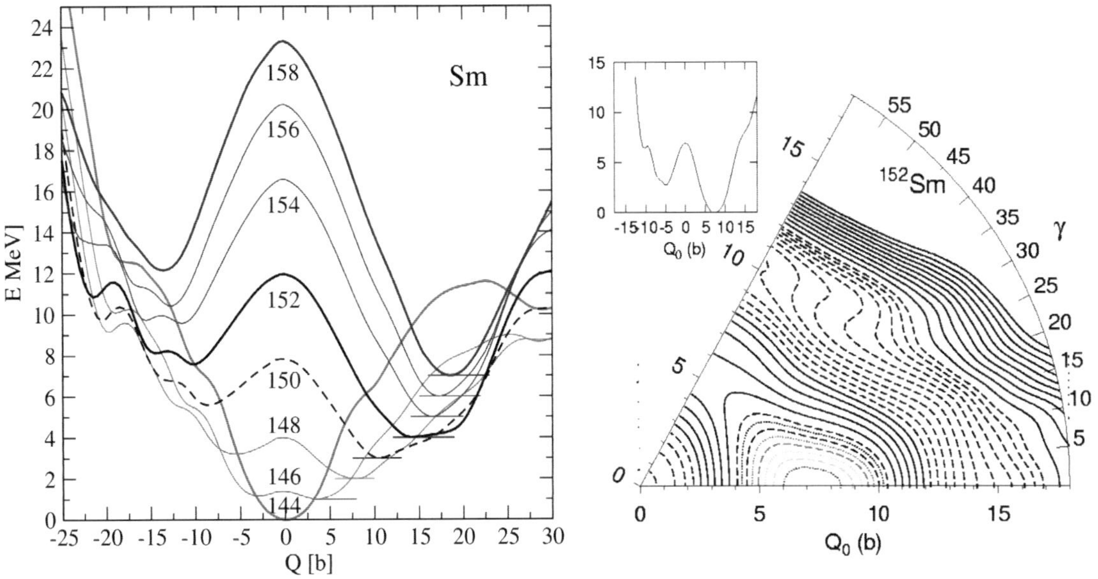

Figure 2. (Left panel) Axial PES for Sm isotopes with SLy4. (Right panel) Contour plot of the PES for ^{152}Sm with the Gogny D1S force (notice that Q is defined as $2Q_0$).

makes this nucleus a reasonably good candidate to show E(5)-like behavior. Similar conclusions are obtained for the other candidates ^{108,110}Pd, ^{128,132}Xe, and $^{130-134}$Ba [4, 5].

In Figure 2 we study the X(5)-like behavior on the example of Sm isotopes, and in particular on ^{152}Sm. Constrained HF+BCS calculations for Sm isotopes with SLy4 and zero-range pairing force are shown in the left panel of Figure 2. The axial plots show a clear shape transition from spherical ^{144}Sm ($N = 82$) to well developed prolate shapes at $^{156-158}$Sm. The isotope ^{152}Sm ($N = 90$) shows a transitional behavior with a shallow minimum on the prolate side and an additional minimum on the oblate side with an energy barrier rather high between these minima. On the right panel, the triaxial contour plot for the X(5) candidate ^{152}Sm shows that for Q_0 values in the vicinity of the prolate minimum, the energy presents a parabolic behavior as a function of γ. This is particularly true for low γ angles but tends to become flat as γ approaches 60°. The oblate axial minimum becomes also a saddle point when looking at the $Q_0 - \gamma$ plane. Thus, the assumption of an infinite square well made for the β-potential to arrive to the X(5) symmetry, is not realized within non-relativistic microscopic calculations that produce systematically potential energy barriers at $\beta = 0$, excluding the spherical configuration from the coexisting shapes. Similar results are obtained for the other X(5) candidates ^{150}Nd, ^{154}Gd, and ^{156}Dy [4, 5]. The results also agree with those obtained from relativistic calculations [8, 9].

We next consider the chain of W isotopes as examples of oblate-prolate shape transitions occurring in this mass region. Figure 3 shows the evolution of the axial shapes with increasing neutron number, calculated from various forces, Skyrme SLy4 and Gogny D1S. The transition takes place at $N = 116 - 118$, where the transitional nuclei show oblate and prolate minima coexisting at close energies. In Figure 4 we can see the contour plots of the PES with SLy4 in the $^{186-196}$W isotopes. We can see in ^{186}W a real prolate minimum and an oblate saddle point. ^{190}W shows a triaxial ground state with the axial prolate and oblate minima transformed into saddle points. There is a very soft behavior of the PES along the γ degree of freedom, developing a very shallow triaxial minimum. Finally, in ^{196}W the oblate minimum has been

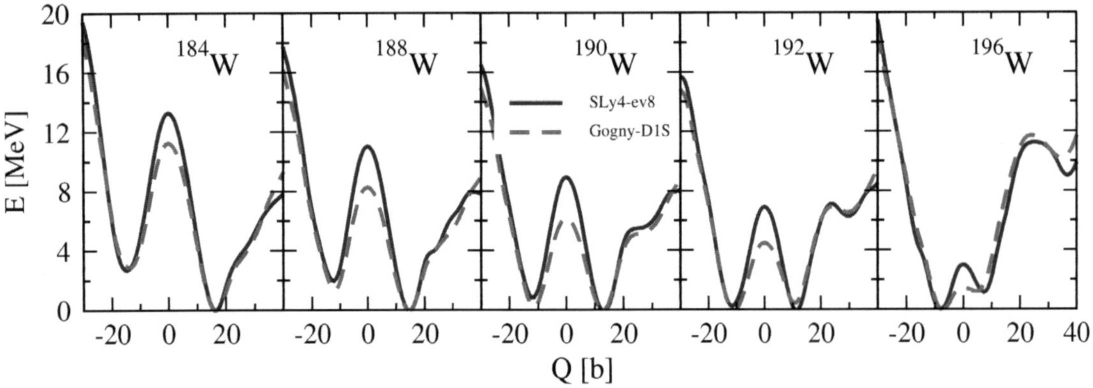

Figure 3. Axial PES for $Z = 74$ isotopes from mean field calculations with various forces.

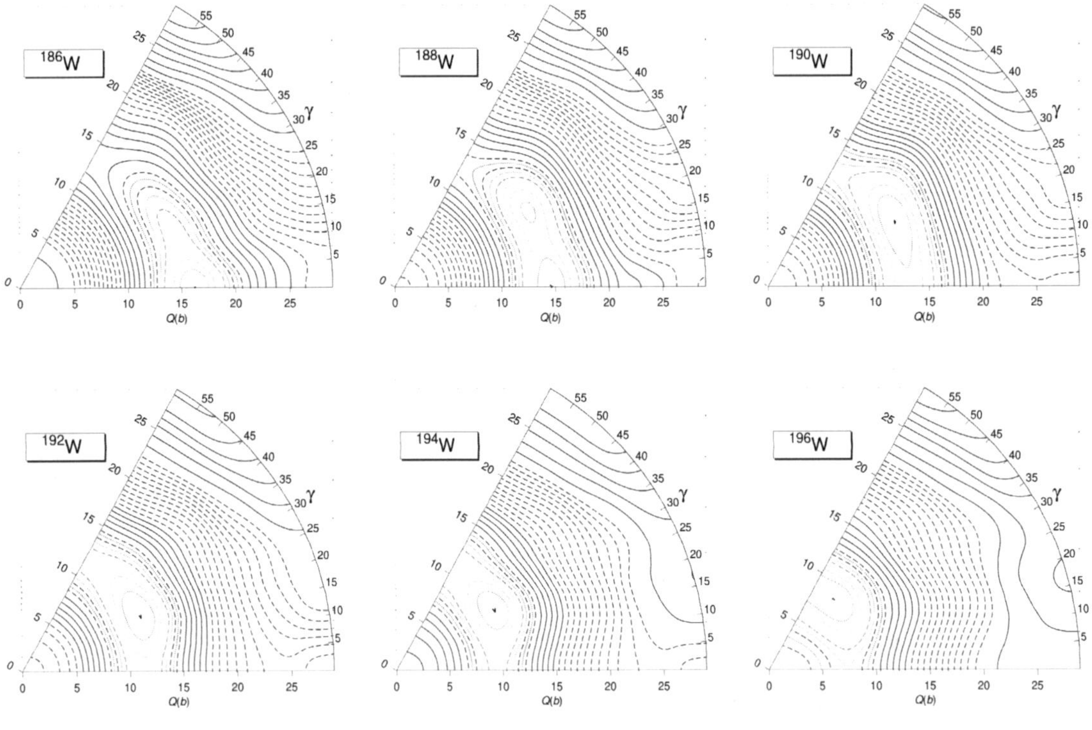

Figure 4. Contour plots of the PES for W isotopes with the Skyrme SLy4 force.

developed, while the prolate one is now a saddle point. Similar results are found in Yb, Hf, Os, and Pt isotopes [6, 7]. In general, for all of these isotopes, the transition from prolate to oblate shapes takes place at $N = 116 - 118$, where the energies of oblate and prolate shapes are nearly degenerate. The triaxial analysis of these nuclei shows that the axial oblate and prolate minima, which are separated by high potential barriers in the β degree of freedom, are linked very softly in the γ degree of freedom, making these minima saddle points in the extended $\beta - \gamma$ plane.

In Figure 5 we can see the single-particle energies (SPE) for the Gogny force corresponding

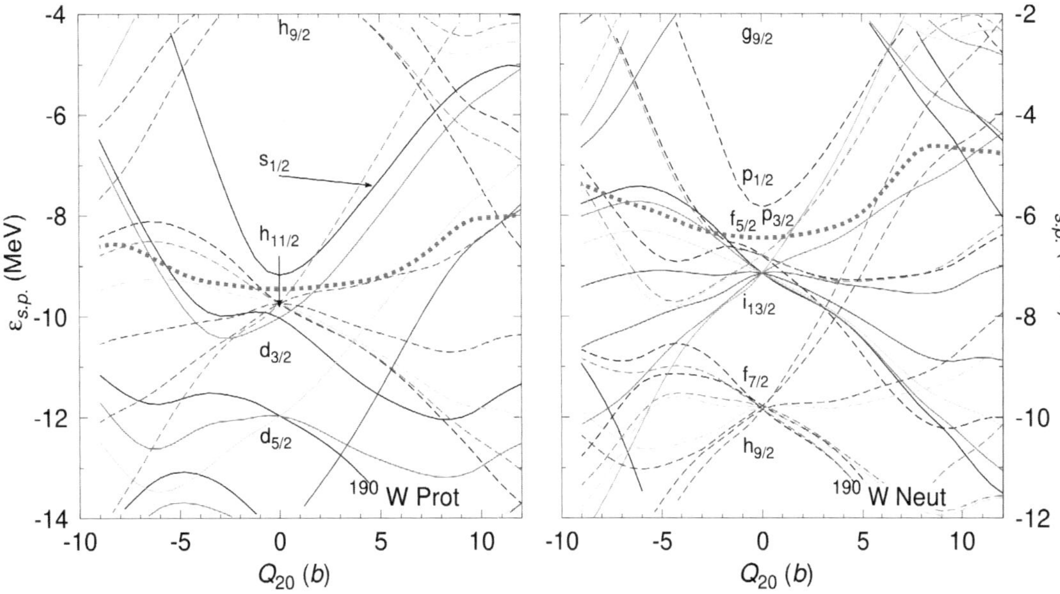

Figure 5. Single-particle energies for protons (left) and neutrons (right) for ^{190}W with the D1S Gogny force.

to ^{190}W. In the plot for protons we observe the presence of the $3s_{1/2}$ level just above the Fermi level and the levels $1h_{11/2}, 2d_{3/2}$ and $2d_{5/2}$ below. For neutrons we have the $3p_{1/2}$ level above the Fermi level and almost degenerate $3p_{3/2}$, $2f_{5/2}$ and $1i_{13/2}$ orbitals just below the Fermi level. A couple of MeV below we find degenerate $1h_{9/2}$ and $2f_{7/2}$ orbitals. Those levels evolve with deformation and at Q_{20} around 6.6 b a gap in the SPE spectrum signaling a region of low level density appears both in the proton and neutron spectra that is responsible for the prolate minimum observed in the axially symmetric potential energy curve. In the oblate side, at Q_{20} = -6.6 b the neutron's Fermi level approaches another gap that is responsible for the oblate minimum observed in the axially symmetric case. Both minima are in fact saddle points as long as the γ degree of freedom is considered. In the case of ^{184}W, the Fermi level for neutrons is lowered and the SPE spectrum shows a gap near the Fermi level for Q_{20} = 8 b. This fact together with the proton's gap also observed in that region of Q_{20} favors the development of the prolate minimum observed. In the oblate side a gap is developed in the neutron's SPE spectrum at Q_{20} = -8 b that is responsible for the observed oblate minimum. For the nucleus ^{196}W, the neutron's SPE spectrum shows a gap around the Fermi level for oblate deformations with Q_{20} in the range between -1 b and -5 b that is responsible for the oblate minimum observed in this case. Therefore the prolate-oblate transition seen at $N = 116$ is a consequence of the two gaps in the neutron's SPE, one in the prolate and the other in the oblate side as the Fermi level crosses them. On the other hand, the proton's SPE spectrum seems to favor the appearance of coexisting oblate and prolate configurations as Z increases that are the precursors of the triaxial instability observed in that case.

We can also look at the onset of deformation in this region by using the ideas developed by Federman and Pittel (FP) [16] in trying to unify the description of deformation both for light nuclei and heavy ones. The argument of Refs. [16] is that deformation is driven by the $T = 0$ neutron-proton interaction and this is particularly intense between spin orbit partners. Next in the range of relevance of the n-p interaction strength we find interactions between orbitals with the same radial quantum number and large orbital angular momenta differing by one unit (i.e., $n_p = n_n$ and $l_p = l_n \pm 1$). By looking at the SPE plots in Figure 5 we find the relevant role of the $1h_{11/2}$ orbital for protons which is very close to the Fermi level for all the nuclei considered in the region. According to FP's argument this orbital could interact with its neutron spin orbit partner, namely the $1h_{9/2}$ orbital but this one is well below the Fermi level and can be considered as inert. Near the neutron's Fermi level we have a $1i_{13/2}$, $2f_{5/2}$ and $3p_{3/2}$. Obviously, it is the first one that fulfills the above criteria of $n_p = n_n$ and $l_p = l_n \pm 1$ and therefore is the strongly interacting one with the $1h_{11/2}$ orbital. For values of N around 110 the $1i_{13/2}$ is in the middle of the Fermi level favoring the observed prolate deformation with well established and deep prolate wells. As N increases the $1i_{13/2}$ gets more and more occupied and at some point it ceases to play a role that is transferred to the $2f_{5/2}$ and $3p_{3/2}$ orbitals. Among them, only the $2f_{5/2}$ can interact with the $2d_{3/2}$ of protons but as the l values are low we do not expect a strong interaction. This explains why as N increases the depth of the deformation wells decreases favoring triaxial deformations.

3. Summary and Conclusions

We have carried out a systematic mean-field exploration of the triaxial PESs in various isotopic chains where nuclear shape transitions occur and where the realization of the critical point symmetries E(5) and X(5) have been proposed. We use the HFB framework based on the finite-range density-dependent Gogny interaction (D1S), as well as the HF+BCS formalism based on the short-range density-dependent Skyrme interaction (SLy4) in connection with a zero-range density-dependent interaction in the pairing channel.

We have found that the assumptions of flat potentials in the quadrupole deformation β and constant potentials in the γ direction, which are characteristics of the E(5) critical point symmetry, are supported by our results in ^{108,110}Pd, $^{128-132}$Xe, and ^{130,134}Ba. In the case of the rare-earth isotopes with $N = 88, 90$, it is shown that they present a transitional behavior between spherical and axially deformed shapes. However, we do not find the characteristic flat potentials of the X(5) symmetry but rather a potential barrier at zero deformation. In these cases we find that, at equilibrium, the potentials in γ exhibit a quadratic behavior centered at $\gamma = 0$ that becomes flat as γ increases. These results, together with those obtained within the relativistic mean-field approach, allow us to conclude that all the available state-of-the-art mean-field approximations, support the assumption of flat potentials compatible with the ones postulated in the E(5) critical point symmetry for some selected isotopes. However, we have not found satisfactory examples for PES showing patterns similar to those of the X(5) symmetry.

We have also studied the oblate-prolate shape transitions in Yb, Hf, W, Os, and Pt isotopes within the same theoretical formalism. We have shown that increasing the proton number in this mass region leads the nuclei to triaxiality. On the other hand, increasing the neutron number, the ground-state shapes in the isotopes studied evolve from axially deformed prolate shapes to axially deformed oblate shapes. The transitional nuclei ($N = 116$) exhibit a γ soft behavior with very shallow triaxial minima. The transition occurs with different degrees of stiffness depending on the isotope. It is rather sharp for low Z isotopes (Yb and Hf) and much broader for W, Os, and Pt, where a region of triaxial ground states develops between the regions of prolate and oblate minima. The analysis of the single-particle energies both for axially symmetric and triaxial configurations demonstrates the role of different gaps showing up in the SPE of both protons and neutrons, as well as the role played by the $T = 0$ proton-neutron interaction.

Acknowledgments

This work was partially supported by MICINN (Spain) under contracts Nos. FIS2008-01301 and FPA2007-66069.

References

[1] Wood J L, Heyde K, Nazarewicz W, Huyse M and Van Duppen P 1992 *Phys. Rep.* **215** 101
[2] Wu C Y *et al.* 1996 *Nucl. Phys.* A **607** 178
Stuchbery A E *et al.* 1996 *Phys. Rev. Lett.* **76** 2246
Wheldon C *et al.* 2000 *Phys. Rev.* C **63** 011304(R)
Podolyák Zs *et al.* 2000 *Phys. Lett.* B **491** 225
Pfützner M *et al.* 2002*Phys. Rev.* C **65** 064604
Caamaño M *et al.* 2005 *Eur. Phys. J.* A **23** 201
Podolyák Zs *et al.* 2009 *Phys. Rev.* C **79** 031305(R)
[3] Jolie J and Linnemann A 2003 *Phys. Rev.* C **68** 031301(R)
Naik Z *et al.* 2004 *Pramana* **62** 827
Stevenson P D *et al.* 2005 *Phys. Rev.* C **72** 047303
Walker P M and Xu F R 2006 *Phys. Lett.* B **635** 286
Rodriguez T R and Egido J L 2008 *Phys. Lett.* B **663** 49
Sun Y, Walker P M, Xu F R and Liu Y X 2008 *Phys. Lett.* B **659** 165
Morales I O , Frank A, Vargas C E and Van Isacker P 2008 *Phys. Rev.* C **78** 024303
Garcia-Ramos J E and Heyde K 2009 *Nucl. Phys.* A **825** 39
[4] Rodríguez-Guzmán R and Sarriguren P 2007 *Phys. Rev.* C **76** 064303
[5] Robledo L M, Rodríguez-Guzmán R and Sarriguren P 2008 *Phys. Rev.* C **78** 034314
[6] Sarriguren P, Rodríguez-Guzmán R and Robledo L M 2008 *Phys. Rev.* C **77** 064322
[7] Robledo L M, Rodríguez-Guzmán R and Sarriguren P 2009 *J. Phys.* G **36** 115104
[8] Fossion R, Bonatsos D and Lalazissis G A 2006 *Phys.Rev.* C **73** 044310
[9] Niksic T, Vretenar D, Lalazissis G A and Ring P 2007 *Phys. Rev. Lett.* **99** 092502
Li Z P, Niksic T, Vretenar D, Meng J, Lalazissis G A and Ring P 2009 *Phys. Rev.* C **79** 054301
[10] Iachello F and Arima A 1987 *The Interacting Boson Model* (England: Cambridge University Press)
[11] Iachello F 2000 *Phys. Rev. Lett.* **85** 3580
Iachello F 2001 *Phys. Rev. Lett.* **87** 052502
[12] Dehargé J and Gogny D 1980 *Phys. Rev.* C **21** 1568
[13] Berger J F, Girod M and Gogny D 1984 *Nucl. Phys.* A **428** 23c
[14] Chabanat E, Bonche P, Haensel P, Meyer J, and Schaeffer R 1998 *Nucl. Phys.* A **635** 231
[15] Bonche P, Flocard H and Heenen P H 2005 *Comput. Phys. Comm.* **171** 49
[16] Federman R and Pittel S 1977 *Phys. Lett.* B **69** 385
Federman R and Pittel S 1979 *Phys. Rev.* C **20** 820

Journal of Physics: Conference Series **205** (2010) 012025 doi:10.1088/1742-6596/205/1/012025

Signatures of shell-effects in collective excitations

V. Werner[1], J. R. Terry[1], M. Bunce[1,2] and Z. Berant[1,3]

[1] Wright Nuclear Structure Laboratory, Yale University, P.O. Box 208120, New Haven, CT 06520, USA

[2] Department of Physics, University of Surrey, Guildford GU2 7XH, United Kingdom

[3] Nuclear Research Center Negev, Beer-Sheva, Israel

E-mail: `volker.werner@yale.edu`

Abstract. Two methods for measuring lifetimes of excited states in nuclei are discussed. One recently developed method involves a plunger device and inverse kinematics Coulomb excitation with heavy beams. The other method covers a different lifetime range and relies on electronic timing. Both methods have recently been applied at Yale University. This paper will give a brief overview of technical aspects of those measurements. A short motivation and preliminary results of the experiments are given.

1. Introduction

This lecture covered methods for lifetime measurements in the picosecond and nanosecond region, using different techniques. One is the well-known plunger technique, which we now put to novel use in inverse kinematics reactions, namely Coulomb excitation. The new technique [1] has been applied to stable beams at the Wright Nuclear Structure Laboratory (WNSL) at Yale University, and has recently also been used by the Cologne group at Jyvaskylä (see contribution by Ch. Fransen). The second technique presented in this paper is the measurement of lifetimes of excited states via fast electronic timing. This method has been known since a long time, but is one of the standard techniques for decay-spectroscopy at existing and upcoming radioactive ion beam facilities.

Both methods have recently been applied at WNSL attacking two different physics themes, which have the common ground of investigating the evolution of the B(E2) excitation strengths of the first excited 2^+ state over series of isotopes. For nuclei near closed shells those lifetimes are usually in the picosecond range, too short for fast electronics measurements, and therefore require the use of Doppler shift techniques. The plunger method that we used will be presented in Section 2. We performed a recoil distance Doppler shift (RDDS) experiment on ^{120}Te, in order to extend the $B(E2)$ excitation strength systematics within the Te isotopic chain to its lightest stable isotope.

Section 3 will address measurements of lifetimes of the first excited state in near mid-shell even-even nuclei. Such lifetimes are usually in the proximity of a nanosecond and are therefore accessible with fast electronics techniques. An experiment on the deformed nucleus ^{172}Hf will be presented. The method will be shown, and the new, preliminary result will be discussed in the framework of saturation of $B(E2)$ strengths toward mid-shell.

2. Inverse kinematics RDDS

2.1. Experimental details

For many even-even nuclei, the lifetimes of the first excited 2^+ states are known from their measured Coulomb excitation cross sections. However, in general this can result in systematic errors due to unknown quadrupole moments or non-observed feeding. The RDDS technique offers a model independent way of measuring lifetimes, independent from the excitation mechanism. Whereas, classically, RDDS measurements have been performed mostly in normal kinematics fusion evaporation or Coulomb excitation experiments, in the era of radioactive beam experiments the use of inverse kinematics is mandatory. We performed a first measurement in inverse kinematics Coulomb excitation using heavy beams.

The measurement was performed at WNSL using the New Yale Plunger Device (NYPD) and eight clover detectors in the SPEEDY array [2]. A 300-MeV beam of ^{120}Te was produced by the 20-MeV ESTU tandem van de Graaff accelerator at WNSL, and impinged on a stretched carbon target with a thickness of 400 μg/cm^2. The beam was Coulomb excited on the carbon target, and then reacted and unreacted beam were stopped in a copper target of 14 mg/cm^2 thickness. The distance between the target and stopper foils were varied over a range of 5 - 1100 μm, therefore varying the probability of in-flight decay of the Coulomb excited 2_1^+ state in ^{120}Te. The beam energy was well below (about 50% at mid-target) the Coulomb barrier, ensuring that excitations of states other than the 2_1^+ state were negligible. This simplified the analysis tremendously, since no feeding needed to be taken into account, and a simple decay function was fitted to the data.

Carbon from the target was forward scattered by the reaction and passed the stopper foil. The carbon particles were subsequently detected in a circular silicon detector, placed directly behind the target. Gamma-rays from the de-excitation of the 2_1^+ state were detected in the clover detectors, which were placed at 41.5° and 138.5° relative to the beam axis, either at Doppler shifted energies in the case of in-flight decay, or at the excitation energy of 560.4 keV of the 2_1^+ state. The trigger for the data acquisition was a particle-gamma coincidence. After random background subtraction the observed peaks appeared on almost no background, and were well separated due to the large v/c value of about 6%. Sample spectra are shown in Fig. 1.

Figure 1 also includes lines from the decays of higher-lying excited states, which are two orders of magnitude smaller than the peak of interest and can therefore be neglected. Statistics in those lines were insufficient to perform an RDDS analysis for those states, however, we were able to perform a Coulomb excitation analysis to extract their $B(E2)$ decay strengths relative to that of the 2_1^+ state.

2.2. RDDS analysis and results

In order to extract a lifetime, we define the ratio

$$P(d) = \frac{I_\gamma^{(shifted)}(d)}{I_\gamma^{(total)}(d)} , \tag{1}$$

of the shifted peak intensity $I_\gamma^{(shifted)}$ over the total peak intensity $I_\gamma^{(total)}$ for each distance d between the two foils. The peak intensities are corrected for detector efficiencies and the angular distribution of gamma-rays, which was obtained from the Coulomb excitation code by Winther and de Boer [3]. Neglecting feeding from higher-lying states, the P ratio should follow a simple decay law:

$$P(d) = 1 - e^{-\lambda(d-d_0)} , \tag{2}$$

where $\lambda = 1/(v\tau)$ contains the lifetime information. The mean velocity v is extracted directly from the mean energies of the Doppler shifted peaks at forward and backward angles. d_0 is left

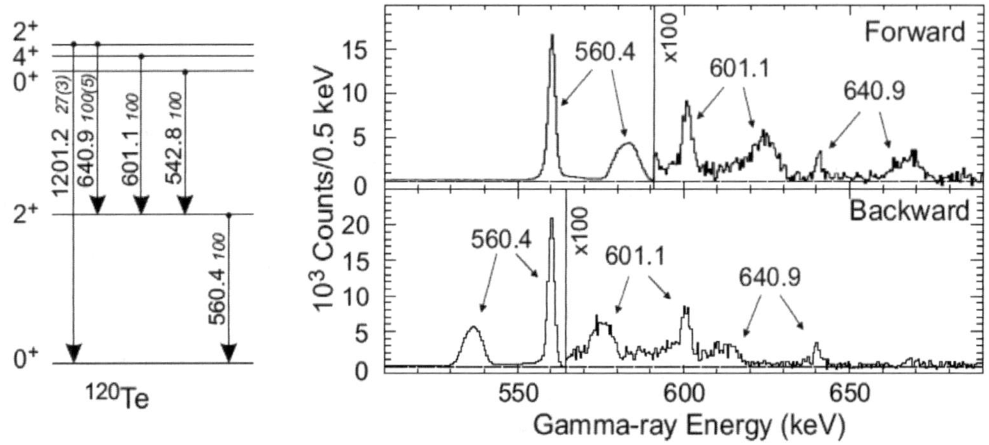

Figure 1. Particle gated and random subtracted sum spectra of ^{120}Te for all plunger distances. The left top and bottom panels show the summed spectra for the $2_1^+ \rightarrow 0_1^+$ transition at forward and backward angles, respectively. The right top and bottom panels correspondingly show other observed transitions, according to the level scheme on the left hand side.

as a free fit parameter that corresponds to the minimum achievable distance between the two foils.

At high recoil velocities, relativistic effects need to be taken into account. As such, the detector angles in the laboratory frame have to be corrected. The Lorentz transformation of the angular distribution consists of a transformation of the emission angle, as well as of a multiplicative factor J from the Jacobian transformation to the center of mass (CM) frame:

$$J = \frac{d\Omega_{CM}}{d\Omega_{lab}} = \frac{1-(v/c)^2}{(1-v/c \cdot \cos\theta_{lab})^2} \ . \tag{3}$$

While the transformation of the emission angle has a negligible effect on the angular distribution of gamma-rays, the Jacobian of the transformation leads to a significant difference between the angular distributions of gamma-rays emitted in-flight or stopped. The correction in our case is on the order of 8%.

The hyperfine interaction between the nucleus and the atomic electrons when recoiling into vacuum causes a de-orientation of the aligned nuclear state [4]. This effect is taken into account by time-dependent attenuation coefficients

$$G_k^{(stopped)}(d) = \exp\left(-\frac{|g|d}{vC_k}\right) \text{ and} \tag{4}$$

$$G_k^{(in-flight)}(d) = \frac{\frac{1}{1+\frac{|g|\tau}{C_k}} \, \frac{1-\exp\left(-(1+\frac{|g|\tau}{C_k})\frac{d}{v\tau}\right)}{}}{1-\exp\left(-\frac{d}{v\tau}\right)} \ , \tag{5}$$

where $|g|$ is the absolute value of the g factor of the excited state. De-orientation due to recoil into vacuum has been extensively studied for tellurium isotopes at $v \approx 0.06c$ [5], where the parameters C_k that appear within the G_k coefficients have been fixed, and therefore could be used for the present analysis. The data points from the ^{120}Te experiment, including all corrections, and the fit to obtain the lifetime of the 2_1^+ state are shown in Fig. 2. The fit results in a mean lifetime of $\tau = 10.4(2)$ ps, which is a significant improvement compared to the currently listed value of $\tau_{lit} = 1.3(27)$ ps [6].

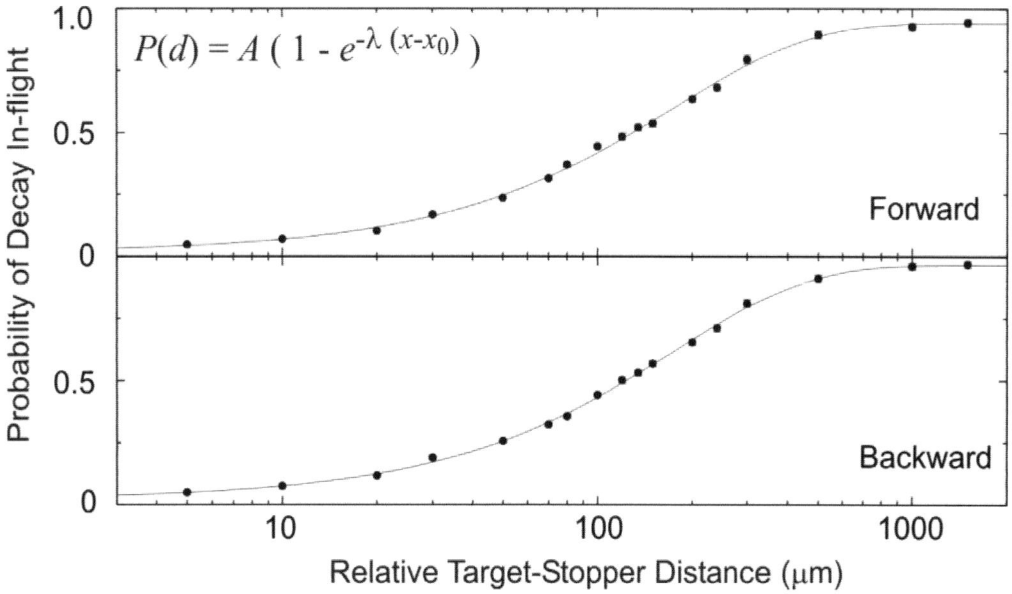

Figure 2. P ratios from eq. (1) and the fitted function from eq. (2) for the 2_1^+ state in ^{120}Te.

2.3. Future use of the plunger method

The RDDS technique as presented above is suitable for experiments with weak beams. In order to test this, only part of the present data has been sorted corresponding to a beam intensity of 10^{5-6} particles per second, which is in the range of new radioactive beam facilities. An analysis of the reduced data set yielded a lifetime result with a statistical error well below 10%. The plunger technique should also allow to measure absolute values of g factors of excited states over series of isotopes. Therefore, the de-orientation from recoil into vacuum needs to be measured by the angular distributions of gamma-rays. The hyperfine parameters in eqs. (4) and (5) need to be fixed from a measurement of a state with known g factor. Since those parameters should not vary within an isotopic chain, for neighbor isotopes the only free parameter in the de-orientation coefficients then is the g factor. This technique is currently being tested at WNSL.

3. Fast electronics lifetime measurements

3.1. Experimental details

In this section a method of measuring lifetimes using electronics is presented. In our case, we have been interested in the lifetime of the 2_1^+ state of the well-deformed rare-earth nucleus ^{172}Hf. Data on the Hf isotopes (as well as other isotopic chains) suggest that the $B(E2)$ strengths in deformed nuclei rise toward mid-shell, but show a saturation effect around mid-shell [7, 8]. Since much of the available data stems from the 1970's, and was often obtained in relatively low statistics experiments, we tested this by a new lifetime measurement.

The safest way to obtain the lifetime of a given state is to have a signal from the direct population of the state, and a signal from its decay, and measure the time difference between both. The signals in our case were given by the gamma-ray transitions $4_1^+ \rightarrow 2_1^+$ at 214.0 keV and $2_1^+ \rightarrow 0_1^+$ at 95.2 keV. The corresponding gamma-rays were detected by four BaF_2 scintillators surrounding the source. The beta-decay from ^{172}Ta populates mostly directly the 4_1^+ state in ^{172}Hf. The mother nuclei ^{172}Ta were produced at the moving tape collector (MTC) setup [9] at WNSL using the reaction ^{165}Ho(^{12}C,5n). Fusion evaporation residues, which were recoiling out

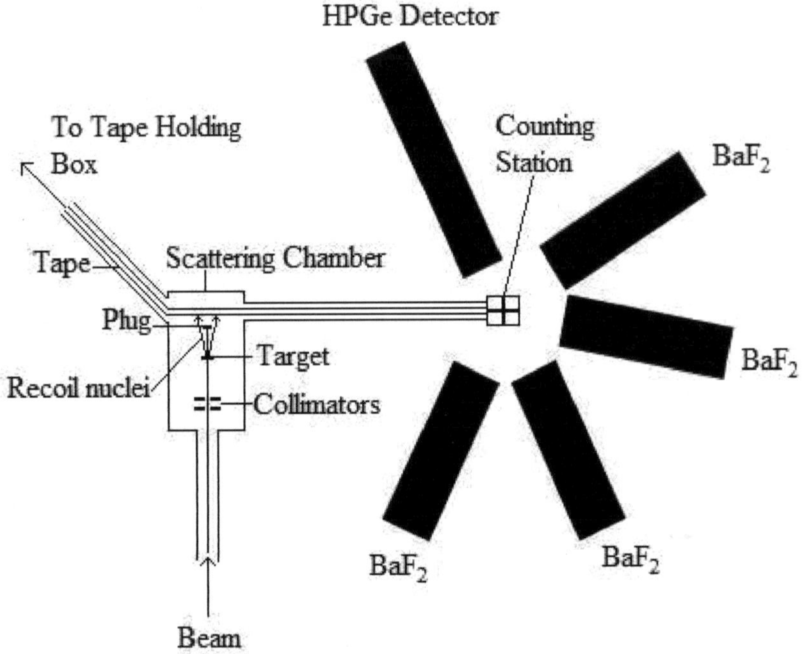

Figure 3. Schematic of the setup at the MTC. Four BaF_2 detectors were used for the lifetime measurement, in addition one Ge detector was used as a monitor.

of the 4 mg/cm^2 self-supporting foil, were implanted on a tape behind the target. The beam itself was stopped by a gold plug in between target and tape, while most recoils passed the plug at angles larger than 2°.

The halflife of ^{172}Ta is about 36 minutes, therefore a tape cycle of about 67 minutes was chosen. That means that activity was collected on tape for 67 minutes, and then moved in between the detectors, while activating the next spot on the tape in parallel to measuring gamma-rays after beta-decay at the well-shielded detector setup. After the detection period, the activated spot was carried away to a tape holding box in order to minimize unwanted background from subsequent beta-decays. A schematic of the MTC setup is shown in Fig. 3.

The array of four BaF_2 detectors allowed for a total of twelve independent detector pairs for the lifetime measurement. For each pair, an energy gate was set on the $4_1^+ \rightarrow 2_1^+$ transition in one detector, and on the $2_1^+ \rightarrow 0_1^+$ in the other detector. The constant fraction time signals from both detectors (with the condition on the energies) were fed into VME based time-to-digital converters (TDCs), and the amplified energy signals were fed into VME based analog-to-digital converters (ADCs). The TDC time resolution was set to its minimum of about 35 ps per channel.

A standard germanium detector was also added to the setup, in order to derive a high-resolution spectrum from the source, and therefore exclude the possibility of contaminants in the peaks of interest. A sample energy spectrum of a BaF_2 detector is shown in comparison to the Ge spectrum in Fig. 4. The energy resolution of the scintillator would not have allowed to exclude contaminants, but no lines other than the expected decays of interest are observed in the Ge detector. The energy gates placed on the BaF_2 spectra are indicated by "A" and "B". The corresponding times of events in regions "A" and "B" of two detectors were subtracted from each other and sorted into time-difference spectra. A sample is shown in the insert of Fig. 4.

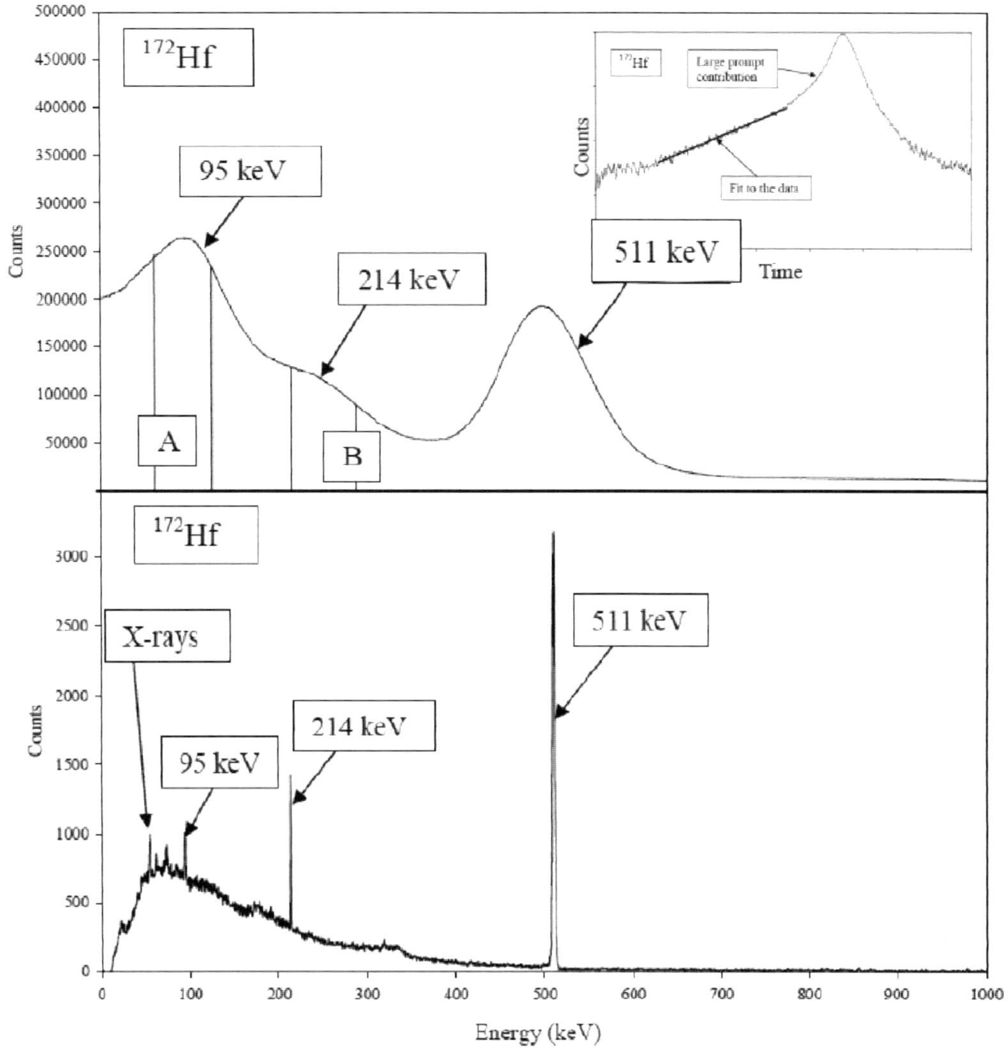

Figure 4. Energy spectrum of a BaF_2 detector (top), compared to the spectrum taken with a Ge detector (bottom). The insert in the top panel illustrates a sample time-difference spectrum (log-scale) between two BaF_2 detectors and a fitted line to the relevant portion.

3.2. Fast timing analysis and results

In order to obtain the lifetime of the 2_1^+ state, we simply fit a line to the portion of the time-difference spectrum that corresponds to the exponential decay of the state. The region in the time-difference spectrum that corresponds to the detector response, therefore the time resolution of the system (prompt peak), has been determined before the actual experiment using prompt coincidences from a ^{60}Co source. In that measurement gates were placed on the Compton background of the ^{60}Co spectra around 95 keV and 214 keV, thus the energy regions of interest. This step is important since the time resolution is a function of gamma-ray energy. The portion of the time-difference spectra which is dominated by the system response function is well excluded from the region of the lifetime fit.

For the final analysis, only 6 out of 12 detector pairs could be used due to problems with some of the electronics branches. Still, we are left with 6 independent values for the 2_1^+ lifetime in ^{172}Hf, which we average to a mean lifetime of $\tau = 2.655(79)$ ns. This is a significant difference to the literature value of $\tau_{lit} = 2.23(14)$ ns [10], and lowers the error bar by a factor of two. Lowering the lifetime of the 2_1^+ state in ^{172}Hf leads to an increase of its $B(E2)$ value, which now does not show a drop as from listed data (*e.g.*, Fig. 1 in Ref. [7]), but rather further increases from the value in ^{170}Hf, which has recently been remeasured [11]. Therefore, it is important to check further isotopes toward mid-shell in order to see to which extent a saturation effect is actually present in the Hf isotopic chain.

3.3. Future prospects of fast lifetime measurements

The largest drawback in our measurements using BaF_2 detectors is the poor energy resolution of those detectors, which is typically in the range of 10%. In our case the obtained spectra were sufficiently clean to avoid problems that may arise from contaminant or overlapping peaks. However, in general it would be appreciable to get a better energy resolution, but a similarly good timing resolution. A scintillator material ($LaBr_3$) which has these properties has recently been developed and will be of particular interest for future studies in decay spectroscopy.

4. Acknowledgments

This work was supported by U.S. DOE under grant no. DE-FG02-91ER40609.

References

[1] Terry J R, Werner V, Berant Z, Casperson R J, Casten R F, Heinz A, Henning G, Lüttke R, McCutchan E A, Qian J, Shoraka B, Williams E and Winkler R 2009 *AIP Conf. Proc.* **1090** 337
[2] Krücken R 2000 *J. Res. Nat. Inst. Stan. Techn.* **105** 53
[3] Winther A and de Boer J 1965 *Tech. rep.* California Institute of Technology
[4] Rascher R, Lieb K P and Uhrmacher M 1976 *Phys. Rev.* C **13** 1217
[5] Stuchbery A E and Stone N J 2007 *Phys. Rev.* C **76** 034307
[6] Kitao K, Tendow Y and Hashizume A 2002 *Nucl. Data Sheets* **96** 241
[7] Zhang J, Casten R F, Wolf A, Berant Z, Cakirli R B, Zamfir N V and McCutchan E A 2006 *Phys. Rev.* C **73** 037301
[8] Bian B-A, Di Y-M, Long G-L, Sun Y, Zhang J and Sheikh J A 2007 *Phys. Rev.* C **75** 014312
[9] Zamfir N V and Casten R F 2000 *J. Res. Natl. Inst. Stand. Techn.* **105** 147
[10] Singh B 1995 *Nucl. Data Sheets* **75** 199
[11] Costin A, Ahn T, Bochev B, Dusling K, Li T C, Pietralla N, Rainovski G and Rother W 2007 *Phys.Rev.* C **76** 024307

Journal of Physics: Conference Series **205** (2010) 012026 doi:10.1088/1742-6596/205/1/012026

Dynamic chirality in the interacting boson model

S Brant[1] **and D Tonev**[2,3]

[1] Department of Physics, Faculty of Science, University of Zagreb, 10000 Zagreb, Croatia
[2] INFN, Laboratori Nazionali di Legnaro, 35020 Legnaro, Italy
[3] Institute for Nuclear Research and Nuclear Energy, BAS, 1784 Sofia, Bulgaria

E-mail: brant@phy.hr

Abstract. The chiral interpretation of twin bands in odd-odd nuclei is investigated in the Interacting Boson Model framework. The analysis of the wave functions shows that the possibility for angular momenta of the valence proton, neutron and core to find themselves in the favorable, almost orthogonal geometry is present, but not dominant. Such behavior is found to be similar in nuclei where both the level energies and the electromagnetic decay properties display the chiral pattern, as well as in those where only the level energies of the corresponding levels in the twin bands are close together. The difference in the structure of the two types of chiral candidates nuclei can be attributed to different β and γ fluctuations, induced by the exchange boson-fermion interaction, i.e. by the antisymmetrization of odd fermions with the fermion structure of the bosons. In both cases the chirality is weak and dynamic. Among the nuclei that are candidates for chiral behavior, the best candidate is probably the odd-even ^{135}Nd, where the β and γ fluctuations are strongly reduced and chirality, although dynamic in origin, is rather close to the static limit.

1. Introduction

The rotation of triaxial nuclei may give rise to pairs of identical $\Delta I = 1$ bands with the same parity in odd-odd nuclei - the chiral doublet bands [1, 2, 3]. The condition for the appearance of chiral doublet bands is that the proton and neutron Fermi levels are located in the lower part of the valence proton high-j (particle-like) and in the upper part of the valence neutron high-j (hole-like) subshells (or vice versa), and the core is triaxial. The angular momenta of the valence particles are aligned along the short and long axes of the triaxial core, while the angular momentum of the rotational core is aligned along the intermediate axis. The three angular momenta can be arranged to form two systems that differ by intrinsic static chirality, a left- and a right-handed system. When this chiral symmetry is broken in the body-fixed frame, the restoration of the symmetry in the laboratory frame results in the occurrence of degenerate doublet $\Delta I = 1$ bands. Chirality in odd-mass nuclei is a novel feature [4, 5]. The conditions for chirality in odd-mass nuclei can be more favorable than in odd-odd nuclei. They can be fulfilled for three-particle configurations of the type one-proton-two-neutrons or one-neutron-two-protons.

Several theoretical models have been applied in a number of articles, like Tilted Axis Cranking Model [1], Two Quasiparticle + Triaxial Rotor Model [1, 6], Core-Particle-Hole Coupling Model [7]. All these models have one assumption in common, they suppose a rigid triaxial core. On the contrary, all nuclei in which twin bands have been observed have another characteristics in common, they are in regions of masses where even-even nuclei are γ-soft, i.e. effectively triaxial but not rigid. It is evident that nuclei in these mass regions can not reach the full requirements needed for the existence of chirality, but they can approach them, or at least retain some fingerprints of chirality.

2. IBFFM calculations

The structure of twin bands in odd-odd nuclei was also described in the Interacting Boson Fermion-Fermion Model (IBFFM) [8, 9], in the version where there is no distinction between proton bosons and neutron bosons. The triaxial equilibrium deformation was generated by the cubic (three-body) term [10, 11] added to the standard IBM-1 core Hamiltonian. In the case of the ^{134}Pr nucleus [12, 13, 14] the calculated IBFFM $B(E2)$ and $B(M1)$ values for twin bands based on the $\pi h_{11/2}\ \nu h_{11/2}^{-1}$ configuration have been in excellent agreement with the experimental values, while the results of the Two Quasiparticle + Triaxial Rotor Model (TQPTR) calculations displayed sizable differences in respect to experimental data (Fig.1). In the IBFFM besides the orientation in space, the deformation of the core is the additional degree of freedom. Valence quasiparticles are coupled to all structures of the boson core that are present in the basis, limited by the total boson number. In the IBFFM calculation we have performed, the core was not rigid. The minimum of potential energy surface was very shallow in the γ-direction and also rather broad in the β-direction.

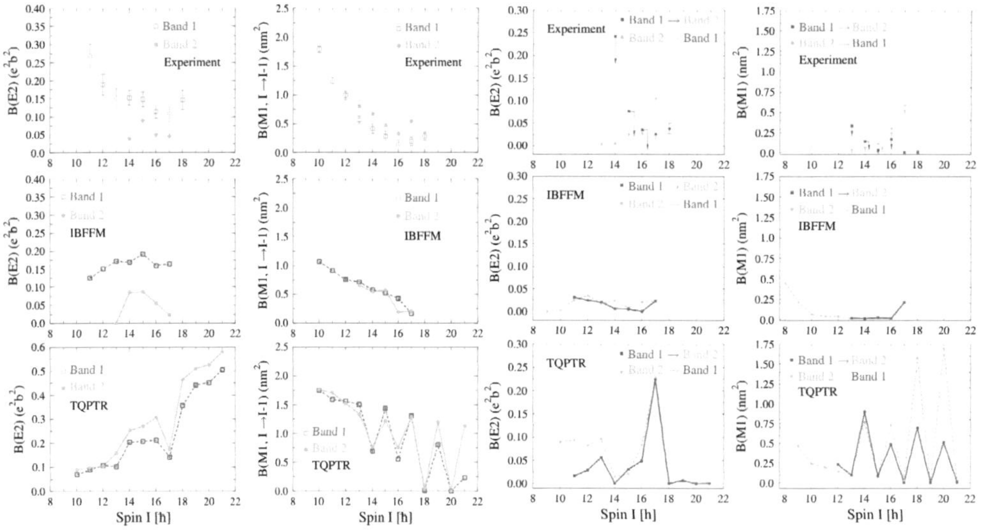

Figure 1. Experimentally determined and theoretically calculated $B(E2)$ and $B(M1)$ values in ^{134}Pr. In the upper panels experimental values are presented. In the middle panels the predictions of the IBFFM are presented. In the panels on the bottom the calculations of the TQPTR are shown.

Due to the underlying symmetry, the pair of chiral twin bands should exhibit systematic properties. The yrast and the side bands should be nearly degenerate. In the angular momentum region where chirality sets in (which for the A$\sim$130 mass region corresponds to $I \geq 12$) the $B(E2)$ values of the electromagnetic transitions de-exciting analogue states of the chiral twin bands should be almost equal. Correspondingly the $B(M1)$ values should exhibit odd-even staggering, being for the $\pi h_{11/2}\ \nu h_{11/2}^{-1}$ configuration much bigger for transitions de-exciting states with odd spins than for transitions de-exciting states with even spins. The $B(M1)$ values for $\Delta I = 1$ transitions connecting the side to the yrast band should have the odd-even staggering out of phase with respect to the $B(M1)$ staggering for transitions de-exciting states in the yrast and the side bands, i.e. for the $\pi h_{11/2}\ \nu h_{11/2}^{-1}$ configuration $B(M1)$ values for transitions de-exciting states with even spins have to be much bigger then for the ones de-exciting states with odd spins. Neither of the conditions for the $B(E2)$ and $B(M1)$ values was found in ^{134}Pr. This nucleus belongs to a class of odd-odd nuclei in which a pair of twin bands is close in excitation energy, but the electromagnetic decay properties do not show the chiral pattern. For shortness they will be denoted as *case A* nuclei.

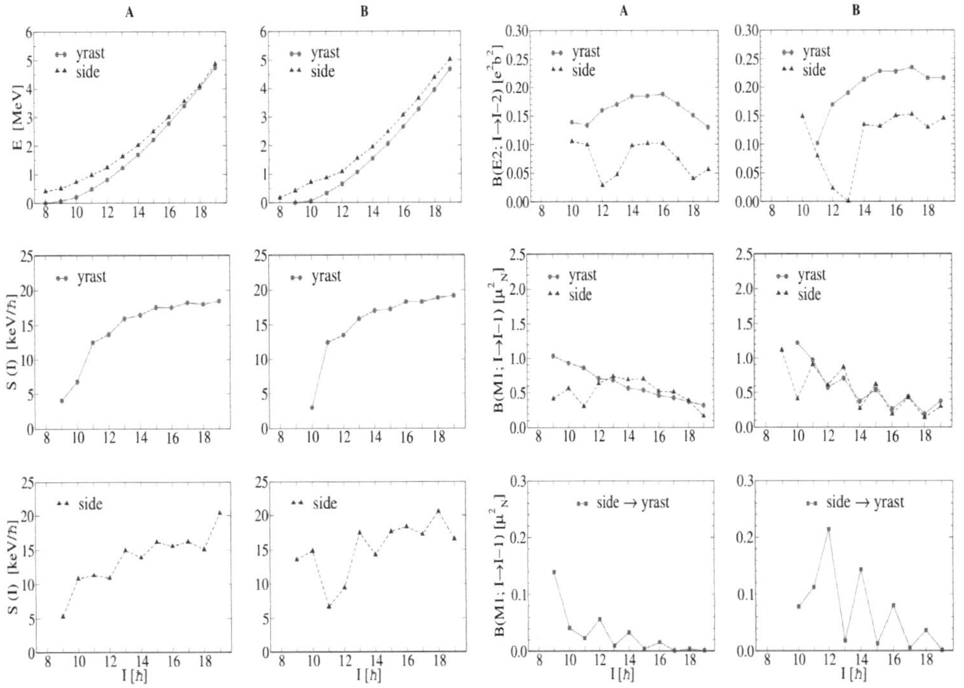

Figure 2. Twin bands calculated in the IBFFM (excitation energies, signature $S(I)$, $B(E2)$ and $B(M1)$ values) for the *case A* (panels in the first and third column) and the *case B* (panels in the second and fourth column).

To investigate the setting up of chirality in a certain nucleus, is crucial to determine the $B(E2)$ and $B(M1)$ values, Unfortunately, such measurements have been done only for few nuclei of interest: ^{134}Pr, ^{132}La and ^{128}Cs. In the case of ^{128}Cs the electromagnetic decay properties display the expected chiral pattern [15]. Odd-odd nuclei where the pair of twin bands is close in excitation energy and the electromagnetic decay properties display the chiral pattern, will be denoted as *case B* nuclei.

Odd-odd nuclei in the A$\sim$130 mass region can be classified as *case A* or *case B* nuclei. In all these nuclei the cores are γ-soft, their odd-proton odd-mass neighbors have also a similar structure and their odd-neutron odd-mass neighbors have a similar structure, too. IBFFM calculations have been performed for such nuclei, taking the same γ-soft core, same fermion occupation probabilities, same strengths of the boson-fermion interactions for the odd proton, and different strengths of the boson-fermion interactions for the odd neutron [16]. The main difference was in the strength of the exchange interaction for the odd neutron: in the *case A* it was strong and in the *case B* was weaker (but still rather strong). The results are presented in Fig.2.

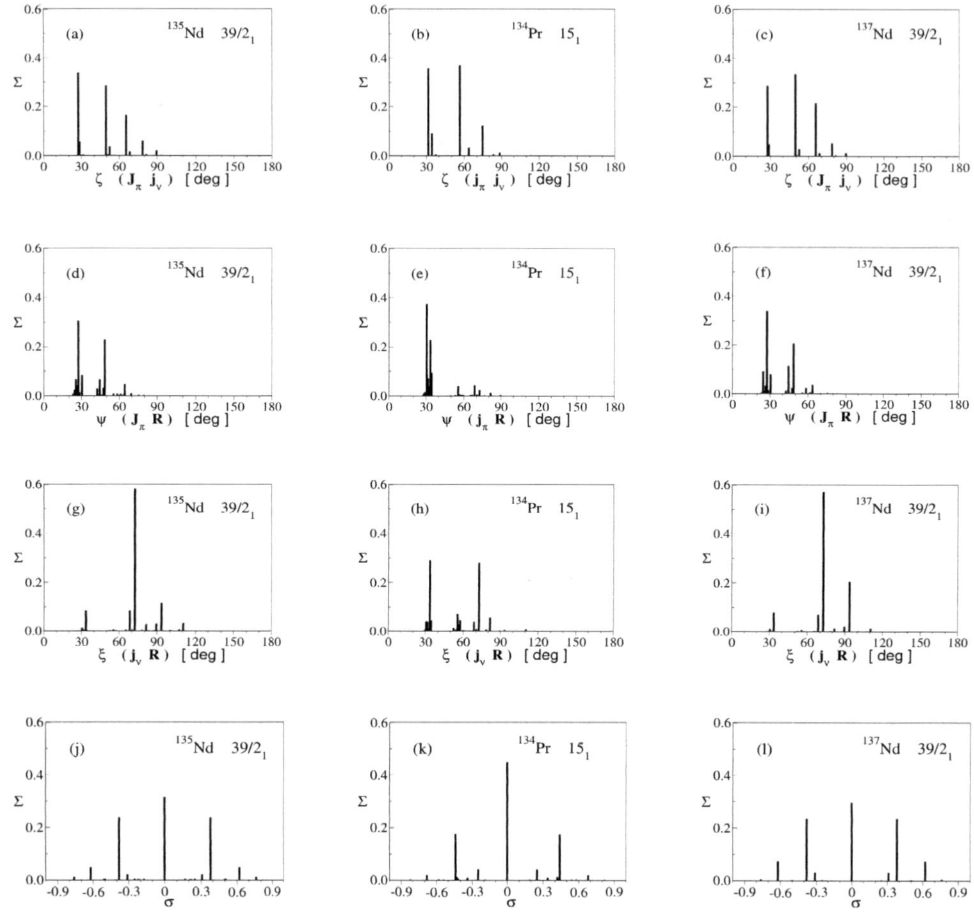

Figure 3. Calculated distributions $\zeta\ (J_\pi\ j_\nu)$ (panel (a)), $\psi\ (J_\pi\ R)$ (panel (d)) and $\xi\ (j_\nu\ R)$ (panel (g)) of the angles between $\vec{J}_\pi$, $\vec{j}_\nu$ and $\vec{R}$, and the σ distribution (panel (j)) for the $39/2_1$ state of the yrast band A in ^{135}Nd. In panels (b), (e), (h) and (k) the corresponding distributions for the yrast $\pi h_{11/2}\ \nu h_{11/2}\ 15_1$ state in ^{134}Pr are presented. In panels (c), (f), (i) and (l) the corresponding distributions for the $39/2_1$ state of the yrast band A in ^{137}Nd are presented.

In the *case A* the composition of the yrast band, in terms of contributions from core states, shows that the yrast band is basically built on the ground-state band of the even-even core. With increasing spin the admixture of the γ-band of the core becomes more pronounced. The side band wave functions contain large components of the γ-band and with increasing spin, of higher-lying collective structures of the core, that near the band crossing become dominant. The analysis of the distributions $\zeta\ (j_\pi\ j_\nu)$, $\psi\ (j_\pi\ R)$ and $\xi\ (j_\nu\ R)$ [16] of the angles between $\vec{j}_\pi$, $\vec{j}_\nu$ and $\vec{R}$, and the σ distribution (details of the procedure can be found in Ref. [14]), found that the presence of configurations with the angular momenta of the proton, neutron and core in the for chirality favorable, almost orthogonal geometry, is substantial but far from being dominant (panels in the second column of Fig.3). The analysis of the distributions $\zeta\ (j_\pi\ j_\nu)$, $\psi\ (j_\pi\ R)$ and $\xi\ (j_\nu\ R)$ for the *case B* has shown no significant differences in respect to the *case A* [16]. The presence of all chiral signatures in the *case B* is not the consequence of (for chirality) more favorable geometry.

Journal of Physics: Conference Series **205** (2010) 012026 doi:10.1088/1742-6596/205/1/012026

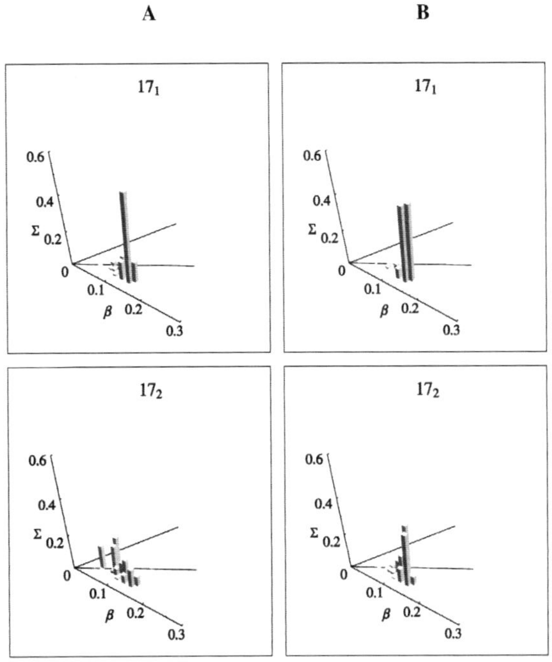

Figure 4. IBFFM β and γ distributions for the *case A* (first column) and the *case B* (second column). In the upper panels the distributions for the yrast 17_1 states and in the panels on bottom for the side band 17_2 states are presented. The axis in the $\beta-\gamma-$ plane have $\gamma = 0^o$ and $\gamma = 60^o$, while the middle line in the $\beta-\gamma-$plane marks $\gamma = 30^o$.

The main, but decisive, difference is revealed in the analysis of distributions in the $\beta-\gamma-$plane [16] (details of the procedure and the definition of the deformation parameters β and γ can be found in Ref. [14]). The β and γ distributions for *cases A* and *B* are sizeably different. The two bands have significantly more similar deformations in the *case B* than in the *case A* (Fig. 4). The yrast bands in both cases are similar, with fluctuations on higher spins being somewhat smaller in the *case B*. The side band on medium and higher spins has far less shape fluctuations and a significant decrease of components with small β in the *case B*. Different deformations and larger fluctuations in the *case A* reflect in $B(E2)$ values, that are much smaller in the side band than in the yrast band. In the *case B*, due to more similar deformations, the $B(E2)$ values for transitions between states in the side band are closer to the $B(E2)$ values in the yrast band. Smaller shape fluctuations and more similar deformations in the *case B* increase the probability that the two bands could be even and odd superpositions of separated left-handed and right-handed configurations, that reflects in the chiral-like behavior of $B(M1)$ values. The distribution of the core structure in the wave functions reveals the dynamical mechanism. In the *case B* the weaker, but still strong, boson-fermion interactions (particularly the exchange interaction) for the neutron (hole-like) quasiparticle are not strong enough to admix big components from higher lying core bands. The ground and the $\gamma-$band components are dominant in the states of the side band. All chiral signatures are present, but large shape fluctuations sizably reduce their magnitude in respect to the full chiral predictions. The $\gamma-$softness and shape fluctuations prevent the angular momenta of the proton, neutron and core to create chiral favorable geometry on the scale they would do if the core is triaxial. For nuclei where the boson-fermion interaction is stronger, other higher lying core structures admix into the states of the side (and partially into the yrast, too) band, increasing the shape fluctuations. This allows the admixture of more contributions from near axial shapes and consequently washes out the $B(M1)$ staggering. The only visible signature of dynamical chirality remains the vicinity of the excitation energies of the two bands.

However, the possible formation of chiral bands is not limited to odd-odd nuclei. Whenever three configurations in a nucleus have large angular momentum components along the three principal axes, the conditions for chirality are present. In odd-mass nuclei this can be fulfilled for three-particle configurations of the type one-proton-two-neutrons or one-neutron-two-protons. In this case the two-particle configuration takes the role of the one-particle configuration (of the same type) in the odd-odd nucleus. The situation in odd-mass nuclei could be even more favorable for chirality, because the two-particle angular momentum is longer than the one-particle angular momentum in odd-odd nuclei. The first observation of a possible three-quasiparticle chiral structure was in ^{135}Nd [4]. The negative-parity ground state band in ^{135}Nd has the one-neutron-quasiparticle $\nu h_{11/2}$ hole-like structure. At spin 25/2 it is crossed by a three-quasiparticle band with the structure interpreted by the $\nu h_{11/2}(\pi h_{11/2})^2$ configuration. The two-proton-quasiparticle configuration $(\pi h_{11/2})^2$ consists of $\pi h_{11/2}$ particle-like quasiprotons and is therefore particle-like. A twin band interpreted by the same $\nu h_{11/2}(\pi h_{11/2})^2$ configuration was observed as the yrare band rather close in excitation energy. In Ref. [4] these bands have been interpreted as a pair of chiral bands. Recently, $B(E2)$ and $B(M1)$ values deexciting states of the twin bands in ^{135}Nd have been measured [5]. The corresponding $B(E2)$ and $B(M1)$ values have been found to be almost identical in both bands, giving a strong argument to interpret this pair of bands as chiral partner bands.

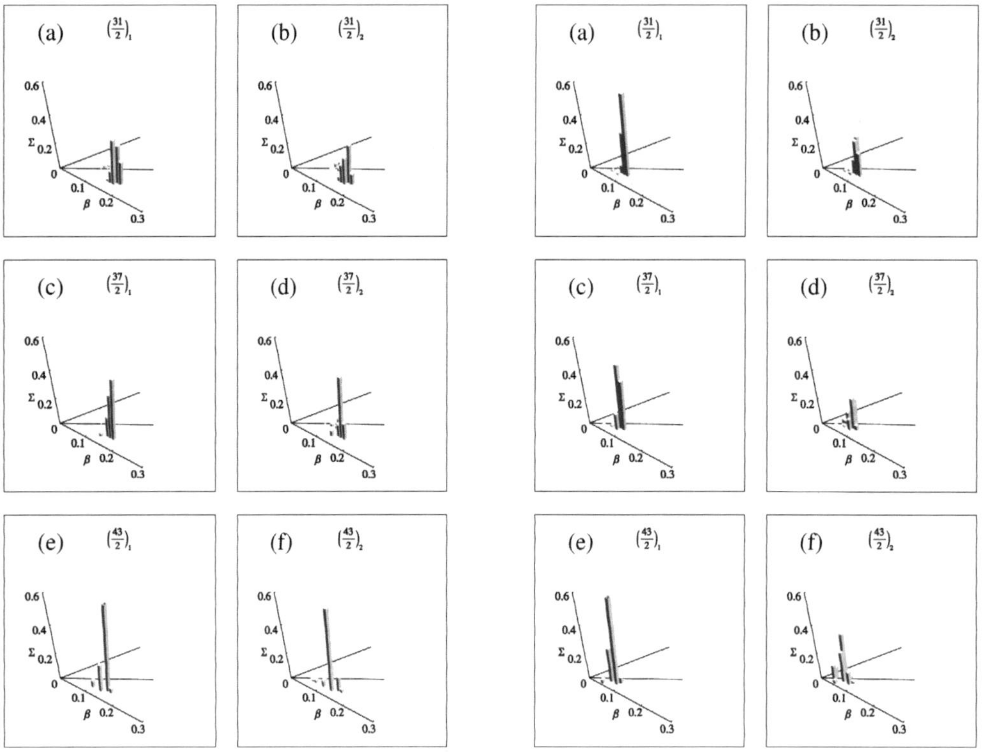

Figure 5. IBFFM β and γ distributions for ^{135}Nd (first and second column) and ^{137}Nd (third and forth column): panels (a) and (b) for the $31/2_1$ and $31/2_2$ states, panels (c) and (d) for the $37/2_1$ and $37/2_2$ states and panels (e) and (f) for the $43/2_1$ and $43/2_2$ states.

The properties of the three-quasiparticle bands in ^{135}Nd have also been investigated [17] within the interacting boson fermion plus broken pairs framework (IBFBPM) [18]. The IBFBPM calculations have shown that both partner bands are basically built on the ground-state band of the even-even core. The admixture of the γ-band components of the core is small. This behavior is different than in the IBFFM wave functions of odd-odd nuclei. The two-proton angular momentum is twice as long as the one-proton angular momentum in the neighboring odd-odd nuclei, providing a better possibility for the formation of chiral structures. The analysis of the distributions ζ $(J_\pi\ j_\nu)$, ψ $(J_\pi\ R)$ and ξ $(j_\nu\ R)$ of the angles between $\vec{J}_\pi$, $\vec{j}_\nu$ and $\vec{R}$ and of the σ distributions shows that in the states of both bands the presence of configurations with the angular momenta of the protons, neutron and core in the for chirality favorable, almost orthogonal geometry, is substantial. It is not so pronounced as if it were for a rigid triaxial core, but is far more pronounced than in the case of neighboring odd-odd nuclei (Fig.3). The partner bands show a structural change at medium spins that is due to shape fluctuations (Fig.5). For states with $I \leq 35/2$ shape fluctuations are of the order of those in the neighboring odd-odd nuclei [14, 16]. At spin 37/2 shape fluctuations are reduced, and for higher spins, although different values of β are present, the distribution shows one dominant peak. The structure changes from the geometry dominated by shape fluctuations into the geometry that is rather close to static chirality. This transition from the vibrational into a static chiral regime has recently been obtained in the calculations of TAC with RPA [5].

The structure of the N = 77 ^{137}Nd nucleus looks very similar to ^{135}Nd. But ^{137}Nd is less deformed and closer to the close shell. The IBFBPM predictions for ^{137}Nd [17] show that shape fluctuations in the yrare band are far more pronounced than in the yrast band. The fluctuations are not reduced for higher spins. On the contrary, in the yrare band the shape fluctuations are enhanced for higher spins and besides triaxial slightly prolate components, in the distribution oblate components appear (Fig.5). This leads to the conclusion that in $N = 77$ nuclei the structure of twin bands is more determined by shape fluctuations and prolate-oblate coexistence than by chirality.

3. Conclusions

The analysis of the wave functions of chiral candidates bands in the framework of the Interacting Boson Model, shows that the possibility for angular momenta of the valence proton, neutron and core to find themselves in the favorable, almost orthogonal geometry is present, but not dominant. The difference in the structure of chiral candidates nuclei can be attributed to different β and γ fluctuations. The coupling of the proton and neutron quasiparticles to the shape degrees of freedom is the dominant mechanism, leading to a weak chirality of dynamical origin. In some cases, in odd-even nuclei, this dynamic chirality can be close to static chirality.

References

[1] Frauendorf S and Meng J 1997 *Nucl. Phys.* A **617** 131
[2] Dimitrov V I, Frauendorf S and Dönau F 2000 *Phys. Rev. Lett.* **84** 5732
[3] Starosta K, Koike T, Chiara C J, Fossan D B and LaFosse D R 2001 *Nucl. Phys.* A **682** 375c
[4] Zhu S *et al* 2003 *Phys. Rev. Lett.* **91** 132501
[5] Mukhopadhyay S *et al* 2007 *Phys. Rev. Lett.* **99** 172501
[6] Ragnarsson I and Semmes P 1988 *Hyperfine Interact.* **43** 423
[7] Starosta K, Chiara C J, Fossan D B, Koike T, Kuo T T S, LaFosse D R, Rohoziński S G, Droste Ch, Morek T and Srebrny J 2002 *Phys. Rev.* C **65** 044328
[8] Brant S, Paar V and Vretenar D 1984 *Z. Phys.* A **319** 355
[9] Brant S and Paar V 1988 *Z. Phys.* A **329** 151
[10] Heyde K, Van Isacker P, Waroquier M and Moreau J 1984 *Phys. Rev.* C **29** 1420
[11] Casten R F, von Brentano P, Heyde K, Van Isacker P and Jolie J 1985 *Nucl. Phys.* A **439** 289
[12] Brant S, Vretenar D and Ventura A 2004 *Phys. Rev.* C **69** 017304
[13] Tonev D *et al* 2006 *Phys. Rev. Lett.* **96** 052501
[14] Tonev D *et al* 2007 *Phys. Rev.* C **76** 044313
[15] Grodner E *et al* 2006 *Phys. Rev. Lett.* **97** 172501
[16] Brant S, Tonev D, de Angelis G and Ventura A 2008 *Phys. Rev.* C **78** 034301
[17] Brant S and Petrache C M 2009 *Phys. Rev.* C **79** 054326
[18] Vretenar D, Bonsignori G and Savoia M 1995 *Z. Phys.* A **351** 289

Journal of Physics: Conference Series **205** (2010) 012027 doi:10.1088/1742-6596/205/1/012027

Testing a Variational Method for Fluctuations

P. D. Stevenson and J. M. A. Broomfield
Department of Physics, University of Surrey, Guildford, Surrey, GU2 7XH, UK
E-mail: p.stevenson@surrey.ac.uk

Abstract. A variational approach for many-body systems due to Balian and Vénéroni, which goes beyond the Harrtree-Fock mean field, has been implemented in the case of nuclear Skyrme effective interactions used in examining large amplitude collective motion. An evaluation of the numerical issues involved with the method is presented, in which it is found that the effect of model parameters is generally under control, but long-time running leads to unphysical results.

1. Introduction

A variational approach which is optimised to reproduce the expectation values of two-body observables while retaining a one-body density matrix representation of the physical state was proposed by Balian and Vénéroni [1, 2, 3] and implemented [4, 5, 6] in the 1980s. The first implementations in nuclear collisions and resonances necessarily featured restrictions in the realisation due to computational limitations. Since then, Time-Dependent Hartree-Fock (TDHF) capabilities, upon which the Balian-Vénéroni technique relies, have become more sophisticated [7, 8, 9, 10]. Recently, we have revisited the approach and implemented it in a symmetry-unrestricted way with full Skyrme forces [11, 12, 13, 14, 15].

Despite the fact that fuller calcualtions are now computationally feasible, there are still many numerical issues of concern, and some of these are addressed in the present paper.

2. The Method

The key point behind the Balian-Vénéroni approach is that the variational principle should include in it any observable whose expectation value is to be optimised, and that one can do this with two-body observables, even if one uses one-body density matrices as the trial states. One still improves upon those variational principles in which such considerations are not made.

The two-body observable studied in our work is the fluctuation of the number operator, which is applied to counting the number of nucleons in a spatial region identified with one nucleus or fragment. We have considered the excitation and subsequent decay of giant resonancs, and heavy ion collisions of a deep inelastic and a fusion-evaporation nature. The application of the Balian-Vénéroni procedure leads to an expression for the fluctuation of a one-body observable, $\hat{N}$, of the form

$$\Delta N^2(t_1) = \lim_{\epsilon \to 0} \frac{1}{2\epsilon^2} \mathrm{Tr} \left[(\rho(t_0) - \sigma(t_0))^2 \right] \tag{1}$$

where

$$\sigma(t_1) = e^{i\epsilon\hat{N}} \rho(t_1) e^{-i\epsilon\hat{N}}. \tag{2}$$

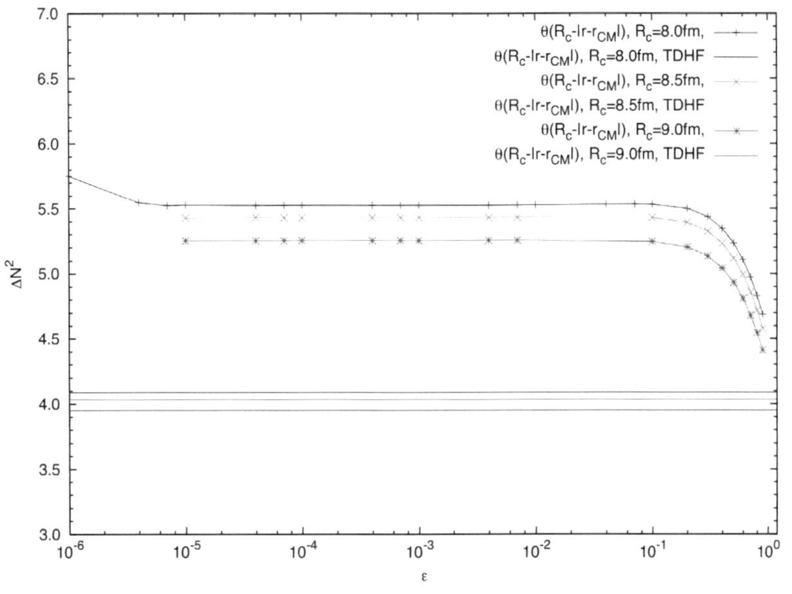

Figure 1. $\Delta N^2(t_1)$ plottes as a function of ϵ for different values of the cutoff radius R_c. The TDHF results, calculated at t_1 (and independent of ϵ) are shown for comparison.

The implementation of this approach consists of running a TDHF code from $t = 0$ to some finite end time t_1, apply the transformation (2) to *only the remnant nucleus and not any dripped flux*, run the code backwards to time $t = 0$ and evaluate (1). The procedure should be performed for a series of values of ϵ and extrapolated back to $\epsilon = 0$.

Various numerical issues arise in this process, the TDHF process being a rather computationally intensive one. It is important to ensure that the process of extrapolation of ϵ is as well-controlled as possible, that the transformation is applied to only the daughter nucleus, that parameters related to the spatial box in which the wavefunctions are represented, and the choice of end time are all as free from giving ambiguous physical results as possible. We explore these issues in the next section.

3. Results

All examples given here use as a test case the excitation of a large-amplitude giant dipole resonance in ^{32}S. As a check of the stability of results as a function of parameters in the procedure, we look at the dependnce on the cutoff radius R_c inside which we consider the daughter nucleus to exist. This has previously been reported in less detail [15]. In figure 1, values of the number fluctuation are given for different cutoff radius sizes; 8fm, 8.5fm and 9fm from the centre of the daughter nucleus. The number fluctuation does change as a function of cutoff radius, as more of the tail of the nuclear wavefunction is included, but only in the same way that there is an increase in TDHF, leaving the percentage change in fluctuation between the TDHF and the (better) Balian-Vénéroni answer quite constant. One should run the code in a region where the sensitivity to cutoff radius is smallest.

In figure 1, the range of sizes of ϵ for which a linear extrpolation can me made is clear. The range is actually quite large, but one must be careful. The flat region is only so large when one evaluates the $\rho(t_0)$ in (1) by evolving the system forward in time to t_1 and then backwards to

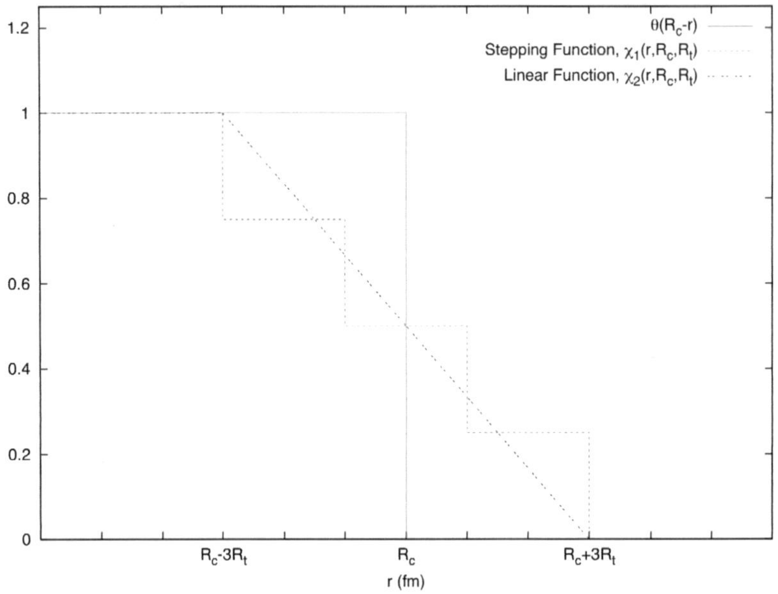

Figure 2. A schematic view of the radial dependence (plotted outwards from the centre-of-mass of the nucleus) of the step function, $\chi_1(|\bar{r}-\bar{r}_{CM}|, R_c, R_t)$, and the linear function, $\chi_2(|\bar{r}-\bar{r}_{CM}|, R_c, R_t)$, used in place of the theta function.

Table 1. The dependence of $\langle N\rangle$, ΔN_{TDHF} and ΔN_{BV} on the form of the spatial cutoff function (and on R_c and R_t). The function $\theta(R_c - |\bar{r}-\bar{r}_{CM}|)$ provides a sharp cutoff at $r = R_c$. The function $\chi_1(r, R_c, R_t)$ provides a cutoff which decreases in steps whilst the function $\chi_2(r, R_c, R_t)$ provides a linear cutoff. R_t controls the sharpness of the cutoff functions $\chi_1(r, R_c, R_t)$ and $\chi_2(r, R_c, R_t)$ as shown schematically in figure 2. These results values were all calculated for $t_1 = 250$ fm/c using the SLy6 parameterisation of the Skyrme force.

	$\mathbf{R_c}$ (fm)	$\mathbf{R_t}$ (fm)	$\langle \mathbf{N}\rangle\|_{t_1}$	$\Delta \mathbf{N_{TDHF}}\|_{t_1}$	$\Delta \mathbf{N_{BV}}\|_{t_1}$	Change
$\theta(R_c - r)$	8.0	-	26.6366	2.0218	2.3506(0)	+16%
$\chi_1(r, R_c, R_t)$	8.0	0.25	26.6244	2.0234	2.3346(0)	+15%
$\chi_2(r, R_c, R_t)$	8.0	0.25	26.6295	2.0226	2.3387(0)	+16%
$\theta(R_c - r)$	9.0	-	26.8975	1.9875	2.2924(0)	+15%
$\chi_1(r, R_c, R_t)$	9.0	0.25	26.9044	1.9877	2.2689(0)	+14%
$\chi_2(r, R_c, R_t)$	9.0	0.25	26.8951	1.9885	2.2758(0)	+14%

t_0 without performing the Balian-Vénéroni transformation on it. This ensures that systematic errors are eliminated from process of finding the difference between two small values in (1).

The form of the spatial cutoff in figure 1 is a sharp step function. It is not obvious that the is the best choice. We have examined other function, such as a multistep function and a linear funcion to lessen the sudden break between nucleus and environment.

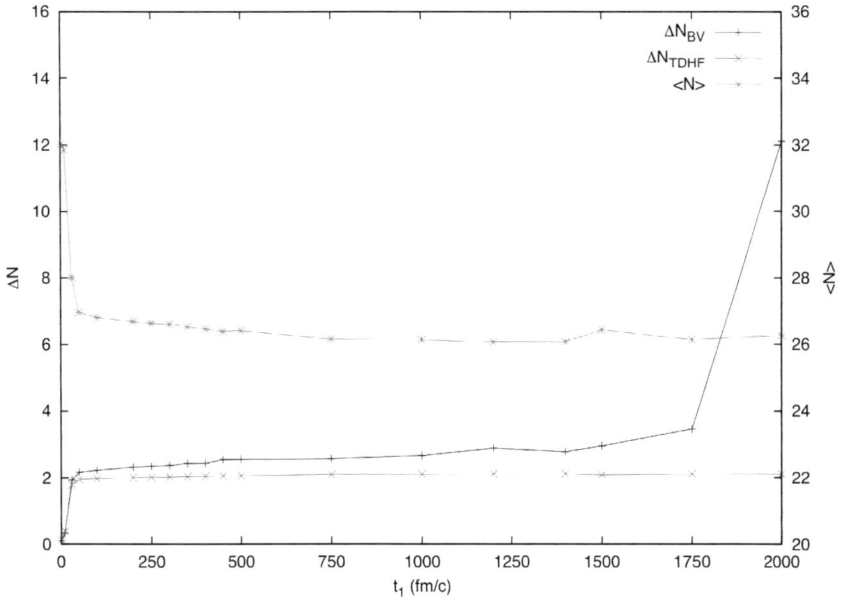

Figure 3. Mean number $\langle N \rangle$ and the fluction in TDHF and BV methods as a function of end time t_1.

Figure 2 shows the spatial form of the different cutoff functions used, and Table 1 the results from using these different functions to split the system between the daughter nucleus (in which the transformation (2) is applied) and the environment (where it is not). The table shows that the differences between the different methods are slight, with the numbers being physically reasonable given that the stepping and linear functions actually feature a little less of the daughter nucleus in them.

Finally, we look at the dependence of the final time, t_1. In principle, one should aim to perform a calculation in which the daughter nucleus has completely decayed and all flux that has left the nucleus has gone far off into the environment away from the nucleus. In practice, one is limited in box size and the customary boundary conditions (reflecting or periodic) result in unphysical interference with the emitted particles and the daughter nucleus. This means that one must pick an intermediate value of t_1; large enough that the system has approached its asymptotic state, but small enough that numerical problems are not starting to rear their head.

Calculations as a function of end time t_1 are shown in figure 3. It shows that the Balian-Vénéroni calculation of the number fluctuation increases significantly at large times. This happens after the number of nucleons in the daughter nucleus starts to rise again due to reflections and is an indication of numerical problems. The results are at least reasonably stable for a large range of times, and the numbers at any of those times can be taken as a good value for the fluction *at that time*. In the long run, better boundary conditions at the edge of the box will be needed to allow for better long time behavior.

Other numerical aspects have been studied too; the effect of variation of time step and spatial discretisation. These variations were negligible.

4. Conclusion

We have presented an analysis of some of the numerical issues related to implenting the Balian-Vénéroni variation procedure in nuclear collective motion. Parameters related to the separation of the system between nucleus and environment give rise to slightly different answers (as they do in TDHF), but in a physical way. Long-time behaviour is problematic and will require improved boundary conditions.

References

[1] Balian R and Vénéroni M 1985 *Ann. Phys.* **164** 334-410
[2] Balian R and Vénéroni M 1988 *Ann. Phys.* **187** 29-78
[3] Balian R and Vénéroni M 1992 *Ann. Phys.* **216** 351-430
[4] Troudet T and Vautherin D 1985 *Phys. Rev.* C **31** 278-9
[5] Bonche P and Flocard H 1985 *Nucl. Phys.* A **437** 189-207
[6] Marston J B and Koonin S E 1985 *Phys. Rev. Lett.* **54** 1139-41
[7] Nakatsukasa T and Yabana K 2005 *Phys. Rev.* C **71** 024301
[8] Simenel C and Chomaz Ph 2003 *Phys. Rev.* C **68** 024302
[9] Umar A S and Oberacker V E 2006 *Phys. Rev.* C **73** 054607
[10] Maruhn J A, Reinhard P-G, Stevenson P D and Strayer M R 2005 *Phys. Rev.* C **74** 027601
[11] Broomfield J M A and Stevenson P D 2008 *J. Phys.* G**35** 095102
[12] Broomfield J M A and Stevenson P D 2009 *AIP Conf. Proc.* **1098** 133
[13] Stevenson P D and Broomfield J M A 2009 *AIP Conf. Proc.* **1165** 177
[14] Broomfield J M A 2009 *Ph.D. Thesis*, University of Surrey
[15] Stevenson P D and Broomfield J M A, *Proc. XXXII Symposium on Nuclear Physics, Cocoyoc, Mexico* (*Preprint* 0903.0130)

Journal of Physics: Conference Series **205** (2010) 012028 doi:10.1088/1742-6596/205/1/012028

Isospin symmetry-breaking corrections for superallowed β decay in relativistic RPA approaches

Haozhao Liang[1,2], **Nguyen Van Giai**[2] **and Jie Meng**[1,3,4]

[1]State Key Laboratory of Nuclear Physics and Technology, School of Physics, Peking University, Beijing 100871, China

[2]Institut de Physique Nucléaire, IN2P3-CNRS and Université Paris-Sud, 91406 Orsay, France

[3]School of Physics and Nuclear Energy Engineering, Beihang University, Beijing 100191, China

[4]Department of Physics, University of Stellenbosch, Stellenbosch, South Africa

E-mail: `nguyen@ipno.in2p3.fr`

Abstract. The isospin symmetry-breaking corrections δ_c obtained with the self-consistent relativistic RPA approaches are presented. It is shown that the proper treatment of the Coulomb mean field is very important for extracting these corrections. The nucleus-independent $\mathcal{F}t$ value, the V_{ud} matrix element and the unitarity of the Cabibbo-Kobayashi-Maskawa matrix are discussed. The effects of neutron-proton mass difference on the isospin symmetry-breaking corrections are also investigated.

1. Introduction

The Cabibbo-Kobayashi-Maskawa (CKM) matrix [1, 2] relates the quark eigenstates of the weak interaction with the quark mass eigenstates. Its unitarity is one of the most important properties of the Standard Model of particle physics. Measuring its matrix elements independently and verifying its unitarity condition provide a rigorous test for the Standard Model. For the leading matrix element V_{ud}, there are four experimental methods to determine its value: nuclear $0^+ \to 0^+$ superallowed β decays [3], neutron decay [4], pion β decay [5] and nuclear mirror transitions [6]. In particular, the first method provides the most precise determination of the $|V_{ud}|$ value [7].

In order to determine the $|V_{ud}|$ value with these nuclear superallowed β transitions, the radiative corrections Δ_R^V, δ'_R, $\delta_{\rm NS}$ and the isospin symmetry-breaking corrections δ_c for the experimental ft values have to be taken into account [3], i.e.,

$$V_{ud}^2 = \frac{K}{2G_F^2(1+\Delta_R^V)\mathcal{F}t}, \tag{1}$$

and

$$\mathcal{F}t = ft(1+\delta'_R)(1+\delta_{\rm NS}-\delta_c), \tag{2}$$

where G_F is the Fermi coupling constant for purely leptonic decays and K is a physical constant related to electron mass [7]. The f and t factors represent the statistical rate function and partial half-life, respectively, and they are obtained through measurements of the Q values, branching ratios, and half-lives of the superallowed decays. With the radiative and isospin symmetry-breaking corrections, the quantity $\mathcal{F}t$ should be nucleus independent.

Since the isospin symmetry is not an exact symmetry mainly due to the presence of the Coulomb forces in nuclei, the superallowed transition strength $|M_F|^2$ from initial state $|i\rangle$ to final state $|f\rangle$ slightly differs from its ideal value $|M_0|^2$ as

$$|M_F|^2 = |\langle f|\, T_\pm \,|i\rangle\,|^2 = |M_0|^2(1-\delta_c), \tag{3}$$

where $M_0 = \sqrt{2}$ for $T = 1$ states having exact isospin symmetry.

Shell model calculations are generally used to evaluate the isospin symmetry-breaking corrections δ_c [8]. Alternatively, self-consistent Random Phase Approximation (RPA) based on microscopic mean field theories is another reliable approach for evaluating the superallowed transition strength M_F. Such calculations have been performed for a few nuclei with the non-relativistic Skyrme Hartree-Fock approach in Ref. [9].

During the last decade, great efforts have been dedicated to developing the charge-exchange (Q)RPA within the relativistic framework [10–12]. In particular, a fully self-consistent charge-exchange RPA based on the relativistic Hartree-Fock (RHF) approach [13] was established recently [12]. A very satisfactory description of spin-isospin resonances was obtained without any readjustment of the energy functional. Therefore, it is appropriate now to re-investigate the isospin symmetry-breaking corrections δ_c for superallowed β decay with the relativistic approaches.

In this paper, the calculations of the isospin symmetry-breaking corrections δ_c with the self-consistent relativistic RPA approaches will be presented. The nucleus-independent $\mathcal{F}t$ value, the V_{ud} matrix element and the unitarity of the CKM matrix will then be discussed. The effects of neutron-proton mass difference on the isospin symmetry-breaking corrections will be also investigated.

2. Self-consistent Relativistic RPA

The basic ansatz of the relativistic Hartree (RH), also known as relativistic mean field (RMF), and relativistic Hartree-Fock (RHF) theories is a Lagrangian density $\mathcal{L}$, where nucleons are described as Dirac spinors that interact via the exchange of σ-, ω-, ρ-, π-mesons and the photon [14–16]. In order to give a satisfactory description of nuclear matter and finite nuclei, the nonlinear self-coupling of mesons [17, 18] or density-dependent meson-nucleon couplings [13, 19] are introduced.

In the Hartree and Hartree-Fock approximations, the total energy E of the system is the trial ground state (Slater Determinant) expectation value of the effective Hamiltonian operator $\hat{H}$, which can be obtained with the general Legendre transformation from the Lagrangian density $\mathcal{L}$. This total energy contains the kinetic energy and the direct (Hartree) and exchange (Fock) potential energies. In the Hartree approximation, the Fock terms are neglected for simplicity, and the parameters of $\mathcal{L}$ are readjusted accordingly.

In the self-consistent RPA calculations, the particle-hole residual interaction $V_{\rm ph}$ in the RPA equations is derived from the second derivative of the same energy functional as that used in the ground-state description. In general, there are additional rearrangement terms in the particle-hole residual interaction when the explicit density dependence of the meson-nucleon couplings are included [20]. However, since the rearrangement terms are due to the dependence on isoscalar ground-state densities, it is easy to see that they are absent in the charge-exchange channels, e.g., in the description of superallowed β decays. The RPA equations can be written as

$$\begin{pmatrix} \mathcal{A}^J_{p\bar{n}p'\bar{n}'} & \mathcal{B}^J_{p\bar{n}n'\bar{p}'} \\ -\mathcal{B}^J_{n\bar{p}p'\bar{n}'} & -\mathcal{A}^J_{n\bar{p}n'\bar{p}'} \end{pmatrix} \begin{pmatrix} X^{J\nu}_{p'\bar{n}'} \\ Y^{J\nu}_{n'\bar{p}'} \end{pmatrix} = \omega_\nu \begin{pmatrix} X^{J\nu}_{p\bar{n}} \\ Y^{J\nu}_{n\bar{p}} \end{pmatrix}, \tag{4}$$

where p and $\bar{p}$ (n and $\bar{n}$) denote unoccupied and occupied proton (neutron) states. It should be emphasized that within the no-sea approximation the unoccupied states include not only the states above the Fermi surface, but also the states in the Dirac sea. With the X and Y amplitudes, the transition probabilities $B^{\pm}_{\nu}$ between the 0^+ parent ground-state and the 0^+ excited states ν in the daughter nuclei ($N \pm 1$, $Z \mp 1$) can be obtained [21].

3. Results and Discussion

For all the calculations presented in this paper, the spherical symmetry is assumed and the filling approximation is applied to the last partially occupied orbital. The radial Dirac equations are solved in coordinate space by the Runge-Kutta method within a spherical box with a box radius R = 15 fm and a mesh size dr = 0.1 fm. The single-particle wave functions thus obtained are used to construct the RPA matrices $\mathcal{A}$ and $\mathcal{B}$ in Eq. (4) with the single-particle energy truncation $[-M, M + 120$ MeV$]$, i.e., the occupied states are the positive energy states below the Fermi surface, whereas the unoccupied states can be either positive energy states above the Fermi surface or bound negative energy states [12]. With these numerical inputs, the model-independent sum rule,

$$\sum_{\nu} B^-_{\nu} - \sum_{\nu} B^+_{\nu} = N - Z, \tag{5}$$

where $B^{\pm}_{\nu}$ corresponds to the transition operator $T_{\pm}$, can be fulfilled up to 10^{-5} accuracy, and the isospin symmetry-breaking corrections δ_c are stable with respect to these numerical inputs at the same level of accuracy.

Table 1. Isospin symmetry-breaking corrections δ_c for the $0^+ \to 0^+$ superallowed transitions obtained by self-consistent RHF+RPA calculations with PKO1 [13] as well as RH+RPA calculations with DD-ME2 [19]. The column PKO1* presents the results obtained with PKO1 without the Coulomb exchange (Fock) term. The results obtained by shell model calculations [8] are listed in the column T&H for comparison. All values are expressed in %.

	PKO1	PKO1*	DD-ME2	T&H [8]
$^{10}C \to {}^{10}B$	0.082	0.148	0.150	0.175(18)
$^{14}O \to {}^{14}N$	0.114	0.178	0.197	0.330(25)
$^{18}Ne \to {}^{18}F$	0.270	0.357	0.430	0.565(39)
$^{26}Si \to {}^{26}Al$	0.176	0.246	0.252	0.435(27)
$^{30}S \to {}^{30}P$	0.497	0.625	0.633	0.855(28)
$^{34}Ar \to {}^{34}Cl$	0.268	0.359	0.376	0.665(56)
$^{38}Ca \to {}^{38}K$	0.313	0.406	0.441	0.765(71)
$^{42}Ti \to {}^{42}Sc$	0.384	0.460	0.523	0.935(78)
$^{26}Al \to {}^{26}Mg$	0.139	0.193	0.198	0.310(18)
$^{34}Cl \to {}^{34}S$	0.234	0.298	0.307	0.650(46)
$^{38}K \to {}^{38}Ar$	0.278	0.344	0.371	0.655(59)
$^{42}Sc \to {}^{42}Ca$	0.333	0.395	0.448	0.665(56)
$^{54}Co \to {}^{54}Fe$	0.319	0.392	0.393	0.770(67)
$^{66}As \to {}^{66}Ge$	0.475	0.571	0.572	1.56(40)
$^{70}Br \to {}^{70}Se$	1.140	1.234	1.268	1.60(25)
$^{74}Rb \to {}^{74}Kr$	1.088	1.230	1.258	1.63(31)

In Table 1, the isospin symmetry-breaking corrections δ_c in Eq. (3) for the $0^+ \to 0^+$ superallowed transitions are shown. The results are obtained by self-consistent RHF+RPA calculations with PKO1 [13] model as well as by self-consistent RH+RPA calculations with DD-ME2 [19] model, and the results obtained by shell model calculations (T&H) [8] are also listed for comparison. The present corrections δ_c range from about 0.1% for the lightest nucleus ^{10}C to about 1.2% for the heaviest nucleus ^{74}Rb, which are two or three times smaller than the T&H results. It is noticed that even smaller estimates of δ_c compared to the shell model calculations have been recently obtained in Ref. [22] using perturbation theory.

In Table 1, it is found that the corrections δ_c of RHF+RPA are systematically smaller than those of RH+RPA. To understand this systematic discrepancy between RHF+RPA and RH+RPA, it must be kept in mind that in RHF+RPA the exchange (Fock) terms of mesons and photon are kept in both the mean field and RPA levels, whereas they are neglected altogether in RH+RPA. Among all the Fock terms, we expect, in particular, the exchange terms of the Coulomb field to play an important role due to the following reason. The isobaric analog state would be degenerate with its isobaric multiplet partners, and it would contain 100% of the model-independent sum rule (5), i.e., $\delta_c = 0$, if the nuclear Hamiltonian commutes with the isospin raising and lowering operators $T_\pm$. This would be true when the Coulomb field is switched off. While this degeneracy is broken by the mean field approximation, no matter the exchange terms of mesons are included or not, it can be restored by the RPA calculations as long as the RPA calculations are self-consistent [23]. Therefore, the Coulomb field is essential for the $0^+ \to 0^+$ superallowed transitions and the Coulomb exchange (Fock) term should be responsible for the different isospin symmetry-breaking corrections δ_c in RHF+RPA and RH+RPA approaches.

In order to verify the above argument, we have performed the following calculations. Using PKO1, the Hartree-Fock calculations are performed by switching off the exchange contributions of the Coulomb field. From the single-particle spectra thus obtained, self-consistent RPA calculations are then performed. The isospin symmetry-breaking corrections δ_c thus obtained are listed in the column denoted as PKO1* in Table 1. It is seen that these results are almost the same as those of RH+RPA calculations with DD-ME2, i.e., by switching off the exchange contributions of the Coulomb field, the corrections δ_c in RHF+RPA calculations recover the results in RH+RPA calculations. Therefore, one can conclude that the proper treatment of the Coulomb field is very important to extract the isospin symmetry-breaking corrections δ_c.

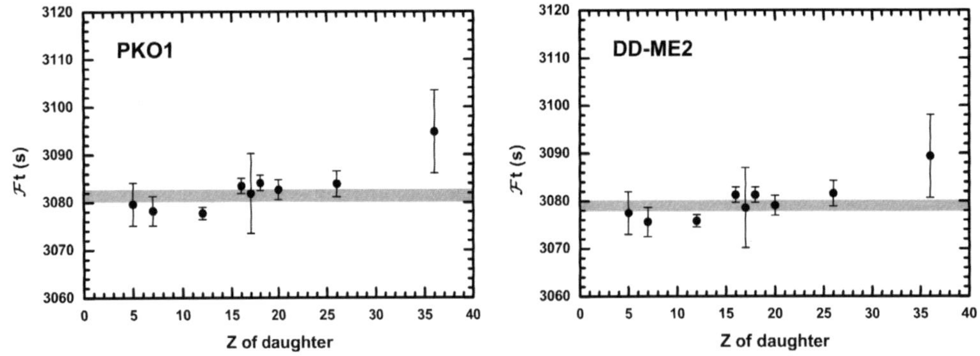

Figure 1. Nucleus-independent $\mathcal{F}t$ values as a function of the charge Z for the daughter nucleus. The values of δ_c are respectively obtained by RHF+RPA calculations with PKO1 (left panel) and by RH+RPA calculations with DD-ME2 (right panel). The shaded horizontal band gives one standard deviation around the average $\overline{\mathcal{F}t}$ value.

Combining the δ'_R and δ_{NS} values from recent calculations [8], and δ_c in Table 1, as well as measured ft values [3], the nucleus-independent $\mathcal{F}t$ values for superallowed β decays are obtained and plotted as a function of the charge Z for the daughter nucleus in Fig. 1, where the uncertainty of δ_c is taken as zero and the shaded horizontal band gives one standard deviation around the average $\overline{\mathcal{F}t}$ value.

It is found that the χ^2/ν is 1.0 $\sim$ 1.1 s, which indicates that the constancy of the nucleus-independent $\mathcal{F}t$ values is satisfied, even though not as well as the shell model calculations in Ref. [3]. It is also found that the $\mathcal{F}t$ value of RHF+RPA is about 2 s larger than that of RH+RPA, which is larger than the difference due to the different effective interactions in either RHF or RH approximations [21]. It again demonstrates that the proper treatment of the Coulomb field plays an important role in this specific subject, and one cannot effectively take care of the Coulomb exchange term by adjusting the parameters in the direct terms of mesons as done in the usual RH approximation.

With the nucleus-independent $\mathcal{F}t$ value, the element V_{ud} of the CKM matrix can be calculated by Eq. (1) with $K/(\hbar c)^6 = 8120.2787(11)\times 10^{-10}$ GeV^{-4}s, $G_F/(\hbar c)^3 = 1.16637(1)\times 10^{-5}$ GeV^{-2} [7], and $\Delta_R^V = 2.361(38)\%$ [8]. Then in combination with the other two CKM matrix elements $|V_{us}| = 0.2255(19)$ and $|V_{ub}| = 0.00393(36)$ [7], one can test the unitarity of the matrix.

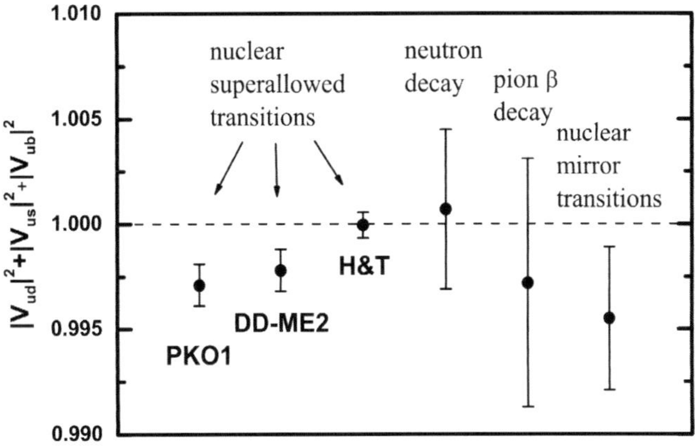

Figure 2. The sum of squared top row elements of the CKM matrix obtained by RHF+RPA calculations with PKO1 and by RH+RPA calculations with DD-ME2 in comparison with those in shell model (H&T) [3] as well as in neutron decay [7], pion β decay [5] and nuclear mirror transitions [6].

In Fig. 2, the sum of squared top row elements of the CKM matrix obtained by RHF+RPA calculations with PKO1 and by RH+RPA calculations with DD-ME2 is shown in comparison with those in shell model (H&T) [3] as well as in neutron decay [7], pion β decay [5] and nuclear mirror transitions [6].

It is interesting to note that the present results agree well with those obtained in neutron decay, pion β decay and nuclear mirror transitions. However, they deviate somewhat from the unitarity condition, which is in contradiction with the conclusion of shell model calculations (H&T) [3]. However, it should be noticed that the uncertainties of the present results are underestimated to some extent as the uncertainty of δ_c is assumed to be zero and the systematic errors are not taken into account. These systematic deviations would come from extracting δ_c with different effective interactions, with the proper neutron-proton mass difference, with deformation, or with pairing correlations.

Finally, we examine the effects of the neutron-proton mass difference on the V_{ud} value. We repeat the self-consistent relativistic RPA calculations with effective interactions PKO1 and DD-

Table 2. Isospin symmetry-breaking corrections δ_c for the $0^+ \rightarrow 0^+$ superallowed transitions. The columns PKO1† and DD-ME2† present the results obtained with effective interactions PKO1 and DD-ME2 but adopting experimental values of neutron and proton masses. The columns $\Delta\delta_c$ show the difference between the results of PKO1† and PKO1, as well as DD-ME2† and DD-ME2. All values are expressed in %.

	PKO1†	$\Delta\delta_c$	DD-ME2†	$\Delta\delta_c$
$^{10}C \rightarrow {}^{10}B$	0.089	0.007	0.161	0.011
$^{14}O \rightarrow {}^{14}N$	0.127	0.013	0.214	0.017
$^{18}Ne \rightarrow {}^{18}F$	0.288	0.017	0.454	0.024
$^{26}Si \rightarrow {}^{26}Al$	0.188	0.012	0.268	0.016
$^{30}S \rightarrow {}^{30}P$	0.531	0.034	0.672	0.039
$^{34}Ar \rightarrow {}^{34}Cl$	0.287	0.019	0.400	0.024
$^{38}Ca \rightarrow {}^{38}K$	0.334	0.021	0.467	0.026
$^{42}Ti \rightarrow {}^{42}Sc$	0.405	0.020	0.549	0.026
$^{26}Al \rightarrow {}^{26}Mg$	0.150	0.011	0.212	0.014
$^{34}Cl \rightarrow {}^{34}S$	0.252	0.018	0.329	0.022
$^{38}K \rightarrow {}^{38}Ar$	0.299	0.020	0.395	0.024
$^{42}Sc \rightarrow {}^{42}Ca$	0.352	0.019	0.472	0.024
$^{54}Co \rightarrow {}^{54}Fe$	0.336	0.017	0.414	0.020
$^{66}As \rightarrow {}^{66}Ge$	0.500	0.026	0.601	0.028
$^{70}Br \rightarrow {}^{70}Se$	1.188	0.048	1.320	0.051
$^{74}Rb \rightarrow {}^{74}Kr$	1.132	0.044	1.308	0.050

ME2, but adopting experimental values of neutron and proton masses. The isospin symmetry-breaking corrections δ_c thus obtained are listed in the columns PKO1† and DD-ME2† of Table 2. The difference between the results of PKO1† and PKO1, as well as DD-ME2† and DD-ME2 are presented in the columns $\Delta\delta_c$ to show the net effects of the neutron-proton mass difference.

It can be seen that the effects of the neutron-proton mass difference on the corrections δ_c range from about 0.01% to about 0.05%, and roughly speaking, larger δ_c leads to larger $\Delta\delta_c$. With these effects, the V_{ud} value predicted by RHF+RPA with PKO1 is changed from $|V_{ud}| = 0.97273(27)$ to $|V_{ud}| = 0.97280(27)$, and the value predicted by RH+RPA with DD-ME2 is changed from $|V_{ud}| = 0.97311(26)$ to $|V_{ud}| = 0.97321(26)$. This indicates the effects of the neutron-proton mass difference are small compared to other kinds of uncertainties, and it is not the main reason why the present results of the sum of squared top row elements of the CKM matrix deviate from the unitarity condition.

4. Summary

In summary, self-consistent relativistic RPA approaches are applied to calculate the isospin symmetry-breaking corrections δ_c for the superallowed β transitions. It is found that the proper treatments of the Coulomb field is very important to extract the isospin symmetry-breaking corrections δ_c. By switching off the exchange contributions of the Coulomb field, the corrections δ_c in RHF+RPA calculations recover the results in RH+RPA calculations. In other words, one cannot effectively take care of the Coulomb exchange term by adjusting the parameters in the direct terms of mesons as done in the usual RH approximation.

With the isospin symmetry-breaking corrections δ_c calculated by relativistic RPA approaches, the values of $|V_{ud}|$ thus obtained agree well with those obtained in neutron decay, pion β decay and nuclear mirror transitions. However, the sum of squared top-row elements seems to deviate from the unitarity condition.

The effects of the neutron-proton mass difference on the isospin symmetry-breaking corrections δ_c have been investigated. It is shown that these effects are small compared to other kinds of uncertainties. It is not the main reason why the present results of the sum of squared top row elements of the CKM matrix deviate from the unitarity condition.

Acknowledgements
This work is partly supported by Major State 973 Program 2007CB815000, the NSFC under Grant Nos. 10435010, 10775004 and 10221003. One of the authors (H.L.) is grateful to the French Embassy in Beijing for the financial support for his stay in France.

References

[1] Cabibbo N 1963 *Phys. Rev. Lett.* **10** 531
[2] Kobayashi M and Maskawa T 1973 *Prog. Theor. Phys.* **49** 652
[3] Hardy J C and Towner I S 2009 *Phys. Rev. C* **79** 055502
[4] Thompson D 1990 *J. Phys. G* **16** 1423
[5] Počanić D *et al.* 2004 *Phys. Rev. Lett.* **93** 181803
[6] Naviliat-Cuncic O and Severijns N 2009 *Phys. Rev. Lett.* **102** 142302
[7] Amsler C *et al.* 2008 *Phys. Lett. B* **667** 1
[8] Towner I S and Hardy J C 2008 *Phys. Rev. C* **77** 025501
[9] Sagawa H, Van Giai N and Suzuki T 1996 *Phys. Rev. C* **53** 2163
[10] De Conti C, Galeão A P and Krmpotić F 1998 *Phys. Lett. B* **444** 14
[11] Paar N, Nikšić T, Vretenar D and Ring P 2004 *Phys. Rev. C* **69** 054303
[12] Liang H, Van Giai N and Meng J 2008 *Phys. Rev. Lett.* **101** 122502
[13] Long W H, Van Giai N and Meng J 2006 *Phys. Lett. B* **640** 150
[14] Serot B D and Walecka J D 1986 *Advances in Nuclear Physics Vol. 16: The Relativistic Nuclear Many Body Problem* (Plenum Press, New York)
[15] Bouyssy A, Mathiot J F, Van Giai N and Marcos S 1987 *Phys. Rev. C* **36** 380
[16] Meng J, Toki H, Zhou S G, Zhang S Q, Long W H and Geng L S 2006 *Prog. Part. Nucl. Phys.* **57** 470
[17] Lalazissis G A, König J and Ring P 1997 *Phys. Rev. C* **55** 540
[18] Long W, Meng J, Giai N V and Zhou S G 2004 *Phys. Rev. C* **69** 034319
[19] Lalazissis G A, Nikšić T, Vretenar D and Ring P 2005 *Phys. Rev. C* **71** 024312
[20] Nikšić T, Vretenar D and Ring P 2002 *Phys. Rev. C* **66** 064302
[21] Liang H, Van Giai N and Meng J 2009 *Phys. Rev. C* **79** 064316
[22] Auerbach N 2009 *Phys. Rev. C* **79** 035502
[23] Engelbrecht C A and Lemmer R H 1970 *Phys. Rev. Lett.* **24** 607

Journal of Physics: Conference Series **205** (2010) 012029 doi:10.1088/1742-6596/205/1/012029

Structure of the low-lying states in some N=80 isotones

D Tarpanov [1], Ch Stoyanov [1], N Lo Iudice[2], Nguyen Van Giai[3], V Voronov [4]

[1]Institute for Nuclear Research and Nuclear Energy, 1784 Sofia, Bulgaria
[2] Dipartimento di Scienze Fisiche, Universitá di Napoli "Federico II" and Istituto Nazionale di Fisica Nucleare, Monte S. Angelo Via Cintia, I-80126 Napoli, Italy
[3]Institut de Physique Nucléaire, IN2P3/CNRS and Université Paris-Sud, F-91406 Orsay Cedex, France
[4]Bogoliubov Laboratory of Theoretical Physics, Joint Institute for Nuclear Research, 141980 Dubna, Moscow Region, Russia

E-mail: tarpanov@inrne.bas.bg

Abstract. The quasiparticle-phonon model using a separable interaction deduced from a Skyrme force is adopted to compute low-lying spectra and the giant dipole resonance in the $N = 80$ isotones. It is shown that the Skyrme interaction not only reproduces the fragmentation of the giant resonance but also yields results comparable to the ones obtained when a Woods-Saxon potential is used in the description of the mixed symmetry states recently discovered in these nuclei. This test is encouraging in view of future extensions of the method to neutron-rich nuclei far from stability.

1. Introduction

Skyrme effective interactions are widely used for the study of nuclear properties. Attention is focused mostly on the bulk properties, like binding energies, mean square radii, energy centroids, and on the strength distributions of Giant Resonances [1, 2]. Recently, a tensor term was incorporated in the Skyrme force [3, 4]. Such a term came out to improve significantly the single particle level spectra.

Recently, methods for obtaining separable approximations to Skyrme forces have been developed. One is anchored to the mean field self-consistency [5], the other is based on a finite rank approximation [6]. Both methods have been applied to deformed [5] as well as spherical nuclei [6, 7, 8]. The finite rank approximation procedure was used to calculate the properties of low-lying quadrupole excitations in the chain of tin isotopes [9].

The Quasiparticle-Phonon Model (QPM) of Soloviev and collaborators [10] has been applied successfully to the study of the low-lying excitations and giant resonances in spherical and deformed nuclei. The main ingredients of the model are phonons obtained in the quasi-particle random-phase approximation ($QRPA$) interacting by means of a residual interaction. The QPM unifies the description of the structure of low-lying states in terms of single and multiple phonon states. Within the model, the consistency between the mean field and residual interaction is ensured by the relation between the radial part of the interaction and the derivative of the mean field potential,

$$f(r) = \frac{dU(r)}{dr}. \quad (1)$$

In the QPM one adopts generally a Woods-Saxon one-body potential. This is a good choice for stable nuclei, where the Woods-Saxon parameters can be inferred from a fit of the experimental quantities, like single-particle energies. This is no longer possible in nuclei far from stability. In neutron rich nuclei, in fact, the proton mean field potential becomes much deeper than the neutron well because of the enhancement of the proton-neutron interaction. For those nuclei, it is safer to generate the mean field self-consistently . This can be done by Hartree-Fock (HF) or Hartree-Fock-Bogoliubov (HFB) methods using Skyrme forces.

A preliminary condition for the reliability of the one-body potentials so obtained is that they reproduce the low-energy properties and spectra of stable nuclei. A meaningful test is provided by the work presented here. We have, in fact, performed a QPM calculation using a two-body interaction of separable form derived from a Skyrme force. Here, a detailed comparison with an analogous calculation using a Woods-Saxon potential shows that the separable version of the Skyrme force is successful in the study of low-lying collective states in spherical nuclei as well as in reproducing the distribution of the strength in the domain of the Giant Dipole Resonance (GDR).

2. Main ingredients of the model

The intrinsic QPM Hamiltonian has the following structure

$$H = H_{sp} + V_{pair} + V_M^{ph} + V_{SM}^{ph} + V_M^{pp}. \quad (2)$$

H_{sp} is the one-body Hamiltonian, V_{pair} is the monopole pairing force, V_M^{ph} and V_{SM}^{ph} are respectively the particle-hole separable multipole and spin-multipole interactions, and V_M^{pp} is the particle-particle interaction.

As pointed out already, most of the QPM calculations have been performed using Woods-Saxon one-body potentials. In the case of spherical symmetry, the parameters of these potentials have been fitted for several domains of mass number [10]. The comparison of a typical Woods-Saxon potential with the corresponding SHF potential, obtained through a $SIII$ parametrization, is shown in Fig. 1 for ^{136}Ba. We have to keep in mind that, in the case of SHF potential, the effective mass m*(r)/m is smaller than 1. Thus, the term of comparison is not the central part of the potential but $\frac{m^*(r)}{m}V_{central}(r)$ [4].

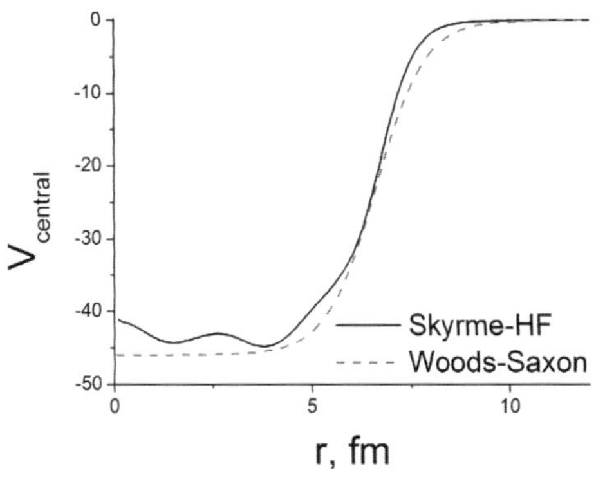

Figure 1. A mean field potential calculated within the Skyrme-HF approximation with $SIII$ parameterization compared to the typical Woods-Saxon potential for ^{136}Ba .

The close similarity between the two potentials has been proved quantitatively in Ref. [11] and exploited for the study of the pygmy resonance. In the mentioned study, the mean field part was calculated by means of the Hartree-Fock-Bogoliubov method and the parameters of Woods-Saxon potentials were fitted to reproduce the single-particle separation energies, the charge radii and the difference between proton and neutron mean square radii.

The residual interaction in the particle-hole form has the following expression,

$$V \approx \sum_{l} \kappa_l Q^{\dagger}_{lm} Q_{lm}, \tag{3}$$

where

$$Q^{\dagger}_{lm} = \sum_{j_1 j_2} \langle j_1 | f(r) Y_{lm}(\Theta, \Phi) | j_2 \rangle a^{\dagger}_{j_1} a_{j_2}, \tag{4}$$

$a^{\dagger}(a)$ being creation(annihilation) particle operators. If $f(\vec{r}) = r^l Y_{lm}(\Theta, \Phi)$ Eq.(4) yields the well known standard multipole operator. The parameters κ_l are chosen according to prescriptions of the QPM [10].

The procedure used here differs from the one adopted in Ref.[11] for studying the pygmy resonance. Here, in fact, we derive the mean field potential directly from the Skyrme-force by a Skyrme-HF method and keep the consistency between the mean field and residual interaction by enforcing the consistency condition expressed by Eq. (1).

3. Results

Recently, the QPM is the only method that allows to perform microscopic calculations in a sufficiently large configuration space and provides a unified description of low-lying and high-energy single- and multi-phonon excitation levels, including the excited states in the domain of Giant Resonances.

In recent years, the QPM has been adopted with success in disclosing the properties of the so-called mixed symmetry states discovered for the first time in Mo [12] and, since then, found in most spherical nuclei in the vicinity of the $N = 50$ and $N = 82$ neutron shell closures [13]. The QPM calculations , using a WS one-body potential, nicely reproduce the low- lying spectra in these nuclei and provide specific signatures for the symmetric and mixed symmetry states [14]. Moreover, the same QPM studies were able to point out the importance of shell structure in determining the splitting of the mixed symmetry mode in selected nuclei around $N = 82$ [15, 16].

In view of the extension of the method to nuclei far from stability, it is of special importance to perform a QPM study of these states using a separable interaction deduced from a Skyrme force. We have done this for the even-even N=80 isotones.

3.1. Single particle schemes

We have examined several Skyrme interactions in order to find the potential that yields the best single-particle spectrum in the N=80 chain. Our main criterion was to choose the force that best reproduces the level ordering and level density around the Fermi level for both proton and neutron subsystems. We have compared the results to those obtained by a Woods-Saxon potential, parameterized for the given mass region, and known to give a good description of the experimental spectra. The comparison for the neutron level scheme is shown in Fig.2. As seen from the figure, the best results are obtained when the SIII parametrization is used. The same SIII interaction yields also for protons a level sequence in good agreement with the one obtained using the WS potential (Fig.3). Thus, we have adopted the SIII parameterization for the QRPA and QPM calculations.

3.2. Mixed symmetry states

As it was shown in [15], the single-particle spectra have a crucial role in the formation of the structure of mixed symmetry states along the N=80 isotones as well as the changing of the structure within the isotonic chain. The nucleus ^{138}Ce reveals an interesting feature. The $M1$ strength is split into two pieces while in ^{136}Ba the corresponding $M1$ strength is concentrated

Figure 2. Neutron single-particle levels obtained with different Skyrme parameterizations, compared to Woods-Saxon single particle spectrum.

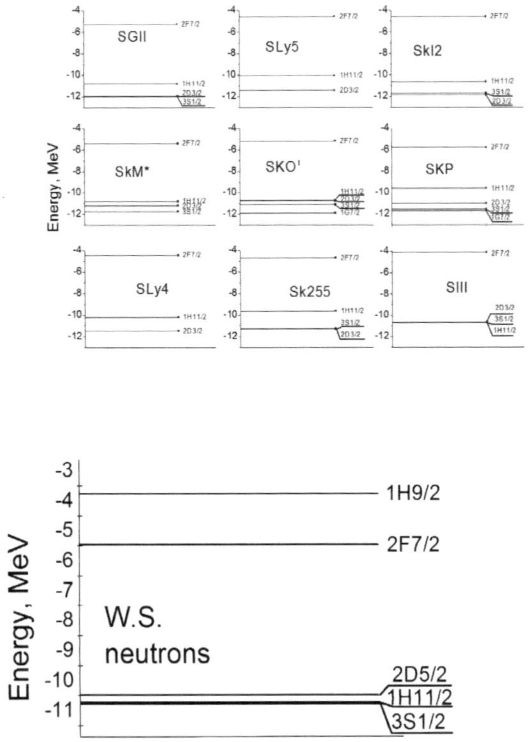

Figure 3. Proton single-particle levels obtained with different Skyrme parameterizations, compared to Woods-Saxon single particle spectrum.

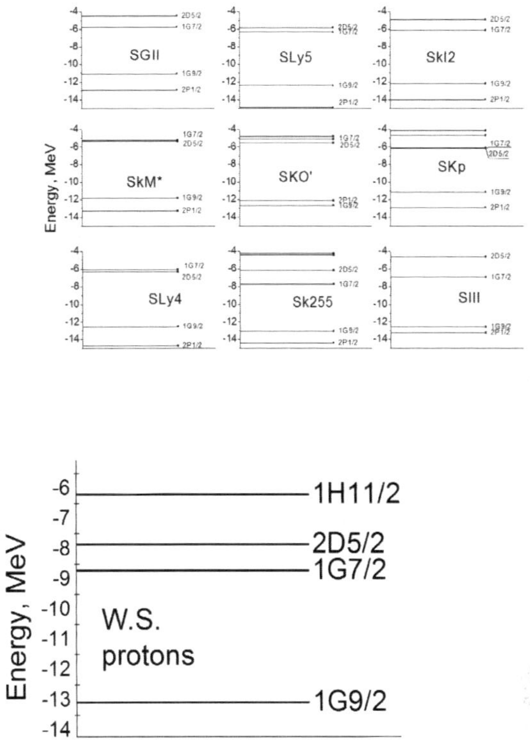

Table 1. Energy of the lowest quasi-particle proton and neutron states of ^{136}Ba

	Woods-Saxon [MeV]	SIII [MeV]
$(1g_{7/2})_p$	1.31	1.06
$(2d_{5/2})_p$	1.57	2.32
$(2d_{3/2})_n$	1.095	1.176
$(1h_{11/2})_n$	1.245	1.164
$(3s_{1/2})_n$	1.55	1.62
$(2d_{5/2})_n$	3.30	3.55

Table 2. Energy and $E2$ decay strengths of the lowest $[2^+]_{RPA}$ states. The numbers in brackets are the RPA values obtained using a WS potential

Nucleus	λ_i^π	$\omega_{\lambda_i^\pi}$ MeV	$B(E2)\downarrow$ W.u.
^{136}Ba	2_1^+	1.06 (1.03)	15 (24.8)
	2_2^+	2.5 (2.12)	1.7(1.7)
	2_3^+	2.89 (2.25)	0.5 (0.07)
^{138}Ce	2_1^+	1.06(1.02)	15(22.2)
	2_2^+	2.6 (2.21)	5 (1.6)
	2_3^+	2.89 (2.31)	0.1(3.96)

on a single excitation level. This property [15] is connected with shell structure and pairing. The important single-particle states around the Fermi level are $2d_{3/2}$, $1h_{11/2}$ and $3s_{1/2}$ for neutrons and $1g_{7/2}$ and $2d_{5/2}$ for protons. The lowest proton and neutron quasi-particle energies in ^{136}Ba are shown in Table 1. One can seen that both potentials give similar values for the quasi-particle

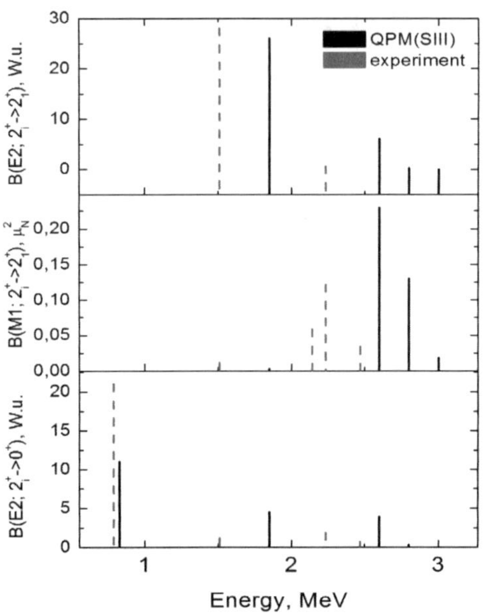

Figure 4. Experimental (dashed lines) versus QPM (solid line) transition strengths in ^{138}Ce. The QPM calculations are done with SIII parameterization of the Skyrme force.

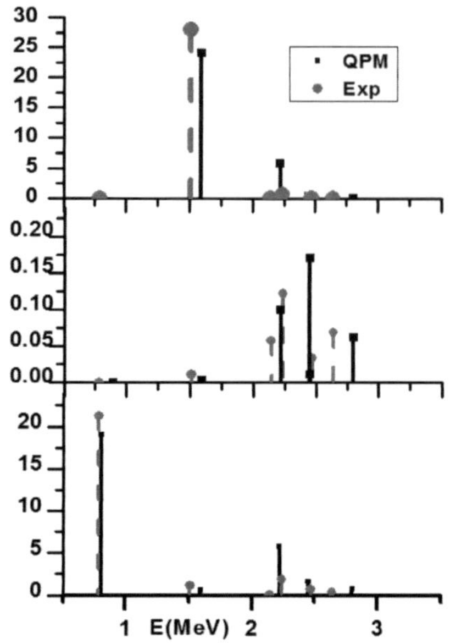

Figure 5. Experimental (dashed lines) versus QPM (solid line) transition strengths in ^{138}Ce. The QPM calculations are done with WS.

energies of the states around the Fermi level. Because of this close correspondence, the low-lying states of ^{136}Ba and ^{138}Ce, calculated by means of $Woods - Saxon$ and $SIII$ potentials must have a similar structure.

The energy and $E2$-strength of the low-lying $(2^+)_{RPA}$ states computed in the $QRPA$ are shown in Table 2. As discussed in [15], because of the different position of the proton chemical potential, the second and third $(2^+)_{RPA}$ states of ^{138}Ce are much more collective than those in ^{136}Ba. The total value of $E2$ transition strengths $\sum_{i=2,3} B(E2; 2_i^+ \to g.s.)$ for ^{138}Ce is much larger than the corresponding value for ^{136}Ba. This is the reason of the splitting of $M1$ strength into two peaks corresponding to the $M1$ transitions connecting the high-lying 2^+ excitations to the first 2_1^+ state. Table 2 shows that the $SIII$ potential gives the same results.

In Table 3, the QPM energies and transition strengths, obtained using the $SIII$ force, are compared with the corresponding QPM values using a WS potential and the experimental quantities. It is seen that large $B(M1)$ values are collected by the transitions connecting the 2_3^+ and 2_4^+ excited states to the 2_1^+ state in ^{138}Ce while only one large $B(M1)$ is obtained for ^{136}Ba. The overall comparison between QPM results and experiments is shown in Fig.4. Though the energy of the excitations is not perfectly reproduced, the Skyrme force provides an overall picture which is consistent with the results obtained using the WS potential and in fair agreement with the experimental data.

We conclude that the Skyrme-HF mean field potential incorporated in QPM is very suitable for structure calculations concerning low-lying excitations.

Table 3. QPM versus experimental strengths of $E2$ and $M1$ transitions. The $E2$ strengths are given in W.u. for ^{138}Ce, and in e^2b^2 for ^{136}Ba. The $M1$ strengths are in μ_N^2.

Nucleus	$J_i \rightarrow J_f$	B(E2)			B(M1)		
		EXP	QPM (SIII)	QPM(WS)	EXP	QPM (SIII)	QPM(WS)
^{136}Ba	$0^+_{gs} \rightarrow 2^+_1$	0.400(5)	0.24	0.33			
	$0^+_{gs} \rightarrow 2^+_2$	0.016(4)	0.09	0.046			
	$0^+_{gs} \rightarrow 2^+_3$	0.045(5)	0.03	0.065			
	$2^+_2 \rightarrow 2^+_1$	0.09(4)	0.12	0.12			
	$2^+_3 \rightarrow 2^+_1$					0.007	
	$2^+_4 \rightarrow 2^+_1$				0.26(3)	0.21	0.27
^{138}Ce	$2^+_1 \rightarrow 0^+_{gs}$	21.2(14)	11	19			
	$2^+_2 \rightarrow 0^+_{gs}$	1.16(8)	4.5	0.60			
	$2^+_3 \rightarrow 0^+_{gs}$		3.9	5.8			
	$2^+_4 \rightarrow 0^+_{gs}$	1.86(16)	0.35	1.6			
	$2^+_2 \rightarrow 2^+_1$	28(2)	26	24	0.011(2)	0.003	0.004
	$2^+_3 \rightarrow 2^+_1$		6.1	5.8	0.058(6)	0.23	0.10
	$2^+_4 \rightarrow 2^+_1$	0.65(10)	0.28	0.32	0.122(10)	0.13	0.17

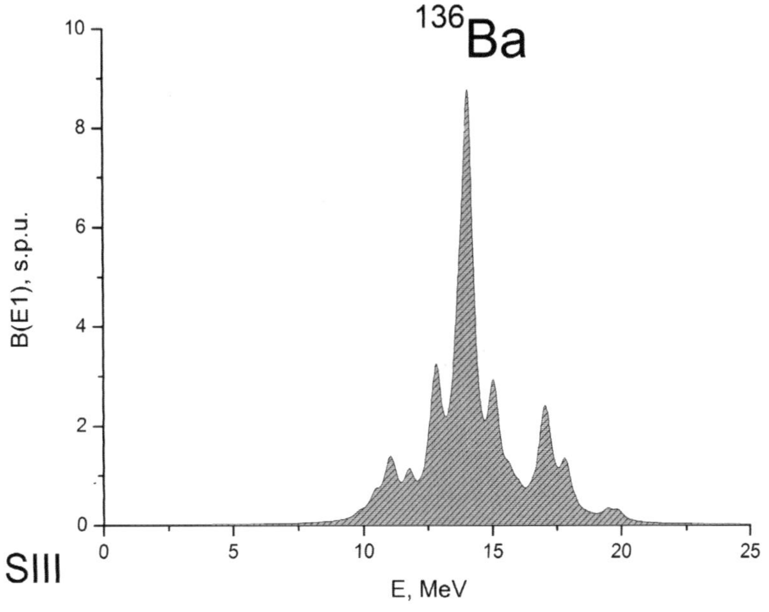

Figure 6. The distribution of the B(E1) strength for ^{136}Ba. A SIII Skyrme parameterization has been used.

3.3. Giant Dipole Resonance

Another test of the single-particle basis and of the value of the fitted QPM parameters, as obtained from the $SIII$ parametrization of the Skyrme force, is provided by QRPA calculations of the $E1$ strength distribution. The is done for ^{136}Ba. As shown in Fig. 6, a realistic distribution of the $E1$ strength in the Giant Dipole resonance domain is obtained. The Energy Weighted Sum Rule is fulfilled by 96% up to 20MeV.

4. Conclusions

The results obtained here show that the mean field generated by Skyrme forces can be successfully incorporated into the numerical scheme of the QPM. In fact, the new set of single particle states generated by these interactions allow a satisfactory description of the structure of the mixed symmetry states in the $N = 80$ isotones. Even the splitting of the $M1$ strength observed in ^{138}Ce is fairly well reproduced. This result gives a very positive message for the application of the QPM to exotic nuclei.

Acknowledgments

This work is partly supported by CNRS (France)- BAS(Bulgaria) PICS Program No 4329, from Bulgarian Nuclear Regulatory Agency - contract with JINR (Dubna) and from a DAAD-D002-25 contract.

References

[1] Hagino K, Giai N V and Sagawa H 2004 *Nucl. Phys.* A **731** 264
[2] Bai C L, Zhang H Q, Zhang X Z, Xu F R, Sagawa H and Colò G 2009 *Phys. Rev.* C **79** 041301
[3] Colò G, Sagawa H, Fracasso S and Bortignon P F 2007 *Phys. Lett.* B **646** 227
[4] Tarpanov D, Liang H, Giai N V and Stoyanov C 2008 *Phys. Rev.* C **77** 054316
[5] Vesely P, Kvasil J, Nesterenko V, Kleinig W, Reinhard P G and Ponomarev V Y 2009 *arXiv* 0907.0923v1
[6] Giai N V, Stoyanov C and Voronov V V 1998 *Phys. Rev.* C **57** 1204
[7] Nesterenko V O, Kvasil J and Reinhard P G 2002 *Phys. Rev.* C **66** 044307
[8] Vesely P, Kvasil J, Nesterenko V, Kleinig W, Reinhard P G and Ponomarev V 2009 *Phys. Rev.* C **80** 031302(R)
[9] Severyukhin A P, Voronov V V and Giai N V 2007 *arXiv* 0711.2423
[10] Soloviev V G 1992 *Theory of atomic nuclei: Quasiparticles and phonons* (Bristol: Institute of Physics Publishing)
[11] Tsoneva N, Lenske H and Stoyanov C 2004 *Phys. Leet.* B **586** 213
[12] Pietralla N, Fransen C, Belic D, von Brentano P, Frießner C, Kneissl U, Linnemann A, Nord A, Pitz H H, Otsuka T, Schneider I, Werner V and Wiedenhöver I 1999 *Phys. Rev. Lett.* **83** 1303–1306
[13] Pietralla N, von Brentano P and Lisetskiy A 2008 *Prog. Part. Nucl. Phys.* **6** 225
[14] Lo Iudice N and Stoyanov C 2002 *Phys. Rev.* C **65** 064304
[15] Lo Iudice N, Stoyanov C and Tarpanov D 2008 *Phys. Rev.* C **77** 044310
[16] Lo Iudice N, Stoyanov C and Pietralla N 2009 *Phys. Rev.* C **80** 024311

Journal of Physics: Conference Series **205** (2010) 012030 doi:10.1088/1742-6596/205/1/012030

New aspects of chiral symmetry breaking in atomic nucleus

J. MENG

School of Physics and Nuclear Energy, Beihang University, Beijing 100191, China
School of Physics and State Key Laboratory of Nuclear Physics and Technology, Peking University, 100871 Beijing, China
Department of Physics, University of Stellenbosch, Stellenbosch, South Africa

E-mail: mengj@pku.edu.cn

S. Q. ZHANG

School of Physics and State Key Laboratory of Nuclear Physics and Technology, Peking University, 100871 Beijing, China

B. QI and S. Y. WANG

School of Space Science and Physics, Shandong University at Weihai, Weihai 264209, China

Abstract. The significant progresses of the chirality in atomic nuclei are briefly reviewed for both experimental and theoretical sides. A particle rotor model is developed which couples several valence protons and neutrons to a rigid triaxial rotor core and applied to investigating the chirality in odd-A nucleus ^{135}Nd. Static chirality has been shown to be a transient phenomenon surrounding by chiral vibrations. It is found that the $B(M1)$ staggering is associated strongly with the characters of nuclear chirality, i.e., the staggering is weak in chiral vibration region while strong in the static chirality region.

PACS numbers: 21.60.Ev, 21.10.Re, 23.20.Lv

Keywords: Particle-Rotor Model (PRM); Tilted Axis Cranking (TAC); static chirality; chiral vibration; multiple chiral bands (MχD)

1. Introduction

Handedness or chirality is a subject of general interest in molecular physics, elementary particles, and optical physics. The occurrence of chirality in nuclear physics was suggested in 1997 by Frauendorf and Meng.[1] This effect is expected to occur in atomic nuclei having triaxial shapes, and in which there are a few high-j valence particles and a few high-j valence holes. For a triaxially deformed rotational nucleus, the collective angular momentum favors alignment along the intermediate axis, which in this case has the largest moment of inertia, while the angular momentum vectors of the valence particles (holes) favor alignment along the nuclear short (long) axis. The three mutually perpendicular angular momenta can be arranged to form two systems with opposite chirality, namely left- and right-handedness. These two systems are transformed into each other by the chiral operator which combines time reversal and spatial rotation of 180°, $\chi = \mathcal{T}\mathcal{R}(\pi)$. The spontaneous breaking of chiral symmetry thus happens in the body-fixed reference frame. In the laboratory reference frame, with the restoration of chiral symmetry due to quantum tunneling, the so-called chiral doublet bands, a pair of separated $\Delta I = 1$ bands (normally regarded as nearly degenerate) with the same parity, are expected to be observed in triaxial nuclei.

2. Fingerprints for chirality in atomic nuclei

After the prediction of the chirality in atomic nuclei, lots of efforts have been made to find out their fingerprints for the experimental observations. Based on one particle and one hole coupled to a triaxial rotor, a set of experimental signatures have been suggested as fingerprints of nuclear chirality, for example, see Refs. [1, 2, 3]. For energy spectra, the chiral bands should contain a pair of nearly degenerate $\Delta I = 1$ bands with the same parity. The selection rule for electromagnetic transitions in the chiral geometry has been proposed in Ref. [2]. Further, in ideal chiral pair bands all corresponding physics properties, such as spin alignments, moment of inertia, and electromagnetic transition probabilities, must be identical or, in practice, very similar.[4, 5]

Nevertheless, one should bear in mind that the fingerprints of the chiral doublet bands mentioned above, in particular $B(M1)$ staggering, are obtained by assuming one proton (neutron) particle and one neutron (proton) hole sitting in a high-j shell coupled with a triaxial rotor with $\gamma = 30°$. It is therefore interesting and necessary to investigate the doublet bands with more general cases where γ deviates from 30°.

3. Experimental progress

During the last decade, lots of experimental efforts have been devoted to search for nuclear chirality. In fact, a pair of $\Delta I = 1$ bands found in ^{134}Pr with the $\pi h_{11/2} \otimes \nu h_{11/2}$ configuration [6] have been reinterpreted in Ref. [1] as a candidate for chiral doubling. Thereafter, in 2001 similar low-lying doublet bands were reported in $_{55}$Cs, $_{57}$La, and

$_{61}$Pm $N = 75$ isotones of ^{134}Pr, and an island of chiral rotation was suggested in the $A \sim 130$ mass region [7]. Up to now, candidate chiral doublet bands have been proposed in a number of odd-odd nuclei in the $A \sim 130$ mass region[7, 8, 9, 10, 11, 12, 13, 14, 15, 16] with the suggested configuration $\pi h_{11/2} \otimes \nu h_{11/2}$, $A \sim 100$ [17, 18, 19] with $\pi g_{9/2}^{-1} \otimes \nu h_{11/2}$, and $A \sim 190$ [20, 21] with $\pi h_{9/2} \otimes \nu i_{13/2}^{-1}$. A few more candidates with more than one valence-particle and hole were also reported in odd-A [22, 24, 23, 25, 26] and even-even [27] nuclei. Furthermore, the doublet bands in even-even ^{106}Mo [28] and ^{110,112}Ru [29] are regarded to be soft chiral vibrations, which may exemplify the general chiral geometry, because the non-planar geometry of rotation cannot be directly related to the alignment of high-j particles and holes with different principal axes [29].

Lifetime measurements are essential to extract the absolute $B(M1)$ and $B(E2)$ transition probabilities. Their importance has stimulated experimental programs aimed at identifying chiral doublet bands [30, 31, 32, 33, 34].

4. Theoretical progress

On the theoretical side, chiral bands were first predicted in the particle-rotor model (PRM) and tilted axis cranking (TAC) approach for triaxially deformed nuclei[1].

Chiral rotation has been studied by the Strutinsky shell correction TAC (SCTAC) method with a hybrid potential which combines the spherical Woods-Saxon single-particle energies and the deformed part of the Nilsson potential [35, 36]. Recently, chiral TAC solutions have also been found in ^{132}La isotones within the self-consistent Skyrme Hartree-Fock cranking model [37, 38]. The cranked relativistic mean field (RMF) theory has been reported only in the contexts of principle axis rotation [39, 40] and planar rotation [41, 42]. The generalization thereof for searching for chiral solutions, i.e., the aplanar rotation, is still under development. The advantage of the cranked mean field approach and its limitation has been extensively discussed in, e.g., refs. [43, 44].

In contrast, the PRM is a quantum mechanical model and describes the system in the laboratory reference frame and yields directly the energy splitting and tunneling between doublet bands. Chirality for nuclei in $A \sim 100$ and $A \sim 130$ regions has been studied with the particle-rotor model for certain particle-hole configurations[45, 46] ,or the core-quasiparticle/core-particle-hole coupling model[13, 47], following the Kerman-Klein-Dönau-Frauendorf method[48]. With the pairing correlations taken into account to simulate the configurations of multi-particles sitting in a high j-shell, particle rotor model with a quasi-proton and a quasi-neutron coupled with a triaxial rotor has been applied to study chiral doublet bands.[49, 50] The (quasi-)particle rotor model can well describe the energy spectra and the ratios $B(M1)/B(E2)$ and $B(M1)_{\rm in}/B(M1)_{\rm out}$ of chiral bands at the whole spin region for many nuclei, for example ^{126}Cs in $A \sim 130$ mass region [50], ^{106}Rh in $A \sim 130$ mass region [51], and ^{198}Tl in $A \sim 190$ mass region [21]. By analyzing the orientation of the angular momentum for the rotor as well as the valence proton and neutron, and the effective angles between these angular momenta, the chiral geometry represented by a remarkable aplanar rotation can be revealed by means of

PRM [49, 51, 21]. It is also noted that the efforts to describe the doublet bands in ^{134}Pr are not so successful, where other effects may take also important roles [4, 30].

In addition to the PRM and TAC approaches, the so-called doublet bands have been investigated in terms of the interacting boson fermi model [52, 53, 54] and a pair truncated shell model [55], etc.

5. Triaxial n-particle-n-hole PRM

As mentioned above, apart from odd-odd nuclei, chiral doublet bands have been observed in odd-A nuclei, an exact version of multi-particle and multi-hole coupled with a triaxial rotor model is highly required to describe the doublet bands based on multi-quasiparticle configurations. Recently we have developed a triaxial n-particle-n-hole PRM to treat more than one valence proton and one valence neutron and applied to the study of nuclear chirality in ^{135}Nd with the $2p1n$ configuration $\pi h_{11/2}^2 \otimes \nu h_{11/2}^{-1}$ [56].

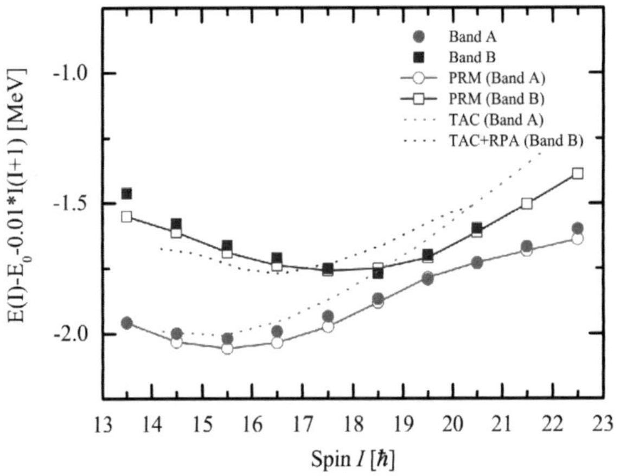

Figure 1. (color online) The excitation energies $E(I)$ for the chiral bands in ^{135}Nd calculated by the triaxial PRM with configuration $\pi h_{11/2}^2 \otimes \nu h_{11/2}^{-1}$ (open symbols) in comparison with the data (filled symbols) [22, 32] and the corresponding TAC+RPA results (dotted lines) [32]. The energies are relative to the band head E_0 of the chiral bands, with a rotor reference subtracted. Take from Ref. [56].

The calculated excitation energy spectra $E(I)$ for the doublet bands A and B in ^{135}Nd are presented in Fig. 1, together with the corresponding data [22, 32]. For comparison, the results from the random phase approximation (RPA) based on TAC [32] are also included.

The energy splitting between two sister bands is also well described in the TAC+RPA calculations for $I \leq 39/2$, which is the vibrational regime, where the RPA is stable. The energy splitting disappears at $I = 41/2$ in TAC+RPA, where the TAC solution attains static chirality while anharmonic quantal tunneling takes the role of energy splitting. The description of the experiment in the region $I > 39/2$ is beyond the realm of RPA based on the planar TAC solution. On the other hand, being a quantum theory that is not restricted to small amplitude vibrations, PRM is able to perfectly reproduce the energy splitting for the whole observed spin region, staying within the configuration $\pi h^2_{11/2} \otimes \nu h^{-1}_{11/2}$ and the triaxial rotor.

In addition to energy spectra, the reduced probabilities $B(M1)$ and $B(E2)$ for in-band as well as for the interband transitions are reproduced excellently by the triaxial n-particle-n-hole PRM. The details can be found in ref. [56].

6. From chiral vibration to static chirality

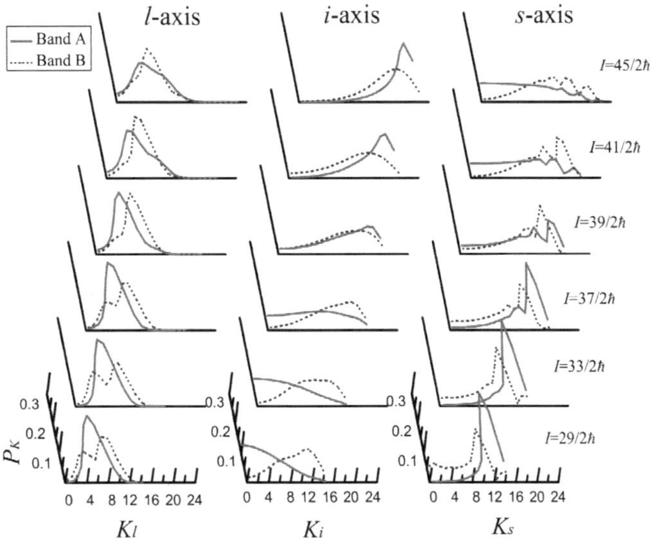

Figure 2. (color online) The probability distributions for projection of total angular momentum on the long (l-), intermediate (i-) and short (s-) axis in PRM for the doublet bands in ^{135}Nd. Take from Ref. [56].

The TAC calculations have suggested that the chiral rotation cannot exist below a certain critical frequency [37]. Before the onset of chirality of the mean field, the precursor of the symmetry breaking occurs as a soft vibration between the right- and left-handed configurations. These chiral vibrations can be studied in the framework of the TAC+RPA and the static chirality is shown to be a transient phenomenon [32, 57].

In Fig. 2, the probability distributions for the projection of the total angular momentum along the l-, i- and s-axes are shown for the doublet bands in ^{135}Nd. For each PRM state, the probability for the projection K of total angular momentum on quantization axis is given by the square sum of all the expansion coefficients with a certain K quantum number. The probability distributions with respect to the three principle axes are obtained by placing γ into different sectors, such that the shape is the same but different principal axis is used for quantization.

For spin $I = 29/2$, near the band head, the probability distribution of two bands differ as expected for a chiral vibration. At spin $I = 39/2$, the probability distributions for band A and B are very similar, which show the characteristics of static chirality. For higher spin, where the energy difference between the chiral partners increases, they attain vibration character again.

7. Reexamination of fingerprints for chirality

Let us come to discuss the fingerprints for chirality, in particular the feature of odd-even staggering of B(M1) values. The electromagnetic transitions of the chiral doublet bands have been discussed for different triaxiality parameter γ in PRM with $\pi h_{11/2} \otimes \nu h_{11/2}^{-1}$ configuration [58]. The $B(M1)$ staggering as well as the resulting $B(M1)/B(E2)$ and $B(M1)_{in}/B(M1)_{out}$ staggering with the triaxiality parameter γ were investigated in order to answer whether or not such staggering is a fingerprint to identify the chiral doublet bands.

After careful reexamination of the $B(M1)$ in ref. [58], one finds that the $B(M1)$ staggering is associated strongly with the characters of chirality. The $B(M1)$ staggering is associated strongly with the characters of nuclear chirality, i.e., the staggering is weak in chiral vibration region while strong in the static chirality region.

This result also qualitatively agrees with the recent lifetime measurements for the doublet bands in ^{128}Cs and ^{135}Nd. The former case was claimed as a good example revealing chiral symmetry breaking in Ref. [31], where the pronounced $B(M1)$ staggering is exhibited. The latter case was suggested to reveal the chiral vibration motions for $I < 41/2\hbar$ region in Ref. [32], where the weak $B(M1)$ staggering is shown.

8. Multiple chiral doublet bands

Is it possible to exist more than one pair of chiral bands in one single nucleus? Such possibilities of multiple chiral doublet (MχD) bands have been demonstrated recently in Ref. [44] with the microscopic triaxial relativistic mean field (RMF) approaches. With the adiabatic and configuration-fixed constrained triaxial RMF theories developed, triaxial shape coexistence, high-j proton hole- and neutron particle-configurations, and possible MχD of ^{106}Rh are suggested.

Ref. [59] has extended the investigation to other rhodium isotopes. The ground states of ^{102,104,106,108,110}Rh are found to be triaxially deformed. Wave functions from the Cartesian basis are transformed to a spherical basis, and the corresponding configurations of the minima are specified by quantum numbers in spherical basis. Based on the triaxial deformation and the corresponding high-j particle and hole configurations, the existence of MχD, are suggested in ^{104,106,108,110}Rh. The investigation provides not only further support for the prediction of MχD in ^{106}Rh, but also presents new experimental opportunity for the observation of MχD in $A \sim 100$ mass region.

In Refs. [44, 59], the time-reversal invariance was assumed from the beginning, i.e., the time-odd fields were neglected. In Ref. [60], the configuration-fixed constrained triaxial relativistic mean-field approach has been extended by including time-odd fields and applied to studying the candidate MχD nucleus ^{106}Rh. It has been found that the calculations including time-odd fields have confirmed the previous prediction of the possible existence of MχD in the configuration-fixed triaxial RMF approach.

9. Summary and perspective

The significant progresses of the chirality in atomic nuclei since 1997 are briefly reviewed. A particle rotor model is developed which couples several valence protons and neutrons to a rigid triaxial rotor core and applied to investigating the chirality in odd-A nucleus ^{135}Nd. Static chirality has been shown to be a transient phenomenon surrounding by chiral vibrations. It is found that the $B(M1)$ staggering as well as the resulting $B(M1)/B(E2)$ and $B(M1)_{in}/B(M1)_{out}$ staggering are sensitive to the triaxiality parameter γ. One cannot exclude the chirality simply by the odd-even staggering of $B(M1)/B(E2)$ values. It is further clarified that for partner bands with near degenerate energy spectra and similar $B(M1)$ and $B(E2)$ transitions, the strong $B(M1)$ staggering can be used as a fingerprint for the static chirality, i.e., the staggering is weak in chiral vibration region while strong in the static chirality region. Nevertheless, the identification of chiral doublet bands is still an open question, which needs more efforts from both experimental and theoretical sides, and continues to be a subject of intense discussions.

Acknowledgments

This work is partly supported by the Major State Basic Research Developing Program 2007CB815000, the National Natural Science Foundation of China under Grant Nos. 10975008, 10875074, and 10775004.

[1] S. Frauendorf and J. Meng, *Nucl. Phys. A* **617**, 131 (1997).
[2] T. Koike, K. Starosta and I. Hamamoto, *Phys. Rev. Lett.* **93**, 172502 (2004).
[3] S. Y. Wang, S. Q. Zhang, B. Qi and J. Meng, *Chin. Phys. Lett.* **24**, 664 (2007).
[4] C.M. Petrache, G.B. Hagemann, I. Hamamoto and K. Starosta, *Phys. Rev. Lett.* **96**, 112502 (2006).
[5] P. Joshi, et. al., *Phys. Rev. Lett.* **98**, 102501 (2007).
[6] C. M. Petrache *et al.*, *Nucl. Phys. A* **597**, 106 (1996).
[7] K. Starosta *et al.*, *Phys. Rev. Lett.* **86**, 971 (2001).
[8] T. Koike, K. Starosta, C.J. Chiara, D.B. Fossan, and D.R. LaFosse, *Phys. Rev. C* **63**, 061304(R) (2001).
[9] A. A. Hecht *et al.*, *Phys. Rev. C* **63**, 051302(R) (2001).
[10] D. J. Hartley *et al.*, *Phys. Rev. C* **64**, 031304(R) (2001).
[11] R. A. Bark, *et al.*, *Nucl. Phys. A* **691**, 577 (2001).
[12] X. F. Li *et al.*, *Chin. Phys. Lett.* **19**, 1779 (2002).
[13] T. Koike, K. Starosta, C. J. Chiara, D. B. Fossan, and D. R. LaFosse, *Phys. Rev. C* **67**, 044319 (2003).
[14] G. Rainovski *et al.*, *Phys. Rev. C* **68**, 024318 (2003).
[15] A. J. Simons *et al.*, *J. Phys. G* **31**, 541 (2005).

[16] S. Y. Wang, Y. Z. Liu, T. Komatsubara, Y. J. Ma, and Y. H. Zhang, *Phys. Rev. C* **74**, 017302 (2006).
[17] C. Vaman, D. B. Fossan, T. Koike, K. Starosta, I. Y. Lee and A. O. Macchiavelli, *Phys. Rev. Lett.* **92**, 032501 (2004).
[18] P. Joshi *et al.*, *Phys. Lett. B* **595**, 135 (2004).
[19] P. Joshi *et al.*, *Eur. Phys. J. A* **24**, 23 (2005).
[20] D. L. Balabanski *et al.*, *Phys. Rev. C* **70**, 044305 (2004).
[21] E. A. Lawrie, *et al.*, *Phys. Rev. C* **78**, 021305(R) (2008).
[22] S. Zhu *et al.*, *Phys. Rev. Lett.* **91**, 132501 (2003).
[23] J. A. Alcántara-Núñez *et al.*, *Phys. Rev. C* **69**, 024317 (2004).
[24] J. Timár *et al.*, *Phys. Lett. B* **598**, 178 (2004).
[25] J. Timár, *et al.*,*Phys. Rev. C* **73**, 011301(R) (2006).
[26] Y. X. Zhao, *et al.*, *Chin. Phys. Lett.* **26**, 082301 (2009).
[27] E. Mergel *et al.*, *Eur. Phys. J. A* **15**, 417 (2002).
[28] S. J. Zhu, *et al.*, *Eur. Phys. J. A* **25**, (Suppl. 1) (2005) 459.
[29] Y. X. Luo, *et al.*,*Phys. Lett. B* **670**, (2009) 307.
[30] D. Tonev *et al.*, *Phys. Rev. Lett.* **96**, 052501 (2006).
[31] E. Grodner *et al.*, *Phys. Rev. Lett.* **97**, 172501 (2006).
[32] S. Mukhopadhyay *et al.*, *Phys. Rev. Lett.* **99**, 172501 (2007).
[33] T. Suzuki *et al.*, *Phys. Rev. C* **78**, 031302(R) (2008).
[34] L. L. Wang *et al.*, *Chin. Phys. C* **33**, Suppl.1, 173(2009).
[35] V. I. Dimitrov, S. Frauendorf and F. Dönau, *Phys. Rev. Lett.* **84**, 5732 (2000).
[36] V. I. Dimitrov, F. Dönau and S. Frauendorf, *Phys. Rev. C* **62**, 024315 (2000).
[37] P. Olbratowski, J. Dobazewski, J. Dudek and W. Plóciennik, *Phys. Rev. Lett.* **93**, 052501 (2004).
[38] P. Olbratowski, J. Dobazewski and J. Dudek, *Phys. Rev. C* **73**, 054308 (2006).
[39] W. Koepf and P. Ring, *Nucl. Phys. A* **493**, 61 (1989).
[40] A. V. Afanasjev, P. Ring and J. König, *Nucl. Phys. A* **676**, 196 (2000).
[41] H. Madokoro, J. Meng, M. Matsuzaki and S. Yamaji, *Phys. Rev. C* **62**, 061301(2000).
[42] J. Peng, J. Meng, P. Ring and S. Q. Zhang, *Phys. Rev. C* 78, 024313 (2008).
[43] S. Frauendorf, *Rev. Mod. Phys.* **73**, 463 (2001).
[44] J. Meng, J. Peng, S. Q. Zhang and S.-G. Zhou, *Phys. Rev. C* **73**, 037303 (2006).
[45] J. Peng, J. Meng, S. Q. Zhang, *Phys. Rev. C* **68**, 044324 (2003).
[46] J. Peng, J. Meng, S. Q. Zhang, *Chin. Phys. Lett.* **20**, 1223 (2003).
[47] K. Starosta *et al.*, *Phys. Rev. C* **65**, 044328 (2002).
[48] A. Klein, *Phys. Rev. C* **63**, 014316 (2000), and references therein.
[49] S. Q. Zhang, B. Qi, S. Y. Wang, and J. Meng, *Phys. Rev. C* **75**, 044307 (2007).
[50] S. Y. Wang, S. Q. Zhang, B. Qi and J. Meng, *Phys. Rev. C* **75**, 024309 (2007).
[51] S. Y. Wang, S. Q. Zhang, B. Qi, J. Peng, J. M. Yao, and J. Meng, *Phys. Rev. C* **77**, 034314 (2008).
[52] D. Tonev *et al.*, *Phys. Rev. C* **76**, 044313 (2007).
[53] S. Brant, D. Tonev, G. de Angelis, and A. Ventura, *Phys. Rev. C* **78**, 034301 (2008).
[54] S. Brant and C. M. Petrache, *Phys. Rev. C* **79**, 054326 (2009).
[55] K. Higashiyama, N. Yoshinaga and K. Tanabe, *Phys. Rev. C* **72**, 024315 (2005).
[56] B. Qi, S. Q. Zhang, J. Meng, S. Y. Wang, and S. Frauendorf, *Phys. Lett. B* **675**, 175 (2009).
[57] D.Almehed and S. Frauendorf, arXiv:0709.0969v1 [nucl-th].
[58] B. Qi, S. Q. Zhang, S. Y. Wang, J. M. Yao, and J. Meng, *Phys. Rev. C* **79**, 041302(R) (2009).
[59] J. Peng, H. Sagawa, S. Q. Zhang, J. M. Yao, Y. Zhang, and J. Meng, *Phys. Rev. C* **77**, 024309 (2008).
[60] J. M. Yao, B. Qi, S. Q. Zhang, J. Peng, S. Y. Wang, and J. Meng, *Phys. Rev. C* **79**, 067302 (2009).

Journal of Physics: Conference Series **205** (2010) 012031 doi:10.1088/1742-6596/205/1/012031

On the 0^+ excited states in even-even deformed nuclei

V.P. Garistov[1,2], A.I. Georgieva[1], A. Solnyshkin[2], J. Adam[2], D. Bogachenko[3], O. Egorov[3], A. Islamov[3], V. Silaev[3]

[1] Institute for Nuclear Research and Nuclear Energy, Bulgarian Academy of Sciences, Sofia, Bulgaria;
[2] Joint Institute for Nuclear Research, Dubna, Russia
[3] Institute for Theoretical and Experimental Physics, Moscow, Russia

E-mail: garistov@inrne.bas.bg

Abstract. The low-lying 0^+ excited states remain an object of particular interest in the nuclear structure physics. Recently long sets of 0^+ excited states were experimentally observed. Our analysis of the experimental data have shown that in even-even nuclei of the rare-earth and actinide regions the energies of all low lying 0^+ excited states with great accuracy can be distributed on parabolic functions of the number of monopole excitations building these states. Along with the classification of the energies of the 0^+ excited states in respect to the number of bosons that build the band heads we analyze the role of their collectivity in the structure and evolution of the yrast bands and the B(E2) transition probabilities within these bands. The experimental determination in ^{160}Dy of the predicted in this way 0^+ excited state with energy 0.6813 MeV is presented.

1. Introduction

The complexity in the structure of excited 0^+ states springs up a variety of approaches ranging from microscopic studies of anharmonic effects [1], consideration of quadrupole and pairing vibrational modes in conversional electrons and internal pair decay [2], or exact diagonalization in the restricted space of collective phonons of different types [3]. On the other hand, the description of the energies and electromagnetic decay properties of the excited states are important tests in the evaluation of the applicability of the different models, like the shell model, cluster-vibrational model, quasi-particle - phonon model, a deformed configuration mixing shell model, the interacting boson approximation and in particular its O(6) limit.

In this work, we suggest an approach for the evaluation of the yrast bands energies and B(E2) probabilities between yrast band states. The basic assumption is, that some of the states with $K^\pi = 0^+$ are the band heads of crossing rotational bands, parts of which form the yrast band and we show the importance of the structure of band head, which determines the value of the moment of inertia of the nucleus in this state. The different behavior of such rotational bands gives us the possibility to consider the yrast band as composed from several crossing rotational bands which start from different band heads with $K^\pi = 0^+$. Hence, our first step is to review in short the energy distribution of the 0^+ excited states in terms of their collectivity (number of boson excitations) within the same nucleus.

2. Energy distribution of the 0^+ excited states

Usually the classifications of experimental data is mainly done from a "horizontal perspective" in sequences of nuclei where the investigated nuclear characteristics are empirically studied as functions of the numbers of their valence nucleons. In contrast we analyze the experimental data on the low-lying spectra in terms of the order of collectivity (number of monopole bosons constructing the corresponding 0^+ excited state) in the same nucleus. Such a classification of large amount of experimental data in terms of integer classification parameter has been done in [4], based on phenomenological collective Hamiltonian for single level approach, written in terms of boson creation and annihilation operators R_+ , R_- and R_0

$$H = \alpha R_+ R_- + \beta R_0 R_0 + \beta \Omega R_0. \qquad (1)$$

The operators in (1) are constructed from pairs of fermion creation and annihilation operators $a^\dagger$ and a on a single $j-$ level.

$$\begin{aligned} R_+ &= \tfrac{1}{2}\sum_m (-1)^{j-m}\alpha^\dagger_{jm}\alpha^\dagger_{j-m} \,, \\ R_- &= \tfrac{1}{2}\sum_m (-1)^{j-m}\alpha_{j-m}\alpha_{jm} \,, \\ R_0 &= \tfrac{1}{4}\sum_m (\alpha^\dagger_{jm}\alpha_{jm} - \alpha_{j-m}\alpha^\dagger_{j-m}) \,. \\ [R_0, R_\pm] &= \pm R_\pm, \qquad [R_+, R_-] = 2R_0 \end{aligned} \qquad (2)$$

Applying the Holstein-Primakoff transformation on the operators (2) (where $\Omega = \frac{2j+1}{2}$)

$$R_- = \sqrt{2\Omega - b^\dagger b}\, b \quad R_+ = b^\dagger \sqrt{2\Omega - b^\dagger b} \quad R_0 = b^+ b - \Omega \qquad (3)$$

we express them in terms of ideal bosons, defined by $\left[b, b^\dagger\right] = 1 \quad [b, b] = \left[b^\dagger, b^\dagger\right] = 0$, where $b|0\rangle = 0 \quad |n\rangle = \frac{1}{\sqrt{n!}}(b^\dagger)^n|0\rangle$ and the initial Hamiltonian (1) becomes:

$$H = \mathbf{a}\, b^\dagger b - \mathbf{b}\, b^\dagger b b^\dagger b. \qquad (4)$$

Thus the energy spectrum raised by the Hamiltonian (4) is a parabolic function of the number n of the ideal monopole bosons

$$E_n = \mathbf{a}\, n - \mathbf{b}\, n^2 \qquad (5)$$

where $\mathbf{a} = (2\Omega + 1)\alpha - \Omega\beta$ and $\mathbf{b} = \alpha - \beta$. There are more realistic models that reproduce similar dependence of the energies of the collective states on the number of collective excitations, for instance, the Interacting vector boson model (IVBM)[5] based on the introduction of two kinds of vector bosons (called p - and n-bosons), that "built up" the collective excitations in the nuclear system. The Hamiltonian, corresponding to one of the dynamical symmetries of the IVBM [5], expressed in terms of the first and second order invariant operators of the different subgroups of its corresponding chain, is obviously diagonal in the basis labelled by their irreducible representations with eigenvalues:

$$E((N,T);KLM;T_0) = a'N + b'N^2 + \alpha_3 T(T+1) + \beta_3 L(L+1) + \alpha_1 T_0^2 \qquad (6)$$

where N is the total number of vector bosons, T, T_0, are the T-spin, which defines the parity of the states and its third projection and L is the angular momentum of each state. For fixed parity and angular momentum we have obviously an expression similar to (5).

Thus we choose the parabolic function to be a rule for the distribution of the $K^\pi = 0^+$ state energies in the space of monopole bosons. On Figure 1. we show the respective new representation of the sets of recent experimental data [6]. Now we can label every $K^\pi = 0^+$

state by an additional characteristic n, the number of monopole bosons determining it's collective structure. The parameters **a** and **b** are evaluated by fitting the experimental energies of the different 0^+ states of a given nucleus to the theoretical ones applying all possible permutations of the classification numbers n and evaluating the distribution corresponding to the minimal value of χ-square. Often the first excited 0^+ state is considered as less collective than the next one in increasing energy. In some even-even nuclei the first in energy excited state is not necessarily the state with the smallest n (degree of collectivity). For instance the 0^+ state with excitation energy 0.2548 MeV ($n = 20$) observed in ^{158}Gd [7] is much more collective than the 0^+ state with energy 0.5811 ($n = 1$) MeV. Similar behavior of the distribution of the energies of 0^+ states, can be seen on Figure 1 for ^{180}W and ^{152}Gd, as well as the high accuracy of the distribution of all the experimental states on the second order curves. In comparison, the calculations of the 0^+ excited states' energies in ^{158}Gd within the semi-microscopic approach of the "quasi-particle - phonon model" [8] are far from the accuracy obtained in our simple approach.

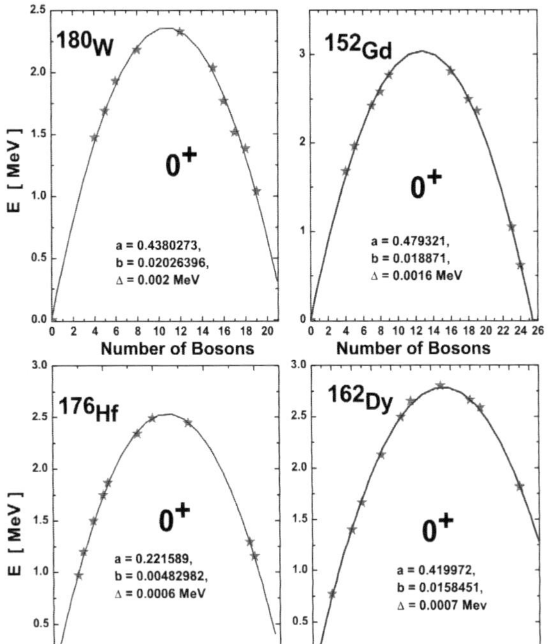

Figure 1. Experimental data [6] distributed on parabolic curves, defined by the parameters **a** and **b** given for each nucleus, in the space of monopole bosons. Δ gives the mean square deviation of the experimental data from the theoretical curve (5) $E_n = \mathbf{a}\ n - \mathbf{b}\ n^2$.

3. Density distribution and moment of inertia

As a first example of the application of the above results we present the description of the yrast bands of non-magic even-even nuclei. For this purpose we take the nuclear surface oscillation model [9] giving the density distribution of the nucleus placed in n-boson excited state $|n\rangle$:

$$\rho_{nn}(r, R_0, S) = \frac{3\rho_{00}(r, R_0, S)}{4\pi R_0\,(3S^2 + R_0^2)} + \sum_{k=1}^{n} \binom{n}{k} \frac{\frac{(\frac{xR_0^2}{8\pi^2})^k \frac{d^{2k}}{dR_0^{2k}}\rho_{00}(r,R_0,S)}{k!}}{\frac{nxR_0^3}{\pi} + \frac{4}{3}\pi\,(3S^2 + R_0^2)\,R_0} \tag{7}$$

Then, having the density distribution as a function of the number of monopole bosons n and choosing any nuclear form, defined by the nuclear size R_0 and the diffuseness parameter S, the moment of inertia can be calculated. As a result it also becomes function of number of bosons, through the density distribution (7). We suppose that the shape of the nucleus is an ellipsoid with a spherical core inside and half-axis c(n)and d(n). The moment of inertia corresponding to this nuclear shape (see Figure 2) is $\Im(c(n),d(n)) = \Im_{ellipce}(c(n),d(n)) - \Im_{sphere}(c(n),c(n))$. In the case of spherical nucleus $d = c$ the moment of inertia $\Im(c(n),c(n)) = 0$.

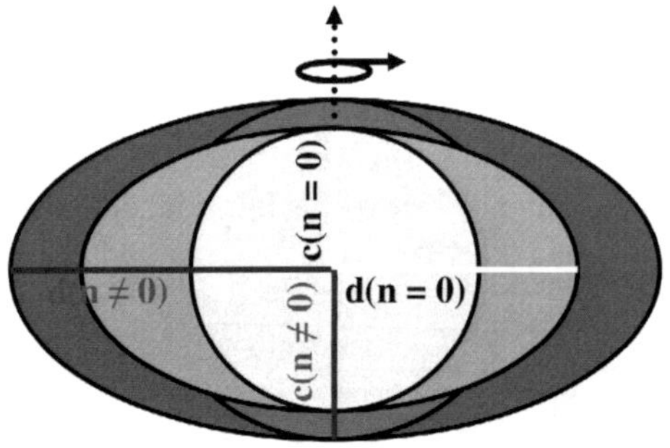

Figure 2. Nuclear shape. For the nucleus in the ground state n = 0 the half-axis c(0) and d(0) do not depend on the number of bosons n

After introducing the oscillations of the nuclear surface and using the density distribution (7) the moment of inertia of the nucleus with this form becomes:

$$\begin{aligned}\Im(c(n),d(n),x,n) &= \frac{M(d(0)^2-c(0)^2)}{6400\pi^4}(27n^2x^2 - 360n^2x^2\pi + 48n\pi^2(25nx - 2\sqrt{15})x \\ &+ 6402\sqrt{15}nx\pi^3 + 1280\pi^4). \qquad (8)\end{aligned}$$

where $x = \frac{E_0}{C_0}$, E_0 - the energy of the one-boson state and C_0 - the nuclear surface compressibility parameter in respect to the monopole excitations. In all our calculations we impose for the ratio of the ellipsoid's half-axis (of any nuclear excited state $|n>$), the condition $\frac{c(n)}{d(n)} = \frac{c(0)}{d(0)} = Const$, where $c(n) = c(0)\left(1+\sqrt{nx(\frac{3}{2\pi}+\frac{9}{40\pi^2})}\right)$ and $d(n) = d(0)\left(1+\sqrt{nx(\frac{3}{2\pi}+\frac{9}{40\pi^2})}\right)$. For the

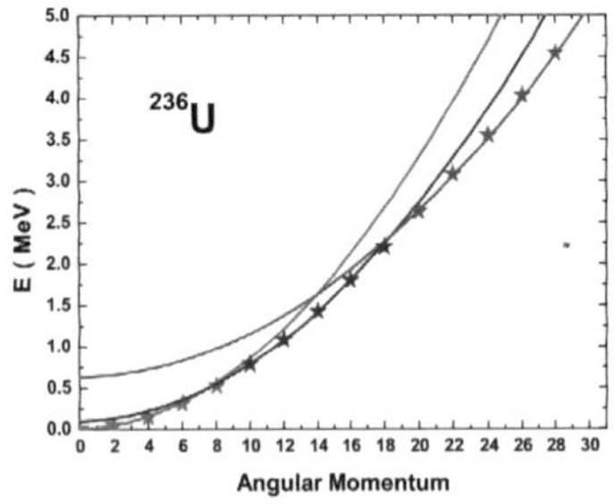

Figure 3. Yrast line of ^{236}U energies described as three crossing rotational bands. $x = 0.054$; red line - ground band, blue line - $n = 3$, green line - $n = 8$.

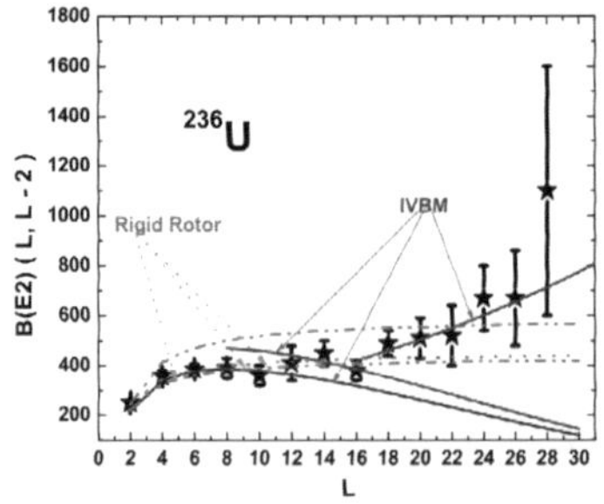

Figure 4. Comparison of B(E2) values of ^{236}U calculated within rigid rotor (dash-dotted lines) and IVBM (full lines)approaches with experiment.

ground band $n = 0$ and the moment of inertia for the rest of the 0^+_n excited bands depends only on the number of bosons that build their heads structures. This leads to the possibility of separation of the yrast band into two or more parts belonging to different rotational bands having different values of the moment of inertia because of the difference in the collective structure of the rotational bands heads (see for example Figure 3). In the next Figure 4. we present the comparison of the calculated and experimental B(E2) values for ^{236}U nucleus, when the yrast line separation is taken into account.

4. Analysis of the experimental situation in ^{160}Dy

Calculated energy distributions of the 0^+ states in the ^{160}Dy nucleus plotted on Figure 5 for the best fit number of monopole bosons show that apart from the 0^+ states with the number of bosons n = 2, 5, 6, and 7 (stars on Figure 5) currently known from the experiments [10], states with the number of bosons n = 1, 3, 4, and 8 (circles in Figure 5) should be observed. In

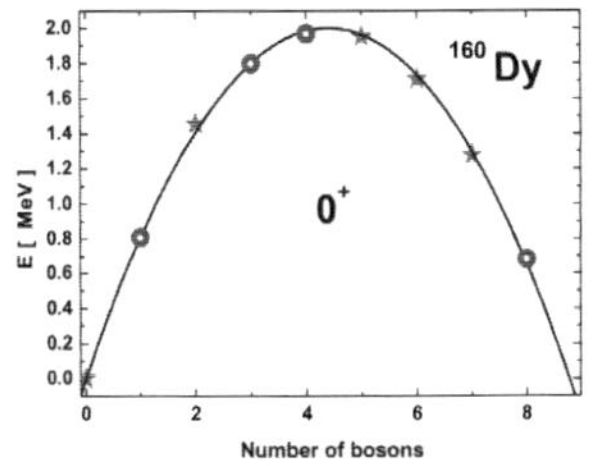

Figure 5. Experimental data for the energies of 0^+ states in the ^{160}Dy distributed on the parabolic curve in the space of monopole bosons.

the light of the above-mentioned theoretical predictions of possible existence of new excited 0^+ states in the ^{160}Dy nucleus new experimental investigations were carried out with the aim to look for new possible E0 transitions between known and predicted 0^+ states in the spectrum of internal conversion electrons (ICE) measured at the β decays $^{160}Ho \rightarrow ^{160}Dy$ and $^{160}Er \rightarrow ^{160}Ho \rightarrow ^{160}Dy$. In order to obtain the ICE spectrum, several photo-emulsion plates (irradiated at JINR DLNP, Dubna, in different years) were measured, using permanent magnet beta-spectrographs [11] at the β decay of Ho and Er fractions. These fractions were separated by means of the chromatographic method [12] from the tantalum target irradiated by a 660 MeV proton beam with an intensity of 2 μA from the DLNP phasotron. The plates were measured by a new technique developed at ITEP (Moscow) and described in detail in the paper [13]. An example of the results of such measurement are illustrated on Figure 6.

The spectrum was calibrated in energy and intensity against the well studied intense single lines of K electrons and three of them are shown on Figure 6. Our more thorough investigations of particular ICE spectrum regions (Figure 7) on the left of the K682.3 keV, corresponding to the known 682.3 keV γ transition in the ^{160}Dy nucleus [10], revealed an earlier unknown line. Analysis of the available experimental data on γ transitions in the ^{160}Dy nucleus, showed that in the γ ray spectrum, γ transitions "responsible" for internal conversion electrons which could give rise to the observed unknown line, do not exist.

Therefore, we assumed that this line may correspond to the E0 transition between a new excited 0^+ state with the energy 681.3 keV and the ground 0^+ state in ^{160}Dy. We were further motivated by the observation that this energy (681.3 keV) is close in value to the predicted by the theoretical model energy for 0^+ state built by n = 8 bosons (see Figure 5). Since population of this 0^+ state directly from the β decay of $^{160g,m}Ho$ is practically impossible because of the

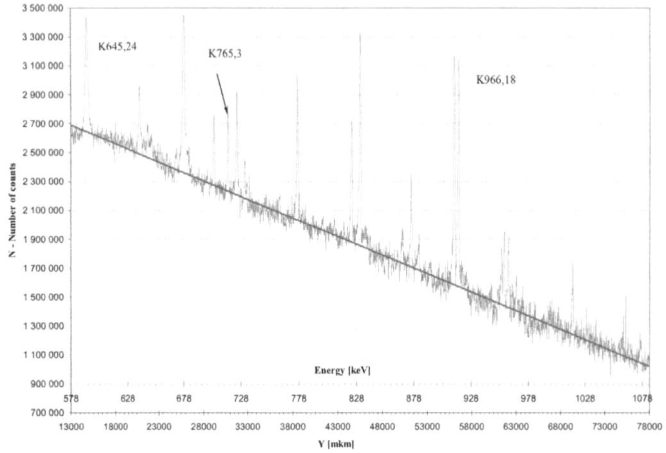

Figure 6. ICE spectrum in the energy region 578 - 1078 keV

quantum characteristics of the ground and isomeric states with $I^{\pi} = 5^{+}$ and $I^{\pi} = 2^{-}$ of the parent nucleus, it may only be populated by transitions from higher lying excited states of ^{160}Dy. We found at least two such transitions. One of them was the E0 transition with the energy K1271.0 keV observed in the ICE spectrum on the left of the known K1271.89 keV line. It's energy exactly corresponds to the energy of the transition between the 1952.3 keV and 681.3 keV 0^{+} states. As the analysis showed, in the γ spectrum [10] there were no γ rays of appropriate energy whose conversion electrons could be responsible for this line. Further analysis made it possible to reveal one more transition which could be treated as a transition populating the

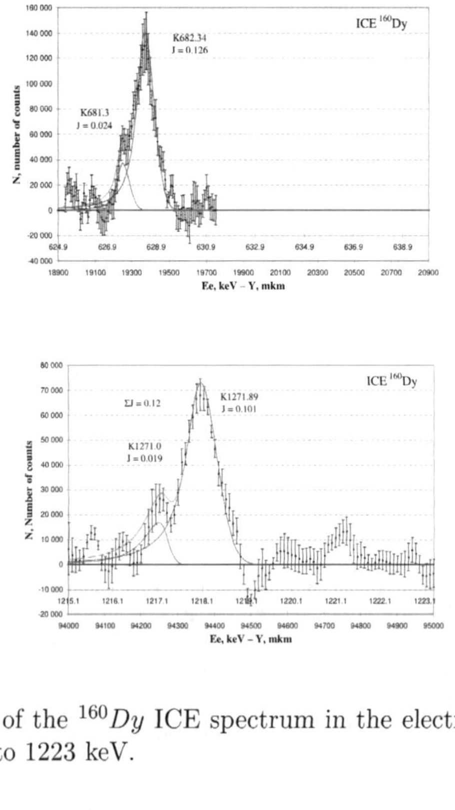

Figure 7. Two fragments of the ^{160}Dy ICE spectrum in the electron energy ranges from 625 to 632 keV and from 1215 to 1223 keV.

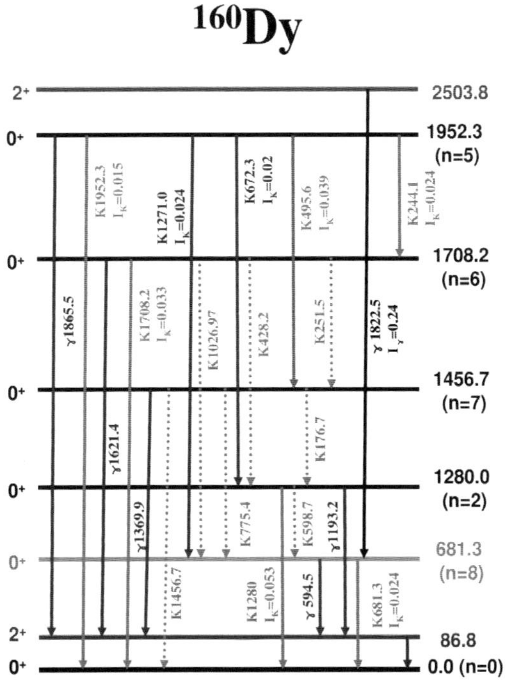

Figure 8. Fragment of the scheme of ^{160}Dy excited states.

681.3 keV 0^+ state. It was the γ transition reliably identified in the experiment [10], but not placed in the $^{160m,g}Ho \rightarrow ^{160}Dy$ decay scheme. This transition with energy 1822.4(3) keV fits best between the quite reliably established 2503.8 keV 2^+ state and the predicted 681.3 keV 0^+ state. As is evident from Figure 8, de-excitation of the 681.3 keV 0^+ state is possible and should only be due to the E0 transition to the ground state and its competing E2 transition to the 86.8 keV 2^+ level of the ground state band in ^{160}Dy. The sum of their intensities should completely compensate the sum of the intensities of the transitions populating the 681.3 keV state. In other words, the balance of total intensities of all transitions to and from this level should be zero. This situation is observed for all known 0^+ states in the ^{160}Dy nucleus [10]. The major part of the intensity in the de-excitation of the 0^+ states in question is attributed to the E2 transitions to the 86.8 keV 2^+ level. Fulfillment of the zero balance condition for the 681.3 keV 0^+ state also requires the E2 transition to the 86.8 keV 2^+ level. The energy of this transition should be 594.5 keV. Based on the aforesaid, we analyzed the energy region involving the energy 594.5 keV in the γ ray spectrum obtained in our measurements. This region is very difficult for analysis because of the insufficient energy resolution, and therefore it is impossible to separate explicitly the 594.5 keV line against the background of two lines with similar energies 593.5 and 595.3 keV. Nor was it done in [10], where only two lines with energies 593.5 and 595.3 keV appear in the list of γ rays for the region in question. However, it seems possible to determine the upper limit for the intensity of the 594.5 keV E2 transition. According to our evaluation, the intensity of this transition should not be higher than 0.3 relative units. This evaluation allowed us to calculate the limit for the dimensionless Rasmussen parameter for the 681.3 keV 0^+ state $X_{calc.} \approx 0.6$ and to compare it with the known $X_{exp.}$ for this nucleus. The values of $X_{exp.}$ are presented in Table 1, where $I_K(E0)$ is the relative intensity of the E0 transition K line, Ω_K is the electron factor, $E_\gamma(E2)$ and $I_\gamma(E2)$ are the energy and relative intensity of the E2 transition. It is evident that

Table 1.

$E(0^+)_{level}$ $[keV]$	$I_K(E0)$ $rel.un.$	Ω_K	$E_\gamma(E2)$ [keV]	$I_\gamma(E2)$ rel.un.	$X_{exp.}$
681.3	0.024	$4.089 \cdot 10^{10}$	594.5	< 0.3	> 0.3
1280.0	0.053	$7.778 \cdot 10^{10}$	1193.2	13.8	0.3
1456.7	-	$9.068 \cdot 10^{10}$	1369.9	40.0	-
1708.2	0.033	$1.103 \cdot 10^{11}$	1621.4	12.7	0.6
1952.3	0.015	$1.310 \cdot 10^{11}$	1865.6	6.1	0.9

our evaluation of $X_{calc.} \approx 0.6$ for the predicted 681.3 keV 0^+ state is in good agreement with $X_{exp.} > 0.3$ for the other 0^+ states in ^{160}Dy.

Summing up, we think it possible to deduce that the 681.3 keV 0^+ state exist in the spectrum of the ^{160}Dy nucleus. Naturally, this conclusion would be more convincing if one or more other E0 transitions from upper 0^+ states to the 681.3 keV level were found. However, if these transitions do exist, their negligibly low intensities make them very difficult to observe. The really low values of the expected intensities follows from the fact that the balance of the intensities of the 681.3 keV level is already almost zero, and any significant addition of intensity from above will result in braking it. Nevertheless, we do not lose interest in searching for E0 transitions and even plan more thorough measurements of ICE spectra.

In conclusion, we hope that our theoretical predictions of vacant places in the distributions of the energies of the 0^+ states (for example, see Figure 1 and Figure 5),the evaluated energies positions of bands heads and the estimations of E0 transition probabilities between different 0^+ states in the same nucleus, may be used by experimentalists in identifying the spins, parities and energies of new excited collective states like in the presented case of the ^{160}Dy nucleus.

Acknowledgments
The Bulgarian authors wish to acknowledge financial support from Contract $\Phi - 1501$ with the National Science Foundation of the Bulgarian Ministry of Education and Science and a grant for scientific collaboration with the Bogolyubov Laboratory for Theoretical Physics of JINR, Dubna.

References

[1] Silvestre-Brac B and Piepenbring R 1977 Phys. Rev. C 16, N4 1638 ; 1978 Phys. Rev. C 17 N1
[2] Passoja A et al. 1983 Physics Letters 124B, N3 4 157
[3] Kantele Jet al. 1986 Physics Letters 171B N2 3 151
[4] Vladimir P Garistov e-Print Archive:nucl-th/0008067(August 2000);nucl-th/0309058(September 2003) ;
[5] Georgieva A, Ganev H G, Draayer J P and Garistov V P 2009 PEPAN, 40, issue 4, 461
[6] Meyer D A, et.al. 2006 Phys. Rev. C 74, 044309
[7] Lesher S R et al. 2002 Phys. Rev. C66 051305 (R); Aprahamian A et al. 2002 Phys. Rev. C 65 031301(R)
[8] Lo Iudice N et al. 2004 Phys. Rev. C 70, 064316
[9] Garistov V P 1995 IJMP-E 4 N2 371
[10] Adam J et al. 2002 Izv. RAN, Ser. Fiz. 66 1384; Reich C W 2005 Nuclear Data Sheets 105, 557
[11] Abdurazakov A A et al. 1972 Permanent Magnet Beta-Spectrographs, Tashkent, FAN
[12] Molnar F et al. 1973 Fiz. Elem. Chastits At. Yadra 4 (4) 1077-1155
[13] Egorov O K et al., JTF 2003 vol. 46 no. 3 p.870

Journal of Physics: Conference Series **205** (2010) 012032 doi:10.1088/1742-6596/205/1/012032

Microscopic Calculations of ^{8}He+p Elastic Scattering Cross Sections

V K Lukyanov[1], E V Zemlyanaya[1], K V Lukyanov[1], D N Kadrev[2], A N Antonov[2], M K Gaidarov[2] and S E Massen[3]

[1]Joint Institute for Nuclear Research, Dubna 141980, Russia

[2]Institute for Nuclear Research and Nuclear Energy, Bulgarian Academy of Sciences, Sofia 1784, Bulgaria

[3]Department of Theoretical Physics, Aristotle University of Thessaloniki, 54124 Thessaloniki, Greece

E-mail: gaidarov@inrne.bas.bg

Abstract. A microscopic approach to calculate the optical potential (OP) with the real part obtained by a folding procedure and with the imaginary part inherent in the high-energy approximation is applied to study the ^{8}He+p elastic scattering at energies of tens of MeV/nucleon (MeV/N). The OP's and the cross sections are calculated using different models for the neutron and proton densities of ^{8}He. The role of the spin-orbit potential is studied. Comparison of the calculations with the available experimental data on the elastic scattering differential cross sections at beam energies of 15.7, 26, 32, 66 and 73 MeV/N is performed. The problem of the ambiguities of the depths of each component of the optical potential is considered by means of the imposed physical criterion related to the known behavior of the volume integrals as functions of the incident energy. It is shown also that the role of the surface absorption is rather important, in particular for the lowest incident energies (e.g., 15.7 and 26 MeV/N). The present approach, which uses only parameters that renormalize the depths of the OP, can be applied along with other methods using microscopically calculated optical potentials.

1. Introduction

The experiments with intensive secondary radioactive nuclear beams have made it possible to investigate the structure of light nuclei near the neutron and proton drip lines as well as the mechanism of scattering of the weakly bound nuclei. A widely used way to study the structure of exotic nuclei is to analyze their elastic scattering on protons or nuclear targets at different energies. Here we would like to mention, for example, the experiments on scattering of helium isotopes on protons at incident energies E_{inc} less than 100 MeV/N, in particular, for ^{8}He at energy 15.7 [1], 26 [2], 32 [3, 4], 66 [3, 4] and 73 MeV/N [3, 4, 5].

The experimental data on differential and total reaction cross sections of processes with light exotic nuclei have been analyzed using a variety of phenomenological and microscopic methods. Among the latter methods we note, e.g. the microscopic analysis based on the coordinate-space g-matrix folding method [6, 7, 8, 9, 10, 11, 12, 13], as well as works where the real part of OP (ReOP) is microscopically calculated using the folding approach (e.g. [14, 15, 16, 17, 18, 19, 20, 21, 22, 23]). Usually the imaginary part of the OP's (ImOP) and the spin-orbit (SO) terms have been determined phenomenologically. Thus, the OP's have a number of fitting parameters.

The main aim of this work is to calculate differential cross sections of elastic ^{8}He+p scattering at energies less than 100 MeV/N studying the possibility to describe the existing experimental data by calculating microscopically not only the real (in a double-folding procedure) but also the imaginary optical potential (instead of using phenomenological one) within the high-energy approximation (HEA) and using a minimal number of fitting parameters. In this respect, we want to establish the limits of the applicability of the HEA OP for calculations of differential elastic cross sections of the ^{8}He+p scattering for different regions of angles and incident energies. The standard folding OP includes an exchange term and, correspondingly, the non-linear effects in calculations of the potential. In the calculations we pay attention to the importance of the contribution of various physical quantities and features, such as microscopically obtained spin-orbit forces and regularization of the NN forces used in folding calculations. Also, we consider effects of renormalization of the real and imaginary parts of microscopic optical potentials, the differences or similarities of various models for the ^{8}He structure which are used in the description of the experimental data on the cross section of the ^{8}He+p elastic scattering. We study the density dependence of the effective NN interaction, as well as the sensitivity of the results to the predictions of different theoretical models for the densities of ^{8}He. In addition, the nuclear surface effects are also studied by introducing an additional surface term in OP. This is related to investigations of the lowest energy limit of the applicability of the HEA OP in ^{8}He+p elastic scattering. Also we pay attention to the energy dependence of the renormalization parameters, as well as to the respective volume integrals. We note the necessity to analyze the differential cross sections estimating simultaneously the values of the total reaction cross section. This would give an additional test of the various ingredients of the approach.

2. Theoretical scheme

2.1. The optical potential

The general expression for the optical potential used to calculate the ^{8}He+p elastic scattering differential cross sections has the form:

$$U_{opt}(r) = N_R V^F(r) + iN_I W^H(r) + 2\lambda_\pi^2 \left\{ N_R^{SO} V_0^F \frac{1}{r}\frac{df_R(r)}{dr} + iN_I^{SO} W_0^H \frac{1}{r}\frac{df_I(r)}{dr} \right\} (\mathbf{l}.\mathbf{s}), \quad (1)$$

where V^F (calculated within the folding approach) and W^H (calculated within the HEA) are the real and imaginary contributions to the OP, respectively, and V_0^F and W_0^H are the depths of the SO optical potential obtained simultaneously from the approximation of the volume real and imaginary OP's by Woods-Saxon (WS) form. In Eq. (1) N_R, N_I, N_R^{SO}, and N_I^{SO} are additionally introduced strength parameters that renormalize the depths of the different components of the OP.

The real part of the nucleon-nucleus OP is assumed to be a result of a single folding of the effective NN potential and the nuclear densities. It involves the direct and exchange parts (for more details, see, e.g. [19, 20, 21] and also [17]):

$$V^F(r) = V^D(r) + V^{EX}(r). \quad (2)$$

In Eq. (2) the direct part (V^D) is composed of the isoscalar (IS) and isovector (IV) contributrions, correspondingly:

$$V_{IS}^D(r) = \int \rho_2(\mathbf{r}_2) g(E) F(\rho_2) v_{00}^D(s) d\mathbf{r_2}, \quad (3)$$

$$V_{IV}^D(r) = \int \delta\rho_2(\mathbf{r}_2) g(E) F(\rho_2) v_{01}^D(s) d\mathbf{r_2}. \quad (4)$$

In Eqs. (3) and (4) the energy dependence of the effective NN interaction is taken in the usually used form:

$$g(E) = 1 - 0.003E. \quad (5)$$

Also, for the NN potentials v_{00}^D and v_{01}^D we use the expression from [21] for the CDM3Y6-type of the effective interaction based on the solution of the equation for the g-matrix, in which the Paris NN potential has been used. The density dependence of the effective interaction is taken in the following form:

$$F(\rho) = C\left[1 + \alpha e^{-\beta\rho(\mathbf{r})} - \gamma\rho(\mathbf{r})\right], \tag{6}$$

where C=0.2658, α=3.8033, β=1.4099 fm^3, and γ=4.0 fm^3.

The isoscalar part of the exchange contribution to the ReOP has the form (see, e.g. [17]):

$$\begin{aligned} V_{IS}^{EX}(r) &= g(E)\int \rho_2(\mathbf{r}_2, \mathbf{r}_2 - \mathbf{s})F\left[\rho_2\left(\mathbf{r}_2 - \frac{\mathbf{s}}{2}\right)\right] \\ &\times\ v_{00}^{EX}(s)j_0[k(r)s]d\mathbf{r}_2. \end{aligned} \tag{7}$$

In Ref. [17] the so-called complex HEA optical potential has been applied to explain the available data on the ^{6}He+p elastic differential cross sections and energies less than 100 MeV/N. The HEA OP was derived in [24] on the basis of the eikonal phase inherent in the optical limit of the Glauber theory. It can be obtained as a folding of form factors of the nuclear density and the NN amplitude $f_{NN}(q)$ [24, 25]:

$$U_{opt}^H = V^H + iW^H = -\frac{\hbar v}{(2\pi)^2}(\bar{\alpha}_{NN} + i)\bar{\sigma}_{NN} \times \int_0^\infty dq q^2 j_0(qr)\rho_2(q)f_{NN}(q). \tag{8}$$

In (8) $\bar{\sigma}_{NN}$ and $\bar{\alpha}_{NN}$ are, respectively, the NN total scattering cross section and the ratio of the real to imaginary parts of the forward NN scattering amplitude both averaged over the isospin of the nucleus.

2.2. The density distributions of ^{8}He

In the calculations of the OP's we use the following point-nucleon density distributions of ^{8}He:

i) the Tanihata densities deduced in [26] by means of comparison of the measured total reaction cross section of ^{6,8}He+^{12}C at 800A MeV with the respective expression from [27] derived in the framework of the optical limit of the Glauber theory;

ii) the LSSM densities calculated in a complex $4\hbar\omega$ shell model space [22, 23] using the WS basis of single-particle wave functions with realistic exponential asymptotic behavior;

iii) the densities obtained in [28, 29] with accounting for the NN central-type short-range Jastrow correlations.

3. Results and discussion

In this Section we present the results of the calculations of the microscopic OP's and the respective ^{8}He+p elastic scattering differential cross sections at energies $E_{inc} < 100$ MeV/N. In principle, the OP's do not contain free parameters, but they depend on the density distribution of the target nucleus. This allows one to test advanced theoretical methods that give predictions for the density distribution. In Fig. 1 in logarithmic and linear scales are shown the proton $\rho_p(r)$, neutron $\rho_n(r)$ and matter $\rho(r)$ densities of ^{8}He obtained in different models. Also, for comparison, the known COSMA densities [30, 31] are presented. We note that among them only the LSSM densities have a realistic exponential asymptotics, whereas the others have a Gaussian one. The results for the JCM densities are given for the value of the correlation parameter β=2.5 fm^{-1} in the Jastrow correlation factor $1 - e^{-\beta^2 r^2}$, where r is the distance between neutrons. It was shown in Refs. [28, 29] that the inclusion of this factor causes a slight increase of the density in the central part of the nucleus.

The discussion that follows is based on the fitting procedure, where the additionally introduced strength parameters N_R, N_I, N_R^{SO}, N_I^{SO} are varied step by step. So, we start

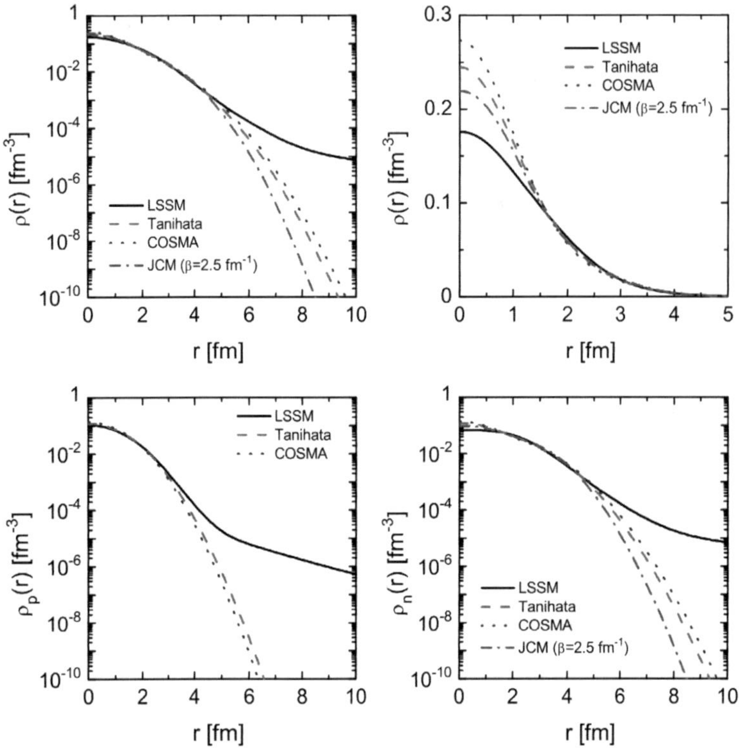

Figure 1. Total ((a) and (a′)), point-proton (b) and point-neutron (c) densities of ^{8}He from the model of Tanihata [26], COSMA [30, 31], LSSM [22, 23] and JCM calculations [28, 29].

from the case N_R=N_I=1, N_R^{SO}=N_I^{SO}=0, then fit successively both coefficients N_R and N_I, and after that the values of N_R^{SO} and N_I^{SO}. First, we give in Fig. 2 the results of our methodical calculations of the cross sections for different energies (15.7, 26, 32, 66 and 73 MeV/N) using the densities of ^{8}He from LSSM, Tanihata and JCM approaches in the case when N_R=N_I=1 and N_R^{SO}=N_I^{SO}=0 (i.e. without spin-orbit interaction). It can be seen that the behavior of the cross sections for a given energy and interval of angles is weakly sensitive to the choice of the model for the density of ^{8}He. In spite of this uncertainty we choose for the further applications the LSSM density since it has a realistic exponential behavior in the peripheral region of the nucleus.

Later, as a next step, we allow the "depth" of each of the parts of the OP (1) in our semi-microscopic models to vary in order to find the optimal values of the parameters N_R, N_I, N_R^{SO} and N_I^{SO} by a fitting procedure to the available experimental data for the cross sections. In Fig. 3 we present the results of our calculations of ^{8}He+p elastic scattering cross sections for various energies and the LSSM density with the fitted values of the parameters N_R, N_I, N_R^{SO} and N_I^{SO}. The values of these renormalization parameters are given in Table 1 together with the predicted total reaction cross sections. The results obtained using the values of the parameters from the first line of this Table for each energy are given by solid line in Fig. 3, while those from the second line for each energy are given by dashed line.

The difficulties in the description of the cross sections at low energies can be overcome by adding the contribution of a surface part of OP to the volume form. For this reason we consider

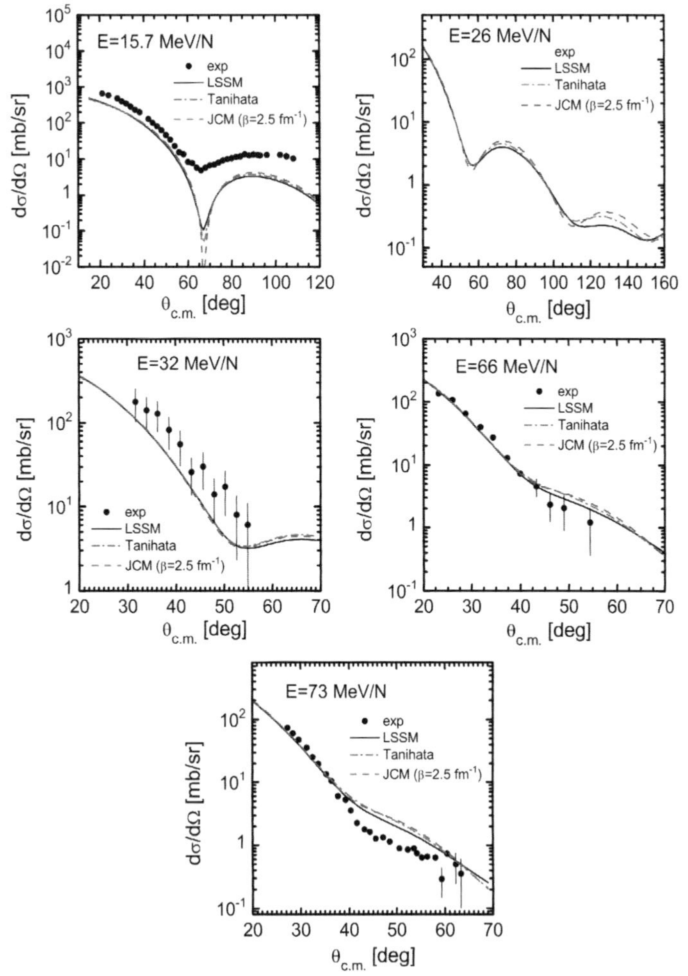

Figure 2. The ^{8}He+p elastic scattering cross sections at different energies calculated using U_{opt} [Eq. (1)] for values of the parameters N_R=N_I=1 and N_R^{SO}=N_I^{SO}=0. The used densities of ^{8}He are LSSM, Tanihata, and JCM (β=2.5 fm^{-1}). Experimental data are taken for 15.7 [1], 26 [2], 32 [3, 4], 66 [3, 4] and 73 MeV/N [3, 4, 5].

also the contribution of the surface potential:

$$U'_{opt}(r) = U_{opt}(r) - i4aN_S\frac{dV^F(r)}{dr}, \tag{9}$$

where the first term in the right-hand side is the expression for the OP given by Eq. (1) (in which the ImOP is taken in the form of $V^F(r)$) and the second term is responsible for the surface effects. We would like to note that, in particular, for the lowest incident energy, the combination of the microscopically folded real and imaginary parts in the form of V^F is more appropriate. In Eq. (9) a is the diffuseness parameter of $V^F(r)$ fitted by the WS form.

We present in Fig. 4 our results for the cross section in the case of E=15.7 MeV/N obtained using the LSSM density of ^{8}He. The calculations are performed by fitting the strength parameters N_R, N_I, N_R^{SO}, N_I^{SO} entering Eqs. (1) and (9) and the depth parameter N_S of the surface term of the OP [Eq. (9)]. It is seen from Fig. 4 that the inclusion of the surface contribution to the imaginary part of the OP improves the agreement with the experimental data, especially for small angles and in the region of the cross section minimum. Obviously, for more successful description of the cross sections at low energies (15.7 and 26 MeV/N) our method has to be modified and improved by an inclusion of virtual excitations of inelastic and decay channels of the reactions.

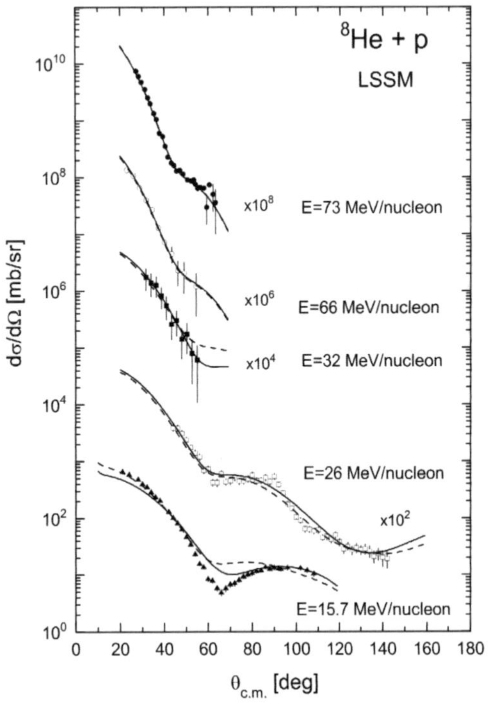

Figure 3. The ^{8}He+p elastic scattering cross sections at different energies calculated using U_{opt} [Eq. (1)] for various values of the renormalization parameters N_R, N_I, N_R^{SO} and N_I^{SO} (presented in Table 1) giving the best agreement with the data. The used density of ^{8}He is LSSM. Experimental data are taken for 15.7 [1], 26 [2], 32 [3, 4], 66 [3, 4] and 73 MeV/N [3, 4, 5].

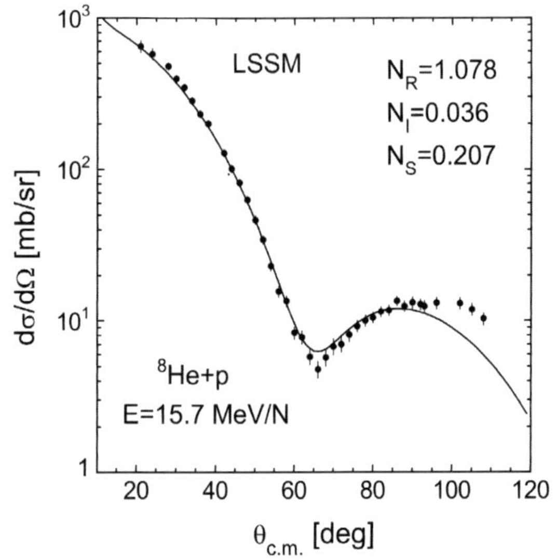

Figure 4. The ^{8}He+p elastic scattering cross section at energy E=15.7 MeV/N using LSSM density of ^{8}He. The fitted values of N_R, N_I, and N_S are given. Experimental data are taken from [1].

4. Conclusions

The results of the present work can be summarized as follows:

i) The optical potentials and cross sections of ^{8}He+p at E=15.7,26, 32, 66 and 73 MeV/N elastic scattering were calculated and comparison with the available experimental data was performed.

(a) The ReOP (V^F) was calculated microscopically using the folding procedure and M3Y effective interaction based on the Paris NN potential.

(b) The ImOP (W^H) was calculated within the HEA.

(c) Different model densities of protons and neutrons in ^{8}He were used in the calculations: Tanihata, COSMA, LSSM and JCM.

Table 1. The renormalization parameters N_R, N_I, N_R^{SO} and N_I^{SO} obtained by fitting the experimental data in Fig. 3 in the case of LSSM density. The energies are in MeV/N and the total reaction cross sections σ_R are in mb.

E	N_R	N_I	N_R^{SO}	N_I^{SO}	σ_R
15.7	1.0	0.236	0	0	603.6
15.7	0.9	0.1	0.107	0.040	693
26	0.422	0.104	0.090	0.010	275.11
26	0.439	0.144	0.087	0.023	377.22
32	0.438	0.036	0.096	0	71.9
32	1.0	0.374	0	0	419.5
66	0.876	0.071	0	0	55.7
66	0.854	0.086	0	0	65.9
73	0.875	0.02	0	0	1.48
73	0.869	0.01	0.010	0.002	1.22

(d) The SO contribution to the OP was included in the calculations.

(e) The cross sections were calculated by numerical integration of the Schrodinger equation by means of the DWUCK4 code using all interactions obtained (Coulomb plus nuclear optical potential).

ii) The results show that the LSSM densities of ^{8}He which have more diffuse tails at larger r than the densities based on Gaussians lead to a better agreement with the data for the ^{8}He+p elastic scattering at different energies.

iii) It was shown that, generally, at energies $E > 25$ MeV/N a good agreement with the experimental data for the differential cross sections can be achieved using OP with calculated both V^F and W^H varying mainly the volume part of the OP and neglecting the SO contribution.

iv) The explanation of the ^{8}He+p cross sections at lower energies ($E < 25$ MeV/N) needs accounting for the effects of the nuclear surface. In this case the use of ImOP of the HEA type is limited. A more successful explanation of the cross section at low energies could be given by inclusion of polarization contributions due to virtual excitations of inelastic and decay channels of the reactions.

v) It was shown that the effects of the Jastrow central short-range NN correlations on the OP's and on the shape of differential cross sections are weak.

vi) The problem of the ambiguity of the values of the parameters N_R, N_I, N_R^{SO}, and N_I^{SO} when the fitting procedure is applied to a limited number of experimental data is considered. Physical criteria imposed in our work [18] on the choice of the values of the parameters N were the known behavior of the volume integrals as functions of the incident energy in the interval $0 < E_{inc} < 100$ MeV/N, as well as the values of the total cross section of scattering and reaction.

Concluding, we would like to note that the microscopic approach presented in our work can be used along with other more sophisticated methods mentioned above.

4.1. Acknowledgments
The work is partly supported on the basis of the Project from the Agreement for co-operation between the INRNE-BAS (Sofia) and JINR (Dubna) and from the Agreement between BAS and Aristotle University of Thessaloniki. Three of the authors (D.N.K., A.N.A. and M.K.G.) are grateful for the support of the Bulgarian Science Fund under Contracts Nos. 02–285 and Φ–1501.

The authors E.V.Z. and K.V.L. thank the Russian Foundation for Basic Research (Grant No. 09-01-00770) for the partial support.

References

[1] Skaza F *et.al.* 2005 *Phys. Lett.* B **619** 82
[2] Wolski R *et al.* 2002 *Nucl. Phys.* A **701** 29c; 1998 *JINR Preprint, Dubna* **E15-98-284**; 1999 *Phys. Lett.* B **467** 8; Stepantsov S *et al.* 2002 *Phys. Lett.* B **542** 35
[3] Korsheninnikov A A *et al.* 1997 *Nucl. Phys.* A **616** 189c
[4] Korsheninnikov A A *et al.* 1997 *Nucl. Phys.* A **617** 45
[5] Korsheninnikov A A *et al.* 1993 *Phys. Lett.* B **316** 38
[6] Dortmans P J, Amos K, Karataglidis S and Raynal J 1998 *Phys. Rev.* C **58** 2249
[7] Karataglidis S, Brown B A, Amos K and Dortmans P J 1997 *Phys. Rev.* C **55** 2826
[8] Amos K, Richter W A, Karataglidis S and Brown B A 2006 *Phys. Rev. Lett.* **96** 032503
[9] Deb P K, Clark B C, Hama S, Amos K, Karataglidis S and Cooper E D 2005 *Phys. Rev.* C **72** 014608
[10] Deb P K, Amos K, Karataglidis S, Chadwick M B and Madland D G 2001 *Phys. Rev. Lett.* **86** 3248
[11] Deb P K and Amos K 2003 *Phys.Rev.* C **67** 067602
[12] Amos K, Dortmans P J, von Geramb H V, Karataglidis S and Raynal J 2000 *Adv. Nucl. Phys.* **25** 275
[13] Arellano H F and Girod M 2007 *Phys. Rev.* C **76** 034602
[14] Avrigeanu M, Anagnostatos G S, Antonov A N and Giapitzakis J 2000 *Phys. Rev.* C **62** 017001
[15] Avrigeanu M, Anagnastatos G S, Antonov A N and Avrigeanu V 2002 *Int. J. Mod. Phys.* E **11** 249
[16] Avrigeanu M, Antonov A N, Lenske H and Stetcu I 2001 *Nucl. Phys.* A **693** 616
[17] Lukyanov K V, Lukyanov V K, Zemlyanaya E V, Antonov A N and Gaidarov M K 2007 *Eur. Phys. J.* A **33** 389
[18] Lukyanov V K, Zemlyanaya E V, Lukyanov K V, Kadrev D N, Antonov A N, Gaidarov M K and Massen S E 2009 *Phys. Rev.* C **80** 024609
[19] Satchler R J and Love W G 1979 *Phys. Rep.* **55** 183; Satchler R J 1983 *Direct Nuclear Reactions* (Clarendon, Oxford)
[20] Khoa D T and von Oertzen W 1993 *Phys. Lett.* B **304** 8; 1995 *ibid* B **342** 6; Khoa D T, von Oertzen W and Bohlen H G 1994 *Phys. Rev.* C **49** 1652; Khoa D T, von Oertzen W and Ogloblin A A 1996 *Nucl. Phys.* A **602** 98; Khoa D T and Than H S 2005 *Phys. Rev.* C **71** 044601; Knyaz'kov O M 1986 *Sov. J. Part. Nuclei* **17** 137
[21] Khoa D T and Satchler G R 2000 *Nucl. Phys.* A **668** 3
[22] Karataglidis S, Dortmans P J, Amos K and Bennhold C 2000 *Phys. Rev.* C **61** 024319
[23] Amos K 2000 *Proceedings of 9th Conf. Nuclear Reaction Mechanisms*, Varenna, Italy, June 2000, ed. by Gadioli E (Ricerca Scientifica, Milano) 51
[24] Lukyanov K V, Zemlyanaya E V and Lukyanov V K (2006) *Phys. At. Nucl.* **69** 240
[25] Shukla P 2003 *Phys. Rev.* C **67** 054607
[26] Tanihata I 1992 *Phys. Lett.* B **289** 261 (and references therein)
[27] Karol P J 1975 *Phys. Rev.* C **11** 1203
[28] Massen S E and Moustakidis Ch C 1999 *Phys. Rev.* C **60** 024005
[29] Moustakidis Ch C and Massen S E 2000 *Phys. Rev.* C **64** 014314
[30] Zhukov M V *et al.* 1993 *Phys. Rep.* **231** 151
[31] Zhukov M V, Korsheninnikov A A and Smedberg M H 1994 *Phys. Rev.* C **50** R1

Journal of Physics: Conference Series **205** (2010) 012033 doi:10.1088/1742-6596/205/1/012033

Quasi-elastic reactions: a survey on recent results

S Szilner

Ruđer Bošković Institute, HR-10 001 Zagreb, Croatia

E-mail: suzana.szilner@irb.hr

Abstract.

Binary reactions at energies close to the Coulomb barrier received recently a significant boost thanks to the advent of the large solid angle magnetic spectrometer PRISMA coupled to the γ array CLARA. In the present paper different aspects of the recent results of nuclear structure and reaction dynamics will be presented, focusing more closely on the reaction mechanism.

1. Introduction

Transfer reactions play an essential role in the study of the structure of nuclei, in particular with light ions they have been instrumental for the construction of the shell model and they provided important data to establish the properties of particle-particle correlations in the nuclear medium. With heavy ions, the transfer studies were further extended. The two colliding partners can exchange a large number of nucleons, thus enabling to probe the pair density in the nuclear medium. From the reaction mechanism point of view, the study of multinucleon transfer reactions provides an insight about which degrees of freedom have to be included in any model in order to describe the evolution of the heavy-ion reaction from the quasi-elastic to the deep inelastic regimes and to fusion [1].

In the last decade the renewed interest in transfer reactions has been mainly due to the realization that multinucleon transfer reactions could be used to populate nuclei moderately rich in neutrons. This renewed interest benefited from the construction of the new generation large solid angle spectrometers based on trajectory reconstruction, with a large gain in overall efficiency while still keeping a good identification of reaction products. The coupling of these spectrometers with large γ arrays allowed the identification of γ rays coming from the decay of weak transfer channels associated with the population of nuclei moderately far from stability.

In this work a discussion on transfer yields, nucleon-nucleon correlation and the connection between multinucleon transfer reactions and other competing reaction channels is presented. An emphasis will be given to the discussion on the recent results of γ-particle coincidence measurements with PRISMA+CLARA. [2, 3, 4].

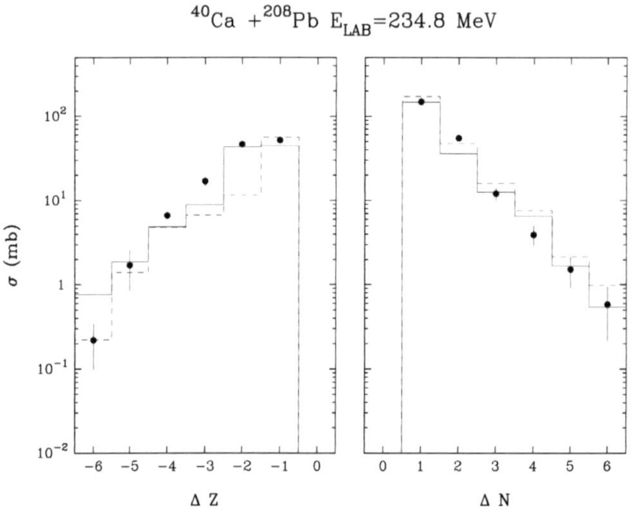

Figure 1. Angle and Q-value integrated cross sections for pure proton stripping and pure neutron pick-up channels of the ^{40}Ca+^{208}Pb reaction at $E_{\rm lab}$ =235 MeV as a function of transferred nucleons. The points and histograms are experimental and theoretical (CWKB) values, respectively. The dashed lines are the results of CWKB calculations using sequential transfer only, while the solid lines also include pair-transfer modes.

2. Nucleon-nucleon correlations

The presently achieved quality of data and of the agreement between theory and measurement is demonstrated in figure 1. The figure shows a comparison of the experimental angle and Q-value integrated cross sections for pure proton stripping and pure neutron pick-up channels of the ^{40}Ca+^{208}Pb reaction as a function of the number of transferred nucleons [5] with semiclassical calculations [6, 7]. A clear identification of the transfer products is obtained up to the pick-up of $\sim$six neutrons and the stripping of $\sim$ six protons. The bombarding energy is close to the Coulomb barrier and the yield, measured at the grazing angle, reflects some of the main characteristics of quasi-elastic processes. The charge and mass distribution of produced nuclei indicates the dominance of a direct mechanism in the population of different fragments. The plot shows that the neutron pick-up drop by almost a constant factor for each transferred neutron as an independent mechanism would suggest. The pure proton transfer cross sections behave differently, with the population of the $-2p$ channel as strong as the $-1p$ channel. This suggests the contribution of processes involving the transfer of proton pairs in addition to the successive transfer of single protons. One observes that the pair-transfer mode gives an essential contribution to the proton transfer channels (left panel), while its

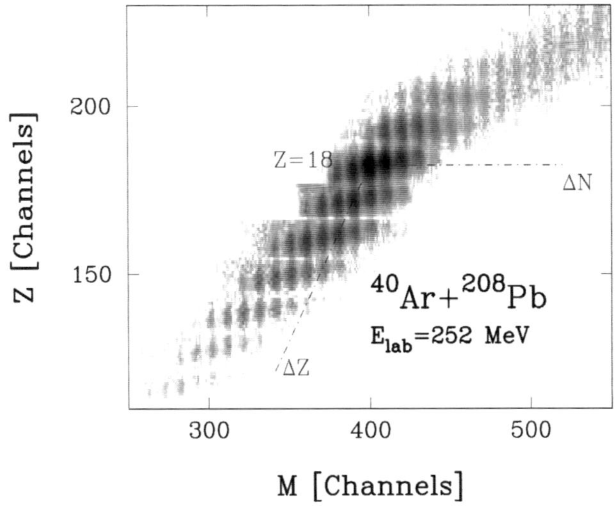

Figure 2. $M - Z$ spectra obtained in multinucleon transfer reactions of ^{40}Ar+^{208}Pb at 250 MeV. The dash-dotted lines correspond to the pure proton stripping (ΔZ) and, pure neutron pick-up (ΔN) channels, crossing at Z=18 and A=40.

contribution is almost negligible for the neutron channels (right panel).

Nuclear structure calculations have shown that the pairing interaction has the same strength for both neutrons and protons [8]. Since the one-neutron transfer cross section is almost one order of magnitude larger than the one-proton transfer, the contribution of a pair-transfer mode is masked, in the neutron sector, by the successive mechanism. However, it is important to point out that the microscopic treatment of pair modes is still a theoretical challenge.

Multinucleon transfer reactions are governed by optimum Q-value consideration, and using stable ions they dominantly populate neutron pick-up and proton stripping channels. In order to study proton-neutron correlation one has to use systems where the population of the np channels is allowed by the Q-value. In particular it is important to study the $+(np)$ channel which can be populated via a direct mechanism, at variance with the $-(np)$ channel which is of complex nature since it can be strongly affected by neutron evaporation mechanism (see next section). Such a goal can be achieved by using neutron rich beams, thus, we have performed a measurement of ^{40}Ar+^{208}Pb by using the PIAVE+ALPI LNL accelerator complex with an ^{40}Ar beam at 250 MeV. The $M - Z$ spectra is plotted in figure 2. The figure clearly shows that the proton pick-up and neutron stripping channels start to be populated. The experimental yields have been compared with GRAZING calculation [9, 10]. The yields have been normalized to the computed +1n channel and the same normalization constant has been kept for other neutron and proton pick-up and stripping channels. The dependence of the cross sections on the number of neutron pick-up and proton stripping channels turns out to be very similar to the one observed in other studied systems [5, 11]. A good agreement between data and theory is obtained for the transfer of pure neutrons and for channels

Journal of Physics: Conference Series **205** (2010) 012033 doi:10.1088/1742-6596/205/1/012033

involving the stripping of one proton $(-1p)$. Noteworthy is the fact that the shape of the mass yield for the $(-1p)$ case is well reproduced, also, on the neutron pick-up side. The proton pick-up channel $(+1p)$ is well reproduced, while these preliminary results indicate that some additional degrees of freedom have to be considered for the $(+1p + xn)$ channels.

Important complementary information can be achieved by looking at the population pattern of individual states, as the effect of the pairing interaction is to redistribute the transfer flux in a way to enhance states of specific structure. The spectra of the neighboring nuclei populated in the reaction comprise partly single particle or hole states and partly states that involve combinations of these states with collective boson. The transfer processes studied here are governed by well known properties of single particle and collective states, and through the excitation of these elementary modes the energy and angular momentum are transferred from the relative motion to the intrinsic degrees of freedom.

For these reasons, the character of the states populated in neutron transfer reactions have been studied in more details. In even Ar isotopes the strongest transitions between the yrast states have been observed. Besides them, the non-yrast feeding of the 2^+ and 0^+ states is also present. In the odd Ar isotopes, besides the transition between the lowest lying levels, a strong population of the $11/2^-$ states have been observed. Recent shell model calculations [12, 13] stress the importance of the coupling of collective boson to single-particle states (i.e. 2^+ to $f_{7/2}$, giving an $11/2^-$ stretched configuration). Multinucleon transfer reactions, in comparison with light ion induced transfer reactions, have larger probabilities for the population of stretched configurations, as the transfer probability has its maximum at the largest angular momentum transfer. It is important to notice that he de-excitation spectra may quite differ from the excitation ones. The quadrupole and octupole matrix elements generally play a major role in the excitation process, while the de-excitation may be dominated by very small components in the wave function. Thus, the structure of the studied nucleus together with the production mechanism strongly affects the observed spectra. Analysis of the all populated nuclei is in progress.

The measured total cross sections shown in figures 1 and 2 refer to the secondary reaction products [11, 14]. The primary yield can be in fact considerably altered by evaporation processes due to large kinetic energy loss components.

3. Evaporation

Making use of the γ-particle coincidence technique and exploiting the binary character of transfer reactions a direct evidence of particle evaporation processes can be outlined. With magnetic spectrometers, by detecting a specific Z and A the velocity vector of the undetected partner can be evaluated and applied for the Doppler correction of its corresponding γ rays. In figure 3 we show as an example the γ spectra for ^{40}Ar (detected) and its associated binary partner in the ^{40}Ca+^{96}Zr reaction [4]. One clearly

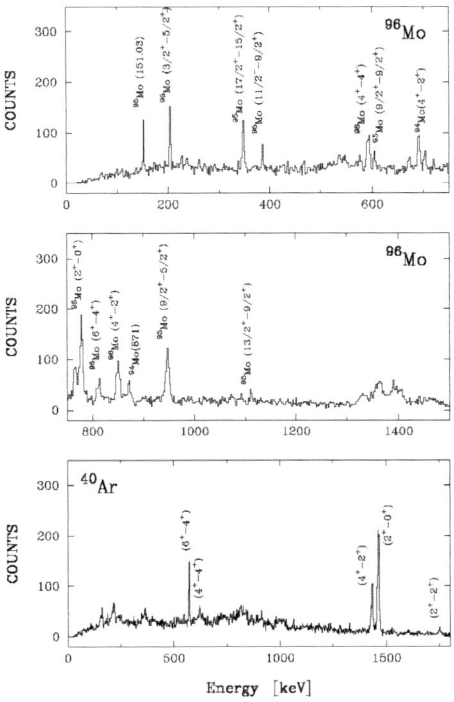

Figure 3. γ spectra for the $-2p + 2n$ channel in the reaction ^{40}Ca+^{96}Zr Doppler corrected for the heavy (top two frames) and light fragments (bottom frame). To have a better identification of the different γ lines for the heavy fragment we used an expanded energy scale.

sees that in the Mo spectra are present not only the γ rays belonging to the primary partner but also those of the nuclei produced after evaporation takes place. The yield associated with these neutron evaporation channels becomes progressively more marked for massive transfer channels. For instance in the $-4p$ channel the primary binary partner has negligible yield. A direct signature of how evaporation modifies the final yield distribution by correlating projectile-like and target-like fragment isotopic yields was outlined in [15] via $\gamma - \gamma$ coincidences. These two experimental techniques give comparable yield for neutron evaporation.

4. Summary

Multi-nucleon transfer reactions have been studied with the large solid angle magnetic spectrometer PRISMA, where ions identification is achieved by reconstructing event-by-event the trajectory of the ions in the magnetic elements. Experimental yields have been compared with semi-classical models that include surface vibrations and single particle transfer modes and taking into account the effect of neutron evaporation. By coupling the PRISMA spectrometer with the high resolution CLARA γ-array it has been shown how the transfer process populates reaction products in regions of angular momentum and excitation energy quite different from fusion-evaporation processes. This has been exploited to gain information on states that can be interpreted along the particle-vibration coupling scheme.

The material presented in this school is the result of a cooperative work of many people of the PRISMA+CLARA collaboration and belonging to different institutions (LNL, Padova, Torino, Zagreb, Bucharest, Strasbourg) which I wish to acknowledge.

References

[1] Corradi L, Pollarolo G and Szilner S 2009 *J. of Phys.* G **36** 113101
[2] Stefanini A M *et al* 2002 *Nucl. Phys.* **A701** 217c
[3] Gadea A *et al* 2004 Eur. Phys. J. A **20** 193
[4] Szilner S *et al* 2007 *Phys. Rev.* C **76** 024604
[5] Szilner S *et al* 2005 *Phys. Rev.* C **71** 044610
[6] Vigezzi E and Winther A 1989 *Ann. Phys. (N.Y.)* **192** 432
[7] Corradi L *et al* 1994 *Phys. Rev.* C **49** R2875
[8] Broglia R A, Hansen O and Riedel C 1973 *Advances in Nuclear Physics* vol 6, ed Baranger M and Vogt E (New York: Plenum)
[9] Winther A 1994 *Nucl. Phys.* **A572** 191; Winther A 1995 *Nucl. Phys.* **A594** 203
[10] Winther A *program GRAZING, htpp:/www.to.infn.it/~nanni/grazing* unpublished
[11] Corradi L *et al* 2002 *Phys. Rev.* C **66** 024606
[12] Caurier E, Martínez-Pinedo G, Nowacki F, Poves A and Zuker A P 2005 *Rev. Mod. Phys.* **77** 427
[13] Lebhertz D *et al*, to be published
[14] Corradi L, Stefanini A M, Lin C J, Beghini S, Montagnoli G, Scarlassara F, Pollarolo G and Winther A 1999 *Phys. Rev.* C **59** 261
[15] Broda R 2006 *J. of Phys.* G **32** R151

Journal of Physics: Conference Series **205** (2010) 012034 doi:10.1088/1742-6596/205/1/012034

Tetrahedral symmetry in nuclei: search for its fingerprints in the Actinide and Rare-Earth regions

D Curien[1], J Dudek[1] and K Mazurek[2]

[1] Institut Pluridisciplinaire Hubert Curien, UMR7178, IN2P3-CNRS
and Université de Strasbourg
23 rue du Loess BP 28 - 67037 Strasbourg Cedex 2, France

[2] Niewodniczanski Institut of Nuclear Physics, Polish Academy of Sciences,
ul. Radzikowskiego 152, Pl-31-342 Krakow, Poland
and Grand Accélérateur National d'Ions Lourds
Bd Henri Becquerel BP 55027 - 14076 CAEN Cedex 05, France

E-mail: Dominique.Curien@ires.in2p3.fr

Abstract. In series of recent theory articles, predictions have been formulated suggesting the existence of nuclear states whose mean-field Hamiltonians and thus the implied geometrical shapes, are characterized by tetrahedral symmetry. Following these publications, series of experiments for the Rare-Earth region have been proposed and performed. In this article, we shortly summarize the theory evolution developed in the original articles and discuss the status of the issue in light of the preliminary results of the ongoing experimental analyses. More recent theoretical results cross checked with extended literature investigations, suggest that the Actinide region might contain the best experimental candidates to investigate the fingerprints of the symmetry. A proposition is made to prove it via lifetime measurements with the so-called "microwave method".

1. Tetrahedral symmetry - theoretical background and experimental signs

The existence of the so-called high-rank symmetries in nuclear structure physics has been predicted in the framework of the realistic mean-field theory with the universal Woods-Saxon potential and confirmed by using Skyrme-Hartree-Fock-Bogolyubov method [1-8]. The term high-rank refers to specific point-groups symmetries: the tetrahedral and octahedral ones. They are the only symmetries leading to four-fold degeneracies of single-nucleonic levels in deformed nuclei: all other symmetries of deformed nuclei studied in nuclear physics lead only to the two-fold Kramer degeneracies. It can be shown that a classical tetrahedral-symmetric surface *(pyramid shape)* is invariant with respect to 24 symmetry elements such as fixed-angle rotations and plane reflections and that the octahedral-symmetric surface *(diamond shape)* is invariant with respect to 48 different symmetry elements. The importance of these symmetries has been deduced from the fact that they can lower the total energy of the nucleus by a few MeV [9].

As a critical consequence, the tetrahedral symmetry helps to produce large shell gaps in the single particle spectra because the degenerate levels "shrink" together [1,2,6] leading to specific tetrahedral shell-closures at the tetrahedral magic numbers: 32, 40, 56, 64, 70, 90, 112, and 136. The presence of those magic numbers produces low-lying minima in the nuclear potential energy surfaces that correspond to tetrahedral shapes. These "central" tetrahedral magic nuclei are accompanied by a large number of neighbouring nuclei that form tetrahedral islands throughout the Periodic Table in analogy to the presence of islands of spherical nuclei distributed around combinations of 8, 20, 28, 50, 82, 126 spherical magic numbers. The increased degeneracies in the mean-field spectra of deformed nuclei may lead to increased stability of the corresponding nuclear shapes: this may have particularly important implications for the exotic-nuclei projects. The discovery of tetrahedral and/or octahedral symmetries would then represent a critical step forward in nuclear structure physics.

The calculation results show clearly that the tetrahedral symmetry is responsible for strong shell effects in the Rare-Earth [9,10] and Actinide regions. It can be shown that nuclei with the exact tetrahedral symmetry have all multipole moments equal to zero (except for the Q_{32}, the "triaxial octupole" one) and in particular all the quadrupole moments $Q_{2\mu}=0$. This implies in particular the existence of very exotic rotational bands: because the tetrahedron is a non-spherical shape, nuclei with such shapes can rotate. The corresponding rotational sequences will be characterized by the "usual" $E\sim I(I+1)$ energy-spin dependence. But unlike in the "traditional" rotational bands, the E2-transitions commonly thought to be synonymous with collective rotation, are predicted either to vanish completely or to be very weak since the quadrupole moments in the exact symmetry limit must vanish. As a consequence the first experimental fingerprints that one can think of are the gamma decay $B_{in}(E2)/B_{out\text{-}out}(E1)$ branching ratios and their evolution with spin along the band. Another important point is that the tetrahedral symmetry is modelled, within the lowest multipolarity spherical harmonics of order $\lambda = 3$ (octupole) and therefore the associated rotational states may carry the negative parity in full analogy to other octupole states.

Figure 1 illustrates the results of the macroscopic-microscopic phenomenological method calculations using the Woods-Saxon Hamiltonian with the "universal-compact" set of parameters as in Refs. [6] and [10]. Comparison shows that the minimum at $\alpha_{32} \approx 0.13$ corresponds to $\alpha_{20} = 0$ and thus to the tetrahedral symmetry. If the measurements show that the quadrupole moment of the underlying rotational band is close to zero then the most likely interpretation would be the tetrahedral rather that pure "pear-shape" octupole deformation since (cf. right-hand of the figure) there is no axial-octupole minimum predicted by theory in this case. Of course, the next step will be to look for unambiguous criteria based on the branching ratios calculated with the help of the group theory methods. But the first condition, necessary although not sufficient, is to verify whether the quadrupole deformations in the discussed states are close to zero.

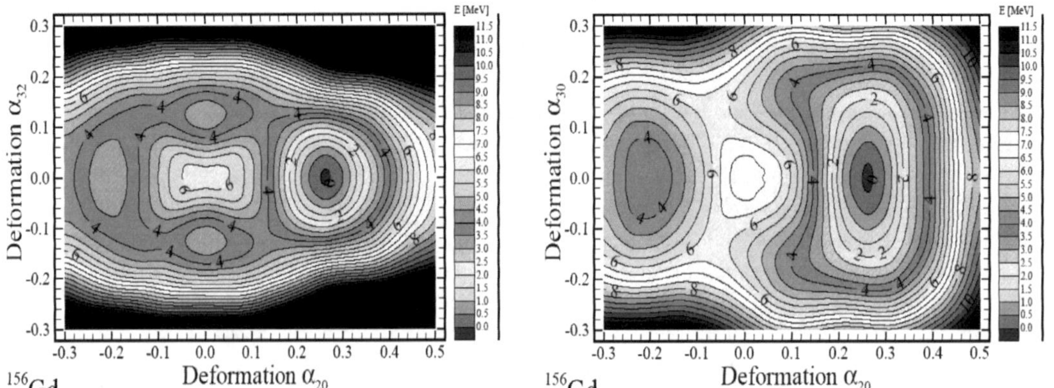

Figure 1. Total energy surfaces in function of the tetrahedral deformation α_{32} (left) and axial "pear shape" octupole deformation α_{30} (right) vs. quadrupole deformation α_{20}. Observe that according to calculations a tetrahedral minimum at $\alpha_{20} = 0$ is not accompanied by a "pear-shape" minimum (no competition between the two "octupole" type deformations).

Traditionally, the negative parity bands that are not identified as based on particle-hole excitations have often been interpreted in terms of the octupole degrees of freedom, the latter understood as α_{30} i.e. axially symmetric deformation. These effects take the form of either pear-shape static-deformed minima on the potential energy surfaces or as pear-shape vibrational-bands (called octupole-vibrational for short). To be more precise, according to such a traditional interpretation the nucleus remains quadrupole-deformed in its ground-state, whereas its one-phonon negative-parity octupole-excitation gives rise to the associated band-head. The latter collects then the rotational transitions from the rotational-states above. According to this scenario, the presence of the quadrupole-deformed equilibrium implies that the quadrupole moments and thus the implied rotational collective E2 transitions associated with those band heads should be present. However, as pointed out above, in many situations such E2 transitions seem to be clearly missing what puts in doubt this traditional interpretation in terms of the octupole-vibrational bands as always valid. Our approach adds a new interpretation without negating the validity of the possible two traditional interpretations of the collective octupole-mode related bands. More detailed calculations (which will be reported elsewhere) show that in fact, when the neutron number increases, in many actinide nuclei the pear-shape susceptibility sets in, in the otherwise quadrupole-deformed ground-state minima, with an increasing chances for the traditional interpretation in terms of the low-lying octupole α_{30}-vibrations. This tendency is accompanied, in the calculations, by a weakening of the tetrahedral symmetry minima through the lowering of the corresponding potential energy barriers (the tetrahedral minima vanish with increasing N while being relatively strong for the lighter isotopes). Detailed calculation results will be published.

In fact the existing literature contains already reports about several collective rotational bands of the above properties in the Rare-Earth and Actinide nuclei in an excellent correspondence to the predictions summarized above. In particular Refs. [10] and [11] suggest that tetrahedral symmetry has probably been seen in experiments in the early 1980's presenting a kind of curiosity whose interpretation was left out then. These rotational bands were simply interpreted in terms of octupole vibrational structures ignoring the unusual vanishing of the E2 intra-band transitions. This was particularly the case in the Rare-Earth region where for nuclei such as ^{156}Gd a negative parity rotational band was observed with vanishing E2's when the spin is decreasing [12]. Historically, this nucleus was the first experimental case that we have investigated. Now it appears to us from the current literature that the most significant cases might be found in the Actinide region and more specifically in some of the Uranium isotopes.

2. Recent experiments in the Rare-Earth region

Our first theoretical results suggest the existence of very stable total-energy minima corresponding to sizeable tetrahedral deformations in several nuclei around Z=62-70 (Sm, Gd, Dy, Er, Yb) with masses A~150-160. In the Gadolinium nuclei the tetrahedral-symmetric states are predicted to lie relatively low above the ground state (between 0.5 - 2.0 MeV approximately). The predicted barriers separating the "new" structures from the "traditional" ones (the quadrupole deformed ground-state minima) are of the order of several MeV, ranging typically from about 1.5 to about 5 MeV or more in various isotopes of the mentioned elements [6]. Following these indications we have started our experimental effort in this region.

2.1. The ^{156}Gd JYFL-JUROGAM experiment

In the ^{156}Gd nucleus a low-lying negative-parity side-band with odd-spins have been reported with no transition of the E2 character below spin 9 [12]. This particular nucleus has been studied in over 20 experiments using varying target-projectile combinations and detection systems [13], but the E2 transitions below spin 9 have never been seen. The presence of the corresponding states has been detected exclusively through the inter-band E1 transitions to the ground-state band. No convincing explanation was given to interpret this observation. Ultimately this band, together with its even-spin partner, was considered as an octupole $K^{\pi}=1^{-}$ vibrational band.

As mentioned earlier, at the ideal symmetry limit the dipole and the quadrupole moments of a tetrahedral-symmetric nucleus vanish. Then, according to theory arguments based on the pure symmetry assumption, the first allowed collective transition has an E3 character (pure octupole transitions) and consequently there should be no E2 or E1 transitions! But this is very unlikely because the ratio of the transition probability between E1 and E3 transitions is proportional to 10^{12}. Therefore at low spin any partial symmetry breaking such as the zero-point motion will favor the electromagnetic decay through E1 transitions. At increasing spins the nuclear rotation will necessarily privilege some particular directions of the total spin and because of the Coriolis interactions, the nucleonic alignment is going to gradually impose the triaxial quadrupole polarization of the core. As the result of those polarization effects we expect that some E2 transitions will become possible above a certain critical spin value in qualitative agreement with the clear observation of weak E2 transitions above spin 11.

The nucleus ^{156}Gd was therefore considered as a very good candidate to test the possibility to use the B(E2)/B(E1) branching ratios as a first fingerprint of the symmetry. This nucleus could be easily produced with a very high fusion-evaporation cross section using an alpha beam that favors a direct feeding of the low spin yrare states of interest. We have run a γ-coincidence experiment at Jyväskylä (FIN) with the JUROGAM array using the reaction ^{154}Sm(α,2n) at a 27.5 MeV bombarding energy that was pre-determined in an OSCAR pilot experiment at IPN-Orsay. A key issue here is that JUROGAM offers the possibility to study ^{156}Gd with triple γ-coincidences. Such a study was never performed on this nucleus in the past with an alpha beam, but it is essential in order to disentangle various gamma multiplets present in these nuclei, unambiguously locate the states of interest and be able to accurately measure the branching ratios in order to be sure that the previous results were not due to some experimental limits.

The preliminary results of the Jyväskylä experiment have been reported in Ref. [11]. These results confirm the pattern of vanishing intra-band E2 transitions below spin 11, however new inter-band transitions between the two negative-parity bands were found. They are very important to pin down the presence of some states of the tetrahedral candidate band that were previously seen only trough their E1 decays-out to the ground-state band and therefore attributed to the negative-parity even-spin band using only the argument of the state's excitation energy. We used the new inter-band transitions also to gate from above the states of interest to extract the branching ratios that are reported in Table 1 (below spin 11 the branching ratio of the odd-spin band are given in terms of upper limits whose estimates take into account the experimental resolving power). They are compared with the available data on ^{222}Th, a nucleus seen in the literature as one of the best "static" octupole deformed nuclei. From this Table we could conclude the following:

- The behaviour of the two negative-parity bands is very different: the even-spin partner has branching ratio one to two orders of magnitude larger than the odd-spin partners that is our candidate band. (Note that in the case of even-spin negative-parity band the inter-band dipole transitions are in fact non-stretched E1 transitions while in the case of the tetrahedral band they are the stretched ones).
- The E2 intra-band transitions in the even-spin band are seen down to spin 4, but only down to spin 11 in the odd-spin band.
- Both bands have decreasing B(E2)*in*/B(E1)*out* branching ratios with spin which is not the case of the octupole band of ^{222}Th.

Consequently the large structural differences seen in the branching ratios between the compared configurations suggest that they may have different origins. Even though there is no justification to claim the existence of the tetrahedral symmetry in ^{156}Gd, we can conclude from our accurate measurements that the first-excited negative-parity side-band with odd-spin is a serious candidate because the usual explanation in terms of octupole vibration appears less and less possible with the impossibility to explain the vanishing E2 intra-band transitions in a comprehensive way. As we will see in the following, the final word will probably come from the direct evaluation of the quadrupole moment extracted from lifetime measurements.

Table 1. Branching ratios R=B(E2)*in*/B(E1)*out* $*10^6$ fm^2 in ^{156}Gd, even and odd-spins, compared to the existing analogous data on ^{222}Th

Spin	**19^-**	**18^-**	**17^-**	**16^-**	**15^-**	**14^-**	**13^-**	**12^-**	**11^-**	**10^-**	**9^-**	**8^-**	**7^-**	**6^-**
^{156}Gd odd	50[a]		16[a]		5		6		<9		<26		<92	
^{156}Gd even		-		-		-		-		640		330		210
^{222}Th[b]	0.3		0.4		0.4		0.3		0.4		0.4		0.4	

[a] data are from Ref. [13], [b] data are from Ref. [14]

2.2. Other experiments in the Rare-Earth region

Three other experiments were performed after the JUROGAM measurement. Two of them aimed at attacking the problem of the vanishing E2 transitions in ^{156}Gd through different methods and nuclear reactions. The third experiment was performed with GAMMASPHERE at ANL to investigate ^{156}Dy, an N=90 isotone as opposed to ^{156}Gd which has N=92 neutrons. All these experiments are still under analysis and we will briefly present here only preliminary results.

2.2.1. ^{156}Gd LNL-GaSp experiment (Legnaro-I). The idea of this experiment [15] was to study the states of interests in ^{156}Gd by feeding them not from above in term of spin as in the case of the Jyväskylä fusion-evaporation experiment, but from below via Coulomb excitations using a ^{58}Ni beam. Therefore these states, if fed, could only be fed via multi-E2 ground-state band transitions followed by a single E3 inter-band transition. This was what we observed since the negative-parity states of the odd-spin band were seen up to spin 17. This fact clearly indicates that the octupole degrees of freedom and possibly the ones leading to tetrahedral symmetry are strongly at work in this nucleus. No E2 transitions below spin 11 are seen and the extracted branching ratios are compatible with the ones in the Jyväskylä measurement within the experimental errors, thus enforcing our previous conclusions. The absolute intensities of the transitions seen in this experiment have been compiled and will be analysed thanks to the GOSIA code in order to extract a value of the quadrupole moment of the tetrahedral-candidate band.

2.2.2. ^{156}Gd ILL-GAMS experiment (Grenoble-F). This experiment [16] is aiming at the measurement of the lifetime of the 5^- state from the γ-ray induced Doppler broadening (GRID technique) of the 5^- to 6^+ E1 inter-band transition. For that purpose a very intense γ-beam produced by the ^{155}Gd(n,γ)^{156}Gd reaction at the centre of the ILL-reactor is focused on the GAMS spectrometer where the Bragg diffraction is studied in the range of the energy of interest. The second goal was to try to observe the "missing" 5^- to 3^- E2 transition. No clear-cut evidence of the existence of this transition is currently seen in the available data. Therefore if we combine the lifetime measurement obtained for the 5^- states together with the lowest intensity value of the resolving power, we are able to extract an upper limit of the quadrupole moment. We end up with a value which is at its maximum estimated limit, close to the one of the ground-state band. If we take this limit as real, this observation would imply both the presence of the quadrupole band-head of the negative-parity band and yet vanishing E2 transitions. This would be another interesting scenario implying, however, the existence of a very strong E1 moment since otherwise it will be difficult to understand why strong quadrupole moments do not result in observation of the implied E2 transitions. We aim now to get a factor of ten in statistics by a dedicated experiment.

2.2.3. ^{156}Dy ANL-GAMMASPHERE experiment (Argonne-USA). In ^{156}Dy the same kind of structure as the one studied in ^{156}Gd has been reported in the literature: a low-lying negative-parity odd-spin band with E2 inter-band transitions seen only down to spin 9 with no intensities reported [17] and this prevents obtaining the branching ratios. GAMMASPHERE with its very high sensitivity was used to observe the decay of the interesting states populated in the ^{148}Nd(^{12}C,4n) reaction [18]. A very weak transition between the 9^- and the 7^- state was observed but even after an extended analysis, no in-band transitions below spin 7 were seen. The extracted $B_{in}(E2)/B_{out}(E1)$ branching ratios show the same pattern and magnitude as in ^{156}Gd with a regular decrease down to spin 9. These results clearly prove that the structure seen in ^{156}Gd is not a single accident but is most probably well spread in this region as predicted by the calculations involving tetrahedral symmetry arguments around the N=90 magic number. The next step with GAMMASPHERE will be to measure the lifetime in this band with a plunger experiment.

3. The Actinide region

In the actinide region our theoretical results available today predict the existence of stable total-energy minima laying at about 500 keV or more, above the ground-state and corresponding to sizeable tetrahedral deformations. The barriers separating these minima from the ground-state minimum are of the order of 1.0 to 2.5 MeV and the effect is even strengthened by the tetrahedral and octahedral symmetries superposing. (Recall that the tetrahedral group is a sub-group of the octahedral one and the superposition in question still results in the tetrahedral symmetry). It is worth emphasizing that these barriers are comparable with-, or even more sizeable, than those predicted as responsible for the prolate/oblate shape coexistence phenomena already studied experimentally, among others, in the Mercury region. We believe that these predictions signify a stronger stability in yet undiscovered tetrahedral nuclei as compared to already discovered prolate/oblate/spherical shape co-existing minima. The 230–234 even-even Uranium isotopes are good examples of tetrahedral candidates. Figure 2 shows the series of even-even Uranium isotopes from ^{226}U to ^{238}U for the ground-state bands and the first negative-parity bands. It is clear that the behaviour of the $^{230\text{-}234}$U isotopes corresponds to what we expect for a pure tetrahedral symmetry: no E2's intra-band transitions and only E1's inter-band transitions decaying from the negative-parity excited-states to the ground-state band members. To the contrary, on the lightest-mass side of this group we have ^{226}U nucleus that shows a "characteristic zigzag" pattern typical for the static-octupole deformation. Our microscopic calculations suggest that the underlying nature of the negative-parity bands in these actinide nuclei changes when the neutron number increases. At low neutron numbers N<136, calculations confirm the earlier interpretation of the band structure in terms of the Y_{30} static-octupole configurations. At heavier isotopes the Y_{32} tetrahedral interpretation is supported by the theory results. For N>146, the tetrahedral minimum is not well formed implying low stiffness against octupole vibrations and suggesting the presence of the low-lying octupole vibrational bands as observed in ^{238}U [19].

Let us observe in passing that all the odd-spin negative-parity bands from 230–238 even-even Uranium isotopes have been usually interpreted in the literature as octupole vibrational bands similarly to many bands in the Rare-Earth nuclei. They are characterized by two rotational branches: the lower-lying zero-phonon positive parity ground-state band and the negative parity one-phonon higher-lying band that usually remains shifted upwards in energy by the phonon's vibrational energy. However it is important to emphasize that the origin of such a shift (as the energy of octupole vibration) cannot be distinguished from the shift caused by the energy difference between the ground-state minimum and the tetrahedral minimum in the case of the new-symmetry interpretation. But the tetrahedral symmetry offers a natural explanation of the disappearing E2 transitions which is not the case when the usual octupole vibrational states are considered.

One important comment concerns the $I^\pi=1^-$ state. This state is a natural band member in the context of octupole vibrational scenario because they are the so-called $K^\pi=0^-$ or 1^- bands. But there is a particular arbitrariness in operating with $I^\pi=1^-$ states in our context of tetrahedral deformation. These 1^- are "very fragile" states in the sense that they may come from numerous and varying structural origins. In particular, in the tetrahedral-candidate bands we have no way to say whether 1^- states belong to the bands as long as nobody has seen the E2 transition from 3^- to 1^-. Consequently, unless we measure them we have to relay on theory. And theory says: if the bands are tetrahedral then 1^- states certainly do not belong to the tetrahedral symmetry because the states with spins I<2 must not carry the tetrahedral symmetry (they carry most likely the axial symmetry, and this deduction is a direct mathematical consequence of the properties of the Wigner functions). On the other hand there are good reasons to have 1^- states that hang around with close-to-good excitation energies without being neither octupole nor, certainly, tetrahedral. This is particularly seen in the Rare-Earth region: for most of the candidate nuclei if one would fit a parabola to the energies of states 3^-, 5^- and 7^- and then see how it extrapolates to spin 1^-, all possible scenarios are found; in other words, anything may happen and there are also sometimes situation where the lowest 1^- states are higher in energy than 3^- states!

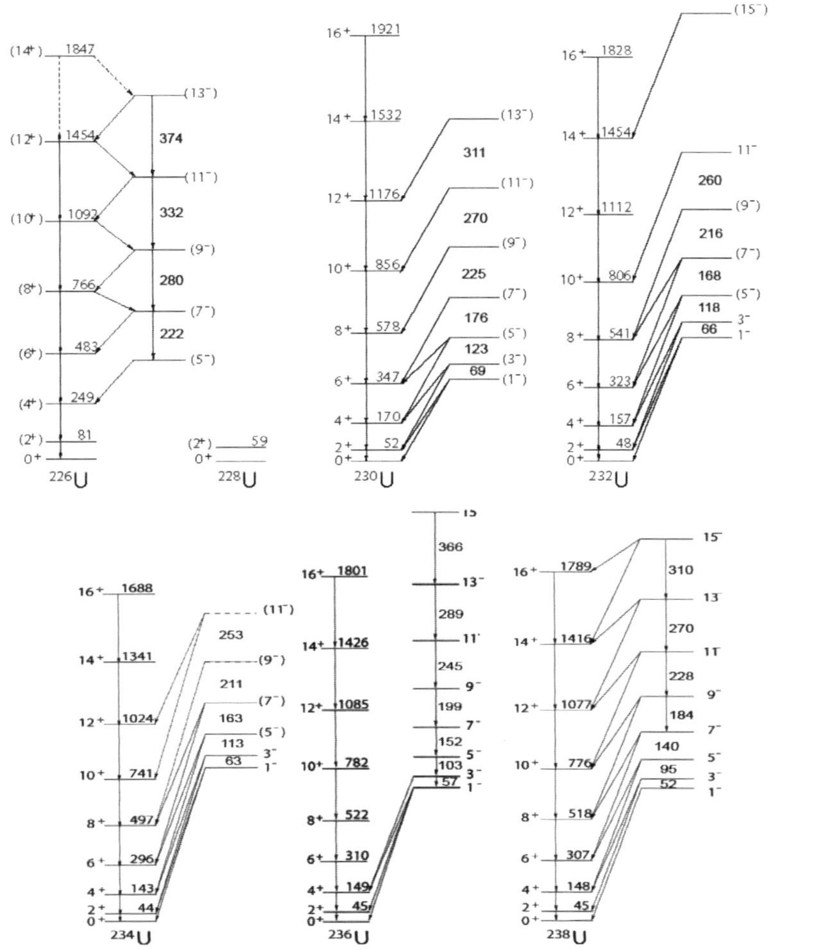

Figure 2. Partial level schemes of the uranium isotopes from ^{226}U to ^{238}U showing only the ground-state bands and the first negative parity bands. Energy differences between the negative parity states are given whether an E2 transition was observed or not. Observe a characteristic "E1 zigzag pattern" in ^{226}U as discussed in the text. Data are taken from Ref. [19-21].

Consequently, unless we measure them we have to relay on theory. And theory says: if the bands are tetrahedral then 1^- states certainly do not belong to the tetrahedral symmetry because the states with spins I<2 must not carry the tetrahedral symmetry (they carry most likely the axial symmetry, and this deduction is a direct mathematical consequence of the properties of the Wigner functions). On the other hand there are good reasons to have 1^- states that hang around with close-to-good excitation energies without being neither octupole nor, certainly, tetrahedral. This is particularly seen in the

Journal of Physics: Conference Series **205** (2010) 012034 doi:10.1088/1742-6596/205/1/012034

Rare-Earth region: for most of the candidate nuclei if one would fit a parabola to the energies of states 3^-, 5^- and 7^- and then see how it extrapolates to spin 1^-, all possible scenarios are found; in other words, anything may happen and there are also sometimes situation where the lowest 1^- states are higher in energy than 3^- states!

One question remains untouched and fully open: are the E2 intra-band transitions missing because the reduce B(E2) transition probabilities are small or, alternatively, because the B(E1) transition probabilities are very large? The first scenario will be a direct consequence of the tetrahedral symmetry hypothesis whereas the second one will favour the traditional octupole vibration explanation under the condition that the absence of E2 transitions in the spectra is well understood. Therefore it is now for us of the up-most importance to address this clear question by measuring the lifetimes of the candidate-band members and get very accurate limits for the branching ratios. We then will obtain the transition probabilities that we will use to extract the values of the transitional quadrupole moment. This will enable us to compare them with the ground-state band values: if they are much lower in magnitude than the ground-state band values, we will approach the tetrahedral hypothesis as an explanation for these structures.

4. Lifetimes and the microwave method to measure quadrupole moments

In the Actinide region taking into account the possible nuclear reactions and the structure of the candidate-bands, it is impossible to employ the usual methods for the lifetime measurements of the excited states of the 230−234 Uranium isotopes. The states of interest in these nuclei could only be produced via fusion evaporation reactions with proton and alpha particles that create the nuclei almost at rest. This totally prevents the use of Doppler measurement techniques such as RDM (Recoil Distance Method) with a plunger. Another possibility would be to use fast-timing measurements but this method implies the possibility to gate on a gamma transition above the measured states. Once again this is not possible in our case because of the absence of the E2 intra-band transitions. Therefore and to the best of our knowledge, we can not use the gamma-rays for this measurement. That being said the only remaining possibility is to use the conversion electrons with a "new" approach, but in fact based on the quite old method that is described briefly below.

The microwave method was originally developed at the Weizmann Institute in 1960 to measure in-beam lifetimes in the range of few tenths of picoseconds [22]. In the early 60's the recoil distance methods were not in use because of the absence of good resolution gamma spectrometers such as the Germanium detectors. A dedicated beam line was then built at the CRN-Strasbourg again in the late 60's to measure the lifetimes in the Rare-Earth region [23]. Basically this method consists of a high-frequency beam chopper driven by a microwave beam sweeping cavity. This cavity produces very short bursts of charged particles impinging on a target. An electromagnetic shutter-like device is used to select the conversion electrons according to their time of emission from the target. That device is another microwave cavity synchronized with the beam chopper. Changing the relative phase between the two cavities will modulate the electron energy in function of time. Therefore a variable time scale is established between the production of the excited nuclear state and its decay. Measuring the conversion electron energy in a beta-ray spectrometer will then give access to the lifetimes of the levels of interest. Combining these measurements with limits of the branching ratios will give upper limits of the quadrupole moments.

5. Conclusions

Our realistic mean-field predictions that take into account both the tetrahedral and the octahedral symmetry have focussed our experimental interest onto two regions of the nuclear chart: the Rare-Earth and the Actinide regions. Specific cases have been selected according to the current literature knowledge that display hints of the symmetry through the absence or the vanishing of E2 transitions in given rotational bands. A series of experiments performed world-wide have brought up preliminary encouraging results in ^{156}Gd and ^{156}Dy nuclei. The Actinide region seems to present purer signs of the symmetry with a possible new type of shape coexistence and shape evolution that we might already see in the Uranium isotopes. We are now looking forward to new lifetime measurements in both regions that will prove, or refute, in a unique way the hypothesis of the existence of the high-rank tetrahedral and octahedral symmetries in nuclear physics.

Acknowledgments
The works reported here are mainly part of the activities of the TetraNuc (Tetrahedral Nuclei) Collaboration involving 18 European and non-European institutions. This informal collaboration aims at searching for the fingerprints of the tetrahedral symmetry throughout the Nuclear Chart. The authors want to warmly thank all the people that have been taken part in these activities or support them and more specifically the Direction of IN2P3, the Direction of the IPN-Orsay and D. Verney, O. Stezowski and Q.T. Doan at IPN-Lyon, J. Gerl at GSI-Darmstadt, R. Julin and P. Jones at JYFL-Jyväskylä, F. Haas, H. Molique, J. Robin, P. Medina and M. Richer at IPHC-Strasbourg, M. Jentschel, B. Lauss and W. Urban at ILL-Grenoble, D. Tonev at INRNE-Sofia, G. de Angelis at LNL-Legnaro, R. J. Singh at IUAC-New Dehli, L. Riedinger at Uni. of Tennessee-Knoxville, D. Hartley at U.S. Naval Academy-Annapolis and last but not least the Direction of ANL-Argonne. Acknowledgement is due also for the support from IN2P3-COPIN collaboration program through the project no. 05-119 and the support under COPIGAL project "Search for the high-rank symmetries in subatomic physics".

References
[1] Li X and Dudek J 1994 *Phys. Rev.* C**94** R1250
[2] Dudek J, Gozdz A, Schunck N, Miskiewicz M 2002 *Phys. Rev. Lett.* **88** 252502
[3] Dudek J, Gozdz A and Rosly D 2001 *Acta Phys. Polon.* B**32** 2625
[4] GozdzA , Dudek D and Miskiewicz M 2003 *Acta Phys. Polon.* B**34** 2123
[5] Dudek J, Gozdz A and Schunck N 2003 *Acta Phys. Polon.* B**34** 2491
[6] Schunck N, Dudek J, Gozdz A and Regan P 2004 *Phys. Rev.* C**69** 061305 (R)
[7] Schunck N and Dudek J, Int. 2004 *Journal of Modern Phys.* E**13** 213
[8] Schunck N, Dudek J and Frauendorf S 2005 *Acta. Phys. Polon.* B**36** 1071
[9] Dudek J, Dobaczewski J, Dubray N, Gozdz A, Pangon V, Schunck N 2007 *Int.J.Mod.Phys.* E**16** 516
[10] Dudek J, Curien D, Dubray N, Dobaczewski J, Pangon V, Olbratowski P and Schunck N 2006 *Phys. Rev. Lett.* **97** 07250
[11] Doan Q T *et al.* 2009 *Acta. Phys. Polon.* B**40**,725; PhD Thesis 2009 *Uni. C. Bernard, Lyon*
[12] Konijn J *et al.* 1981 *Nucl. Phys.* A**352** 191
[13] Reich C W 2003 *Nuclear Data Sheets* **99** 753
[14] Akovali Y A 1996 *Nuclear Data Sheets* **77** 271
[15] Singh R J 2008 *LNL Annual Report* and 2009 *private communication*
[16] Jentschel M 2008 *private communication*
[17] Riley M A *et al.* 1998 *Nucl. Phys.* A**486** 456
[18] Riedinger L 2009 *private communication*
[19] Ward D *et al* 1996 *Nucl. Phys.* A**600** 88
[20] Zeyen P *et al.* 1987 *Z. Phys.* A**328** 399; Ackermann B *et al.*1993 *Nucl. Phys.* A**559** 61
[21] Greenless P *et al.* 1998 *J. Phys. G: Nucl. Part. Phys.* **24** 63
[22] Goldring G 1961 *NIM* **11** 29
[23] Gerber J *et al.* 1969 *Revue de Physique Appliquée* **4** 218; Armbruster R *et al.* 1970 *Nuc. Phys.* A**143** 315; Dar Y *et al.*1971 *NIM* **97** 251 and reference therein.

High-energy excited states in ^{98}Cd

A Blazhev[1], N Braun[1], H Grawe[2], P Boutachkov[2], B S Nara Singh[3], T Brock[3], Zh Liu[4], R Wadsworth[3], M Górska[2], J Jolie[1], F Nowacki[5], S Pietri[2], C Domingo-Pardo[2], I Kojouharov[2], L Cáceres[2], T Engert[2], F Farinon[2], J Gerl[2], N Goel[2], J Grębosz[6], R Hoischen[2,7], N Kurz[2], C Nociforo[2], A Prochazka[2], H Schaffner[2], S Steer[8], H Weick[2], H-J Wollersheim[2], A Ataç[9], L Bettermann[1], K Eppinger[10], T Faestermann[10], F Finke[1], K Geibel[1], C Hinke[10], A Gottardo[4], G Ilie[1], H Iwasaki[1], R Krücken[10], E Merchan[11], J Nyberg[12], M Pfützner[13], Zs Podolyák[8], P Regan[8], P Reiter[1], S Rinta-Antila[14], D Rudolph[7], C Scholl[1], P-A Söderström[12], N Warr[1], P Woods[4]

[1]Universität zu Köln, Cologne, Germany
[2]GSI, Darmstadt, Germany
[3]University of York, York, UK
[4]University of Edinburgh, Edinburgh, UK
[5]IPHC, Strasbourg, France
[6]IFJ PAN, Krakow, Poland
[7]Lund University, Lund, Sweden
[8]University of Surrey, Guildford, UK
[9]Ankara University, Ankara, Turkey
[10]Technische Universität München, Garching, Germany
[11]Universidad Nacional de Colombia, Bogota, Colombia
[12]Uppsala University, Uppsala, Sweden
[13]Warsaw University, Warsaw, Poland
[14]University of Liverpool, Liverpool, UK

E-mail: a.blazhev@ikp.uni-koeln.de

Abstract. In ^{98}Cd a new high-energy isomeric γ-ray transition was identified, which confirms previous spin-parity assignments and enables for the first time the measurement of the $E2$ and $E4$ strength for the two decay branches of the isomer. Preliminary results on the ^{98}Cd high-excitation level scheme are presented. A comparison to shell-model calculations as well as implications for the nuclear structure around ^{100}Sn are discussed.

1. Introduction

Studies of isomerism in the neutron-deficient $N \simeq Z$ nuclei around ^{100}Sn give important insights into the role of proton-neutron pairing, serve as testing grounds for nuclear models and give input for understanding the astrophysical *rp*-process [1, 2]. Previous studies of ^{98}Cd [3, 4] have revealed the existence of a high-energy core-excited state of (12^+) decaying by a 4207 keV $E4$ transition to the (8^+) state [4]. Large-scale shell-model (LSSM) calculations done in the *gds*-model space could reproduce the energy of the (12^+) state [4]. The LSSM calculations have also suggested the existence of a 14^+ state below the 12^+ state resulting in a spin-gap isomer, which is supposed to decay by β or proton emission, and a 10^+ state lying above the 12^+ isomer [4]. These predictions together with the long-standing prediction of the 16^+ and $25/2^+$ spin-gap isomeric states in ^{96}Cd and ^{97}Cd, respectively, [5] have been motivation to continue studies in this very exotic region of the nuclidic chart.

2. Experimental details

In summer 2008 an experiment on ^{96,97,98}Cd was performed at the GSI Darmstadt using the FRS fragment separator [6] and the RISING germanium array in the "Stopped Beam" configuration [7]. These exotic nuclei of interest were produced using fragmentation of a 850 MeV/u ^{124}Xe beam on a 4 g/cm^2 ^{9}Be target and finally implanted into an active stopper consisting of nine double-sided silicon strip detectors [8]. The active stopper was surrounded by the RISING Ge-array with a photopeak efficiency of ~10% at 1.3MeV [7], which collected γ rays correlated with ion implantation and/or particle decays registered in the active stopper. Using the FRS detectors all ions were identified on an event-by-event basis allowing for a clean selection of the fragment of interest.

3. First results

While the analysis of this complex experiment is still in progress first results are already available. In several isotopes like ^{98}Cd, ^{96}Ag [9] and ^{94}Pd [10] new γ-ray sub-μs isomeric transitions were discovered. This work reports preliminary results in ^{98}Cd where a new high-energy isomeric transition was identified.

In the γ-ray spectrum sorted in delayed coincidence with implanted ^{98}Cd ions, beside the known transitions [3, 4] with energies 4207(2), 147, 198, 688 and 1395 keV representing the cascade $(12^+) \to (8^+) \to (6^+) \to (4^+) \to (2^+) \to 0^+$ in ^{98}Cd, a new transition with an energy of 4157(3) keV was identified. Figure 1 presents a spectrum gated on the new γ ray, showing that this transition is in coincidence with the $(8^+) \to (6^+) \to (4^+) \to (2^+) \to 0^+$ cascade in ^{98}Cd.

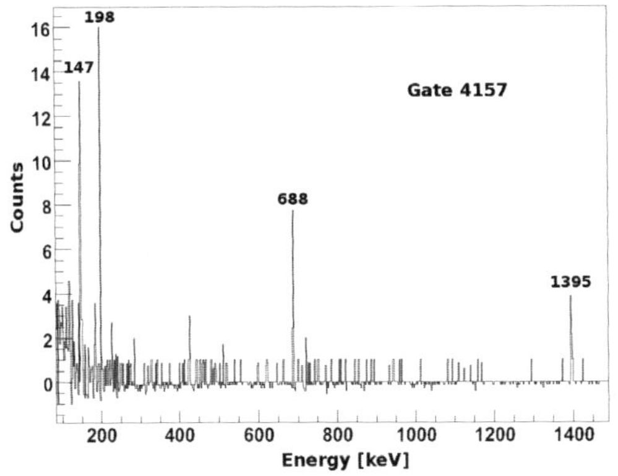

Figure 1. Spectrum in coincidence with ^{98}Cd ions and 4157 keV γ rays.

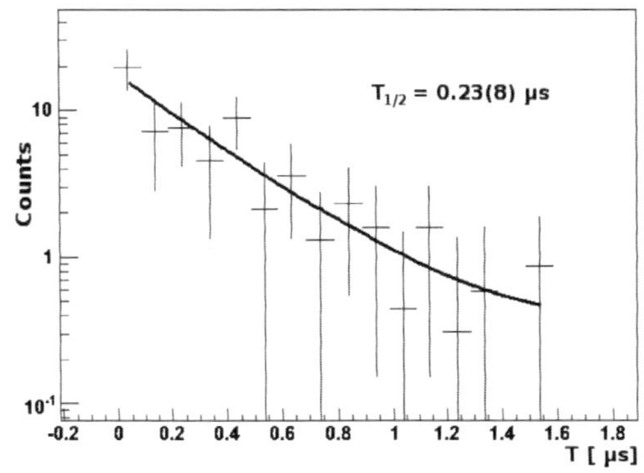

Figure 2. Time spectrum of the 4157 keV γ ray fitted by a single exponential decay plus a constant background.

A time spectrum of the 4157 keV γ-ray transition is shown in Fig. 2. A least-squares fit was performed using a single exponential decay plus a constant background (see Fig. 2). The preliminary result for the halflife is $T_{1/2} = 0.23(8)$ μs, which agrees with the known halflife $T_{1/2} = 0.23(^{+4}_{-3})$ μs of the (12^+) isomeric state [4].

In our previous work [4] the assignment of the $E4$ character to the 4207 keV transition as well as the (12^+) spin and parity to the 6635 keV state was discussed at length. Also in that work [4]

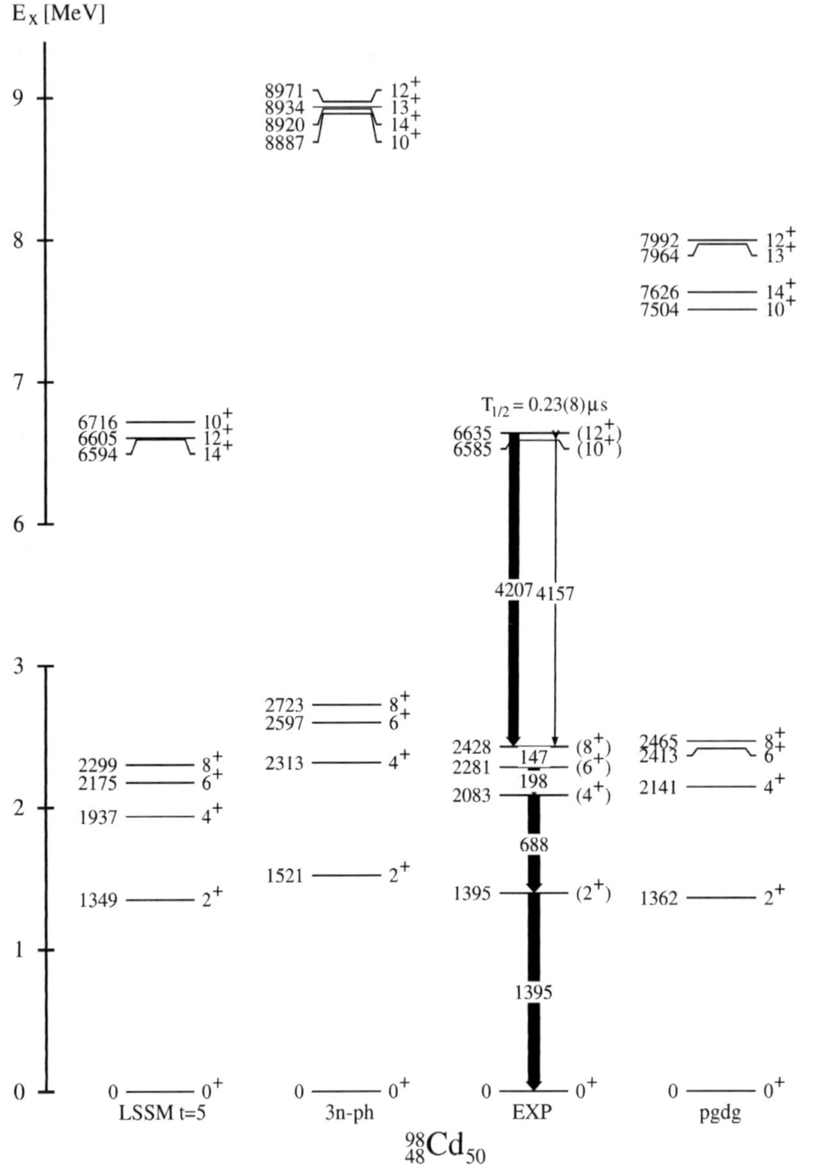

Figure 3. Experimental level scheme of ^{98}Cd (EXP) in comparison to the t=5 ($5p5h$ excitations of the ^{100}Sn core) LSSM spectrum [4] labeled "LSSM t=5" and two revised calculations labeled "3n-ph" and "pgdg". See text for details.

the lower observational limit for $E2$ transitions was 80 keV, and consequently we concluded that a hypothetical (10^+) state would lie less than 80 keV below or above the (12^+) state not changing the $E4$ assignment to the 4207 keV transition.

Both the γ-γ-coincidence relations as well as the halflife suggest that the new transition is part of an alternative cascade connecting the (12^+) state with the (8^+) state. Therefore, we suggest the existence of a state with an energy of 6585 keV, tentatively assign spin and parity of (10^+) to this new state, and suggest an $E2$ character to this new 4157 keV high-energy transition (see Fig. 3 column "EXP"). This also implies an experimentally non-observed 50 keV $E2$ isomeric transition depopulating the (12^+) state and feeding the (10^+) state. With this assignment from the measured halflife and the observed branching the preliminary transition strengths $B(E2; 12^+ \to 10^+) = 2.1(13)$ W.u. and $B(E4; 12^+ \to 8^+) = 3.0(10)$ W.u. are inferred.

4. Shell-model calculations

On the basis of the above discussed experimental evidence for a (10^+) state lying 50 keV below the excited (12^+) state in ^{98}Cd the deficiencies of the previous SM calculations [4] became obvious (see Fig. 3, column "LSSM t=5"). Therefore, new SM calculations have been performed in order to account for the experimental level scheme. Figure 3 shows the new experimental level scheme of ^{98}Cd (EXP) compared to two revised SM calculations, "3n-ph" and "pgdg", which reproduce the correct ordering of the 10^+ and 12^+ states.

The 12^+ is a core-excited state with a main component of the neutron configuration $\nu g_{9/2}^{-1}\, \nu d_{5/2}$. On the other hand the leading configuration of the 10^+ state is the $\nu g_{9/2}^{-1}\, \nu g_{7/2}$. Thus changing the relative position of the neutron $d_{5/2}$ and $g_{7/2}$ orbitals with respect to each other may help in reproducing the experimental observation. The absolute single particle energies of the $\nu d_{5/2}$ and $\nu g_{7/2}$ orbitals relative to ^{100}Sn are not known and up to now their values have only been extrapolated. According to Ref. [11] in ^{101}Sn the $\nu d_{5/2}$ orbital is 80 keV lower than $\nu g_{7/2}$. Therefore, a small variation in these single-particle energies can result in the reversal of the order of these two orbitals. A recent experiment on ^{101}Sn has reported a γ-ray transition of 172 keV [12], which most probably connects the $5/2^+$ and $7/2^+$ states. The authors discuss and give good arguments that the $5/2^+$ state should be the ground state of ^{101}Sn [12]. In the extreme single-particle shell-model this would mean that $\nu d_{5/2}$ is 172 keV lower than $\nu g_{7/2}$, but when accounting for the configuration mixing, in this case especially core-excitations, the situation changes.

The SM calculation labeled "3n-ph"(see Fig. 3) is performed with the m-scheme code NuShell@MSU [13, 14] in the SNE model space $\pi\nu(f_{5/2}, p_{3/2}, p_{1/2}, g_{9/2}, g_{7/2}, d_{5/2}, s_{1/2}) + \nu(h_{11/2})$ using the SNET interaction [14, 15]. The following truncation was used: $\pi(f_{5/2}, p_{3/2})$ and $\nu(f_{5/2}, p_{3/2}, p_{1/2})$ were considered fully occupied, allowing for proton $\pi(p_{1/2}, g_{9/2})^{10}$ configurations and up to 3 neutron particle-hole excitations ("3n-ph") from the $\nu g_{9/2}$ to $\nu(d_{5/2}, g_{7/2})$ orbitals. As shown in Fig. 3 the "3n-ph" calculation reproduces the experimental ordering of the 10^+ and 12^+ states in ^{98}Cd and their close position relative to each other. The calculation "3n-ph" presents a case where the single-particle orbital $\nu d_{5/2}$ is 50 keV higher than $\nu g_{7/2}$, i.e. "reversed" order. That means that in the case of ^{101}Sn, one neutron outside the closed N=50 shell, and no core-excitations, i.e. the "0n-ph" calculation, the $5/2^+$ state is 50 keV above the $7/2^+$. But once the neutron ph-excitations are taken into account the ordering of the calculated levels $5/2^+$ and $7/2^+$ in ^{101}Sn is again "normal". At "1n-ph" the calculated $7/2^+$ state is about 50 keV above the $5/2^+$, at "2n-ph" - 110 keV, and at "3n-ph" - 112 keV above the $5/2^+$ state, showing a converging trend and fairly reproducing what is believed to be experimentally observed [12]. Calculations with higher number of neutron ph core-excitations were not performed due to computational limitations.

The calculation labeled "pgdg" (see Fig. 3) is performed with another m-scheme code OXBASH [15] in the SN model space $\pi\nu(f_{5/2}, p_{3/2}, p_{1/2}, g_{9/2}, g_{7/2}, d_{5/2}, s_{1/2})$ using the GF interaction [16] for $\pi\nu(p_{1/2}, g_{9/2})$ and the SNA interaction [15] for the rest. The calculation was truncated to $\pi(p_{1/2}, g_{9/2})$ and $1p1h$ $\pi g_{9/2}^{-1}\, \pi d_{5/2}, g_{7/2}$ and $\nu g_{9/2}^{-1}\, \nu d_{5/2}, g_{7/2}$ excitations across $Z, N = 50$ shell gaps. A "normal" sequence of the neutron $d_{5/2}$ and $g_{7/2}$ orbitals was assumed, i.e. $d_{5/2}$ below the $g_{7/2}$, and the proton-proton monopole part of the interaction was tuned to reproduce proton separation energies in ^{88}Sr, ^{90}Zr and ^{100}Sn. Meaning that in the "pgdg" stronger proton particle-hole excitations across the shell gap are allowed and as shown in

Fig. 3(right) this calculation also reproduces the relative position of the 10^+ state below the 12^+, although currently the calculated splitting considerably exceeds the experimentally observed one. The existence of stronger proton-core excitations across the $Z = 50$ shell gap has been discussed also in the work of Vaman *et al.* [17] when trying to explain the increased $B(E2)$ strength in the yrast $2^+ \rightarrow 0^+$ transitions in the very neutron-deficient Sn isotopes. Therefore, the experimental $B(E2)$ and $B(E4)$ strengths from the core-excited state in ^{98}Cd provide a crucial test of the SM calculations.

A detailed assessment of the level scheme and transition rates will be given in a forthcoming paper after the experimental analysis and the universal shell-model interaction tuning are completed.

Acknowledgments

The excellent work of the GSI accelerator division is acknowledged. The support under grants 06KY205I and 06MT238 of the BMBF (Germany), and support by STFC(UK) and the Swedish Science Council is acknowledged.

References

[1] Grawe H *et al.* 2006 *Eur. Phys. J* A **27**(S01) 257
[2] Schatz H *et al.* 1998 *Phys. Rep.* **294** 167
[3] Górska M *et al.* 1997 *Phys. Rev. Lett.* **79** 2415
[4] Blazhev A *et al.* 2004 *Phys. Rev.* C **69** 064304
[5] Ogawa K 1983 *Phys. Rev.* C **28** 958
[6] Geissel H *et al.* 1992 *Nucl. Instr. and Meth.* B **70** 286
[7] Pietri S *et al.* 2007 *Nucl. Instr. and Meth.* B **261** 1079
[8] Kumar R *et al.* 2009 *Nucl. Instr. and Meth.* A **598** 754
[9] Boutachkov P *et al. in preparation*
[10] Wadsworth R *et al.* 2009 *Acta Phys. Pol.* B **40** 611
[11] Grawe H and Lewitowicz M 2001 *Nucl. Phys.* A **693** 116
[12] Seweryniak D *et al.* 2007 *Phys. Rev. Lett.* **99** 022504
[13] Rae W D M 2007 Unpublished URL `http://knollhouse.org/`
[14] Brown B A and Rae W D M 2007 MSU-NSCL report
[15] Brown B A *et al.* 2004 MSU-NSCL report number 1289
[16] Gross R and Frenkel A 1976 *Nucl. Phys.* A **267** 85
[17] Vaman C *et al.* 2007 *Phys. Rev. Lett.* **99** 162501

Spin and Parity Assignments to Dipole Excitations of the Odd-Mass Nucleus ^{207}Pb from Nuclear Resonance Fluorescence Experiments with Linearly-Polarized γ-Ray Beams

N. Pietralla[1], T.C. Li[2], M. Fritzsche[1], M.W. Ahmed[3], D. Savran[1], A.P. Tonchev[3], W. Tornow[3], H.R. Weller[3], and V. Werner[4]

[1]Institut fur Kernphysik, Technische Universitat Darmstadt, 64289 Darmstadt, Germany
[2]Nuclear Structure Laboratory, SUNY at Stony Brook, Stony Brook, NY 11794-3800, U.S.A.
[3]Triangle Universities Nuclear Laboratory (TUNL), Duke University, Durham, NC 27708, U.S.A.
[4]A.W. Wright Nuclear Structure Laboratory (WNSL), Yale University, New Haven, CT, U.S.A.

E-mail: pietralla@ikp.tu-darmstadt.de

Abstract. Pb($\vec{\gamma},\gamma'$) photon scattering reactions were studied [1] with the nearly monochromatic, linearly polarized photon beams at the High Intensity γ-ray Source (HIγS) at the DFELL. Azimuthal scattering intensity asymmetries measured with respect to the polarization plane of the beam have been used for the rst time to assign both the spin and parity quantum numbers of dipole excited states of ^{206,207,208}Pb at excitation energies in the vicinity of 5.5 MeV. Evidence for dominant particle-core coupling is deduced from these results along with information on excitation energies and electromagnetic transition matrix elements.

Nuclear Resonance Fluorescence (NRF) is well established as a powerful method for investigation of nclear dipole excitation (see [2–5] and References therein). However, assignments of spin and parity quantum numbers to excited states of odd-mass nuclei from NRF data has remained difficult up to now. Information on spin and parity quantum numbers must be obtained from anisotropies in angular intensity or polarization distributions.

For odd-A nuclei the anisotropies are typically small, in particular, when an unpolarized -ray beam is used for the experiments. The angular distribution function $W(\theta, \phi)$ can be expanded in terms of Legendre polynomials. In the absence of parity violation and third order multipole contributions for the $J_0^{\pi_0}(L_1, L_1^{'})J^{\pi}(L_2, L_2^{'})J_f^{\pi_f}$ cascade with unpolarized photons it can be written as:

$$W(\theta) = \sum_{\nu=0}^{even} B_\nu(\ _1)A_\nu(\ _2)P_\nu(cos\theta) \tag{1}$$

where P_ν are the ordinary Legendre Polynomials of the νth order. The expansion coefficients A and B can be written, using the phase convention of Krane, Ste en, and Wheeler [6] for the multipole mixing ratios δ_1 and δ_2 and in terms of tabulated F-coefficients [6] as:

$$A_\nu(\ _2) = \frac{1}{1+\delta_2^2}[F_\nu(L_2, L_2, J_f, J) + 2\delta_2 F_\nu(L_2, L_2^{'}, J_f, J) + \delta_2^2 F_\nu(L_2^{'}, L_2^{'}, J_f, J)] \quad (2)$$

$$B_\nu(\ _1) = \frac{1}{1+\delta_1^2}[F_\nu(L_1, L_1, J_0, J) \quad 2\delta_1 F_\nu(L_1, L_1^{'}, J_0, J) + \delta_1^2 F_\nu(L_1^{'}, L_1^{'}, J_0, J)] \quad (3)$$

Dipole transitions are the dominant mode of excitation in NRF processes. While in even-even nuclei (with 0^+ ground state) a dipole excitation directly leads to a unique spin assignment $J = 1$ for the excited state, such an unambiguous assignment is not so straightforward in the case of an odd-mass nucleus with half-integer ground-state spin. Let us consider the case of a spin-1/2 ground state. Then, dipole excitations can excite states with spin quantum numbers $J = 1/2$ or $J = 3/2$. Figure 1 shows the angular distribution function (1) for pure dipole NRF cascades $1/2 \rightarrow 1/2 \rightarrow 1/2$ and $1/2 \rightarrow 3/2 \rightarrow 1/2$ using unpolarized -rays in the entrance channel. Unfortunately the di erences of the maxima are not very signi cant in case of an odd-mass-nucleus, which is also mentioned in [7].

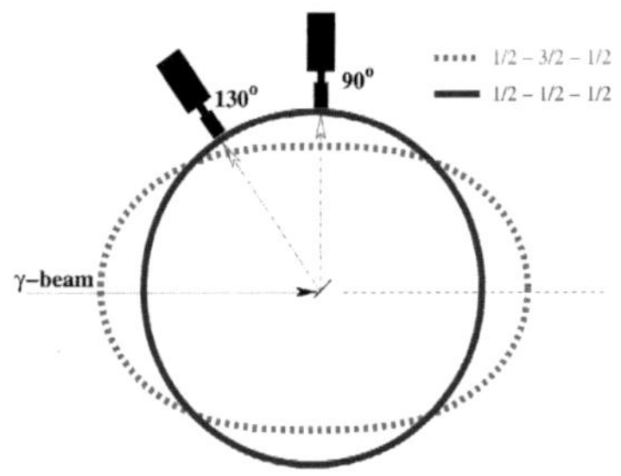

Figure 1. Angular intensity distribution for unpolarized bremsstrahlung in case of the possible two dipole transitions in odd nuclei with $J = 1/2$ ground state. The di erences in the intensity maxima are quite small, making spin assignments difficult.

An alternative for assigning spin and parity quantum numbers in odd-nuclei is to investigate the target with linearly polarized photons. The intensity distribution function for polarized photons is given by

$$W(\theta,\phi) = \sum_{\nu=0}^{even} B_\nu(\vec{\ }_1)A_\nu(\ _2)P_\nu(cos\theta) + (\pm)_{L_1} cos(2\phi) \sum_{\nu=0}^{even} B_\nu^{'}(\vec{\ }_1)A_\nu(\ _2)P_\nu^2(cos\theta)\ . \quad (4)$$

Where P_ν, A, B and δ are the same as in equation (1), $P_\nu^{(2)}$ is the unnormalized associated Legendre polynomial and $(\pm)_{L_1}$ takes the value $+1(\ 1)$ in case of electric (magnetic) character of the leading multipole L_1 of the rst transition. $B^{'}$ can also be written in terms of the tabulated F-coefficients [6]:

$$B_\nu^{'}(\vec{\ }_1) = \frac{1}{1+\delta_1^2}[\kappa_\nu(L_1, L_1)F_\nu(L_1, L_1, J_0, J) + \\ + 2\delta_1^2\kappa_\nu(L_1, L_1^{'})(L_1, L_1^{'}, J_0, J) \quad \delta_1^2\kappa_\nu(L_1^{'}, L_1^{'})F_\nu(L_1^{'}, L_1^{'}, J_0, J)] \quad (5)$$

The κ coefficients can be found in Ref. [8]. For the most important cases of dipole and quadrupole radiation their values are [8] $\kappa_2(1,1) = \ 1/2, \kappa_2(1,2) = \ 1/6, \kappa_2(2,2) = 1/2$ and

Journal of Physics: Conference Series **205** (2010) 012036 doi:10.1088/1742-6596/205/1/012036

$\kappa_4(2,2) =\ \ 1/12$. A $1/2^{\pi_0} \rightarrow 1/2^{\pi} \rightarrow 1/2^{\pi_0}$ NRF cascade has always an isotropic angular distribution, thus $W(\theta,\phi)_{1/2\rightarrow 1/2\rightarrow 1/2} \equiv 1$. In the following we want to discuss the odd-mass nucleus ^{207}Pb with a ground-state spin $J_0^{\pi_0} = 1/2$. An excited spin 1/2 state of the odd-mass nucleus ^{207}Pb therefore radiates isotropically and hence its angular distribution ratio at any two angles is 1. The intensity distribution function of a $1/2\ \ \rightarrow 3/2^{\pi} \rightarrow 1/2$ photon scattering cascades using a polarized incident beam is given by

$$W(\theta,\phi) = 1 + \frac{1}{4}[P_2(\cos\theta)\ \ \frac{1}{2}\ \ \cos(2\phi)P_2^{(2)}(\cos\theta)], \quad (6)$$

with $P_2^{(2)}$ being the unnormalized associated Legendre polynomial of second order and being the parity quantum number of the excited state. For an excitation with positive parity, an E1 excitation, a maximum of NRF intensity shows up perpendicular to the polarization plane of the horizontally polarized photon beam. For an excitation with negative parity (M1), a maximum of intensity occurs in the polarization plane.

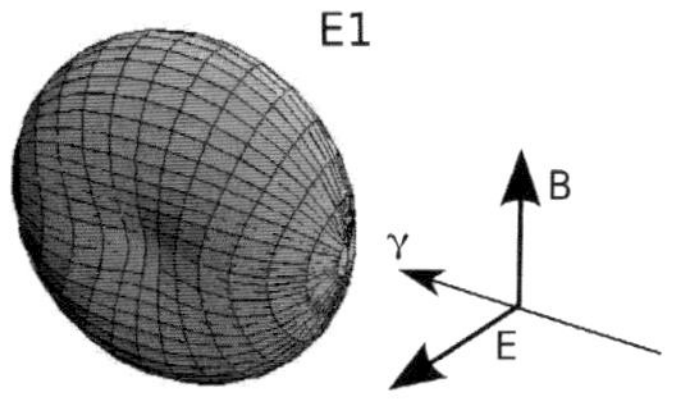

Figure 2. Angular distribution of the scattered photons in case of an E1 excitation in odd-mass nuclei for a horizontally polarized photon beam. The maxima of the intensity distribution occur at $\phi = 0$ and $\phi = 180$.

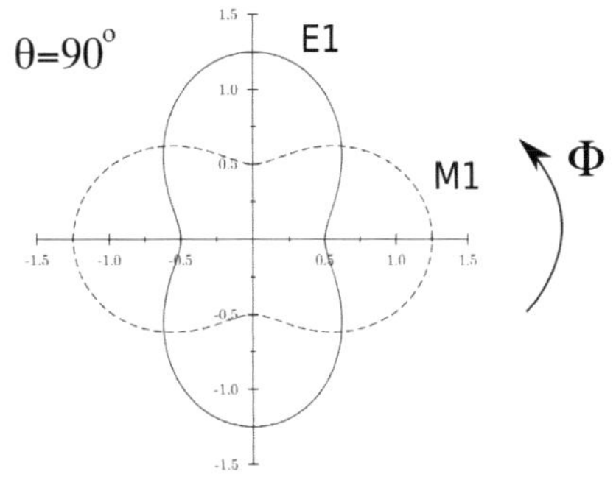

Figure 3. Angular distribution of the scattered photons at $\theta =$ 90 . Depending on the character of the excitation, one obtains an azimuthal NRF asymmetry of 0.4.

Fig. 2 shows the angular distribution for $1/2\ \ \rightarrow 3/2^+ \rightarrow 1/2$ NRF cascade in an odd mass nucleus in case of a horizontally polarized photon beam. The analyzing power is maximum at a polar angle $\theta = 90$. The maxima of NRF intensity occur for $\phi = 0$ and $\phi = 180$. In case of an E1 transition, the angular distribution is shown by the continuous curve, an M1 transition is shown by the dashed curve. Using a linearly polarized -ray beam for NRF studies of an odd-mass nucleus and placing detectors at the above-mentioned angles should make spin and parity assignments possible. The two states of ^{207}Pb observed at 5.490 and 5.597 MeV are of special interest. The assignment of their spin and parity quantum numbers represents a novel use of a linearly polarized photon-beam and, hence, it is discussed in more detail below.

Considering dipole excitations of the $J_0^{\pi} = 1/2$ ground state of ^{207}Pb, the spins and parities of the two excited states could be $J^{\pi}=1/2^{\pm}$, $3/2^{\pm}$ only. The corresponding dipole excitations of ^{207}Pb were previously investigated, using unpolarized bremsstrahlung at the High Intensity Photon Setup (HIPS) at S-DALINAC [7].

The measurements to assign the spin and parity quantum numbers in the 5.490 and 5.597 MeV states in ^{207}Pb have been performed at the High Intensity Gamma Ray Source (HI S) at Duke

Free Electron Laser Laboratory (DFELL) using its nearly monoenergetic linearly polarized -ray beams. The HI S [9] is based on the Duke/OK-4 Storage Ring FEL. Head-on collisions of the FEL photons and the relativistic electrons in the storage ring generate -rays in the laboratory system by the Compton e ect. The wavelength of the FEL photons and the energy of the electron beam in the ring are tunable over a large range of values, which allows the production of rays at a corresponding range of energies. An on-axis collimator selects a narrow cone of linearly polarized, nearly monoenergetic rays from 180 Compton back-scattering processes. That beam is tunable in energy in the MeV range and is available for nuclear resonance uorescence reactions on target nuclei placed in the -ray beam. Fluorescence -quanta result from the decay of resonantly photo-excited nuclear levels and can be observed with high energy resolution using large volume HPGe semiconductor detectors. Details of the set-up and the method have been published previously in Refs. [9, 10]. The OK-4 FEL was tuned to lase at wavelengths around 450 nm. Mean energies of the resulting -ray beams were E_γ=5.5 and 5.6 MeV with a width of about 200 keV. The collimator size was 2.5 cm and the energy spread of the -beam was around 100 keV. The -beam intensity on target was of the order of 10^7 per second. Data was taken for about 6 hours. The target consisted of a 5.65-g/cm^2-thick natural Pb sample with 25.4 mm diameter. (total mass 28 g). It was surrounded by four coaxial HPGe detectors with efficiencies of about 60% relative to a standard 3"x3" NaI detector at 1.3 MeV and they were mounted at a mean polar angle of $\theta = 90$ and at azimuthal angles of $\phi = 0$, 90 , 180 , 270 with respect to the (horizontal) polarization plane of the incident beam.

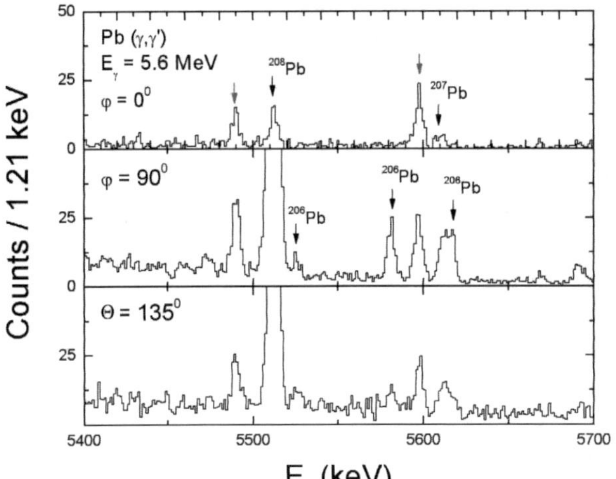

Figure 4. Spectra of the Pb($\vec{}$, $'$) reaction at a mean polar angle $\theta = 90$ relative to the incident photon beam and azimuthal angles $\phi_{\parallel} = (0\ )(top)$ and $\phi_{\perp} = (90\ )$ (middle) relative to the polarization plane of the 5.6±1 MeV -ray beam at HI S. Big arrows (red) mark the two ground state transitions of ^{207}Pb at 5.490 and 5.597 MeV. The di erence of the intensity ratios of these two -rays is clearly visible. Other observed ground state decays are indicated by smaller arrows with the corresponding Pb isotopes labeled on top. The right most peak in the lower panel is a doublet caused by the 5.611 and 5.616 MeV lines from ^{207}Pb and ^{207}Pb, respectively. The spectrum at the bottom shows the mentioned states under $\theta = 135$ and $\phi = 0$

Fig. 4 displays (gain-matched and added) photon scattering spectra at azimuthal angles $\phi_{\parallel} = (0\ , 180\ )$ in the polarization plane of the photon beam (top) and perpendicular to it

Table 1. Excitation energy E_x, spin and parity quantum numbers J^π from this work and from Ref. [7,11] of excited states of ^{206,207,208}Pb. Values for E_x are taken from Ref. [7] unless otherwise noted.

	E_x (MeV)	J^π This work	$J^{\pi 0}$ (h)
^{206}Pb	5.5251(3)	1	$1^{\pm}$
	5.5811(3)	1	1
	5.6161(3)	1	$1^{(\ )}$
^{207}Pb	5.4897(3)	$\frac{3}{2}^+$	$\frac{1}{2}\ , \frac{3}{2}^{\pm}$
	5.5974(3)	$\frac{1}{2}^+$	$\frac{1}{2}^{\pm}, \frac{3}{2}^{\pm}$
	5.611(2)	$\frac{3}{2}^+$	$(\frac{1}{2}, \frac{3}{2})$
^{208}Pb	5.5121(3)	1	1

$\phi_\perp$ = (90 ,270) (middle) using a linearly polarized incident photon beam with 5.6(1) MeV energy. The peaks at 5.490, 5.512, 5.597 and 5.611 MeV in the spectra at $\phi_\parallel$ and $\phi_\perp$ correspond to ground state decay transitions of excited states at these energies in ^{207,208}Pb. They are indicated by red arrows in the $\phi = 0$ spectrum. In the spectrum at $\phi_\perp$, the peaks at 5.525, 5.580 and 5.617 MeV represent ground state decay transitions of excited states at these energies in ^{206}Pb. These peaks are invisible in the $\phi_\parallel$ spectrum, which proves their $E1$ character. To identify the spin and parity quantum numbers of the two states in ^{207}Pb, we form the double ratio (DR) of NRF intensities observed within the polarization plane ($\phi_\parallel$) and perpendicular to it ($\phi_\perp$). Neglecting small di erences in the energy dependence of the detector efficiencies at these two nearby energies, the experimental DR is proportional to the DR of the angular distribution for one of the four possible scattering cascades $\frac{1}{2}\ \rightarrow \frac{1}{2}^{\pm} \rightarrow \frac{1}{2}$ and $\frac{1}{2}\ \rightarrow \frac{3}{2}^{\pm} \rightarrow \frac{1}{2}$. The theoretical DR is de ned as

$$\mathrm{DR}_{J_1 J_2} = \frac{W_{J_1}(\phi_\parallel)}{W_{J_1}(\phi_\perp)} : \frac{W_{J_2}(\phi_\parallel)}{W_{J_2}(\phi_\perp)}, \tag{7}$$

where $W_J(\phi)$ for any $\frac{1}{2}\ \rightarrow \frac{3}{2}^{\pm} \rightarrow \frac{1}{2}$ photon scattering cascades are given by $W(90\ , \phi)$ in Eq. (6). Looking at the angular distributions at Figure 3, the maxima of the intensity for an E1 transition lies at 1.25 at $\phi = 0$ and $\phi = 180$. The relative intensity at $\phi = 90$ and $\phi = 270$, the minimum of intensity, amounts to 0.5. The DR for an E1 transition amounts to DR=0.4.

Table 2. Given is a total of 16 possible spin and parity combinations for the two states of ^{207}Pb, which correspond to di erent values of the double ratio (DR). Combinations restricted to $E1$ excitations of the two states are speci ed on the right.

DR	(J_1^π, J_2^π)	for $E1$ only
0.16	$(\frac{3}{2}^+, \frac{3}{2}\ \ )$	–
0.40	$(\frac{3}{2}^+, \frac{1}{2}^{\pm}\ ), (\frac{1}{2}^{\pm}, \frac{3}{2}\ \ )$	$(\frac{3}{2}^+, \frac{1}{2}^+\ )$
1.00	$(\frac{3}{2}^+, \frac{3}{2}^+\ ), (\frac{3}{2}\ \ , \frac{3}{2}\ \ ), (\frac{1}{2}^{\pm}, \frac{1}{2}^{\pm})$	$(\frac{3}{2}^+, \frac{3}{2}^+\ ), (\frac{1}{2}^+, \frac{1}{2}^+\ )$
2.50	$(\frac{3}{2}\ \ , \frac{1}{2}^{\pm}\ ), (\frac{1}{2}^{\pm}, \frac{3}{2}^+\ )$	$(\frac{1}{2}^+, \frac{3}{2}^+\ )$
6.25	$(\frac{3}{2}\ \ , \frac{3}{2}^+\ )$	–

Table 2 lists all possible DR values and the corresponding spin and parity combinations of the two states. Experimentally one nds

$$\mathrm{DR}_{\mathrm{expt}} = \frac{I_{5.490}(\phi_{\|})}{I_{5.490}(\phi_{\perp})} : \frac{I_{5.597}(\phi_{\|})}{I_{5.597}(\phi_{\perp})} = 0.39(4)\ .$$

Therefore, the quantum numbers $(J^{\pi}_{5.490}, J^{\pi}_{5.597})$ are restricted to the combinations $(3/2^{+}, 1/2^{\pm})$ and $(1/2^{\pm}, 3/2\ \ )$, only.

From the transition rates of the two states at 5.490 and 5.597 MeV, the most likely multipolarity of their ground state decays is E1. Then the measured DR_{expt} and Table 2 leaves $J^{\pi} = \frac{3}{2}^{+}$ for the state at 5.490 MeV and $\frac{1}{2}^{+}$ for the state at 5.597 MeV.

In order to verify our conjecture of $E1$ radiation character, we nally deduce the angular distribution perpendicular and parallel to the polarization plane for both transitions individually. With (6) one obtains

$$\frac{W(90\ \ ,\phi_{\perp})}{W(90\ \ ,\phi_{\|})} = \begin{cases} 0.4 & J^{\pi} = \frac{3}{2} \\ 1 \quad \text{for} & J^{\pi} = \frac{1}{2}^{\pm} \\ 2.5 & J^{\pi} = \frac{3}{2}^{+} \end{cases} \ . \tag{8}$$

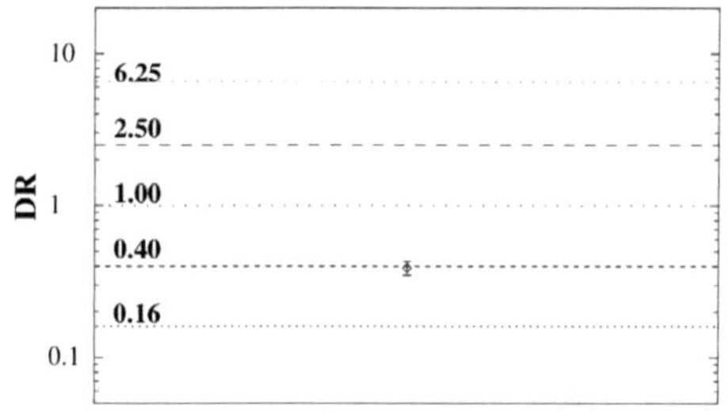

Figure 5. Double ratio (DR) for the two states of ^{207}Pb at 5.490 and 5.597 MeV on a logarithmic scale. Possible values of the DR are labeled on the horizontal lines in the gure. From the corresponding spin and parity combinations $(J^{\pi}_{5.490}, J^{\pi}_{5.597})$ for each DR, listed in Table 2, the measured DR=0.39(4) indicates that $(J^{\pi}_{5.490}, J^{\pi}_{5.597})$ must be $(3/2^{+}, 1/2^{+})$ if $E1$ radiation is considered, only.

Figure 6 displays the azimuthal intensity distribution ratios for the -ray transitions. The values are 2.35(66) and 0.92(25) for the states at 5.490 and 5.597 MeV, respectively. Comparing with (8) the data con rm the $(J^{\pi}_{5.490}, J^{\pi}_{5.597}) = (\frac{3}{2}^{+}, \frac{1}{2}^{\pm})$ assignment to better than two standard deviations. The isotropy of the angular distribution for spin-1/2 states prevents us from making parity assignments to these cases. For clarity we, therefore stress, that one out of these four quantum number assignments, namely the $= (+)$ assignment for the J=1/2 state at 5.597 MeV, is not based on angular distribution of photon scattering intensity but on the large decay rate of the level which leaves $E1$ multipolarity as the only likely possibility. These are the assignment that we adopt in Table 1. The level energies agree within 22 keV and 85 keV, respectively, with the excitation energy of the predominant bound $E1$ excitation of the neighboring doubly-closed shell nucleus ^{208}Pb at 5.512 MeV. I.e., the energy splitting of the doublet $E_{\nu(p^{-1}_{1/2})}(5.512) = E_{3/2^{+}}\ \ E_{1/2^{+}} =\ \ 107$ keV is less than 2% of the centroid value. This is comparable to the relative splitting of the well-established $3\ \ \otimes \nu(p^{\ 1}_{1/2})$ weakly coupled doublet at 2.6 MeV [12,13]. However, literature values for transition rates indicate more pronounced deviation from weak coupling at 5.5 MeV. Like the 1 state of ^{208}Pb at 5.512 MeV, that doublet of individual quantum states of ^{207}Pb dominates its dipole excitation strength distribution, here, with $E1$ decay rates to the ground state of 18(4) and 29(4) mW.u. [7]. While

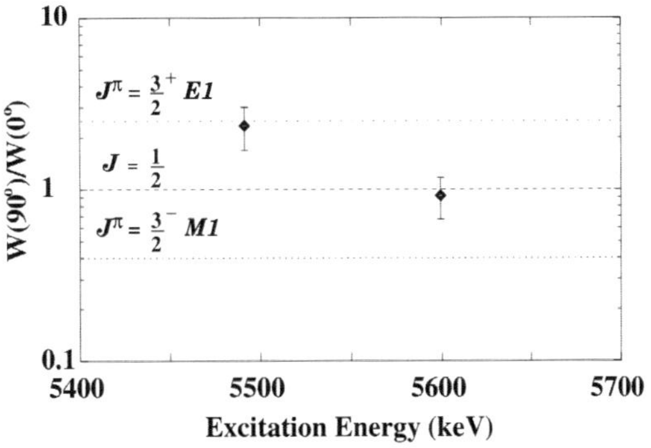

Figure 6. Azimuthal NRF intensity distributions for the two states of ^{207}Pb of 5.490 and 5.597 MeV. The horizontal lines mark the ratios for possible spin quantum numbers 1/2 and 3/2.

these are the largest $E1$ strengths from the ground state to excited states up to 6.2 MeV, which is the energy up to where data on ^{207}Pb are available, these values are a factor of three smaller than in the ^{208}Pb core indicating a substantial deviation from a pure weak coupling.

In summary, we have demonstrated a new method for spin and parity quantum number assignments to highly excited particle-bound dipole excitations of odd-mass nuclei based on the measurement of azimuthal nuclear resonance uorescence-intensity asymmetries about an incident linearly polarized -ray beam. This method has been used to identify doublet of energy levels of ^{207}Pb that predominantly originates in the weak coupling of the $\nu(3p_{1/2}^{1})$ neutron hole to the strongest neutron-bound electric dipole excitation of the doubly closed shell ^{208}Pb core at 5.51 MeV. The observed energy splitting of this $J = 1/2, 3/2$ doublet at an excitation energy of more than $30A^{\ 1/3}$ MeV amounts to less than 2% of its excitation energy with the $J = 3/2$ state being the lower lying level. Deviations from the pure weak-coupling scenario show up in the absolute transition rates due to particle-core interaction.

We thank the sta at the HI S facility at Duke University, especially J. Li, I.V. Pinayev, Y.K. Wu for the support during the experiment. Also we thank T. Ahn, A. Costin, J. Enders, P. von Neumann-Cosel, S. Müller, T. Saito, V. Yu. Ponomarev, and A. Zilges for their contributions. This work was supported by the U.S.-DOE under grant Nos. DE-FG02-04ER41334, DE-FG02-97ER41033, by the German DFG under grant SFB634 and by the DAAD within the German-Bulgarian exchange programme under grant Nos. D/06/05918 and DAAD-09.

References

[1] Pietralla N, Li T C, Fritzsche M, Ahmed M W, Ahn T, Costin A, Enders J, Li J, Muller S, von Neumann-Cosel P, Pinayev I, Ponomarev V Y, Savran D, Tonchev A P, Tornow W, Weller H R, Werner V, Wu Y K and Zilges A 2009 *Phys. Lett. B* **681** 134
[2] Metzger F R 1959 *Progress in Nuclear Physics, published by O. R. Frisch Pergamon, New York* **7**
[3] Berg U E P 1984 *J. Phys.(Paris), Colloq.C4* 359
[4] Kneissl U, Pitz H H and Zilges A 1996 *Prog. Part. Nucl. Phys.* **37** 349
[5] Kneissl U, Pietralla N and Zilges A 2006 *J. Phys. G: Nucl. Part. Phys.* **32** 217
[6] Krane K S, Ste en R M and Wheeler R M 1973 *Nucl. Data Tab.* **11** 351
[7] Enders J *et al.* 2003 *Nucl. Phys. A* **724** 243
[8] Fagg L W and Hanna S S 1959 *Rev. Mod. Phys.* **31** 711
[9] Litvinenko V N *et al.* 1998 *Nucl. Instrum. Methods Phys. Res., Sect. A* **407** 8
[10] Pietralla N, Berant Z *et al.* 2002 *Phys. Rev. Lett.* **88** 12502
[11] Martin M J 1993 *Nucl. Data Sheets* **70** 315
[12] Hausser O, Khanna F C and Ward D 1972 *Nucl. Phys. A* **194** 113
[13] Ungrin J, Diamond R M, Tjom P O and Elbek B 1971 *Mat. Fys. Medd. Dan. Vid. Selsk.* **38** 8

E0 transition strengths from X(5) to the rigid rotor

A Krugmann[1], J Bonnet[1], N Pietralla[1] and R V Jolos[2]

[1] Institut für Kernphysik, Technische Universität Darmstadt, Darmstadt, Germany
[2] Joint Institute for Nuclear Research, Dubna, Russia

E-mail: krugmann@ikp.tu-darmstadt.de

Abstract. Relative and absolute $E0$ transition strengths $[\rho^2(E0)]$ on the transitional path between the X(5) solution and the rigid rotor limit have been evaluated [1] within the framework of the Confined β-Soft (CBS) rotor model. Relative $E0$ transition strengths between the β-vibrational band and the ground state band decrease with increasing angular momentum for a given potential stiffness. The Z-independent quantity $X \propto \rho^2(E0; 0_2^+ \to 0_1^+)/B(E2; 0_2^+ \to 2_1^+)$ has been traced between X(5) and the rigid rotor. It reaches the value $4\beta_M^2$ at the rigid rotor limit, as previously derived by Rasmussen. A new Inter-Band $E0 - E2$ correlation observable $Y \propto \rho^2(E0; 0_2^+ \to 0_1^+)/B(E2; 0_2^+ \to 2_1^+)^2$ has be en proposed, which is independent on the absolute nuclear deformation and solely depends on the nuclear stiffness. Available data for X and Y are in satisfactory agreement with the CBS model.

1. Introduction

As a function of particle number, heavy nuclei can undergo rapid shape phase transitions with respect to their deformation. These shape phase transitions have been studied experimentally and theoretically for many years [2, 3]. This discussion intensified when Iachello proposed analytical solutions of the geometrical Bohr Hamiltonian near the critical points of various nuclear shape phase transitions [4, 5, 6]. The solutions labeled E(5) and X(5) have attracted a great deal of interest and initiated extensive research. Besides the characteristic excitation energy ratios $R_{4/2} = E(4_1^+)/E(2_1^+) = 2.20$ [for E(5)] and $R_{4/2} = 2.90$ [for X(5)] or the evolution of the $E2$ transition rates as a function of spin along the ground state band, the properties of the quadrupole-collective, excited 0^+ states including its $E2$ decays are considered as the identifying signatures of these models. Several experimental studies [7, 8, 9, 10] have dealt with these signatures.

Another prominent decay modes, in particular for 0^+ states, are $E0$ transitions. Brentano *et al.* [11] pointed out that in the interacting boson model the $E0$ transition rates to the ground state increase at the shape phase transitional point and continue to have larger values up to the SU(3) or the O(6) dynamical symmetries.

$E0$ transition strengths have not yet been reported for the X(5) solution. We have recently studied [1] $E0$ transition rates in the Confined $\beta-$Soft (CBS) rotor model [12]. This model interpolates between the X(5) solution and the rigid rotor limit as a function of one structural parameter. Using the CBS rotor model we have obtained relative $E0$ transition rates between bands with one, two or three nodes in the β wave function.

2. CBS Rotor Model

Recently, we have studied [1] the evolution of $E0$ transition strengths for axially symmetric quadrupole deformation from Iachello's X(5) solution [5] towards the rigid rotor limit within the Confined β-Soft rotor model. The CBS rotor model represents an approximate analytical solution to the Bohr Hamiltonian [13]

$$H = -\frac{\hbar^2}{2B}\left[\frac{1}{\beta^4}\frac{\partial}{\partial\beta}\beta^4\frac{\partial}{\partial\beta} + \frac{1}{\beta^2\sin 3\gamma}\frac{\partial}{\partial\gamma}\sin 3\gamma\frac{\partial}{\partial\gamma} - \frac{1}{4\beta^2}\sum_k \frac{Q_k^2}{\sin^2(\gamma - \frac{2}{3}\pi k)}\right] + V(\beta,\gamma) \tag{1}$$

in the quadrupole shape parameters β and γ. Assuming a separable potential $V(\beta,\gamma) = u(\beta) + v(\gamma)$ the wave equation approximately separates into $\Psi(\phi,\theta,\psi,\beta,\gamma) = \xi_L(\beta)\,\eta_K(\gamma)\,D^L_{M,K}(\theta_i)$, where $D^L_{M,K}$ denotes the Wigner functions with θ_i being the Euler angles for the orientation of the intrinsic system. η_K denotes the appropriate wave function in γ. For sufficiently axially symmetric prolate nuclei one might consider a steep harmonic oscillator in γ [5]. $\xi_L(\beta)$ describes the part of the wave function depending on the deformation variable β. The CBS rotor model assumes for prolate axially symmetric nuclei an infinite square well potential $u(\beta)$, with boundaries at $\beta_M > \beta_m > 0$. For this potential the Schrödinger equation is analytically solvable. The ratio $r_\beta = \beta_m/\beta_M$ parameterizes the width of this potential, that is the stiffness of the nucleus in the β degree of freedom. For $r_\beta = 0$ the X(5) limit is obtained with large fluctuations in β. The rigid rotor limit without fluctuations in β corresponds to $r_\beta \to 1$. The full solution of the Bohr Hamiltonian can be analytically written and calculated in terms of Bessel functions of first and second kind.

The CBS rotor model well describes the evolution of low-energy 0^+ bands [12], ground bands of strongly deformed nuclei [14], and the dependence of relative moments of inertia as a function of spin in deformed transitional nuclei [15].

3. E0 Transition Rates

Electromagnetic transition rates can be calculated from the wave functions given above. For $E0$ transitions we obtain with Eqs. (3,4):

$$\rho^2_{\rm if}(E0) \;=\; \left(\frac{3Z}{4\pi}\right)^2 |\langle\psi_{\rm f}|\hat\beta^2|\psi_{\rm i}\rangle|^2 \tag{2}$$

between states that only differ in the β-dependent part of the wave function, *i.e.*, $\langle D_{\rm f}|D_{\rm i}\rangle = \langle\eta_{\rm f}|\eta_{\rm i}\rangle = 1$. As a function of the choice of the potential, *i.e.*, as a function of the nuclear deformation (β_M) and the nuclear stiffness against centrifugal stretching (r_β), we calculated the $E0$ transition rates [1].

4. E0 transitions in quadrupole collective models

The $E0$ transition strength is defined by

$$\rho^2_{\rm if}(E0) = \frac{|\langle\Psi_{\rm final}|\hat T(E0)|\Psi_{\rm initial}\rangle|^2}{(eR^2)^2} \tag{3}$$

with the $E0$ transition operator to lowest order in the deformation parameter

$$\hat T(E0)_{tr} = \frac{3}{4\pi}ZeR^2\,\hat\beta^2 \tag{4}$$

Data on $E0$ transitions have been reviewed by Wood *et al.* [16]. The $E0$ transition operator is similar in structure to the β-dependent part of the $E2$ transition operator

$$\hat T(E2)_{\Delta K=0} = \frac{3}{4\pi}ZeR^2\hat\beta\,, \tag{5}$$

again in the axially symmetric case with $\gamma = 0$.

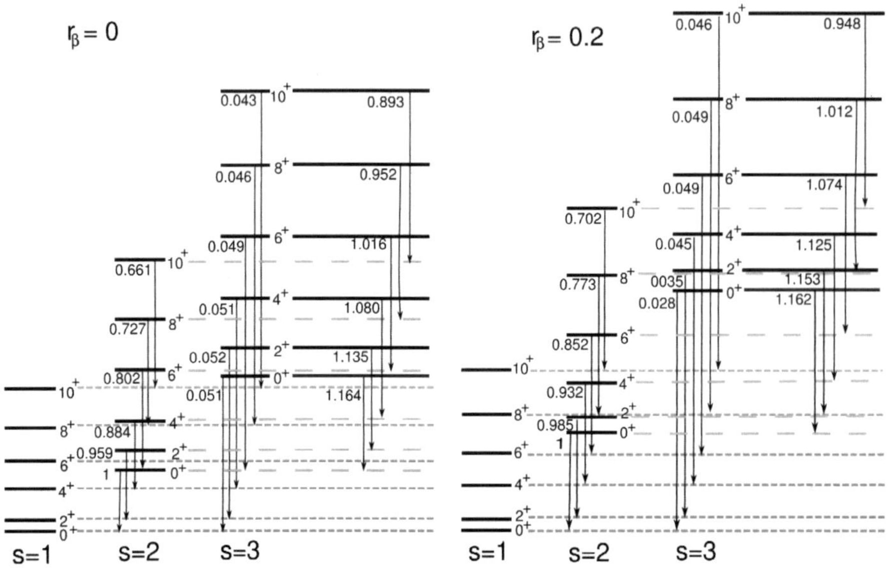

Figure 1. Partial level schemes for low-energy $K = 0$ bands obtained with the CBS rotor model with relative $E0$ transition strengths $\rho^2(E0, J_i \rightarrow J_f)/\rho^2(E0, 0_2^+ \rightarrow 0_1^+)$
left: for X(5) [r_β=0], $R_{4/2} = 2.90$
right: for r_β=0.2, $R_{4/2} = 3.10$

4.1. Stiffness-dependence of relative E0 transition strengths
First we derive the $E0$ transition strengths in the X(5) solution, *i.e.*, for the structural parameter $r_\beta = 0$. The width of the potential well, β_M, is considered as the sole free parameter and thus we normalize the $E0$ transition rates relative to the $0^+_{s=2} \rightarrow 0^+_{s=1}$ $E0$ transition. The left hand side of Fig. 1 shows these relative $E0$ transition rates for $r_\beta = 0$, which is X(5). Some of the values are also tabulated in Table 1. We note that the $\rho^2(E0)$ values are comparable in size for all those $E0$ transitions that have $\Delta s = 1$, while $\Delta s > 1$ $E0$ transitions are suppressed by an order of magnitude. The right hand side of Fig. 1 shows the relative $E0$ transition rates normalized again to the $0^+_{s=2} \rightarrow 0^+_{s=1}$ transition for the CBS model parameter $r_\beta = 0.2$, which corresponds to an intermediate situation with an excitation energy ratio $R_{4/2} = 3.10$. Again, we observe an approximate selection rule for $E0$ transitions between bands with $\Delta s > 1$.

4.2. Angular momentum dependence of relative E0 transition strengths
Next we compare relative $E0$ transition strengths for different width parameters r_β and different angular momenta J for transitions between states with $s = 2$ (β-band) and $s = 1$ (ground state band). The results are shown in Fig. 2. Here, the $E0$ transitions are again normalized to the $0^+_{s=2} \rightarrow 0^+_{s=1}$ transition for every width parameter r_β, separately. For increasing angular momentum, the model predicts a decreasing $E0$ transition rate. For X(5) the $10^+_{s=2} \rightarrow 10^+_{s=1}$ $E0$ transition rate is reduced by 38% relative to the corresponding $0^+_{s=2} \rightarrow 0^+_{s=1}$ transition. Notice also, that for rising potential stiffness, the relative $E0$ transition strengths rapidly approaches a constant value as a function of spin, which is typical for the rigid rotor limit.

4.3. Evolution of absolute E0 rates
Next we focus on absolute transition strengths. Fig. 3 shows $E0$ transition strengths for different angular momenta depending on the width parameter r_β. All transition strengths are normalized

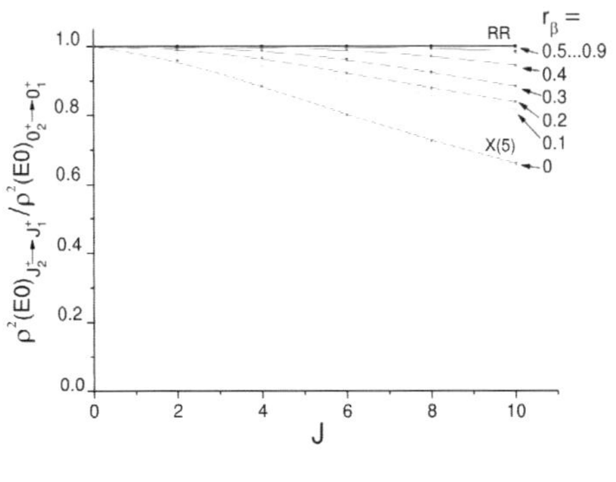

Figure 2. (Color online) Relative $E0$ transition strengths for various nuclear stiffnesses with r_β ranging from 0 to 0.9 in steps of 0.1.

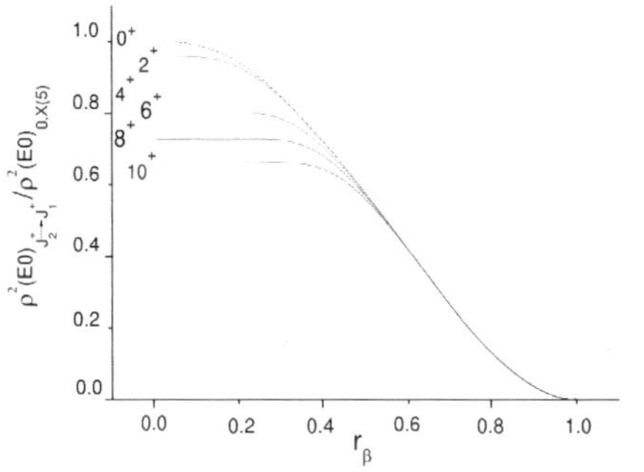

Figure 3. (Color online) $E0$ transition rates from s=2 to s=1 between states with angular momenta $J = 0 - 10$ as a function of r_β. The theoretical values are normalized to the $0_2^+ \to 0_1^+$ $E0$ transition strength in X(5), $\rho^2_{0,X(5)} \equiv \rho^2(E0, 0_2^+ \to 0_1^+; r_\beta = 0)$.

Table 1. $E0$ transition strengths in convenient unitsfor r_β = 0, 0.2 and 0.4.

transition	$10^3\,\frac{\rho^2(E0)}{Z^2\beta_M^4}$								
	$s=2 \to s=1$			$s=3 \to s=1$			$s=3 \to s=2$		
$J_f \to J_i$	X(5)	$r_\beta = 0.2$	$r_\beta = 0.4$	X(5)	$r_\beta = 0.2$	$r_\beta = 0.4$	X(5)	$r_\beta = 0.2$	$r_\beta = 0.4$
0^+	1.81	1.70	1.31	0.092	0.048	0.013	2.10	1.98	1.52
2^+	1.73	1.68	1.30	0.093	0.059	0.015	2.05	1.96	1.52
4^+	1.60	1.59	1.30	0.092	0.076	0.019	1.95	1.92	1.52
6^+	1.45	1.45	1.28	0.088	0.084	0.027	1.84	1.83	1.51
8^+	1.31	1.31	1.23	0.083	0.083	0.037	1.72	1.72	1.49
10^+	1.19	1.20	1.16	0.078	0.078	0.047	1.61	1.61	1.46

to the transition $0_2^+ \to 0_1^+$ for X(5), *i.e.*, we plot the quantity $\frac{\rho^2(E0, J_f \to J_i; r_\beta)}{\rho^2(E0, 0_2^+ \to 0_1^+; r_\beta = 0)}$ for a fixed deformation β_M. For r_β approaching unity all $E0$ transition strengths drop to zero. The evolution of this decrease as a function of r_β depends on angular momentum as it was discussed above. $E0$ transition strengths from 0^+ states monotonically decrease with increasing r_β. $E0$ transition strengths between states with larger angular momenta are significantly effected only above a certain value of r_β, which depends on angular momentum. This behavior originates in the phenomenon of centrifugal stretching for states at higher spins, which causes for higher rotational frequencies a lower sensitivity to the potential at small values of β [12].

For higher stiffness ($r_\beta \gtrsim 0.5$) the decrease of $E0$ transition rates as a function of r_β is entirely dictated by the narrowing of the square-well potential. Independent on low enough spin, the $E0$ transition rates decrease almost linearly with increasing r_β. Table 1 provides an overview over the values of $E0$ transition strengths for the width parameters of r_β = 0, 0.2 and 0.4. In order to present some global values, the $E0$ transition strengths in Table 1 are divided by the appropriate powers of the nuclear charge Z and the scale of its deformation, β_M. A comparison of $E0$ transition strengths, predicted by the CBS rotor model, to experimental values is given

Journal of Physics: Conference Series **205** (2010) 012037 doi:10.1088/1742-6596/205/1/012037

Table 2. Comparison of experimental values for $\rho^2(E0)$ for ^{152}Sm [17] with the CBS rotor model with stiffness parameter $r_\beta = 0.14$. The second line contains absolute predictions (see text). The values given in the 3rd line are normalized to the $E0$ strength of the $0_2^+ \to 0_1^+$ transition in ^{152}Sm.

transition	$0_2^+ \to 0_1^+$	$0_3^+ \to 0_1^+$	$0_3^+ \to 0_2^+$
$\rho^2(E0)$ experiment	0.058(6)	0.0007(4)	0.022(9)
$\rho^2(E0)$ CBS from $E2$	0.3041	0.0114	0.3534
$\rho^2(E0)$ CBS renormalized	0.058	0.0022	0.0674

in Table 2. Data on ^{152}Sm where the strengths of the $E0$ transition rates between the 0_1^+, 0_2^+ and the 0_3^+ states are known are used for example and were taken from Ref. [17]. The CBS values were calculated using a stiffness parameter $r_\beta = 0.14$ obtained in Ref. [12] for a good reproduction of the level scheme of ^{152}Sm.

Reproduction of the $E2$ excitation strength $\mathrm{B}(E2, 0_1^+ \to 2_1^+) = 3.47 \pm 7\, e^2\mathrm{b}^2$ results in a width parameter $\beta_M = 0.46$ and an average deformation of $\langle\beta\rangle = 0.29$. This parameter-free absolute prediction of the $0_2^+ \to 0_1^+$ $E0$ transition strength for ^{152}Sm coincides with the data within a factor of about five when no effective charges are considered.

Data and theory, furthermore, agree on the approximate selection rule $\Delta s = 1$, which suppresses the $0_3^+ \to 0_1^+$ $E0$ transition by one order of magnitude. Normalizing the theory to the $0_2^+ \to 0_1^+$ transition in the spirit of using an effective $E0$ charge yields a good description for the $E0$ decay rates of the 0_3^+ state. This is quite satisfactory.

5. Inter-band $E2$-$E0$-correlation observables

Finally we discuss observables that relate $E0$ and $E2$ transition strengths between the β-band and the ground state band. The Z-independent quantity X was introduced in the 1960s by Rasmussen [18]. He defined the ratio of the $E0$ transition strength $\rho^2(E0, 0_2^+ \to 0_1^+)$ and $B(E2, 0_2^+ \to 2_1^+)$ value between the β-band and the ground state band

$$X_{\mathrm{Rasmussen}} = \frac{\rho^2(0_2^+ \to 0_1^+)\, e^2 R^4}{B(E2, 0_2^+ \to 2_1^+)} = 4\beta_{RR}^2 \tag{6}$$

within the rigid rotor model, where β_{RR} denotes the deformation of the nucleus in the rigid rotor limit. This result has lead to a wide-spread misconception about the evolution of $E0$ transition rates in deformed nuclei. Eq. (6) shows the increase of the $E0/E2$-ratio between ground state band and β-band as a function of the nuclear deformation. It does *not* express an increase of the $E0$ transition strength to the β-band for a sequence of nuclei approaching the Rigid Rotor Limit, *i.e.*, with increasing $R_{4/2}$ value ($\to 3.33$). In contrast, our previous results (see Fig. 3 and Table 1) show that the $E0$ strengths decrease with increasing stiffness of the nuclear quadrupole deformation. *I.e.*, for a typical deformation $\beta \lesssim 0.5$ we must expect the following consequence: the better the nucleus resembles the Rigid Rotor ($R_{4/2} \approx 3.33$), the larger is its stiffness, the smaller are the $E0$ and $E2$ transitions between the β-band and the ground state band.

In the CBS model we can trace the evolution of X as a function of nuclear stiffness. Therefore we consider the quantity

$$X_{\text{CBS}} = \frac{\rho^2(0_2^+ \to 0_1^+)\, e^2 R^4}{B(E2, 0_2^+ \to 2_1^+)} \quad (7)$$

$$= \frac{\langle \Psi_{0_1^+} |\, \mathfrak{b}^2 \,| \Psi_{0_2^+} \rangle^2 \, \beta_M^2}{\langle \Psi_{2_1^+} |\, \mathfrak{b} \,| \Psi_{0_2^+} \rangle^2} \quad (8)$$

with $\mathfrak{b} = \frac{\beta}{\beta_M}$.

As plotted in Fig. 4, the ratio X converges to $4\beta_M^2$ in agreement with the limit obtained by Rasmussen in the rigid rotor model. Data on X exist for the nuclides ^{152}Sm and ^{156}Gd. They are given in Table 2 and included in Fig. 4. The predicted increase of X as a function of nuclear stiffness is obvious. However, the CBS overpredicts the data by a factor of about two. The ratio X/β_M^2 includes the scaling factor β_M^2 and therefore is not determined completely by data. It is costumary to extract β_M^2 from the $B(E2, 2_1^+ \to 0_1^+)$ in a consistent, albeit model-dependend way. The deviations between theory and data for X/β_M^2 seen in Fig. 4, may then be attributed to the usage of both, data on inter-band (X) and intra-band ($\beta_M^2 \propto B(E2, 2_1^+ \to 0_1^+)$) transitions.

In order to fully avoid this dependence on the scaling factor β_M and, thus, any model-dependence, we define the quantity

$$Y = \frac{\rho^2(0_2^+ \to 0_1^+)\,(e^2 R^4 Z)^2\,(\frac{3}{4\pi})^2}{B(E2, 0_2^+ \to 2_1^+)^2}$$

$$= \frac{\langle \Psi_{0_1^+} |\, \mathfrak{b}^2 \,| \Psi_{0_2^+} \rangle^2}{\langle \Psi_{2_1^+} |\, \mathfrak{b} \,| \Psi_{0_2^+} \rangle^4} \quad (9)$$

as the ratio between two appropriate powers of inter-band transition strengths. This observable theoretically depends solely on the stiffness parameter r_β. Due to the monotonic relation between r_β and the $R_{4/2}$ ratio [12], the CBS model uniquely relates the observable Y to the $R_{4/2}$ ratio, which itself is a fundamental, easily obtainable characteristical observable. Figure 5 shows plots of Y as a function of r_β (left) and $R_{4/2}$ (right). Y increases monotonically with the stiffness parameter r_β and with the $R_{4/2}$ ratio, too. Y diverges in the rigid rotor limit. Experimental data points for ^{152}Sm and ^{156}Gd have been included in figures 4 and 5 for example. They have been obtained by evaluating the quantities X and Y as a function of the easily obtainable branching ratio Γ_{E0}/Γ_{E2} for the decays of the 0_2^+ state to the 0_1^+ and 2_1^+ members of the ground state band and the lifetime $\tau_{0_2^+}$ of the 0_2^+ state.

6. Summary

The analytical wave functions in the deformation coordinate β of the Confined β-Soft rotor model have been used to evaluate $E0$ transition strengths between low-energy $K = 0$ bands in axially symmetric nuclei [1]. As a function of the only structural parameter within the CBS rotor model, *i.e.*, r_β, which interpolates from X(5) to the rigid rotor, absolute $E0$-transitions strengths decrease with increasing r_β. The relative $E0$ strengths for a given value of the structural parameter r_β decrease with increasing angular momentum. The Z-independent quantity $X \propto \rho^2(E0; 0_2^+ \to 0_1^+)/B(E2; 0_2^+ \to 2_1^+)$ has been traced between X(5) and the rigid rotor. It reaches the value $4\beta_M^2$ at the rigid rotor Limit, as previously derived by Rasmussen. A new observable $Y \propto \rho^2(E0; 0_2^+ \to 0_1^+)/B(E2; 0_2^+ \to 2_1^+)^2$ has been proposed, which is independent on the absolute nuclear deformation and solely depends on the nuclear stiffness. Available data are in satisfactory agreement with the CBS model.

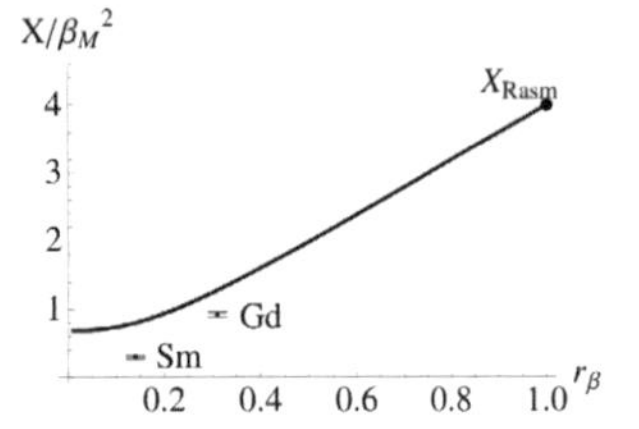

Figure 4. Ratio X_{CBS}/β_M^2 plotted as a function of the stiffness parameter r_β. A value of $X/\beta_M^2 = 4$ has been derived by Rasmussen [18] in the rigid rotor limit. The data points show the experimental values for ^{152}Sm [17] ($r_\beta = 0.14$) and ^{156}Gd [19, 20]($r_\beta = 0.31$).

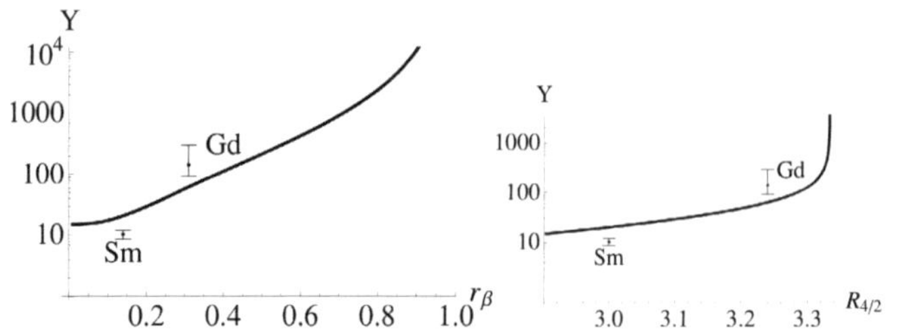

Figure 5. Ratio Y over r_β and $R_{4/2}$. The data points represent the experimental values of Y for $^{152}\mathrm{Sm}(R_{4/2} \approx 3.00)$ [17] and for ^{156}Gd ($R_{4/2} \approx 3.23$) [19, 20].

Acknowledgments

We thank F. Iachello, K. Heyde, P. von Brentano, G. Rainovski, O. Möller and V. Werner for stimulating discussions. Support by the DFG under Grant No. SFB 634, by the Helmholtz International Center for FAIR and by the DAAD within the German-Bulgarian exchange programme under Grant Nos. D/06/05918 and DAAD-09 is gratefully acknowledged.

References

[1] Bonnet J, Krugmann A, Beller J, Pietralla N and Jolos R V 2009 *Phys. Rev. C* **79** 034307
[2] Bohr A and Mottelson B 1975 *Nuclear Structure* vol 2 (Reading: W. A. Benjamin)
[3] Casten R F 2000 *Nuclear Structure from a Simple Perspective* second edition ed (Oxford: Oxford Univ. Pr.)
[4] Iachello F 2000 *Phys. Rev. Lett.* **85** 3580
[5] Iachello F 2001 *Phys. Rev. Lett.* **87** 052502
[6] Iachello F 2003 *Phys. Rev. Lett.* **91** 132502
[7] Clark R M *et al.* 2003 *Phys. Rev. C* **68** 037301
[8] Clark R M *et al.* 2004 *Phys. Rev. C* **69** 064322
[9] Casten R F and Zamfir N V 2001 *Phys. Rev. Lett.* **87** 052503
[10] McCutchan E A and Zamfir N V 2005 *Phys. Rev. C* **71** 034309
[11] von Brentano P, Werner V, Casten R F, Scholl C, McCutchan E A, Krücken R and Jolie J 2004 *Phys. Rev. Lett.* **93** 152502
[12] Pietralla N and Gorbachenko O M 2004 *Phys. Rev. C* **70** 011304
[13] Bohr A 1952 *Mat. Fys. Medd. K. Dan. Vid. Selsk.* **26**
[14] Dusling K and Pietralla N 2005 *Phys. Rev. C* **72** 011303
[15] Dusling K *et al.* 2006 *Phys. Rev. C* **73** 014317
[16] Wood J L, Zganjar E F, Coster C D and Heyde K 1999 *Nucl. Phys.* 323
[17] Artna-Cohen A 1996 *Nuclear Data Sheets* **79** 1
[18] Rasmussen J O 1960 *Nucl. Phys.* **19** 85
[19] Reich C W 2003 *Nuclear Data Sheets* **99** 753
[20] Klora J *et al.* 1993 *Nucl. Phys.* 1

Journal of Physics: Conference Series **205** (2010) 012038 doi:10.1088/1742-6596/205/1/012038

Spectroscopic studies with the PRISMA-CLARA set-up

E Fioretto[1], D Bazzacco[2], S Beghini[2], L Corradi[1], G de Angelis[1], E Farnea[2], A Gadea[3], S M Lenzi[2], S Lunardi[2], P Mason[2], D Mengoni[2], G Montagnoli[2], D R Napoli[1], E Sahin[1], F Scarlassara[2], R Silvestri[1], A M Stefanini[1], C A Ur[2], J J Valiente-Dobón[1], G Pollarolo[4], S Szilner[5], N Marginean[6] and the PRISMA and CLARA collaborations

[1] INFN - Laboratori Nazionali di Legnaro, Viale dell'Università 2, Legnaro (PD), I-35020, Italy
[2] Dipartimento di Fisica dell'Università di Padova and INFN, Via Marzolo 8, Padova, I-35131, Italy
[3] Instituto de Fisica Corpuscolar, CSIC-Universidad de Valencia, Valencia, E-46071, Spain
[4] Dipartimento di Fisica Teorica dell'Università di Torino and INFN, Via P. Giuria 1, Torino, I-10125, Italy
[5] Ruđer Bošković Institute, Bijenička 54, Zagreb, HR-10001, Croatia
[6] Horia Hulubei National Institute of Physics and Nuclear Engineering, Bucharest, R-76900, Romania

E-mail: enrico.fioretto@lnl.infn.it

Abstract. The large solid angle magnetic spectrometer for heavy ions PRISMA, installed at Laboratori Nazionali di Legnaro (LNL), was operated up to the end of March 2008 in conjunction with the highly efficient CLARA set-up. It allowed to carry out nuclear structure and reaction mechanism studies in several mass regions of the nuclide chart. Results obtained in the vicinity of the island of inversion and for the heavy iron and chromium isotopes are presented in this contribution. The status of the new focal plane detectors specifically designed for light ions and slow moving heavy ions is also reported.

1. Introduction

Multinucleon transfer between heavy ions is a suitable mechanism to populate moderately neutron-rich nuclei making feasible the in-beam γ-spectroscopy of these nuclei. This process ranges from the quasi-elastic (i.e. few nucleon transfer and low total kinetic energy loss (TKEL)) to the deep- inelastic (i.e. many nucleon transfer and large TKEL) regime. In thick targets experiments, where one integrates the transfer flux over a broad range of angles and TKEL, γ-γ coincidences are routinely used to build the decay scheme of neutron-rich nuclei, as shown for instance in [1] and in subsequent experiments.

With the PRISMA-CLARA set-up one mainly exploits a regime closer to quasi-elastic processes [2], where a more selective identification of excited states can be obtained via γ-particle coincidences. By properly combining γ-γ coincidences in thick target experiments and γ-particle coincidence measurements performed with magnetic spectrometers (thin target experiments), weak γ transitions belonging to weakly populated transfer channels can be studied [3].

2. The PRISMA-CLARA set-up

PRISMA [4, 5, 6] is a large acceptance magnetic spectrometer designed to be used with heavy-ion beams accelerated at energies up to $E = 10A$ MeV by means of the Tandem/PIAVE-ALPI accelerator complex of LNL. Its optical design consists of a magnetic quadrupole followed by a 60° bending angle magnetic dipole (120 cm curvature radius). The spectrometer is equipped with position-sensitive entrance and focal plane detectors that allow the complete identification and trajectory reconstruction of the reaction products by using the Time of Flight (TOF) and ΔE-E techniques and the position information at the entrance and the focal plane, respectively. The relevant features of the spectrometer are a large solid angle (≈ 80 msr) and momentum acceptance ($\Delta p/p = \pm 10\%$), and a mass resolution up to $\Delta A/A \approx 1/200$. The entrance detector is based on large-area Micro-Channel Plates (MCP), mounted in a Chevron configuration [7], providing time and position signals. The Focal Plane Detector (FPD) is composed of a large area Multi-Wire Parallel Plate Avalanche Counter (MWPPAC) followed by a transverse field multi-anode Ionization Chamber (IC) at about 50 cm downstream [8]. The MWPPAC provides timing signals for both TOF measurements and Data Acquisition trigger, and position information. The IC has an active depth of about 120 cm and provides the energy (E) and the energy loss (ΔE) signals for the identification of the atomic number of the fragments with a Z resolving power $\Delta Z/Z \approx 1/60$ and an overall energy resolution $\Delta E/E \approx 1.5\%$.

CLARA [9] is an array of 25 Compton-suppressed Clover detectors. The Clovers are placed 29.5 cm from the target covering the azimuthal angles from 98° to 180°. The performance of CLARA for E_γ = 1.3 MeV are the following: photopeak efficiency ≈ 3%, peak/total ratio ≈ 45% and energy resolution ≈ 0.6% for v/c = 10%. The latter value is obtained, despite the broad range of product velocities in typical multinucleon transfer reaction, thanks to the excellent event-by-event definition of the recoil velocity provided by PRISMA.

The operation of the large acceptance magnetic spectrometer PRISMA in conjunction with large γ arrays (such as CLARA or the AGATA Demonstrator) is the result of joint efforts between INFN groups involved in γ-spectroscopy and reaction mechanisms studies in collaboration with several European institutions. The experimental campaign of the PRISMA-CLARA set-up started in 2004 and has been completed at the end of March 2008 making use of about 50% of the total beam-time available at the Tandem/PIAVE-ALPI accelerator complex of LNL. Experiments were mainly addressed to obtain information on the shell evolution and the onset of new regions of deformation (collectivity, critical point symmetries) in medium-mass moderately neutron-rich nuclei.

3. Spectroscopic studies of neutron-rich nuclei

The most critical ingredient in determining the properties of a nucleus from a given effective interaction is the overall number of nucleons and the ratio N/Z of neutrons to protons. Changes in the density and the size of nuclei are expected for increasing N/Z ratios with the consequent modification of the average field experienced by a single nucleon. The orbital rearrangement in neutron-rich nuclei causes the breaking of shell gaps and the appearance of new magic numbers.

3.1. Around the island of inversion

In order to investigate the evolution of the $N = 20$ shell gap, neutron-rich nuclei in the region of the "island of inversion" were populated in deep-inelastic processes produced by bombarding a 300 μg/cm^2 ^{208}Pb target with a ^{36}S beam at E_{lab} = 215 MeV. Projectile-like reaction products were identified by means of the PRISMA spectrometer, placed around the grazing angle corresponding to $\theta_g = 56°$, in coincidence with the de-excitation γ-rays detected in CLARA. A wide range of fragments, from Na (Z = 11) to Mn (Z = 25), were populated in this experiment. The γ-ray spectrum in coincidence with ^{36}Si is shown in figure 1. Three γ rays with energies of 1408, 1442 and 842 keV have been clearly observed. However, γ-γ coincidences were not possible because of the low statistics in such a weak channel. Consequently, the level scheme has been built revisiting data from a previous thick target deep-inelastic experiment which used the same projectile. By double gating on transitions that are observed for the first time in this experiment (1442 and 842 keV), the yrast sequence in ^{36}Si has been reconstructed for the first time up to the 6^+ state [10]. Useful information on the evolution of the shape (collectivity) has been extracted from the ratio $R4/2$ between the excitation energies of the 4^+ and 2^+ yrast states, shown in Fig. 2 for even-A Mg and Si isotopes with neutron numbers from 12 to 22. In particular, one can see that at $N = 22$, ^{34}Mg lies close to the rotational limit while ^{36}Si (having only two more protons) shows a nearly spherical shape behavior (vibrational limit), as confirmed by the B(E2) values predicted by shell model calculations.

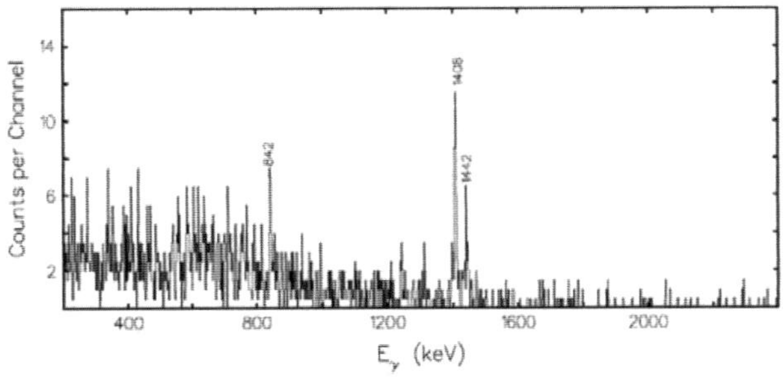

Figure 1. γ-ray spectrum measured with CLARA for ^{36}Si.

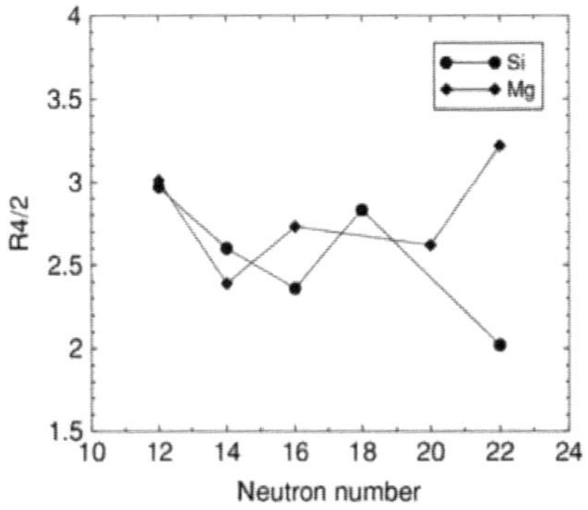

Figure 2. Systematics of the $E(4^+)$ to $E(2^+)$ ratio, $R4/2$, for Mg and Si isotopes from $N = 12$ to $N = 22$.

3.2. The onset of deformation in $A \sim 60$ neutron-rich nuclei

Besides inducing new shell and sub-shell closures, the neutron excess in nuclei can lead to the appearance of new regions of deformation through a spherical-to-deformed transition which has been interpreted as a shape phase transition. In this path, critical points (such as the $E(5)$ and $X(5)$ dynamical symmetries [11]), where nuclei have well-defined properties, could be encountered. Neutron-rich nuclei in the region of heavy Cr and Fe isotopes ($A \sim 60$) constitute an ideal ground for this kind of studies since their structure can be described by large-scale shell model calculations and they become deformed going towards $N = 40$. The experimental information on these nuclei, populated with the ^{64}Ni+^{238}U reaction at $E_{\rm lab} = 400$ MeV, was significantly extended with the PRISMA/CLARA set-up. In particular, yrast states up to $I^\pi = 8^+$ have been observed for the first time in the neutron-rich nucleus ^{58}Cr. Based on γ-γ coincidences and considering that for all known even–even nuclei we observed mainly yrast transitions, we proposed for ^{58}Cr the level scheme shown in figure 3. The experimental values of the ratio $R(4/2)$ between the excitation energies of the 4^+ and 2^+ yrast states for even-A chromium and iron isotopes are shown in figure 4. This ratio for the heavy Fe isotopes is very close to the 2.50 value characteristic of γ-soft rotors ($O(6)$). The ratios between the experimental energies of ^{58}Cr yrast states [12, 13] follow the expectations of the dynamical symmetry $E(5)$ for a nucleus at the critical point of the shape phase transition from a spherical vibrator ($U(5)$) to a γ-soft rotor ($O(6)$). This dynamical symmetry predicts, in a parameter-free analytical way, the following energy ratios for a nucleus at the critical point: $E(4^+)/E(2^+) = 2.20$, $E(6^+)/E(2^+) = 3.59$, $E(8^+)/E(2^+) = 5.17$. The experimental values in ^{58}Cr are 2.20 (figure 4), 3.66 and 5.22, respectively.

A critical test to decide if a nucleus stands in the $E(5)$ critical point is given by the B(E2) values normalized to $B(\mathrm{E2} : 2^+ \to 0^+)$. Although lifetimes in ^{58}Cr are not known, except the lifetime of the first 2^+ state, one can compare the $E(5)$ predictions with the results of large-scale shell model calculations, which give normally a good agreement with the experiment in this region. All theoretical predictions agree quite well in the description of the transition probabilities for the yrast states giving a further hint for the realization of the $E(5)$ critical point (a dynamical symmetry observed so far only in a few stable, heavier nuclei) in ^{58}Cr. Further challenging investigations devoted to the measurement of lifetimes in ^{58}Cr will provide the additional information to confirm the $E(5)$ dynamical symmetry in this nucleus.

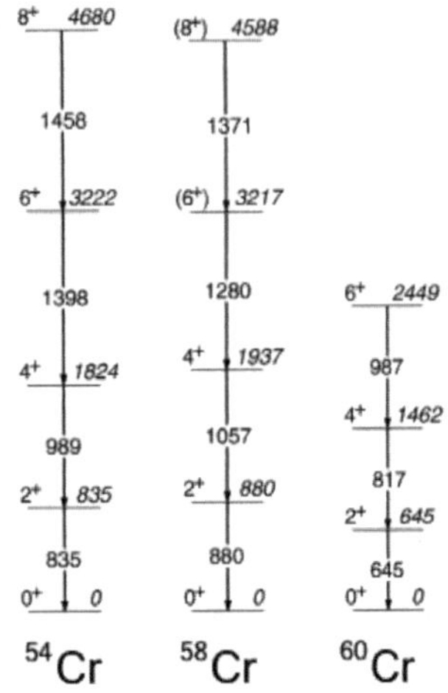

Figure 3. Level schemes of even-*A* heavy chromium isotopes.

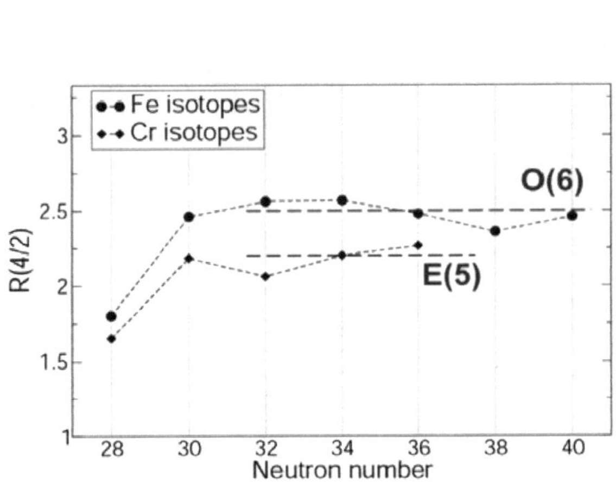

Figure 4. Systematics of $R(4/2)$ ratio in even-*A* Fe and Cr isotopes from $N = 28$ to $N = 40$.

3.3. The neutron-rich iron isotopes

In the ^{64}Ni(400 MeV)+^{238}U reaction it was possible to collect valuable information on the even-*A* neutron-rich iron isotopes up to ^{66}Fe. A second experiment was performed in order to enhance the population of heavy iron isotopes, using the ^{70}Zn(460 MeV)+^{238}U reaction [14]. The comparison of the γ-ray spectra for ^{66}Fe obtained in the two experiments, shown in figure 5, clearly proves that in the second measurement the production cross section for heavy iron isotopes was greatly enhanced.

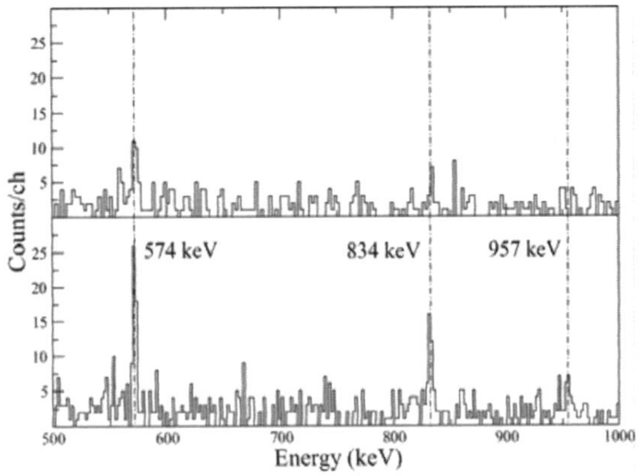

Figure 5. Spectra in coincidence with ^{66}Fe isotopes identified in PRISMA, respectively for the ^{64}Ni(400MeV)+^{238}U (top) and the ^{70}Zn(460MeV)+^{238}U reaction (bottom). In the former case, only the $2^+ \rightarrow 0^+$ ground-state transition (574 keV) was firmly assigned, while in the latter case also the $6^+ \rightarrow 4^+$ (957 keV) and the $4^+ \rightarrow 2^+$ (834 keV) transitions were identified, proving that the population of ^{66}Fe is enhanced.

The decay schemes for the even-*A* neutron-rich iron isotopes populated in this experiment is presented in figure 6 together with the result from large-scale shell model calculations. The effective interaction used is called *fpg* and is described in Ref. [15]. Its ingredients are an inert core of ^{48}Ca, with a valence space including the whole *fp* shell for the protons and the $p_{3/2}$, $f_{5/2}$, $p_{1/2}$, and $g_{9/2}$ orbitals for the neutrons [16]. Suitable truncations were imposed to limit the dimensions of the matrices.

Iron isotopes evolve towards more collective structures when approaching $N = 40$. Inspection of figure 6 shows in fact that the excitation energy of the 2^+ state (usually considered a first fingerprint of nuclear collectivity) decreases with increasing neutron number in the even-*A* iron isotopes. In particular, at $N = 40$ (that for nickel isotopes acts as a sub-shell closure [17]), the drop in the 2^+ excitation energy is large, which suggests increasing collectivity. This can be understood in terms of a decrease of the energy gap between the *fp* shell and the intruder $g_{9/2}$ orbital when the $1f_{7/2}$ proton shell is not completely filled and more neutrons are occupying the upper shell [18]. The fact that the $f_{7/2}$

shell is not fully occupied weakens both the effect of the attractive monopole tensor interaction with the neutrons in the $f_{5/2}$ orbital and the repulsion with those in the neutron $g_{9/2}$ orbital.

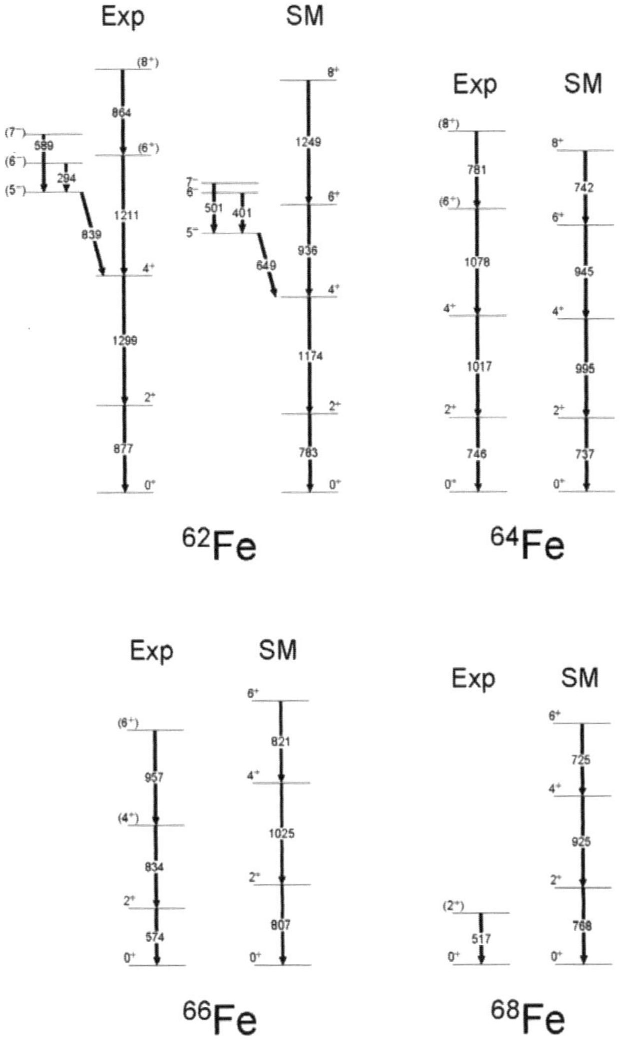

Figure 6. Decay schemes for the even-A iron isotopes ^{62,64,66,68}Fe obtained during the PRISMA-CLARA campaign, compared with the results from large-scale shell model calculations performed with a *fpg* interaction. See text for details.

The theoretical description of the even-A isotopes ^{62,64}Fe (figure 6) is quite satisfactory, although the calculated level schemes are more compressed than the experimental ones. Furthermore, in the case of ^{66}Fe, the excitation energy of the 2^+ state is predicted too high. The dramatic decrease of the energy of the 2^+ state in this nucleus suggests that the $d_{5/2}$ orbital has to be considered as well, as suggested in Ref. [16].

4. Ongoing developments

4.1. Parallel Plate Avalanche Counter for light ions

The MWPPAC, presently installed at the focal plane of PRISMA, was designed for the detection of medium-heavy ions. As expected, low intrinsic efficiencies (below 1%) have been obtained in the detection of light ions. To overcome this limitation a more efficient Parallel Plate Avalanche Counter (PPAC) has been developed (figure 7). It has the same structure of the present MWPPAC. The main difference is the smaller diameter of wires (10 μm instead of 20 μm) used for the construction of the X

Journal of Physics: Conference Series **205** (2010) 012038 doi:10.1088/1742-6596/205/1/012038

anode plane. The position efficiency in this configuration is increased by the higher reduced electric field reached around the *X* wires that implies higher gas gains. In-beam tests were performed at LNL in December 2007. Different ion beams delivered by the XTU-Tandem (^{32}S at energies ranging from 100 MeV to 160 MeV and ^{16}O from 60 MeV to 90 MeV) were elastically scattered from a 200 μg/cm^2 thick ^{197}Au target. Cathode efficiency close to 100% has been measured for ^{32}S and ^{16}O ions over the whole explored energy range. The relative efficiency ε_r (shown in figure 8) of the *X* anode plane with respect to the cathode for ^{16}O at different energies has been estimated from the ratio between the number of counts in the *X* position spectra and the TOF (cathode) ones. The ε_r values for ^{12}C (and for the lowest ^{16}O energy) have been estimated from the measured efficiencies on the basis of the energy losses in the active volume of the detector. An increase in the relative efficiency ε_r up to ≈ 40% has been measured for 50 MeV ^{16}O ions.

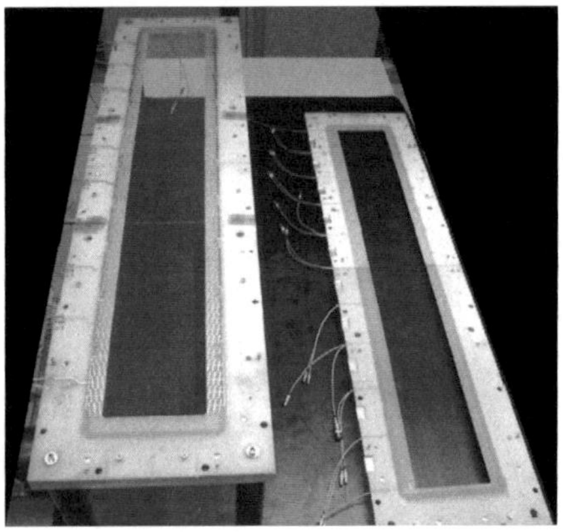

Figure 7. Electrodes of the new PPAC for light ions.

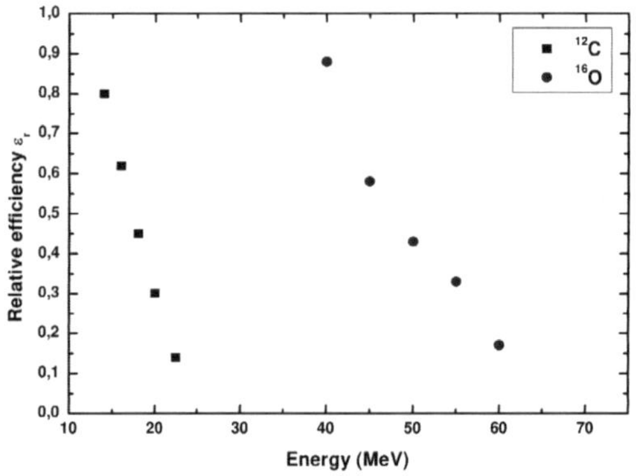

Figure 8. Relative efficiency ε_r for ^{12}C and ^{16}O ions as a function of their energy.

4.2. Secondary electron detector for low-energy heavy ions

The large thickness of the focal plane detector, mainly due to the MWPPAC windows (two mylar foils with a total thickness of 3 μm), limits its performance for low-energy heavy ions for which large energy losses and, consequently, large angular and energy straggling are expected. To overcome this limitation a large area Secondary Electron Detector (SED) is being developed. The working principle of this device is very similar to that of a Micro-Channel Plate. Heavy-ions passing through a thin (0.6 μm) plastic foil produce secondary electrons which are accelerated and transported, by means of electric and magnetic fields, toward the off-beam electron detector composed of a low pressure Multi-Wire Proportional Counter (MWPC). The installation of this new device at the focal plane of PRISMA will allow to decrease the detector thickness from ≈ 420 μg/cm^2 to ≈ 80 μg/cm^2. A small area (7x7 cm^2) prototype (shown in figure 9) has been designed and constructed by the University of Manchester. Gassiplex (GAS) based ASIC electronics has been developed at Daresbury Laboratory for the individual wire readout and data acquisition. The prototype was successfully tested in Manchester and Legnaro with two Saclay designed GAS32 (32 channels - embedding a preamplifier, a shaping time amplifier and a Track & Hold per channel) ASIC boards daisy chained together and coupled to a Daresbury designed V4FADC (fast analogic-to-digital converters) board (figure 10). Position resolutions of the order of 1 mm in the *X* and *Y* coordinates have been measured during the in-beam tests performed at LNL in October 2007 by using ^{80}Se ions at 250 MeV elastically scattered from a 100 μg/cm^2 thick ^{12}C target.

For the large area (1 m wide) SED design, a more size suitable ASIC board named GAS16 (16 channels) has been redesigned. The readout of all anode wires will require a total of 38 GAS16 boards. A new V4FADC module has been also designed to suit a wide range of ASIC electronics.

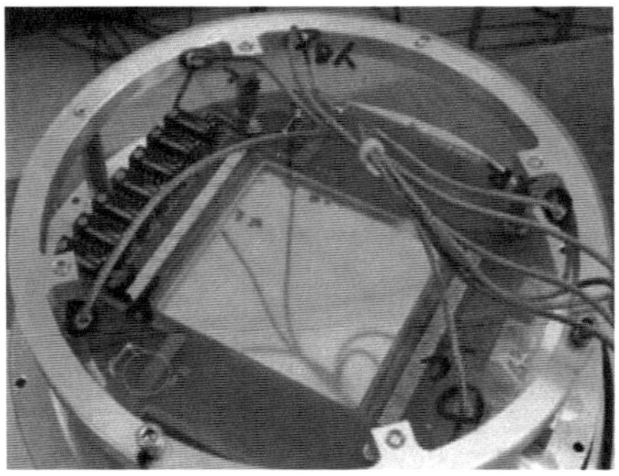

Figure 9. The small area SED based on a low pressure MWPC.

Figure 10. The V4FADC module which can control up to eight GAS16 boards.

5. Summary

The high efficiency and excellent mass resolving power of the PRISMA-CLARA set-up allowed to obtain new and significant information concerning the nuclear structure of neutron-rich nuclei in various regions of the nuclide chart. Moreover, the analysis of γ-γ coincidences from previous thick-target experiments performed with large γ arrays may benefit from the unambiguous attribution of new transitions to specific nuclei obtained with the PRISMA-CLARA set-up. The ongoing construction of new focal plane detectors for the PRISMA spectrometer will allow to overcome the limitations of the presently installed device for light ions and low-energy heavy ions.

Acknowledgments

Most of experiments performed with the PRISMA-CLARA set-up as well as the development of new focal plane detectors for the PRISMA spectrometer have been partially supported by the European Commission within the Sixth Framework Programme through I3-EURONS (contract no. RII3-CT-2004-506065).

References

[1] Broda R *et al* 1990 *Phys. Lett.* **B251** 245
[2] Corradi L *et al* 2009 *J. Phys.* **G36** 113101
[3] Fornal B *et al* 2008 *Phys. Rev.* **C77** 014304
[4] Stefanini A M *et al* 2002 *Nucl. Phys.* **A701** 217c
[5] Scarlassara F *et al* 2004 *Nucl. Phys.* **A746** 195c
[6] Corradi L *et al* 2007 *Nucl. Phys.* **A787** 160c
[7] Montagnoli G *et al* 2005 *Nucl. Instr. and Meth.* **A547** 455
[8] Beghini S *et al* 2005 *Nucl. Instr. and Meth.* **A551** 364
[9] Gadea A *et al* 2004 *Eur. Phys. J.* **A20** 193
[10] Liang X *et al* 2006 *Phys. Rev.* **C74** 014311
[11] Iachello F 2000 *Phys. Rev. Lett.* **85** 3580
Iachello F 2001 *Phys. Rev. Lett.* **87** 052502
[12] Marginean N *et al* 2006 *Phys. Lett.* **B633** 696
[13] Fioretto E *et al* 2007 *Proc. Int. Symp. on Exotic Nuclei, (Khanty-Mansiysk, Russia, 12-17 July 2006)* vol 912 ed Yu E Penionzhkevich and E A Cherepanov (Melville, New York: AIP Conference Proceedings) p 412
[14] Lunardi S *et al* 2007 *Phys. Rev.* **C76** 034303
[15] Sorlin O *et al* 2002 *Phys. Rev. Lett.* **88** 092501
[16] Lenzi S M *et al* 2008 *Report INFN-LNL-***222**(2008) 15
[17] Broda R *et al* 1995 *Phys. Rev. Lett.* **74** 868
[18] Caurier E, Nowacki F and Poves A 2002 *Eur. Phys. J.* **A15** 145

Experimental studies of proton-neutron mixed symmetry states in the mass A ≈ 130 region

G Rainovski[1], N Pietralla[2], J Leske[2], T Ahn[2], L Coquard[2], C Bauer[2], Th Möler[2], K Gladnishki[1], M Danchev[1], A Damyanova[1], D Karagyozov[1], R Topchiyska[1], C J Lister[3], R V F Janssens[3], M P Carpenter[3], S Zhu[3]

[1]Faculty of Physics, St. Kliment Ohridski University of Sofia, 1164 Sofia, Bulgaria
[2]Institut für Kernphysik, Technische Universität Darmstadt, D-64289 Darmstadt, Germany
[3]Argonne National Laboratory, 9700 South Cass Avenue, Argonne, Illinois 60439, USA

E-mail: rig@phys.uni-sofia.bg

Abstract. Considerable progress has been achieved recently in the experimental investigation of quadrupole-collective isovector excitations in the valence shell, the so called mixed-symmetry states (MSSs), in the mass $A \approx 130$ region. This is due to a new experimental technique for study MSSs which is based on the observation of low-multiplicity γ-ray events from inverse kinematics Coulomb excitation with the large 4π spectrometer, such as Gammasphere. The obtained experimental information for the MSSs of stable $N = 80$ isotones indicates that for low-collective vibrational nuclei the underlying single-particle structure can be the most important factor for preserving or fragmenting the MSSs through the mechanism of **shell stabilization**. The evolution of the MSSs from ^{134}Xe to ^{138}Ce is also used to determine the local strength of the proton-neutron interaction derived for first time from states with symmetric and antisymmetric nature.

1. Introduction

Atomic nuclei are quantum, many body systems which consist of two equivalent entities (protons and neutrons). In this respect the atomic nuclei are examples of mesoscopic two-fluid quantum systems which nature is defined by the collectivity, the isospin symmetry and the shell structure. The question how these three aspects coexist, interplay and compete comprises one of the main questions of the quantum theory and the atomic nuclei present a unique laboratory where this question can be studied experimentally. Nuclear phenomena that reflect these three general aspects are nuclear collectivity, shell structure, and the isospin degree of freedom. Of particular importance for studying the mutual balance of these properties are those excitations that are related to the collective two-fluid character of nuclei and to their shell structure. Quadrupole-collective isovector valence shell excitations, so-called Mixed-Symmetry States (MSSs) [1], are the best studied examples of this class of phenomena. A special type of MSSs, the $J^\pi = 1^+$ scissors mode has first been discovered in nuclei [2] and has then been reported or suggested to exist in Bose-Einstein condensates [3] and metallic clusters [4].

States with proton-neutron mixed symmetry have first been defined in the framework of the interacting boson model with proton-neutron degree of freedom (IBM-2) [1]. The concept of proton-neutron mixed symmetry is formalized by the F-spin quantum number [5], which is the isospin analogue for bosons. Within this concept the fully symmetric states have $F = F_{max} = (N_\pi + N_\nu)/2$ ($N_{\pi,\nu}$ denote the proton/neutron boson numbers), while MSSs are those states with $F \leq F_{max} - 1$ [5]. In other words, the F-spin quantum number counts the number of protons and neutrons pairs which are in phase in the quantum state. The IBM-2 states with maximum F-spin quantum number are called Full Symmetry States (FSSs). The F-spin is an approximate quantum number for low-lying collective states of heavy nuclei. The lowest states in a given nucleus are those formed by the FSSs. The $M1$ transitions between these states are forbidden and indeed they are observed to be small on an absolute, single-particle scale. With little modifications due to symmetry restrictions, the MSSs in the IBM-2 repeat the multiplet structure observed for the FSSs albeit at higher energy and with different decay properties.

The most distinct feature of MSSs is the existence of allowed F-vector ($\Delta F = 1$) $M1$ transitions to FSSs. This comprises most prominent experimental signature of MSSs - due to their isovector character they decay to FSSs by strong $M1$ transitions with matrix elements of the order of 1 μ_N [1]. However, finding this signature is a major experimental challenge because it requires full spectroscopic information, *i.e.* the spin-parities of these highly excited non-yrast states, their lifetimes, branching and multipole mixing ratios of their γ-decay have to be determined. Until recently obtaining all this information was possible for a hand-full of stable nuclei only and it required a series of experiments to be measured. Available information on MSSs of vibrational nuclei has recently been summarized in a review article [6]. The best examples of MSSs in stable nuclei are found in the mass $A \approx 90$ region [7, 8, 9], while there are only a few reported cases in the $A \approx 130$, region below the $N = 82$ shell closure, *e.g.* [10, 11, 12]. Apparently, to investigate MSSs, new experimental methods, in particular those that could potentially be applied to radioactive isotopes, are needed. We have shown recently that the observation of low-multiplicity γ-ray events from inverse kinematics Coulomb excitation with the large 4π spectrometers, such as Gammasphere, is a powerful tool to study MSSs. It is worth noting that this experimental technique can straightforwardly be applied to Radioactive Ion Beams (RIBs).

2. Experimental method

The technique which combines the observation of low-multiplicity γ-ray events from inverse kinematics Coulomb excitation with the large 4π spectrometers was first used to study the one-phonon MSS of ^{96}Ru [8] and then exploited to its full capacity in the case of ^{138}Ce [13] which is a low-abundant nucleus. We will use this experiment here to illustrate the method. The experiment was carried out at Argonne National Laboratory. The ^{138}Ce beam with intensity of ≈ 1 pnA ($\sim 6 \times 10^9$ ions/s) was delivered by the ATLAS accelerator. The 480 MeV beam was incident on a 1 mg/cm^2 ^{12}C target. The deexcitation γ rays, following the Coulomb excitation of the projectile, were detected with the Gammasphere array [14] which consisted of 98 HPGe detectors arranged in 15 rings. Gammasphere was used in singles mode resulting in an average counting rate of 4000 counts-per-second (cps), while the room background was producing about 600 cps. A total of 2.4×10^8 events of γ-ray fold 1 or higher was collected in about 14 hours. The contribution of the room background was eliminated in the off-line sort by correlating the γ rays with the accelerator radio-frequency (RF) signal. The final spectrum, which is a difference between the "beam-on" (with respect to the RF) spectrum and the "beam-off" spectrum, scaled to eliminate the 1461-keV room background transition in ^{40}K, is shown in Fig. 1. All γ rays in the spectrum originate from ^{138}Ce nuclei recoiling with $v/c \approx 6.9\%$. The spins of the levels were assigned on the basis of an angular distribution analysis [15].

In this experiment, we have observed the decay of the first six $2^+_{1,2,3,4,5,6}$ states up to an excitation energy of 2.7 MeV. The relative γ-ray yields with respect to the 2^+_1 state measure the relative Coulomb excitation (CE) cross-sections. These data were fitted to the Winther-De

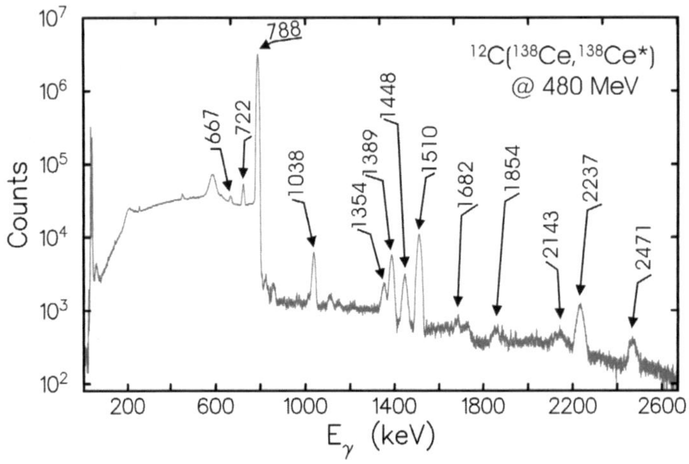

Figure 1. Background-subtracted, Doppler-corrected γ-ray spectrum in ^{138}Ce observed with Gammasphere after Coulomb excitation on a carbon target. Taken form Ref. [13].

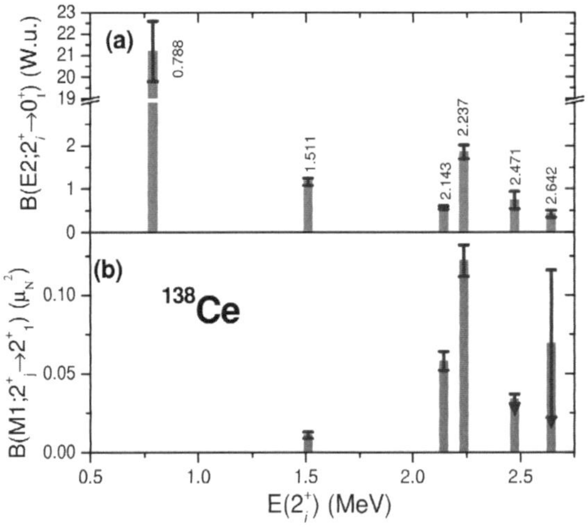

Figure 2. (a) E2 and (b) M1 transition strength distributions for all observed 2^+ states below 2.7 MeV in ^{138}Ce. The arrows indicate upper limits. Taken from Ref. [13].

Boer theory [16] using a multiple CE code [17] and taking into account the energy loss of the beam in the target. Absolute cross-sections were derived using the previously measured value for the B(E2;$0^+_1 \rightarrow 2^+_1$) = 0.450(30) e^{2}b^2 [18, 19]. An unambiguous set of matrix elements $|\langle I^\pi_{\rm f} \parallel E\lambda \parallel 0^+_1\rangle|$ for one-step excitations of interest was obtained. The matrix elements for the excited 2^+_i states together with observed decay branching ratios $I_\gamma(2^+_i \rightarrow 2^+_1)/I_\gamma(2^+_i \rightarrow 0^+_1)$ and deduced $E2/M1$ mixing ratios for the $2^+_i \rightarrow 2^+_1$ transitions provide the $B(E2)$ and $B(M1)$ transition strengths (see Fig. 2).

3. One-phonon MSSs in N=80 isotones

In ^{138}Ce the 2_4^+ state at 2.237 MeV dominates the $2_i^+ \rightarrow 2_1^+$ $M1$ strength distribution (see Fig. 2). This identifies this level as the major fragment of the one-phonon $2_{1,\text{ms}}^+$ state in ^{138}Ce. Its one-phonon character is further corroborated by the fact that the 2_4^+ state exhibits the largest $E2$ strength to the ground state after the 2_1^+ state. However, some parts of the $2_{1,\text{ms}}^+$ strength are spread over nearby 2^+ levels. In particular, the 2_3^+ state at 2.143 MeV acquires a considerable $B(M1; 2^+ \rightarrow 2_1^+)$ value. For analysis of the mixing of the $2_{1,\text{ms}}^+$ state with symmetric configurations we have considered a two-state mixing scenario between the $2_{1,\text{ms}}^+$ MSS and a close-lying fully symmetric state (FSS). This scenario leads to a mixing matrix element of 44(3) keV.

One might expect that excited states in the ^{136}Ba and ^{138}Ce $N = 80$ isotones are quite similar. The energies of the first three 2^+ states indeed correspond to each other within 4%. The 2_4^+ state at 2.129 MeV in ^{136}Ba has previously been identified as a $2_{1,\text{ms}}^+$ state [12] and its excitation energy agrees to within 5% with the excitation energy of the $2_{1,\text{ms}}^+$ state in ^{138}Ce. However, in ^{136}Ba the transition strength $B(M1; 2_4^+ \rightarrow 2_1^+) = 0.26(3)\,\mu_N^2$ [12] is larger than the total $M1$ strength $\sum B(M1; 2_{3,4}^+ \rightarrow 2_1^+) = 0.180(13)\,\mu_N^2$ shared between the 2_4^+ and 2_3^+ states in ^{138}Ce. Moreover, if the above mixing scenario is applied to ^{136}Ba, the resulting mixing matrix element is < 10 keV. This drastic change in the properties of predominantly collective states points precisely to their sensitivity to the underling sub-shell structure. Since the neutron configuration is not expected to differ much for the isotones ^{136}Ba$_{80}$ and ^{138}Ce$_{80}$, it is suggested that this difference in size of the F-spin mixing matrix elements is related to the proton configurations. Ground state spins for proton-odd N = 80 isotones and the shell model indicate the $\pi g_{7/2}$ sub-shell closure for cerium isotopes at proton number Z = 58. While the leading one-phonon 2^+ proton configuration already requires promotion of protons to the $\pi d_{5/2}$ sub-shell in $^{138}_{58}$Ce, the corresponding configuration for $^{136}_{56}$Ba can still be formed within the $\pi g_{7/2}$ sub-shell [20]. Thus, the one-phonon $2_{1,\text{ms}}^+$ state of ^{136}Ba is expected to consist of considerably simpler configurations than the closely lying predominantly symmetric states that surround it at an excitation energy of about 2 MeV. This prevents strong mixing between the $2_{1,\text{ms}}^+$ state and nearby 2^+ states in ^{136}Ba in contrast to the situation in ^{138}Ce. This mechanism is considered as a *shell-stabilization of mixed-symmetry structures* near sub-shell closures [13].

According to the mechanism of shell stabilization one might expect that the one-phonon state of ^{134}Xe will be a single isolated state, while the one-phonon MSS of ^{136}Ce will be fragmented. In order to check this hypothesis we have performed two experiments with Gammasphere at Argonne National Laboratory using the technique described above. Spectra form these experiments are presented in Fig. 3. The $B(E2)$ and $B(M1)$ transition strengths distribution from the observed 2^+ states in ^{134}Xe [21] are presented in Fig. 4. From the $M1$ distribution the 2_3^+ state of ^{134}Xe at 1947 keV can unambiguously be identified as the one-phonon MSS. This result is in agreement with the mechanism of shell stabilization. However, the preliminary results for ^{136}Ce [22] show that the observed spectrum is quite similar to the one of ^{138}Ce but the $M1$ strength is almost completely concentrated in the 2_4^+ state at 2155 keV as the neighboring 2_3^+ state takes almost no $M1$ strength. This fact is in contradiction with the mechanism of shell stabilization. At this point we have to conclude that even though we have found some strong indication that the properties of MSSs reflect the underling sub-shell structure through the effect of shell stabilization, the available experimental data are still not sufficient to allow for any conclusion on the generic nature of this effect.

The experimental efforts, presented above have revealed the evolution of the one-phonon MSSs in the stable $N = 80$ isotones. This has inspired extended microscopic calculations of the structures of the MSSs of the $N = 80$ isotones [23]. It has been demonstrated that a simultaneous description of the properties of the MSSs in the $N = 80$ isotonic chain, including

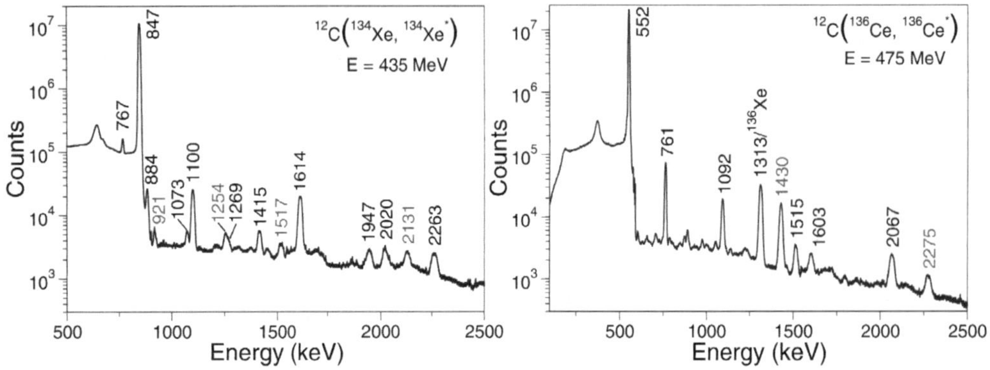

Figure 3. Background-subtracted, Doppler-corrected γ-ray spectra in ^{134}Xe and ^{136}Ce observed with Gammasphere after Coulomb excitation on a carbon target. Taken from Ref. [21, 22].

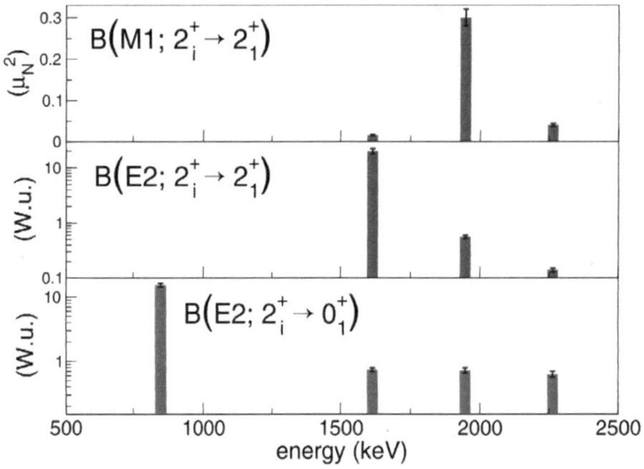

Figure 4. $E2$ and $M1$ transition strength distributions for all observed 2^+ states below 2.5 MeV in ^{134}Xe. Taken from Ref. [21].

the fragmentation of the one-phonon MSS in ^{138}Ce, can be achieved [23] within the frame work of the qasiparticle-phonon model [24]. The QPM calculations show that the splinting of the $M1$ strength in ^{138}Ce is a genuine shell effect due to the specific shell structure of this nucleus and the pairing correlations [23].

Looking at the $N = 80$ isotones, an increase in the energy splitting between the 2_1^+ and $2_{1,\mathrm{ms}}^+$ states can be seen [21]. This separation in energy was empirically fitted using a simple two-state mixing scheme where the interaction between the two states increases in strength as $V(N_\pi) = \beta\sqrt{N_\pi N_\nu}$. As a result of the fit, the relative strength of the proton-neutron quadrupole interaction was derived for the first time from the properties of both, symmetric and mixed-symmetric one-phonon states [21]. These calculations also suggest that a 2^+ state in ^{132}Te at an excitation energy of $\sim$1.6 MeV should be the dominant fragment of the $2_{1,\mathrm{ms}}^+$ state.

4. Summary and outlook

We have proven that the technique of projectile-Coulomb excitation is an efficient tool to study the MSSs of low abundant nuclei because it allows complete spectroscopy to be performed within one experiment. Using this technique we have identified the MSSs of several nuclei from the

A = 130 region. The new experimental data reveals some exciting new effects but to prove and understand them more completely it is obvious that MSSs of unstable nuclei have to be identified. For example, the experimental identification of the one-phonon MSS of ^{132}Te will show how precise is the prediction for the energy of this state [21]. To study further the shell stabilization effect, the MSSs of ^{140}Nd and ^{140}Ba have to be identified. All these task face the question: can MSSs of radioactive nuclei be identified on the basis of large absolute $M1$ transition rates? In other words, can we measure the lifetimes of these highly excited, off-yrast states in radioactive nuclei? The technique of projectile-Coulomb excitation may provide the answer to this question. Experiments we have performed so far using this technique were aimed at stable isotopes that are either rare, with natural abundances $<$1%, or that are noble gases which both complicate the production of massive isotopically enriched targets for traditional scattering experiments. Simple scaling of the available data on stable beams with intensities of the order of 10^9 ions/sec to beam intensities of 10^6 ions/sec, achievable for a large number of radioactive isotopes at present facilities such as REX-Isolde, suggests that this technique offers the potential for identifying and studying MSSs in radioactive isotopes. Given the sensitivity of MSSs to details of the local shell structure, they can become an important tool for testing and refining modern quantitative nuclear structure theories.

Acknowledgments

This work is supported by the US NSF within contract PHY-0245018, by the US Department of Energy, Office of Nuclear Physics, under contract DE-AC02-06CH11357 and under grant DE-FG02-04ER41334, by the SU NSF within contract 249/09, by the BG NSF within contract DO 02-219, by the German-Bulgarian exchange program under grants D/06/05918 and DAAD-09, and by the DFG under grants SFB 634 and Pi 393/2-1.

References

[1] Iachello F 1984 *Phys. Rev. Lett.* **53** 1427
[2] Bohle D *et al.* 1984 *Phys. Lett.* **B137** 27
[3] Marag O M *et al.* 2000 *Phys. Rev. Lett.* **84** 2056
[4] Lipparini E and Stringari S 1990 *Phys. Rev. Lett.* **63** 570 Nesterenko V O *et al.* 1999 *Rev. Lett.* **53** 87
[5] Otsuka T, Arima A and Iachello F 1978 *Nucl. Phys.* A309 1
[6] Pietralla N, von Brentano P and Lisetskiy A F 2008 *Prog. Part. Nucl. Phys.* **60** 225
[7] Pietralla N *et al.* 1999 *Phys. Rev. Lett.* **83** 1303
[8] Pietralla N *et al.* 2001 *Phys. Rev.* C **64** 031301
[9] Werner V *et al.* 2002 *Phys. Lett.* **B550** 140
[10] Molnár G *et al.* 1988 *Phys. Rev.* C **37** 898 Fazekas B *et al.* 1992 *Nucl. Phys.* **A548** 249
[11] Wiedenhöver I *et al.* 1997 *Phys. Rev.* C **56** 751
[12] Pietralla N*et al.* 1998 *Phys. Rev.* C **58** 796
[13] Rainovski G *et al.* 2006 *Phys. Rev. Lett.* **96** 122501
[14] Lee I Y *Nucl. Phys.* 1990 **A520** 641c
[15] Yamazaki T *Nucl. Data* 1967 **A3** 1
[16] Alder K and Winther A *Coulomb Excitation* 1966 (New York: Academic Press)
[17] Ower H, Gerl J and Scheit H computer program CLX
[18] Lo Bianco G *et al.* 1989 *Z. Phys.* **A 332** 103
[19] S. Raman *et al.* 2001 *Atom. Data Nucl. Data Tab.* **78** 1
[20] Lo Iudice N and Stoyanov C 2002 *Phys. Rev.* C **65** 064304
[21] Ahn T *et al.* 2009 *Phys. Lett.***B679** 19
[22] Ahn T 2009 private communication
[23] Lo Iudice N, Stoyanov C and Tarpanov D 2008 *Phys. Rev.* C **77** 044310
[24] Soloviev V G *Theory of Atomic Nuclei: Quasiparticles and Phonons* 1992 (Bristol: Institute of Physics Publishing)

Journal of Physics: Conference Series **205** (2010) 012040 doi:10.1088/1742-6596/205/1/012040

One-phonon excitations in ^{92}Zr from electron scattering

Abdulrahman Scheikh Obeid, O Burda, M Chernykh, A Krugmann, P von Neumann-Cosel, N Pietralla, I Poltoratska, V Ponomarev, and C Walz

Institut für Kernphysik, Technische Universität Darmstadt, 64289 Darmstadt, Germany

E-mail: scheikh@ikp.tu-darmstadt.de

Abstract. Low-lying collective vibrational excitations of ^{92}Zr have been investigated with electron scattering at the S-DALINAC. The form factors of isospin polarized one-quadrupole phonon states of ^{92}Zr at the Z = 40 proton subshell closure have been measured and the momentum-transfer dependence of the form factors for the one-quadrupole phonon states have been compared to the prediction of the Quasiparticle Phonon Model. The E2 transition strengths of the one-quadrupole phonon states and the E3 transition strength of the one-octupole phonon state have been extracted. A comparison to the data on ^{94}Mo and previous spectroscopic information on mixed-symmetric states (MSSs) of ^{92}Zr will be given.

1. Introduction

Collectivity, isospin symmetry and shell structure are generic features of many-body quantum systems that consist of two components (protons and neutrons) like heavy atomic nuclei. The study of collective valence-shell excitations is a useful way to understand how these features coexist, interplay and compete. The investigation of MSSs is an important source of information on the effective proton-neutron interaction in the valance shell [1]. MSSs have been defined in the framework of the Interacting Bosons Model IBM-2 [2]. In analogy to isospin, the symmetry of a multi-boson wave function formed by N_π proton bosons and N_ν neutron bosons with respect to proton and neutron boson labels is quantified by the F-spin. States with $F < F_{max} = (N_\pi + N_\nu)/2$ have wave functions that contain at least one pair of proton and neutron bosons which is antisymmetric under the exchange of proton and neutron labels. They are called mixed-symmetry states. Those states with $F = F_{max}$ are symmetric under the pairwise exchange of the nucleon labels. They are fully symmetric states. The signature for MSSs are (i) strong $M1$ transition to fully symmetric states with matrix elements of about $1\mu_N$ in size and (ii) weakly-collective $E2$ transition to fully symmetric states.

The prediction of the IBM-2 with respect to a multi-phonon structure of MSSs in vibrational nuclei was well confirmed ten years ago by the observation of large $M1$ transitions between low-energy states of ^{94}Mo [3, 4, 5]. The mixed-symmetric 2^+ states were also investigated with electron scattering experiments at the 130 MeV superconducting electron accelerator S-DALINAC [6]. The evaluation of the measured form factors as a function of the momentum-transfer had supported the one-phonon interpretation of the 2^+_1, 2^+_3 states. In the neighbouring even-even isotone ^{92}Zr which is formed by N = 52 neutrons with two valence neutrons and

Z = 40 with no valence protons occupying the $\pi(g_{9/2})$ sub-shell a stronger configurational isospin polarization of the one-phonon states than in ^{94}Mo was expected. This has recently been discussed in the literature [7, 8, 9, 10]. In order to deepen the discussions and to verify this expectation a new electron scattering experiment at the S-DALINAC has been performed on the nucleus ^{92}Zr. A calculation of the momentum-transfer dependence of the form factors for the one-quadrupole phonon states has been done in the framework of the Quasiparticle Phonon Model.

2. Experiment and analysis

2.1. Experimental setup and data

The experiment has been carried out at the Darmstadt superconducting electron linear accelerator S-DALINAC which provides electron beams with energy up to 130 MeV and beam current up to 3 μA. The 33° - 169° spectrometer with momentum acceptance of $\pm 2.1\%$ was used. The spectrometer has in its focal plane a detector system based on four single-strip silicon detectors, each provides 96 strips with thickness of 500 μm and a pitch of 650 μm.
We have used an incident beam energy $E_e = 63$ MeV with beam currents from 0.5 to 1 μA. A ^{92}Zr target with an enrichment of 94.57% and a thickness of 9.75 mg/cm^2 was used. Data were taken at five different scattering angles $\Theta_e = 69°, 81°, 93°, 117°, 165°$ covering the maximum of the $E2$ form factor. Examples of electron spectra are shown in Fig. 1. The salient peaks correspond to the elastic line, the weakly collective one-phonon 2_1^+ state ($E_x = 0.934$ MeV), The 3_1^- state ($E_x = 2.339$ MeV) and the one-phonon 2_2^+ state ($E_x = 1.847$ MeV).

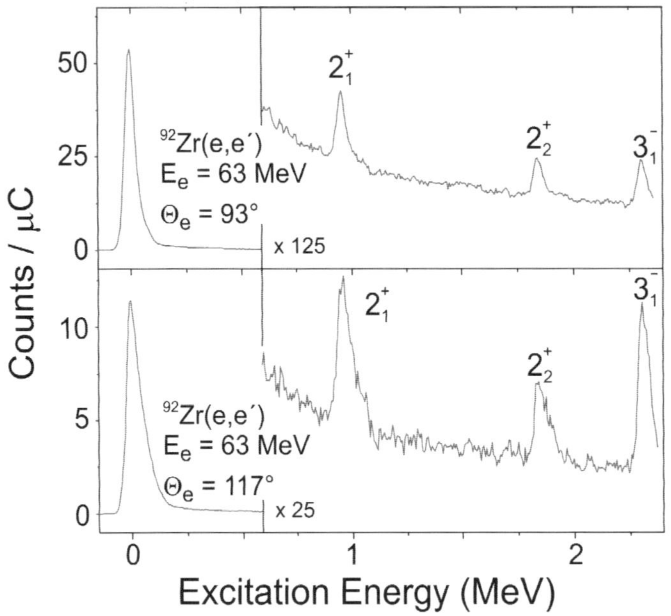

Figure 1. Two electron scattering spectra of the ^{92}Zr(e,e$'$) reaction at $E_e = 63$ MeV and $\Theta_e = 93°, 117°$.

The spectra taken in the ^{92}Zr(e,e$'$) electron scattering reaction were energy calibrated with the excitation energies of the 0_1^+, 2_1^+, 2_2^+, 3_1^- states given in [9].

2.2. Data analysis

The experimentally extracted cross sections for the excited states were normalized to the cross section of the elastic line. The absolute values of the differential cross sections were calculated from the corresponding peak areas taking into account the radiative corrections and the dead time of the electronics according to

$$\left(\frac{d\sigma}{d\Omega}\right) = A_{in}^{exp} \cdot \frac{1}{\Delta\Omega} \cdot \frac{e}{It} \cdot \frac{\mu_{\text{mol}}}{n_{\text{eff}} N_A} \tag{1}$$

were $\Delta\Omega$ is the spectrometer solid angle [sr], It the accumulated charge [C], μ_{mol} the molecular mass, [g/mol], n_{eff} the effective areal density of the target [g/cm^2] and N_A the Avogadro number [1/mol].
The squared form factor is given as:

$$|F(E_0, \theta)|^2 = \frac{(d\sigma/d\Omega)_{exp}}{(d\sigma/d\Omega)_{Mott}} \tag{2}$$

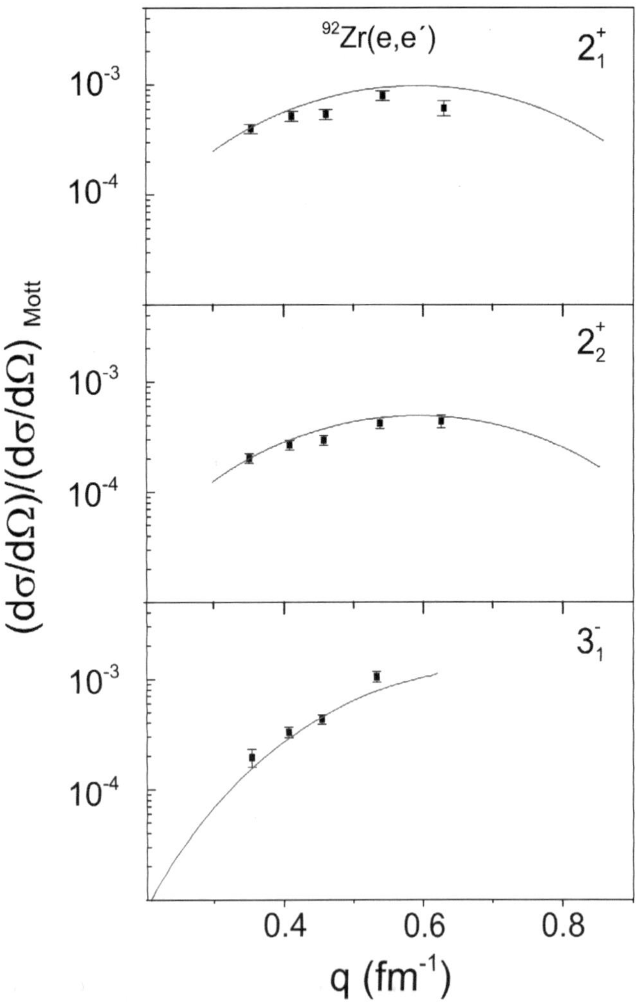

Figure 2. Momentum-transfer dependence of the form factor of the one-phonon FS 2_1^+ state (top), MS 2_2^+ state (middle) and 3_1^- state (bottom) of ^{92}Zr from electron scattering. The data (full squares) are compared to the QPM (solid lines).

with the Mott cross section:

$$\left(\frac{d\sigma}{d\Omega}\right)_{Mott} = \left(\frac{Ze^2}{2E_0}\right)^2 \frac{cos^2(\theta/2)}{sin^4(\theta/2)} \tag{3}$$

Calculations of the momentum-transfer dependence of the form factor of the 2^+ states and of the 3_1^- state have been performed within the Quasiparticle Phonon Model (QPM). Excited states of ^{92}Zr have been described by a wave function which includes one-, two-, and three-phonon configurations. The (e,e′) cross sections have been calculated within the Distorted Wave Born Approximation using the code of Heisenberg and Blok [11] and employing the radial transition charge densities calculated by the QPM. The E2 and E3 strengths can be extrapolated to the photon point, $q = E_x/\hbar c$.

Table 1. E2 strengths from (e,e′) in comparison to the known values from (n,n′γ) [5, 9] and the E3 strength from (e,e′) in W.u.

	This work: ^{92}Zr(e,e′) Eλ strength in W.u. Form factor scaled to QPM	^{92}Zr(n,n′γ) [9]	^{94}Mo(n,n′γ) [5]
$B(E2;2_1^+ \to 0_1^+)$	6.2±0.3	6.4±0.5	16.0±0.2
$B(E2;2_2^+ \to 0_1^+)$	3.3±0.2	3.4±0.4	2.2±0.2
$B(E3;3_1^- \to 0_1^+)$	18.3±1.1		

The $B(E2;2_1^+ \to 0_1^+)$ and $B(E2;2_2^+ \to 0_1^+)$ values that were extracted using the form factor calculated with the QPM are in good agreement with the literature values [9]. The EM strengths are compared to the literature and to the ones of the symmetric and mixed-symmetric states of ^{94}Mo in Tab. 1. The extracted value of the $B(E2;3_1^- \to 0_1^+)$ is 18.1 ± 1.1 W.u. and it compares well to the values obtained from proton scattering experiments $(14.7, 16.2, 18.9, 21.3, 23.6)$ W.u in the review article by Spear [12].

3. Discussion

The (e,e′) electron scattering reaction is predominantly a one-step process. Fig. 1 shows that the 2_2^+ state of ^{92}Zr is almost as strongly populated in (e,e′) as the one-phonon 2_1^+ state suggesting a large one-phonon component being present in the wave function of the 2_2^+ state either. Figure 2 presents the form factors of these 2^+ states and of the 3_1^- state as a function of the momentum transfer q. The similarity of the momentum-transfer dependence of the form factors between the 2^+ states further supports the one-phonon nature of the 2_2^+ state of ^{92}Zr.
The theoretical results shown in Fig. 2 are absolute and not scaled to the experimental data. The QPM results provide a good description of the (e,e′) form factor in this interval of the momentum transfer. The main configurations of the wave functions used to calculate the form factors are given as following:

$$\begin{aligned} \psi_{2_1^+} &= \sqrt{0.59}\ \nu\left(2d_{5/2}^2\right) + \sqrt{0.14}\ \pi\left(1g_{9/2}^2\right) \\ \psi_{2_2^+} &= -\sqrt{0.38}\ \nu\left(2d_{5/2}^2\right) + \sqrt{0.27}\ \pi\left(1g_{9/2}^2\right) \end{aligned} \quad (4)$$

The QPM calculations predict the main proton and neutron components in the wave functions of the one-phonon 2^+ states to have the opposite relative signs which confirms the picture of symmetric and mixed-symmetric character in the valance shell. The 2_1^+ state, because of the large neutron component, is a predominantly symmetric albeit isospin polarized state and the 2_2^+ state is a mixed-symmetric one with a weak isospin polarization.
The E2 strengths extracted in this work are close to the values, $6.4^{+0.6}_{-0.5}$ and 3.4(4) for the 2_1^+ and 2_2^+ respectively, that were obtained from a QPM calculetions performed in Ref. [10]. However,

the wave functions published for the 2^+ states in Ref. [10] show stronger isospin polarisation of the 2^+_1 and 2^+_2 states than what we have found from our calculation.

Acknowledgments
We thank Ch Stoyanov, N Lo Iudice, V Werner, C Fransen, G Rainovski and D Tarpanov for discussions. Support by the DFG under Grant No. SFB 634, by the Helmholtz International Center for FAIR and by the DAAD within the German-Bulgarian exchange program under grant Nos. D/06/05918 and DAAD-09 is gratefully acknowledged.

References

[1] Pietralla N, von Brentano P and Lisetskiy A F 2008 *Prog. Part. Nucl. Phys.* **60** 225
[2] Iachello F and Arima A 1987 *The Interacting Boson Model* (Cambridge University Press.)
[3] Pietralla N *et al.* 1999 *Phys. Rev. Lett.* **83** 1303
[4] Pietralla N, Fransen C, von Brentano P, Dewald A, Fitzler A, Frießner C and Gableske J 2000 *Phys. Rev. Lett.* **84** 3775
[5] Fransen C *et al.* 2003 *Phys. Rev* C **67** 024307
[6] Burda O *et al.* 2007 *Phys. Rev. Lett* **99** 029503
[7] Werner V *et al.* 2002 *Phys. Lett.* A **550** 140
[8] Holt J D, Pietralla N, Holt J W, Kuo T T S and Rainovski G 2007 *Phys. Rev.* C **76** 034325
[9] Fransen C *et al.* 2005 *Phys. Rev* C **71** 054304
[10] LoIudice N and Stoyanv C 2006 *Phys. Rev.* C **73** 037305
[11] Heisenberg J and Blok H P 1983 *Annu. Rev. Nucl. Part. Sci.* C **33** 0569
[12] Spear R H 1989 *Atom. Dat. And Nuc. Dat. Tab* **42** 55

Journal of Physics: Conference Series **205** (2010) 012041 doi:10.1088/1742-6596/205/1/012041

Experimental Investigation of the Warm Rotation in the Order-to-Chaos Region by Gamma-Spectroscopy in the Quasi-Continuum

S Leoni
University of Milano and INFN, Via Celoria 16, 20133 Milano, IT

E-mail: silvia.leoni@mi.infn.it

Abstract. The collective nuclear rotation, at high spins and as a function of temperature T, is investigated in the transition region between the regular ordered motion at T=0 and the chaotic compound nucleus regime around the particle binding energy. The experimental analysis, based on statistical techniques, is carried out in nuclear systems characterized by specific quantum numbers (such as K) and deformation (as for example superdeformed nuclei). Data from high statistics EUROBALL experiments are discussed, focusing on nuclear structure effects at the onset of the chaotic regime. The experimental findings are compared to prediction from a Montecarlo simulation of the γ-decay flow based on microscopic cranked shell model calculations at finite temperature.

1. Introduction

The properties of the atomic nucleus at the limits of angular momentum, temperature and neutron/proton number are currently investigated in great details with selective γ-spectroscopy measurements. In this contribution we focus on the collective response of the nuclear system, in particular on the warm rotation which allows to obtain nuclear structure information in the transition region between the ordered system at temperature T = 0, and the chaotic systems at the compound nucleus level. In fact, the γ-decay of excited rotational nuclei, produced at high spins and moderate excitation energy by fusion reactions between heavy ions, provides information on nuclear structure properties as a function of angular momentum and excitation energy: it probes the region of high level density where a strong mixing between states takes place, leading to the formation of compound nucleus levels [1].

Experimentally, the γ-cascades from the excited nucleus can be measured, with great selectivity, by the use of high-efficiency HpGe-array spectrometers (such as EUROBALL or AGATA in the near future), consisting of more than 100 Ge crystals in 4π geometry around the reaction center, usually combined with other types of ancillary detectors.

After a brief review of the experimental techniques used to extract quantitative nuclear structure information from γ-coincidence spectra (presented in section 2), we will discuss some of the basic results obtained in the study of the warm rotation, such as i) the fundamental role played by the two-body residual interaction in the description of nuclear structure properties already at few MeV of excitation energy above the ground state (section 3); ii) the onset of the chaotic regime, as a weakening of selection rules on specific quantum number such as K (section 4); the evidence for a persistence of nuclear structure effects in the order-to-chaos region, in connection with the warm rotational motion of superdeformed nuclei (section 5).

2. Statistical Analysis of Rotational Spectra

By constructing γ–γ spectra from the measured E_γ energies of high-fold γ cascades (produced by the de-excitation of warm nuclei at high spins), one can distinguish the contribution from different regimes: the γ-decay from the COLD region of discrete bands at T=0 forms ridges in the two-dimensional spectrum, while the excited rotational motion from the WARM region where the decay becomes more irregular/fragmented over many final states gives rise to a rather uniform distribution, filling also the central valley region. This is shown in panel a) of Figure 1, in the case of the normal deformed nucleus ^{168}Yb: the ridge-valley landscape is a direct consequence of the constant spacing between E2 transitions in a regular rotational band, which can not produce counts along the $E_{\gamma 1} = E_{\gamma 2}$ valley region, if the moment of inertia of the band stays constant.

The very existence of many, closely spaced, excited bands results in a quasi-continuum (QC) ridge-valley spectrum, i.e. in a distribution of γ-transitions which can not be resolved experimentally due to the finite detector resolution, although generated by a discrete energy spectrum. As a consequence, the fluctuations of the events collected in the γ–γ spectrum acquire a statistical meaning, providing a powerful tool to estimate the number of bands populating the ridge-valley structures [2]. In particular, the effective number of bands (named paths, N_{path}) both in the ridge and valley region, can be estimated through the relation

$$N_{path} = \frac{N_{eve}}{\frac{\mu_2}{\mu_1} - 1} \times P^{(2)}$$

being N_{eve}, μ_1 and μ_2 the number of events and the first and second moment of the distribution of counts in a given sector [2]. In the previous expression $P^{(2)}$ is a correction factor taking into account the finite energy resolution of the detection system. The analysis is typically performed in a two-dimensional energy region where a rotational band contributes in average with one transition, namely $\Delta E_\gamma \times \Delta E_\gamma = 4\hbar^2/\Im^{(2)} \times 4\hbar^2/\Im^{(2)}$, being $\Im^{(2)}$ the effective dynamic moment of inertia of the rotational bands. Figure 1, panels b) and c), shows the average spectrum "Fit" (obtained as a smooth polynomial interpolation of the two-dimensional ridge-valley distribution shown in panel a)) and the resulting μ_2/μ_1 fluctuation spectrum, which strongly emphasizes the original ridge-valley landscape: large fluctuations are observed along the ridges, pointing to a limited number of discrete regular bands similar to the yrast band, while the small fluctuations observed along the valley suggest the existence of a large number of bands in the warm region of fragmented decay.

The fluctuations of the spectrum events can also provide information on the similarities/correlations between two different event distributions. This can obtained by studying the spectrum covariance [3], defined as

$$\mu_{2,\text{cov}}(A;B) = \frac{1}{N_{ch}} \sum_j (M_j(A) - \langle M \rangle_j (A)) \times (M_j(B) - \langle M \rangle_j (B))$$

where *M(A)* and *M(B)* represent the two event distributions, while *<M>* denotes, again, an average "Fit" spectrum. To normalize the covariance and determine the degree of correlations between spectrum A and B, the correlation coefficient *r(A,B)* is defined as

$$r(A,B) = \frac{\mu_{2,\text{cov}}(A,B)}{\sqrt{(\mu_2(A) - \langle \mu_1 \rangle (A))(\mu_2(B) - \langle \mu_1 \rangle (B))}}$$

The subtraction of the average value of the first moment (over the region N_{ch}) in the denominator corrects for the contribution to μ_2 from counting statistics. From the previous definition one finds that the r correlation coefficient becomes 1 for identical distribution, 0 for completely different spectra. Illustration of the basic idea behind the covariance technique is shown in the right column of Figure 1, in the case of a limited region of a γ–γ spectrum.
The fluctuations and covariance techniques have been employed in the study of quasi-continuum ridge-valley spectra in different region of mass and deformation, as discussed in the next sections.

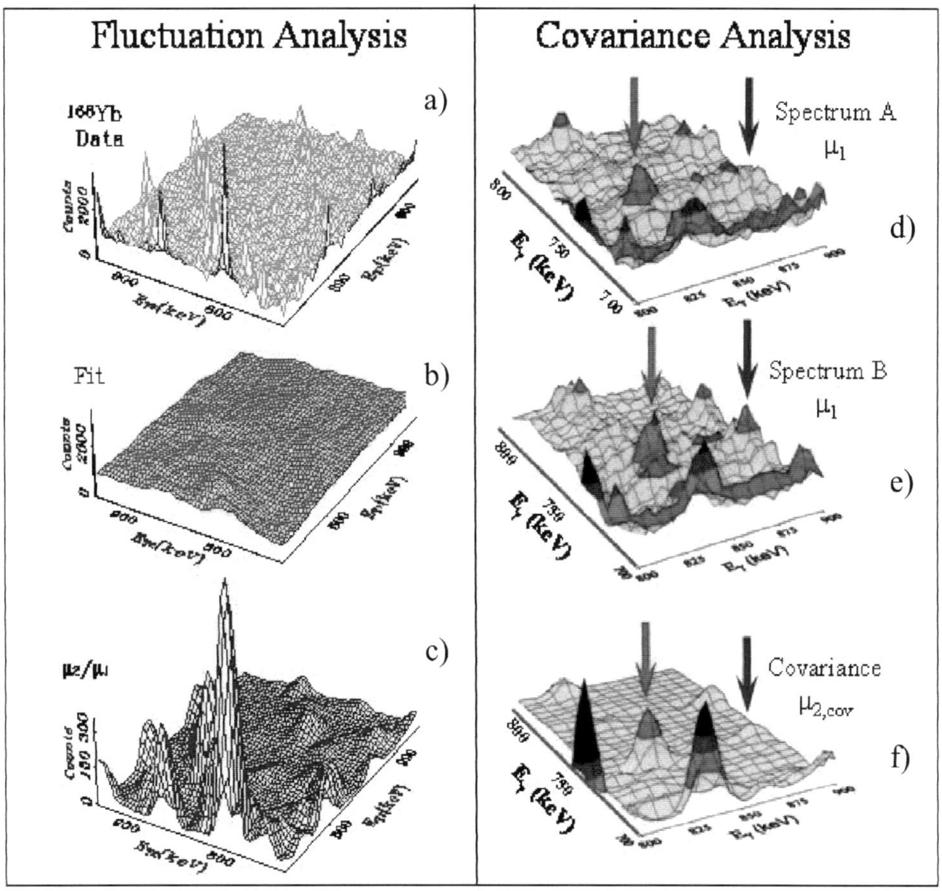

Figure 1: Example of spectral distributions employed in the statistical analysis of γ–γ correlated matrices. The left column refers to the *fluctuation analysis*, from which the effective number of decay paths, $N^{(2)}_{path}$, is extracted, starting from the events distribution (panel a)) and the second moment fluctuation spectrum μ_2/μ_1 (panel c)). The last one is calculated relative to the average spectrum (Fit), shown in the middle panel b). The right column illustrates the basic idea behind the *covariance* technique. The top and middle panels show two sections of γ–γ spectra (spectrum A and B) with the red and blue arrows indicating the positions where similar/different structures are present. In the bottom panel the corresponding covariance spectrum $\mu_{2,cov}$ is given, with emphasized/suppressed structures in correspondence with similar/different distributions (see text for details).

3. Evidence for a band mixing regime

One of the first, basic results obtained from the analysis of quasi-continuum spectra with statistical techniques, is the evidence for the fundamental role played by the two-body residual interaction in the description of the warm rotation [4]. As shown in Figure 2, the number of decay paths extracted from the analysis of the ridge and valley structure for a typical deformed nucleus of mass A=160, such as ^{168}Yb, is of the order of 30 and 10^4-10^5, respectively. It is found that these experimental results can be well explained by microscopic cranked shell model calculations, only including a two-body residual interaction V_{res}, of surface delta type (red solid lines in each panel). This gives rise to a fragmentation of the rotational decay above ~ 1 MeV of excitation energy above yrast, which largely enhances the number of decay paths the nucleus can follow in its warm decay. Moreover, it significantly reduces the number of discrete regular bands in the cold regime. On the contrary, a pure cranked shell model (mean field, MF) description of the excited energy spectrum would provide too many discrete regular bands populating the ridges, and too few (~ 10^3) decay paths populating the valley region, in agreement with simple estimates based on pure level density calculations.

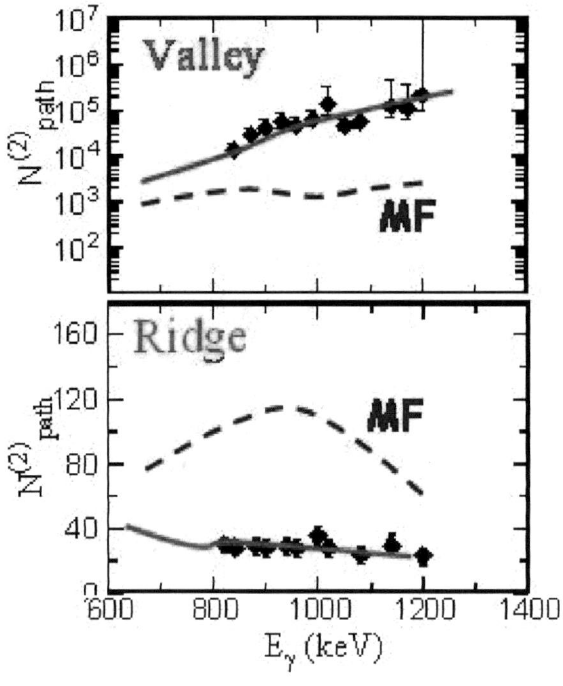

Figure 2. The effective number of path $N^{(2)}_{path}$ extracted from the analysis of the first ridge (bottom) and central valley (top) of the measured and calculated γ–γ coincidence spectra of ^{168}Yb. Symbols refer to data, dashed lines to simulations based on non-interacting bands (mean field description, MF), solid lines to simulations based on bands mixed by the two-body residual interaction (adapted from ref. [4]).

4. Onset of chaos in the atomic nucleus

The statistical analysis of the ridge-valley landscape can also be used to investigate the transition between order and chaos in the atomic nucleus. This can be done by looking at the gradual vanishing of the selection rules associated with the quantum numbers at T = 0, such as K (the projection of the angular momentum on the symmetry axis). The deformed nucleus ^{163}Er is an ideal case, being characterized by a number of rotational bands having low-K (K=5/2) and high-K (K=19/2) values [5]. Following a high statistics EUROBALL experiment, γ–γ matrices in coincidence with low-K and high-K structures have been constructed and analyzed by fluctuation and covariance techniques. It is found that the fluctuation analysis give evidence for a total number of ~40 discrete excited bands, populating the ridge structures of ^{163}Er, half of which of high-K nature. On the contrary, many more

Journal of Physics: Conference Series **205** (2010) 012041 doi:10.1088/1742-6596/205/1/012041

bands, of the order of 10^3-10^5, are found to populate the valley region, with large differences between low-K and high-K states, being the latter ~10 times fewer. This suggests that the K-quantum number is at least partially conserved up to moderate excitation energies, of the order of ~1.5 MeV above yrast, where the rotational motion is fragmented. At higher excitation energies more similar number of bands are obtained for low-K and high-K gated spectra, pointing to a vanishing of selection rules on K and to the onset of a chaotic regime, in which quantum numbers and selection rules loose their meaning. The previous results are strongly supported by the covariance analysis of the spectrum fluctuations [5], which provides specific information on the correlations between spectra gated by different K-states, as shown in Figure 3. In particular, the ridge and valley covariance between spectra gated by low-K states leads to a correlation of the order of 20%, as a consequence of the cross talk by E1 transitions (panel a) and c)). On the contrary, in the case of high-K and low-K covariances, no correlations are found from the analysis of the ridge structures (panel b)), while from the valley region one obtains strong similarities/correlation between high-K and low-K gated distributions only at the highest transition energy values (panel d)). This suggests a gradual transition to a chaotic regime around 1.5 MeV of internal energy. The experimental results shown in Figure 3 are well reproduced by a shell model which combines a cranked mean-field and a residual two-body interaction, together with a terms taking into account the angular momentum carried by the K-quantum number [6]. According to the model, K-mixing is induced by the interplay of the Coriolis and residual interaction, and it is found to gradually increase until a complete violation of the K-quantum number is reached above 2-2.5 MeV of internal energy.

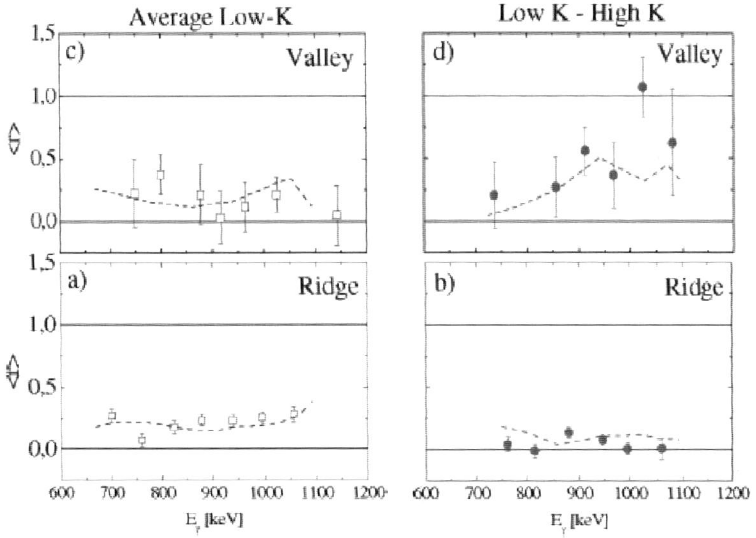

Figure 3. Results of the covariance analysis of ridge/valley structures of ^{163}Er (bottom/top panels). Symbols refer to data, lines to theoretical predictions, as obtained from the analysis of simulated spectra. Panels a) and c) give the correlation coefficient r obtained by averaging over pairs of γ–γ spectra gated by low-K configurations, while results for the low-K versus high-K matrices are shown in panels b) and d) [5].

5. Nuclear structure effects at the onset of chaos in superdeformed nuclei

The study of the thermally excited, rapidly rotating atomic nucleus becomes particularly challenging both experimentally and theoretically, in the case of very elongated systems. The study of superdeformed (SD) nuclei requires, in fact, high selectivity and high statistics, in order to focus on the small fraction of the decay (~few %) which is finally trapped into the SD well and populates discrete rotational bands. A step forward in the understanding of the population and decay of warm SD nuclei has been recently obtained by a comparative study of the nuclei ^{151}Tb and ^{196}Pb, populated with high statistics in two EUROBALL IV experiments [7-9]. The analysis is performed in terms of intensities and fluctuations of several independent observables extracted from quasi-continuum (QC) γ-coincidence spectra, providing quantitative information on the dynamics of the γ decay flow and of the tunneling through the potential barriers between SD and normal deformed (ND) excited states, over a wide spin range. A key point is the comparison of the data with predictions based on a "microscopic" Monte Carlo simulation of the γ-decay in the SD well, i.e. on simulations based on discrete levels calculated microscopically and on microscopic potential barriers between the ND and SD well. It is found that such a detailed study of the order-to-chaos region reveals the influence of nuclear structure effects. In particular, an enhancement in the E1 strength around 1–2 MeV, which could be related to octupole vibrations, is found to be needed to reproduce the intensity of the γ-decay flow directly feeding into the SD band. This is shown in Figure 4, in connection with the population of the SD yrast (panel a)) and of the first ridge/E2 QC in coincidence with the SD yrast (panel b) and c)).

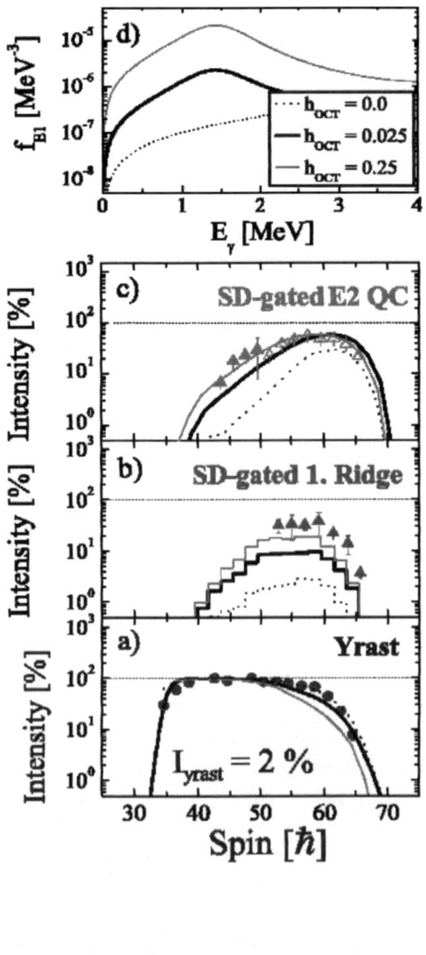

Figure 4: Influence of the low-energy component of the E1 strength (associated with octupole vibrations) on the population of the SD yrast (panel a)), SD-gated 1st ridge (panel b)) and SD-gated E2-QC (panel c)) of ^{151}Tb. In each panel, symbols refer to data, while dashed lines correspond to simulations with standard GDR strength. Thick (thin) solid lines are obtained from calculation including an extra component in the total E1 strength f_{E1}, as shown in the top panel [9].

Furthermore, the simulated intensities and fluctuations can be used to probe the tunneling probability between the SD and ND well, which strongly depends on the shape and height of the microscopically calculated potential barriers, and on the collective mass of the system. This latter quantity is one of the most uncertain parameter, being strongly driven by pairing. It is found that the decay-out properties of nuclei of mass A=190 are particularly sensitive to the magnitude of the collective mass. This is illustrated in Figure 5 where we show the results of the simulation of ^{196}Pb for different value of the parameter C_m , which is here used to scale the collective mass. As one can see, both the intensity of the yrast and of excited bands (panels a) and b)) and the number of discrete excited bands obtained by the fluctuation analysis of the ridge structure (panel c)) are very sensitive to the C_m value. In particular, the analysis shows the need for a strong renormalization of the collective mass ($C_m \sim 2.5$) in order to obtain a reasonable agreement with the data, both for the yrast and excited bands.

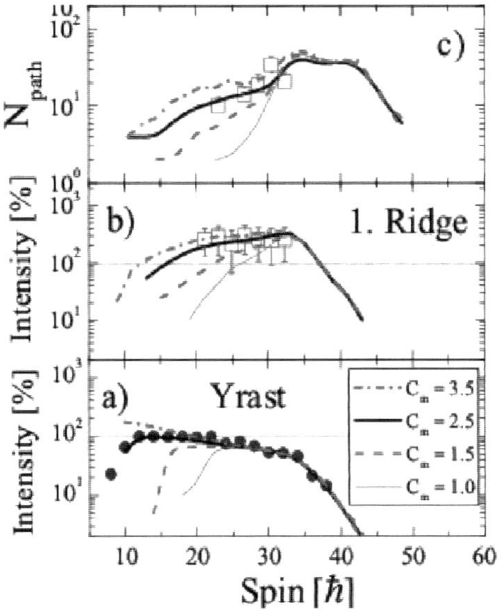

Figure 5: Dependence of the results of the simulation of ^{196}Pb on the scaling factor C_m of the mass parameter. Panels a) and b) refer to the intensity of the SD yrast and of the 1st ridge, panel c) to the number of paths populating the 1st ridge. Symbols refer to the corresponding experimental data (see text for details).

6. Conclusions

We have presented selected examples of nuclear structure studies at high spins and temperature, focusing on the γ-decay of the warm rotating nucleus. It is shown how γ-spectroscopy in the quasi-continuum provides a unique tool for investigating in details the transition between order and chaos, which takes place ~1.5 MeV of excitation energy above yrast. The experimental study is based on statistical analysis techniques and the comparison with theory requires a simulation of the γ-flow based on microscopic quantities (e.g. energy levels and potential barriers). These studies provide a stringent test to the conservation of selection rules, configuration changes and pairing with temperature.

7. References

[1] Bracco A and Leoni S 2002 *Rep. Prog. Phys.* **65** 299
[2] Døssing T et al. 1996 *Phys. Rep.* **268** 1
[3] Leoni S et al. 2000 *Nuc. Phys. A* **672** 71
[4] Herskind B et al. 1992 *Phys. Rev. Lett.* **68** 3008
[5] Leoni S et al. 2005 *Phys. Rev. C* **72** 034307
[6] Matsuo M et al. 2004 *Nuc. Phys. A* **736** 223
[7] Leoni S et al. 2008 *Phys. Rev. Lett.* **101** 142502
[8] Leoni S et al. 2009 *Phys. Rev. C* **79** 064306
[9] Leoni S et al. 2009 *Phys. Rev. C* **79** 064307

Journal of Physics: Conference Series **205** (2010) 012042 doi:10.1088/1742-6596/205/1/012042

Structure effects in collisions induced by halo and loosely bound nuclei

Alessia Di Pietro

INFN- Laboratori Nazionai del Sud, via S. Sofia 62, I-95125, Catania, Italy

E-mail: `dipietro@lns.infn.it`

Abstract.
Reactions induced by halo and loosely bound stable nuclei at low bombarding energies, arouse a lot of interest in the last fifteen years. Many experiments have been performed so far, using both stable and radioactive beams. These experiments are driven by theoretical expectations concerning effects, at sub-barrier energies, of coupling between relative motion and intrinsic excitation of the colliding nuclei which for the weakly bound systems lies into the continuum. In this paper some results of experiments concerning the study of reaction mechanisms with halo and loosely bound stable nuclei at energies around the Coulomb barrier will be discussed.

1. Introduction

In the last years several papers have been published where reaction mechanisms, in particular fusion at low bombarding energies, were investigated, in reactions induced by halo and loosely bound nuclei. Weakly bound nuclei and more specifically halo nuclei, are characterized for having the ground state very close to the particle emission threshold, therefore, in theoretical calculations, in order to proper describe the reaction dynamics, coupling to bound and break-up states must be considered. It is well known in fact, that in order to reproduce the behavior of the cross-section for the different reaction processes at very low energies, dynamical effects, such as the coupling of the relative motion of projectile and target to their intrinsic excitations or to other reaction channels, as for instance transfer, must be included in the calculations. In particular, since weakly bound nuclei have no or at most one bound excited state, coupling to inelastic channels implies coupling to the continuum. This coupling is calculated, within a Coupled Channel approach by performing Continuum Discretized Coupled Channel (CDCC) calculations where the continuum is considered as a sum of discrete states. CDCC calculations predict that, in reactions induced by halo nuclei, dynamical effects enhance the sub-barrier total fusion cross-section with respect to the no-coupling case. In addition, due to the break-up, a suppression of the complete fusion cross-section is predicted by CDCC calculations at energies above the barrier [1]. Conversely, different models as for example [2], based on a time-dependent wave-packet method predicts that the cross-section in the neutron halo case is slightly suppressed than in the non-halo case.

Low energy fusion cross-sections measurement with radioactive halo beams is a difficult task owing to the low intensities of the beams and the small cross-sections involved. Therefore, information on the effect of coupling to break-up can be more easily obtained either by measuring elastic scattering, which has a large cross-section, or using stable weakly bound beams which have much higher intensities than radioactive beams. In this paper an overview of the progress made in reaction studies with halo and weakly bound stable nuclei is presented.

2. Elastic scattering and direct processes in reactions induced by halo nuclei

Most of the existing data concerning low energy reaction studies with halo nuclei, have been performed using ^{6}He as a beam. Elastic scattering and reaction data on several targets are nowadays available and this has stimulated a lot of theoretical work as well. Some data, such as quasi elastic (QE) angular distribution and fusion excitation function, exist also for collisions induced by ^{11}Be beam and, very recently, elastic scattering data induced by the proton halo ^{8}B on medium mass ^{58}Ni have been published [3].

Low energy ^{6}He induced elastic scattering on medium and heavy mass targets, shows a reduction of the elastic cross-section in the rainbow region see e.g. [4, 5] as well as at large scattering angles. In order to describe properly the reduced cross-section within the Optical Model (OM), large diffuseness parameters of the imaginary term of the optical potential must be used.

Also CDCC calculations of ^{6}He elastic scattering on different targets have recently been published [6, 7, 8]. According to these calculations, the reduction of the elastic cross-section at the rainbow is originated by the coupling to states into the continuum. ^{6}He is a three body system, since it can be considered as formed by an ^{4}He core plus two neutrons. In order to simplify the very complicated CDCC calculations of a four body reaction process, two body representations of ^{6}He (^{4}He core plus a di-neutron) are in many cases considered e.g. [6]. More recently, four-body (n-n-alpha+target) CDCC calculations have been developed [7, 8] and good agreement with the data is obtained in some cases. Also these calculations show that coupling to the continuum suppresses the elastic cross-section at the rainbow. In [9], coupling to single neutron stripping and its importance in describing the low-energy scattering of a halo nucleus was investigated. According to the authors f Ref. [9] it has a significant effect on the elastic scattering of ^{6}He and should not be neglected. Coupling to the 2n stripping, which was found to have a large experimental cross-section [10, 11], larger than the 1n stripping and also break-up, could have a major role on the elastic scattering. Inclusive measurements, of reaction processes with ^{6}He beam, have shown that a large cross-section (about 80% of the total reaction cross-section), is due to direct processes (see e.g. [4]). Recently, Chatterjee et al. [10] measured the coincidences between outgoing α-particles, neutrons and γ-rays in the reaction ^{6}He+^{65}Cu. From this experiment the cross-section for the 1n, 2n transfer and the break-up was derived. The cross-section for the 2n transfer was found to be dominating the direct reaction processes (a factor of ten larger than the 1n transfer). This result is in agreement with [11] where the outgoing α-particle was measured in coincidence with two neutrons. In the ^{6}He+^{65}Cu experiment of [10], the elastic scattering cross-section was also measured. Coupled Reaction Channel calculations, which included coupling to 1n and 2n transfer, were performed. These calculations show that coupling to the 2n transfer has a stronger effect than the 1n transfer coupling. The inclusion of the 2n transfer allows to reproduce the magnitude of the elastic cross-section, even if the shape of the cross-section is not reproduced.

Quasi elastic scattering of ^{11}Be+^{209}Bi have been measured by Mazzocco et al. [12]. This experiment was performed at Riken by using a fragmentation beam degraded in energy. The beam energy resolution was insufficient to measure elastic scattering and, as above mentioned, the QE cross-section was measured instead. The QE cross-section included the 1st excited state of ^{11}Be, which is bound and is only 320 keV from the ground state, and the target excited states up to 2.6 MeV. The quality of the data does not allow to see whether elastic scattering induced by the halo ^{11}Be nucleus shows a suppression in the rainbow region as observed in the elastic scattering induced by ^{6}He. On the QE data, Coupled Channel calculations were performed, in order to extract the total reaction cross-section. The extracted total reaction cross-section was

compared to the reaction cross-section for the weakly bound ^{9}Be isotope onto the same ^{209}Bi target, which was measured by the same group [13]. The total reaction cross-section, for the two Be isotopes, is found to be very similar. The authors had also previously measured the fusion excitation function for ^{9,11}Be on ^{209}Bi target, which had also a similar cross-section in the whole energy range explored [14]. The authors concluded that since the absorption is mainly due to fusion and break-up, the breakup process, as the fusion process, must have comparable cross-sections in both ^{9,11}Be nuclei. No difference was found in the behavior of the halo ^{11}Be nucleus and the non-halo ^{9}Be nucleus. This result was in disagreement with what previously found with ^{6}He. However, since both ^{9}Be and ^{11}Be are weakly bound, the break-up channel could play a similar role for the two systems.
Recently, we measured the elastic scattering angular distribution of ^{9,10,11}Be on a medium mass ^{64}Zn target, at the same center of mass energy of 25.4 MeV, using ISOL ^{10,11}Be beams. The experiment with ^{9}Be beam was performed at the 14 MV Tandem of Laboratori Nazionali del Sud (LNS) in Catania. The experiments with the radioactive ^{10,11}Be isotopes were performed at REX-ISOLDE at CERN. Preliminary results show that the total reaction cross-section deduced from OM analysis is $\sigma_R \approx 1$ b for ^{9}Be and ^{10}Be whereas $\sigma_R \approx 2$ b for ^{11}Be. Contrary to what it has been found by [12], a much larger total reaction cross-section is measured in the halo nucleus case. This conclusion disagrees with the result of [12] but agrees with the results obtained with ^{6}He.
The elastic scattering induced by the proton halo ^{8}B nucleus on a ^{58}Ni target was measured at the Twinsol radioactive beam facility of the University of Notre Dame at five beam energies [3]. The total reaction cross-section was extracted from the Optical Model analysis of the elastic data. The results show that similarly to what observed in the ^{6}He data, and also in our ^{11}Be data, the ^{8}B induced reaction has a much larger total reaction cross-section if compared with the reaction cross-section of weakly-bound non-halo nuclei. The authors of [3] attribute this large reaction cross-section to the contribution of break-up and not to transfer as was experimentally observed in the case of ^{6}He induced reactions.

3. Elastic scattering induced by loosely bound nuclei: threshold anomaly effect

Important information about dynamical effects played by the coupling with break-up states, can be obtained by studying reactions induced by loosely bound projectiles. Recently, many experiments have been performed in order to study the threshold anomaly in elastic scattering of ^{6,7}Li and ^{9}Be on different targets. The optical potential extracted from the analysis of elastic scattering of heavy ions, involving tightly bound nuclei, it is known to show a rapid variation with energy near the Coulomb barrier. This energy dependence of the OM potentials is called threshold anomaly, see e.g. [15]. The usual threshold anomaly has been ascribed to the coupling of the elastic scattering to the nonelastic channels. The real part of the OM potential shows a bump at the barrier energy; correspondingly, the imaginary potential decreases. The decrease of the imaginary potential can be understood as being originated by the closure of nonelastic channels at energies near the Coulomb barrier. In weakly bound nuclei the breakup channel is expected to be a relevant reaction channel even at low energies, below the Coulomb barrier. The coupling to the breakup produces a repulsive polarization potential and the usual threshold anomaly may disappear. It has recently been suggested that a new type of anomaly is present in the scattering of weakly bound nuclei, named as 'breakup threshold anomaly' [16]. In this case the strength of the imaginary potential increases as the incident energy decreases, below the Coulomb barrier. The effects of the breakup channel on the elastic scattering of ^{6,7}Li have been studied for different targets [17, 18, 19, 20, 21, 22, 23, 24, 25, 26]. However, conclusions concerning the possible absence of the usual threshold anomaly are contradictory. Keely et al. [17] have investigated elastic scattering of ^{6,7}Li+^{208}Pb at energies around the Coulomb barrier. It was found that the energy dependence of the optical potential is consistent with the existence of the usual threshold anomaly for the ^{7}Li projectile and its absence for the ^{6}Li projectile. According to CDCC calculations [27] the difference for these systems is mainly due to the difference in their breakup thresholds, $S_\alpha = 1.48$ MeV for ^{6}Li and $S_\alpha = 2.45$ MeV for ^{7}Li. The data for the ^{6}Li+^{208}Pb system were reanalyzed by Hussein et al. [16] and the results for the energy dependence of the optical potential have been interpreted as an evidence of a new type of threshold anomaly, named as mentioned above, 'breakup threshold anomaly'. Maciel at al. [18] measured the elastic scattering of ^{6,7}Li+^{138}Ba at near barrier energies. The results of the optical model analysis are similar to those for the ^{208}Pb target [17]. The presence of the usual threshold anomaly for the ^{7}Li elastic scattering has been interpreted as the effect of the strong couplings with the 1st excited state of ^{7}Li (0.48 MeV) as well as other inelastic and direct channels. The ^{6}Li nucleus has no bound excited state, for this reason the breakup channel is expected to be the dominant direct reaction channel. The repulsive polarization potential produced by breakup may compensate the attractive polarization potential produced by other direct and inelastic channels. Gomes *et al.* [24] reanalyzed the ^{6,7}Li+^{138}Ba data and the results suggest the presence of the breakup threshold anomaly for both projectiles. Elastic scattering of ^{6,7}Li nuclei on light targets ^{28}Si and ^{27}Al was measured by Pakou et al. [19, 20] and Figueira et al. [21, 22], respectively. In the case of the ^{28}Si target, no usual threshold anomaly was observed [19, 20]. Similar results were obtained for the ^{7}Li+^{27}Al system [21]. On the other hand, the ^{6}Li+^{27}Al scattering suggests the presence of the breakup threshold anomaly [22].

The study of the threshold anomaly in the elastic scattering of ^{6,7}Li on the medium mass target ^{59}Co has been performed by Souza *et al.* [23]. The experimental results suggest the presence of the usual threshold anomaly for both systems. On the contrary elastic scattering of ^{6}Li on medium mass nuclei ^{58,64}Ni [25] and ^{90}Zr [26] is consistent with the presence of the breakup threshold anomaly. Summarising, for the ^{6}Li scattering on several targets, the results seem to indicate the presence of a break-up threshold anomaly, whereas this is not clear for ^{7}Li where in some analysis the optical potential seem to show the regular threshold anomaly. The different result has been justified as due to the higher break-up threshold of ^{7}Li and to the coupling to the 1st excited state which, in ^{7}Li is bound and in ^{6}Li is unbound. The behavior the of ^{9}Be elastic scattering, should be very similar to that of ^{6}Li. In fact, they have similar break-up thresholds and both do not have bound excited states. However the results for ^{9}Be are rather controversial as for ^{7}Li [28, 29].

At low bombarding energies, below the Coulomb barrier, the elastic cross-section deviates very little from the Rutherford cross-section and the OM analysis is very sensitive to small variation of the cross-section. Therefore, one has to be sure that all systematic errors in the elastic scattering angular distribution measurement are minimized. We recently measured at LNS in Catania, the elastic scattering ^{6}Li+^{64}Zn in the energy range 10 MeV$\leq E_{c.m.} \leq$ 20 MeV. Particular care was taken in order to reduce all source of systematic errors. The OM analysis was performed using a double folding potential. The same shape of the real and imaginary potential was used and the parameters of the fit were the two normalization constants N_R and N_I. The results of the analysis are consistent with the so called 'break-up threshold anomaly' thus confirming the previous results of ^{6}Li measurements.

4. Fusion cross-section in halo and weakly bound induced reactions

Typical intensities of radioactive beams are 10^5-10^7 pps, these are several orders of magnitude smaller than the ones of stable beams. For this reason, in experiments with radioactive beams it is extremely difficult to measure at sub-barrier energies where cross-sections are very small. In particular the fusion cross-section decreases exponentially as the energy decrease below the barrier. Therefore, due to the low measured yield, fusion excitation function do not extend down to deep sub-barrier energies. An open issue is how to evidence, static effects due to the halo as well as the above mentioned dynamical effects due to the coupling to the continuum.

The ideal would be having realistic CDCC calculations to be compared with the experimental data. These, especially in the case of ^{6}He, are still missing owing to the complicate three body structure of this nucleus. The experimental data have been compared with calculations of single barrier penetration, standard CC calculations, or with data involving reactions induced by well bound nuclei on the same target or by comparing data of a reaction with different projectile and target but forming the same compound nucleus [30, 31, 4, 32, 33]. However the result (enhancement/suppression of the cross-section) will depend upon the type of comparison. As an example of discrepancy originated by the different procedure of comparison, in the case of ^{6}He+^{64}Zn of [4, 32] the data are compared with the reaction ^{4}He+^{64}Zn. From that comparison no differences of the two measured fusion excitation function are found in the energy range explored. In [34] the same ^{6}He+^{64}Zn data, are compared with CC calculation without coupling to the continuum. From that comparison, a suppression of the fusion cross-section of the order of 50% at energies above the barrier is claimed and no effect at energies below the barrier is found.
Experiments performed using weakly bound stable nuclei may help to investigate the specific role played by the break-up channel. Reactions induced by the least bound stable nuclei (^{6,7}Li, ^{9}Be) have been studied on several targets see e.g. [35, 36, 37, 38]. The quality of the data is, of course, much better than the one obtained with the radioactive beams. In reactions on heavy targets a reduction of the complete fusion (CF) cross-section is observed at energies above the barrier, whereas no clear effect is observed below the barrier. The reduction was attributed to the projectile break-up in the strong Coulomb field of the target nucleus, followed by incomplete fusion (ICF). No effect was found in the total fusion cross-section, which agrees with the CC calculations [35].
Reactions on medium mass targets have produced different results. Owing to the lower charge of the target nucleus, the ICF process is expected to be small (within the experimental errors). For ^{6,7}Li and ^{9}Be on ^{64}Zn [38] the total fusion cross-section seems not to be affected by the break-up process. For ^{9}Be+^{64}Zn collision the total reaction cross-section was found almost equal to the total fusion cross-section at energies around the barrier. Conversely, the reaction induced by ^{6,7}Li on the same ^{64}Zn target has shown a total reaction cross-section much larger than the total fusion cross-section [38] which is unexpected. However, in a recent publication by Gomes et al. [39] where a universal fusion function (UFF) is derived in order to search for systematic trends in the total fusion excitation function, it is argued that there are problems with the ^{6,7}Li+^{64}Zn data. We recently measured the ^{6}Li+^{64}Zn total fusion excitation function at near and sub-barrier region. The experiment was performed using the same activation technique used to measure ^{6}He+^{64}Zn [4]. This technique is particularly suitable to measure fusion cross-sections when the evaporation residues (E.R.) are so slow that can get stopped into the target. To use this technique, of course, the produced E.R. must be radioactive, and this is the case for ^{6}Li+^{64}Zn total fusion reaction. In figure 1 our results are compared with the data of [38]. As one can see from the figure there is a disagreement between our data and the previously published data. This discrepancy could be originated by the time of flight technique used by Gomes at al. [38] to obtain the fusion cross-section. The low energy heavy fragment produced in fusion or incomplete fusion processes are partially stopped into the target, threshold effect can be responsible for the lower cross-section measured in [38].

5. Summary

In conclusion, reaction studies with halo and weakly bound nuclei at energy around the barrier are currently being extensively performed and a lot of data are becoming available. Elastic scattering of halo nuclei showed a reduced cross-section in the rainbow region. Such a suppression is due to long range absorption that within the CDCC approach can be reproduced once the coupling to the continuum is considered. Coupling to break-up seems also responsible for the

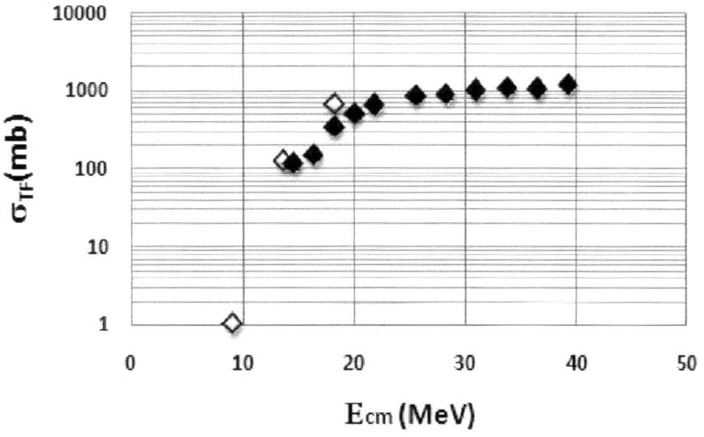

Figure 1. Total fusion excitation function for $^{6}Li+^{64}Zn$. Closed rhombs this work, open rhombs data from [38].

absence of the threshold anomaly in the OM potentials in reactions with weakly bound nuclei. Fusion studies with halo nuclei still shows controversy about possible enhancement/suppression effects, depending upon how the experimental data are compared to extract the searched effect. With weakly bound beams very detailed measurements exist with heavy targets where a suppression of the complete fusion cross-section was found at energies above the barrier. On medium mass targets, measurements do not extend down to energies below the barrier and in some cases it seems that the measured cross-section is underestimated. The anomalously large elastic break-up cross-section observed in $^{6,7}Li+^{64}Zn$ seems indeed to be related to a problem with the previously measured data. Our recent measurement, where a different experimental approach was used, do not show this effect.

6. References

[1] A. Diaz-Torres *et al.* 2002 *Phys. Rev.C* **65** 024606.
[2] M. Ito *et al.* 2006 *Phys.Lett. B* **37** 53.
[3] J.J. Kolata 2009 *Phys. Rev. C* **79**, 027603.
[4] A. Di Pietro *et al.* 2004 *Phys. Rev. C* **69** 044613.
[5] A. M. Sánchez-Benítez *et al.*, 2008 *Nucl. Phys. A* **803** 30
[6] A. Moro *et al.* 2007 *Phys. Rev. C* **75**, 064607.
[7] T. Matsumoto *et al.* 2006 *Phys. Rev. C* **73** 051602R.
[8] M. Rodríguez-Gallardo *et al.* 2008 *Phys. Rev. C* **77**, 064609.
[9] N.Keeley and N. Alamanos 2008 *Phys. Rev. C* **77** 054602.
[10] A. Chatterjee *et al.* 2008 *Phys. Rev. Lett.* **101**, 032701.
[11] P.A. De Young *et al.* 2005 *Phys. Rev. C* **71** 051601(R).
[12] M. Mazzocco *et al.* 2006 *Eur. Phys.J. A* **28**, 295.
[13] C. Signorini *et al.*2002 *Nucl. Phys. A* **701** 23c.
[14] C. Signorini *et al.* 1998 *Eur. Phys. Jour. A* **2** 227.
[15] G.R. Satchler 1991 *Phys. Rep.* **199**, 147.
[16] M.S. Hussein *et al.* 2006 *Phys. Rev. C* **73**, 044610.
[17] N. Keeley *et al.* 1994 *Nucl. Phys. A* **571**, 326.
[18] A.M.M. Maciel *et al.*1999 *Phys. Rev. C* **59** 2103.
[19] A. Pakou *et al.* 2003 *Phys. Lett. B* **556** 21.

[20] A. Pakou *et al.* 2004 *Phys. Rev. C* **69** 054602.
[21] J.M. Figueira *et al.* 2006 *Phys. Rev. C* **73** 054603.
[22] J.M. Figueira *et al.* 2007 *Phys. Rev. C* **75** 017602.
[23] F.A. Souza *et al.* 2007 *Phys. Rev. C* **75** 044601.
[24] P.R.S. Gomes *et al.* 2005 *J. Phys. G* **31** S1669.
[25] M. Biswas *et al.* 2008 *Nucl. Phys. A* **802** 67.
[26] H. Kumawat *et al.* 2008 *Phys. Rev. C* **78** 044617.
[27] N. Keeley and K. Rusek 1998 *Phys. Lett. B* **427** 1.
[28] C. Signorini *et al.* 2000 *Phys. Rev. C* **61**, 061603.
[29] R.J. Woolliscroft *et al.* 2004 *Phys. Rev. C* **69** 044612.
[30] J.J. Kolata *et al* 1998 *Phys.Rev.Lett.* **81** 4580.
[31] R.Raabe *et al.* 2004 *Nature* **431**823.
[32] A. Dipietro 2007 *Eur.Phys.J. Special Topics* **150** 15.
[33] Yu.E. Penionzhkevich *et al.* 2007 *Eur. Phys. J. A* **31** 185.
[34] E. Crema *et al.* 2007 *Phys. Rev. C* **75** 037601 .
[35] M. Dasgupta *et al.* 2004 *Phys. Rev. C* **70** 024606 and reference therein.
[36] C. Beck *et al.* 2003 *Phys. Rev. C* **67** 054602.
[37] A. Mukherjee *et al.* 2006 *Phys. Lett. B* **636** 91.
[38] P.R.S. Gomes *et al.* 2005 *Phys. Rev. C* **71** 034608.
[39] P.R.S. Gomes *et al.* 2009 *Phys. Rev. C* **79**, 027606.

Journal of Physics: Conference Series **205** (2010) 012043 doi:10.1088/1742-6596/205/1/012043

New developments on the Recoil Distance Doppler-Shift Method

C Fransen[1], A Dewald[1], T Baumann[2], D Bazin[2], A Blazhev[1], B A Brown[2,3], A Chester[2,3], A Gade[2,3], T Glasmacher[2,3], P T Greenlees[4], M Hackstein[1], S Harissopulos[5], U Jakobsson[4], J Jolie[1], P M Jones[4], R Julin[4], S Juutinen[4], S Ketelhut[4], T Konstaninopoulos[5], A Lagoyannis[5], M Leino[4], P Nieminen[4], M Nyman[4], P Petkov[1,6], P Peura[4], T Pissulla[1], P Rahkila[4], W Rother[1], P Ruotsalainen[4], J Saren[4], C Scholey[4], J Sorri[4], K Starosta[2,3,7], A Stolz[2], J Uusitalo[4], P Voss[2,3], D Weisshaar[2]

[1] Institut für Kernphysik, Universität zu Köln, 50937 Köln, Germany
[2] National Superconducting Cyclotron Laboratory, Michigan State University, East Lansing, Michigan 48824, USA
[3] Department of Physics and Astronomy, Michigan State University, East Lansing, Michigan 48824, USA
[4] Department of Physics, University of Jyväskylä, Jyväskylä FIN-40014, Finland
[5] INP, N.C.S.R. Demokritos, 153.10 Aghia Paraskevi, Athens, Greece
[6] INRNE, Bulgarian Academy of Sciences, 1784 Sofia, Bulgaria
[7] Department of Chemistry, Simon Fraser University, Burnaby, B.C., Canada V5A 1S6

E-mail: `fransen@ikp.uni-koeln.de`

Abstract. Absolute transition probabilities are fundamental observables for nuclear structure. The recoil-distance-Doppler-shift (RDDS) technique, also called plunger technique, is a well established tool for the determination of these important experimental quantities via the measurement of lifetimes of excited nuclear states. Nowadays nuclear structure investigations are concentrated on exotic nuclei which are often produced with extremely small cross sections or with very low beam intensities. In order to use the RDDS technique also for the investigation of very exotic nuclei this method has to be adapted to the specific needs of these special reactions. This article gives an overview on recent RDDS measurements with the new differential plunger in combination with particle detectors and recoil spectrometers. These were done with projectile multistep Coulomb excitation at low beam energies ($\approx 5\,\mathrm{MeV/u}$) and at intermediate beam energies ($\approx 100\,\mathrm{MeV/u}$) using one step Coulomb excitations and knockout reactions.

1. Introduction

Present day nuclear structure physics focuses on the investigation of exotic nuclei far away from the valley of stability. Absolute transition strengths of excited states of these nuclei provide crucial information on the nuclear structure. This allows to extend the knowledge on the evolution of nuclear properties to extremely proton or neutron rich nuclei and gives the opportunity to test nuclear models in these regions. Absolute transition strengths can be determined from lifetimes of excited nuclear states. However, the measurement of lifetimes in exotic nuclei is very challenging because such nuclei can often be produced only with low production rates due to extremely low cross sections or beam intensities.

The recoil distance Doppler-shift method (RDDS) is a very valuable method for the determination of lifetimes of excited nuclear states and it was discussed in detail, e.g., in [1]. The advantage of this method is that it does not depend on the excitation mechanism. In this article we will describe the adoption of the RDDS method to the measurement of lifetimes of excited nuclear states of very exotic nuclei. These nuclei are typically investigated in experiments with inverse kinematics because the beam production is done, e.g., with in-flight fragmentation and the identification of the exotic beams before and after the secondary target, where the nuclei of interest are excited, is required. Different reactions like fusion, Coulomb excitation, or knockout reactions can be used. In section 2 we will briefly describe the experimental method. In section 3 a lifetime measurement on the stable nucleus ^{128}Xe with multistep Coulomb excitation in inverse kinematics is discussed. This served mainly as a test of the novell experimental method which is expected to be efficiently used in future radioactive beam experiments at energies of about 5–10 MeV/u. The following section 4 gives an overview on the measurements performed in the recent past on radioactive nuclei with the Cologne differential plunger at the National Superconducting Coupled Cyclotron Laboratory (NSCL) at Michigan State University, where both Coulomb excitation and knockout secondary reactions were used.

2. The Cologne differential plunger

The recoil distance Doppler-shift method (RDDS) is a standard method to measure lifetimes of excited nuclear states in the picosecond range. The beam hits a thin target foil and after the nuclear reaction the recoils leave the target foil with the average speed v. Excited states in these nuclei decay via γ decay both after the recoil is stopped in a stopper foil with an unshifted γ ray with energy E_u and in flight whereas (for the non-relativistic case) a Doppler-shifted γ ray with energy

$$E_s = E_u \cdot (1 + \frac{v}{c} \cdot \cos\theta) \tag{1}$$

is observed in a detector positioned under an angle θ relative to the beam axis (see Fig. 1).

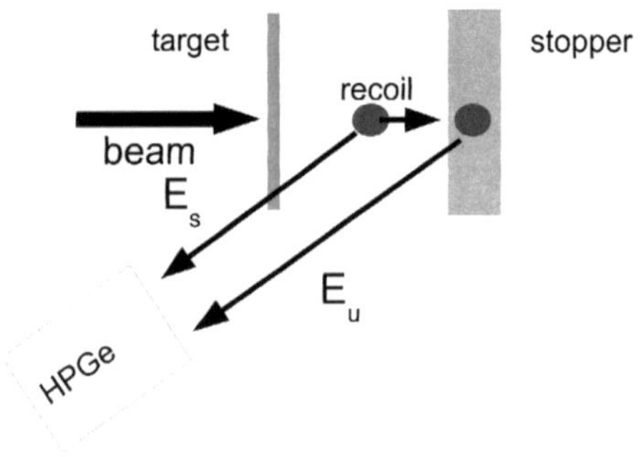

Figure 1. Principle of the RDDS method. E_s and E_u are the Doppler-shifted and unshifted γ-ray energies, respectively, observed with a high-purity Germanium detector.

From the intensities I_s and I_u of shifted and unshifted components for different target to stopper distances and known reaction kinematics one can deduce the lifetime of an excited state independent of the reaction mechanism via the differential decay curve method (DDCM) [2, 3]. Typically, an excited state is populated both from the continuum after a nuclear reaction and also from discrete higher lying states with finite lifetimes. Thus the decay curve of the state of interest involves the population and depopulation history of all states that decay directly or indirectly to this state. The fundamental differential equation follows from the radioactive decay law $\dot{n}_i(t) = -\frac{1}{\tau_i} n_i(t) = -\lambda_i n_i(t)$ and the superposition of all involved states

$$\dot{n}_i(t) = -\lambda_i n_i(t) + \sum_h b_{hi} \lambda_h n_h(t), \tag{2}$$

where $n_h(t)$ is the number of nuclei in a feeding level l_h at time t and b_{hi} is the decay branching ratio of the transition from level l_h to l_i.

After integration and transformation of this equation one can deduce the fundamental equation of the DDCM for lifetime determination, where N_h, N_i are the numbers of nuclei that decayed after a time t from level h, i, respectively

$$\tau_i(t) = \frac{-N_i(t) + \sum_h b_{hi} N_h(t)}{\frac{d}{dt} N_i(t)}. \tag{3}$$

N_i and N_h are proportional to the flight and stop peak intensities. One gets an independent value $\tau_i(t)$ for the lifetime of the level of interest for every target to stopper distance d_i, whereas $t_i = d_i/v$. If the values for $\tau_i(t)$ are plotted as a function of the distance d_i a constant value is expected. A deviation from this is an indicator for a systematical effect that has to be taken into account, like, e.g., unobserved sidefeeding from higher lying levels or deorientation. The latter results from hyperfine interaction of the spins of the recoils with the electron shell during the flight and causes a change of the nuclear orientation with time and thus a change of the angular distribution of the emitted γ rays. This effect is discussed in section 3. The analysis of $\gamma\gamma$ coincidences can circumvent this problem as is pointed out in [4], but due to the aforementioned low production cross sections for radioactive ion beams resulting in low statistics of the measured γ rays usually only singles spectra can be obtained.

For measurements with radioactive ion beams typically recoil identification is necessary because the reaction produces a cocktail beam. In that case often a mass or fragment separator is used to identify the recoils in coincidence with the emitted γ rays. Therefore, the stopper foil of the "standard" plunger has to be replaced by a degrader foil that slows down the recoils rather than stopping them. The degrader foil has to be chosen in a way that a clear separation of the two Doppler-shifted γ-ray lines, one emitted between target and degrader ("flight peak") and one emitted after the degrader ("degraded peak"), becomes possible.

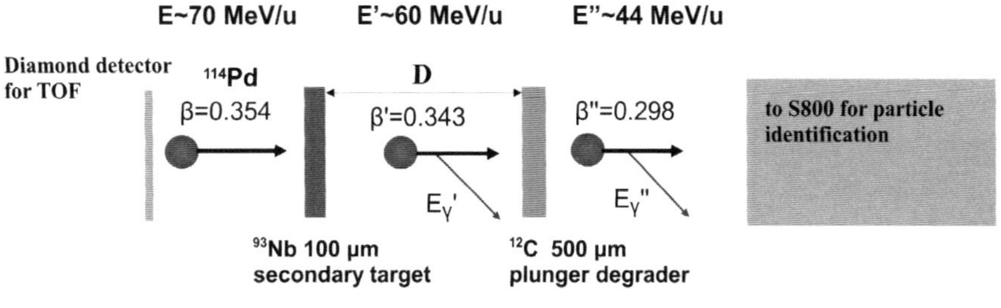

Figure 2. Example of a measurement with the differential plunger. The stopper is replaced by a degrader to allow an identification of the recoils. Given are the energies of a ^{114}Pd beam before and after the target and after the degrader and the correspondig velocities.

Figure 2 depicts the principle of the differential plunger. As an example, the data of a measurement with Coulomb excitation in inverse kinematics on ^{114}Pd at NSCL with the S800 mass separator to identify the ^{114}Pd recoils in a cocktail beam [5] is included.

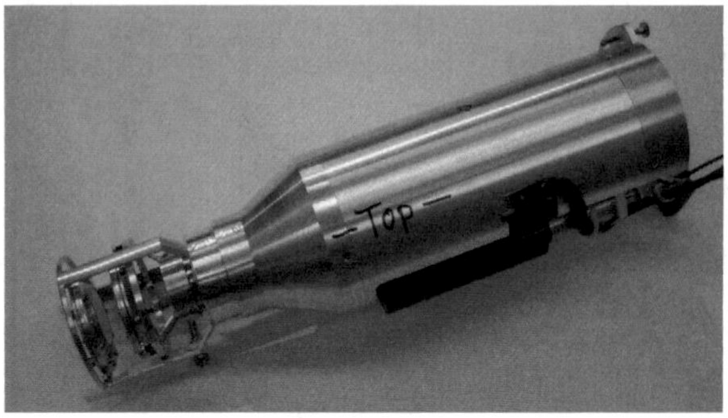

Figure 3. Photograph of the new Cologne differential plunger.

Figure 3 shows a photograph of the new Cologne differential plunger as it was used for the experiments with radioactive ion beams at the NSCL. There are several modifications as compared to the "standard" Cologne plunger for stable beams. Target and degrader foils with diameters up to 4 cm can be used. The foils can be moved to a maximum distance of 2.5 cm with a precision of 1 μm. This large separation is necessary to perform measurements at intermediate beam energies. In the following sections we will give an overview on the recent experiments with the differential plunger.

3. Experiment on ^{128}Xe

An experiment on the stable nucleus ^{128}Xe was performed at the University of Jyväskylä, Finland, with the Cologne plunger and the JUROGAM Germanium detector array. The measurement was done using Coulomb excitation in inverse kinematics. The main aim was the proof of the method of combining the RDDS technique with projectile multistep Coulomb excitation because such measurements are of high interest for future radioactive beam experiments. Further, $E2$ transition strengths in ^{128}Xe were to be measured since this nucleus has been a considered candidate for an E(5) nucleus, i.e., a nucleus at the critical point of the shape phase transition from a spherical vibrator to a γ-soft nucleus [6].

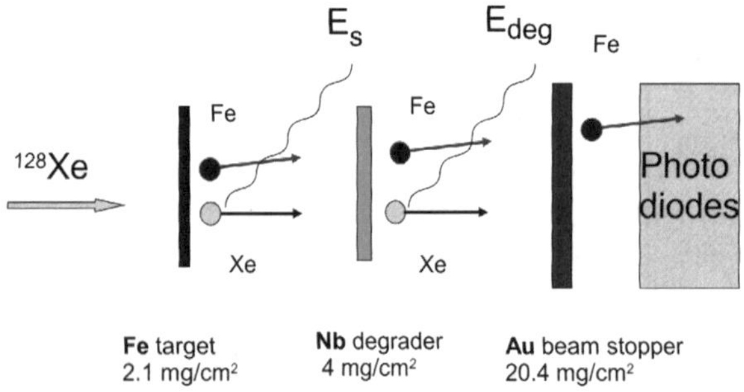

Figure 4. Sketch of the setup of the Coulomb excitation experiment on ^{128}Xe with the differential plunger.

A sketch of the setup is given in Fig. 4. The ^{128}Xe nuclei with an energy of $E(^{128}\mathrm{Xe}) =$ 525 MeV are Coulomb excited on an Fe target foil and Fe nuclei are knocked out of the target. As a degrader, a 4 mg/cm^2 Nb foil was used. However, ^{128}Xe nuclei are also Coulomb excited on the degrader foil. To separate projectile Columb excitation on target and degrader the γ rays from ^{128}Xe are observed in coincidence with Fe target recoils detected with photo diodes mounted behind a gold beam stopper foil that also stops the Nb degrader recoils. Furthermore, the photo diodes are arranged such to allow a separation of different recoil angles to improve the energy resolution of the emitted γ rays. Otherwise, the large angle spread of the ^{128}Xe nuclei after Coulomb excitation would prevent a clear separation of flight and degraded components of the γ-ray lines of interest.

As mentioned before, deorientation is an effect that has to be taken into account when analyzing a RDDS measurement in γ-ray singles spectra. The hyperfine interaction causes the original nuclear spin alignment to diminish in flight as a function of the interaction time.

Figure 5 shows the intensities of flight and degraded component of the $4_1^+ \rightarrow 2_1^+$ transition in ^{128}Xe. Obviously the sum of these intensities is not constant for different distances due to deorientation (most upper curve). Thus a correction had to be performed for the γ-ray intensities at different distances between target and degrader. The upper curves for both the flight and degraded component give the corrected intensities used for the lifetime determination. Details on these corrections are given in [7].

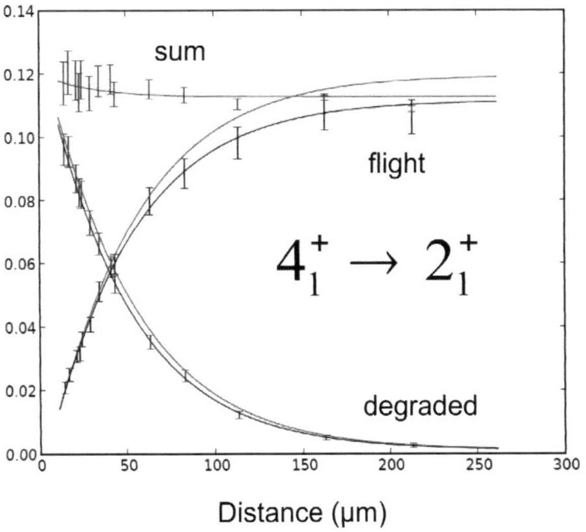

Figure 5. Effect of the deorientation on the intensities of the $4_1^+ \rightarrow 2_1^+$ γ-ray transition including the uncorrected (lower curves for flight and degraded intensities) and corrected curves (upper ones for both cases).

Our experiment yielded a total of six lifetimes of excited nuclear states and eleven $E2$ transition strengths of the lowest collective excitations in ^{128}Xe. Lifetimes of the yrast states up to the 6_1^+ state and of the 2_2^+, 4_2^+, and 3_1^+ states were determined. The data for the yrast states are in quite good agreement with the literature values for these states [8]. More details on this experiment and the interpretation of the results can be found in [7] and will be published in a forthcoming article.

4. Experiments on radioactive isotopes

In this section we give an overview about recent lifetime measurements on radioactive isotopes with the new Cologne differential plunger at NSCL. The experiments used the combination of the NSCL Coupled Cyclotron Facility for the acceleration of the primary beams, the A1900 mass separator for rare isotope selection [9], the diamond timing detector for particle identification [10], the SeGA Germanium detector array for observing the Doppler-shifted γ rays [11], and the Cologne differential plunger mounted at the target position of the S800 spectrograph for identification of the reaction products [12] (see also [13]).

4.1. Knockout reactions

The aim of this experiment, described in detail in [13], was the measurement of lifetimes of the lowest excited states in ^{64}Ge and ^{62}Zn in single neutron knockout reactions. A cocktail beam of 5% ^{65}Ge, 35% ^{64}Ga, 52% ^{63}Zn, and 8% ^{63}Cu was produced via in-flight fragmentation of ^{78}Kr at 150 MeV/u [14]. A 500 μm thick ^{nat}C target and a 250 μm thick ^{93}Nb degrader were used. This resulted in velocities of the reaction producs of $\beta_{tar} \approx 0.39$ after the target and $\beta_{deg} \approx 0.35$ after the degrader. The Doppler-shifted γ rays were observed with the SeGA array with two rings at laboratory angles of 30° and 140° and at three target to degrader distances of 0, 200, and 500 μm. Fig. 6 shows the Doppler-shifted $2_1^+ \rightarrow 0_1^+$ decay transition in ^{62}Zn for illustration. The lineshapes also included in Fig. 6 were calculated from the velocities after target and degrader, considering relativistic effects, and only lifetime and normalization factor were taken as free parameters.

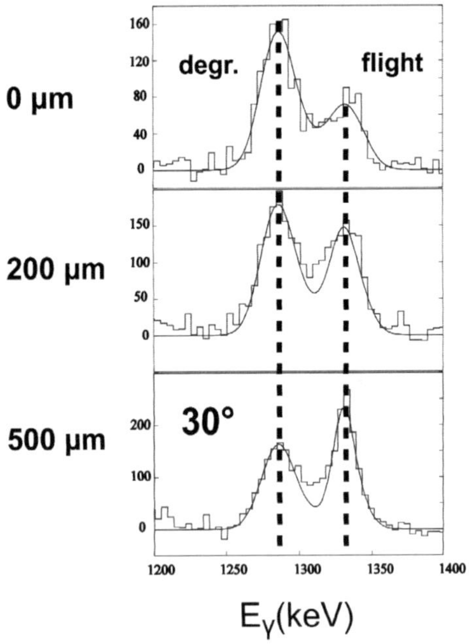

Figure 6. Doppler-shifted $2_1^+ \rightarrow 0_1^+$ transition in ^{62}Zn for three target to degrader distances.

For ^{62}Zn a lifetimes of the 2_1^+ state of $\tau = 4.2(7)$ ps was measured in perfect agreement with a value from the literature of 4.2(3) ps [15]. For the $N = Z$ nucleus ^{64}Ge the results for the first measurements of the lifetimes of the 2_1^+ and 2_2^+ states are 3.3(5) and 8^{+4}_{-2} ps, respectively [13]. A comparison of these results with GXPF1A shell model calculations can be also found in [13].

4.2. Investigation of heavy Palladium isotopes with Coulomb excitation

Absolute transition strenghts of the 2_1^+ states in ^{110,114}Pd were measured with the setup described in the previous section with the RDDS technique following Coulomb excitation at intermediate energies [5]. The ^{110}Pd data were used to check the data analysis and this novel experimental technique. The motivation of the measurement of ^{114}Pd is the fact that the known

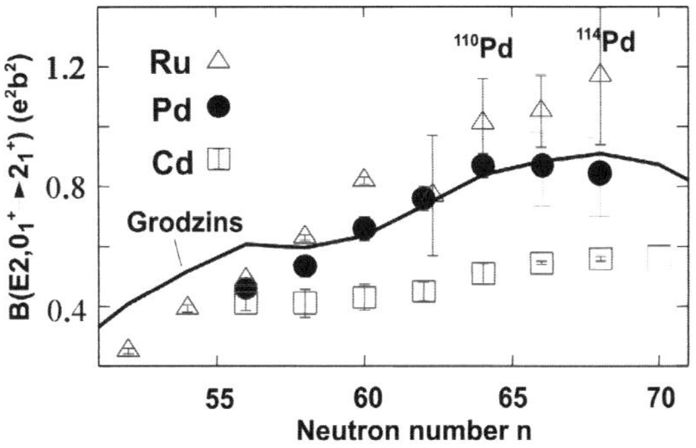

Figure 7. $E2$ excitation strenghts to the 2_1^+ states in Pd isotopes including the new value for ^{114}Pd from [5]. The data is compared to that of the neighboring Ru and Cd isotopic chains and to a prediction with the Grodzins rule.

$B(E2, 2_1^+ \rightarrow 0_1^+)$ value [16] strongly deviated from predictions of the Interacting Boson Model [17] or from the Grodzins relation [18].

The intensities of the ^{110}Pd and ^{114}Pd beams were 11 000 pps and 1500 pps, respectively. They impinged on a 100 μm ^{93}Nb target and a 500 μm thick carbon degrader. The velocities and energies before and after the target and after the degrader are given in Fig. 2. The experiment yielded lifetimes of the 2_1^+ states in ^{110}Pd and ^{114}Pd of τ = 67(8) ps and τ = 118(20) ps, respectively. The value for ^{110}Pd is in a good agreement with the average literature value which gives us confidence in the method. The value for ^{114}Pd is more than two times smaller than the previous result [16]. The resulting $E2$ strengths are plotted in Fig. 7 in comparison to the prediction of the Grodzins relation and to the neighboring isotopes. The new $B(E2)$ value for ^{114}Pd fits nicely in the systematic trend deduced for the ligher isotopes.

This work was supported by the DFG under Contract DE1516/-1 and partly under JO391/4-1, by the U.S. NSF under Grant PHY-0606007, and partly by the National Science Fund of Bulgaria, Ministry of Education and Science under Contract RIC-02/2007.

References

[1] Petkov P, Tonev D, Gableske J, Dewald A, Klemme T and von Brentano P 1999 *Nucl. Instr. Meth.* A **431** 208
[2] Dewald A, Harissopulos S and von Brentano P 1989 *Z. Phys.* A **334** 163
[3] Böhm G, Dewald A, Petkov P and von Brentano P 1993 *Nucl. Instr. Meth.* A **328** 248
[4] Petkov P, Dewald A and von Brentano P 2001 *Nucl. Instr. Meth.* A **457** 527
[5] Dewald A *et al.* 2008 *Phys. Rev.* C **78** 051302(R)
[6] Iachello F 2000 *Phys. Rev. Lett.* **85** 3580
[7] Rother W 2009 Diploma thesis, University of Cologne (unpublished)
[8] Kanbe M and Kitao K 2001 *Nuclear Data Sheets* **94** 227
[9] Morrissey D, Sherrill B, Steiner M, Stolz A and Wiedenhoever I 2003 *Nucl. Instr. Meth.* B **204** 90
[10] Stolz A *et al.* 2006 *Diam. Relat. Mater.* **15** 807
[11] Mueller W *et al.* 2001 *Nucl. Instr. Meth.* A **466** 492
[12] Bazin D *et al.* 2003 *Nucl. Instr. Meth.* B **204** 629
[13] Starosta K *et al.* 2007 *Phys. Rev. Lett.* **99** 042503
[14] Stolz A *et al.* 2005 *Nucl. Instr. Meth.* B **241** 858
[15] Junde H and Singh B 2000 *Nuclear Data Sheets* **91** 317
[16] Blachot J 2002 *Nuclear Data Sheets* **97** 593
[17] Kim K H *et al.* 1996 *Nucl. Phys.* A **604** 163
[18] Grodzins L 1962 *Phys. Lett.* **2** 88

Fast timing: Lifetime measurements with $LaBr_3$ scintillators

I Deloncle[1], B Roussière[2], M A Cardona[3], D Hojman[3], J Kiener[1], P Petkov[4], D Tonev[4] and Ts Venkova[4]

[1]Centre de Spectrométrie Nucléaire et de Spectrométrie de Masse, CNRS/Université Paris-Sud XI, 91405 Orsay Campus,
[2]Institut de Physique Nucléaire,CNRS-IN2P3/Université Paris-Sud XI,
[3]Departamento de Fisica, CNEA, Buenos Aires, Argentina,
[4]Institue of Nuclear Research and Nuclear Energy, BAS, Sofia, Bulgaria.

E-mail: Isabelle.Deloncle@csnsm.in2p3.fr, roussier@ipno.in2p3.fr, cardona@tandar.cnea.gov.ar, hojman@tandar.cnea.gov.ar, petkov@inrne.bas.bg, Dimitar.Tonev@lnl.infn.it,venkova@inrne.bas.bg

Abstract. In order to perform half-life measurements at Alto, we developed a fast-timing set-up using fast scintillators : $LaBr_3$. The main components of our setup and their characteristics (timing properties) are presented. The resulting time resolution, obtained after optimization, and measured off-line with radioactive sources is discussed. This set-up was tested on-line by measuring using the slope method half-lives of $^{137-139}$Cs levels fed by the β^- decay of $^{137-139}$Xe. The neutron-rich Xe nuclei were produced by ^{238}U photofission using, at low intensity, the electron beam and the line of the future facility ALTO. Our results show the ability of such a fast-timing system to measure half-lives lower than the total time resolution. Off-line and on-line results are discussed.

1. Introduction

At the Orsay Tandem, the ALTO project [1] will soon open to nuclear structure studies a large number of exotic nuclei. This project consists in inducing the fission of ^{238}U by the photons emitted during the slowing down of an electron beam in a thick UC target (included in a MK5 ISOL target/source ensemble). The PARRNe-2 device is then used to separate the fragments that exited the source. The LEP-injector Linac has been settled at the Tandem in Orsay to provide the 50MeV electron beam at 10μA intensity. This will lead to a fission rate at ALTO of 10^{11} fission/s, comparable to the ones obtained at ISOLDE, OSIRIS and HRIB. With such a fission rate a large number of Lanthanide nuclei ($^{138-152}$La, $^{143-154}$Ce, $^{144-156}$Pr,$^{145-158}$Nd, $^{148-160}$Pm,$^{152-164}$Sm, $^{155-165}$Eu, $^{158-167}$Gd, $^{161-169}$Tb and $^{166-170}$Dy) [2] should be produced with a yield above 10 atoms per second, allowing, for some of them for the first time, decay spectroscopy to be performed. A lot of new interesting results are expected. Indeed, in this neutron-rich part of the lanthanide region a multitude of different phenomena are predicted, allowing to test our understanding of many nuclear physics aspects. For example, the p-n interaction, due to the orbitals involved, is expected to be strong and to give rise to very large deformation. Reciprocally, in this lanthanide region, a large number of 0^+ excited states are observed that are not expected in deformed nuclei. The measurement of deformation of neutron-rich lanthanides is thus an important goal to be achieved that can be reached by determining the 2_1^+ lifetimes using the fast-timing method with $LaBr_3$ scintillators [3].

2. $LaBr_3$: detectors for direct lifetime measurements

Contrarily to indirect lifetime measurement methods, such as Doppler Shift Attenuation Method, that determine lifetimes from the Doppler shift (or/and the reshaping) of γ-ray lines emitted by recoiling nuclei – and require detectors with extremely good energy resolution –, the fast-timing method is the direct measurement of a time difference between two fast detector (scintillator) signals. The start is given when a populating transition of the interesting level is detected in one detector, and the stop when a deexciting transition of this level is detected in another detector. The determination of this time difference is limited by the time resolution of both detectors: start and stop signals should be well separated, identified as different moments. Therefore, as the decay of a quantum system level is governed by an exponential distribution toward high decay time (stop), lifetime below the time resolution of fast-timing set-u ps can be measured (cf section 6). Very good timing properties are not the only condition for fast-timing detectors. Indeed, the quality of the start and stop depends strongly on the quality of the populating and depopulating γ-ray energy selections and energy resolution is then also an important parameter for such a technique. The fast timing method, developed at first with BaF_2 detectors [4] knows nowadays a renewal with the recent discovery [5] of the $LaBr_3$ scintillators that present good properties on both time and energy response as presented in Table 1 and Table 2.

Table 1. Comparison of main properties of Ge and $LaBr_3$ detectors of same volume.

	Ge	$LaBr_3$
Energy range	keV to MeV	keV to MeV
Abs. Eff.(at 2,54cm from source)[6]	10^{-1} to 10^{-2}	2.10^{-1} to 5.10^{-2}
Rel. Eff. (@1.33MeV)	75%	143%
Energy resolution	1% to 1‰	20% to 3%
Typical time response	250 ns (risetime)	16ns (scint. decay time)
Time resolution	20 ns to 5ns [7]	200 ps(@511keV) [10, 11]

In Table 1 the main charateristics and the evolution of some properties for incident γ energy going from keV to MeV are compared for Ge and $LaBr_3$ crystals with the same volume (cylinders 3"x3"). One can first notice the very high efficiency of the $LaBr_3$ which is at least twice larger than the one of the Ge detector and nearly 40% greater than that of a NaI with the same volume. Moreover, the time resolution of $LaBr_3$ is enhanced by a factor of at least 25 compared to that of a Ge-detector, allowing to measure lifetime of the order of few tens of ps. Of course, Ge detectors are semiconductors and, even cooled to Nitrogen temperature, the gap between valence and conducting bands is small, inducing a high electron yield (300 per keV), and then impressive energy resolution – it goes as the inverse of the square root of the yield – of the order of ‰ at 1MeV. But, as it can be seen in Table 2 in which main properties of most popular scintillators are given (at room temperature), thanks to its high light yield, the $LaBr_3$ has a reasonable energy resolution, the best of the commonly used scintillators, ten times better in particular than that of BaF_2, which is the fastest scintillator nowadays. Nevertheless, the very short decay time of BaF_2 scintillation process (20 times better then for $LaBr_3$) does not fully profit to fast timing set-ups. First, the standard deviation in the timing distribution of the photons emitted by scintillation is proportional to the ratio of the scintillation decay time to the light yield [8, 9], ratio which is then considered as the figure of Merit (F.o.M) of scintillators for timing measurements. From the similar F.o.M. values for BaF_2 and $LaBr_3$ crystals reported in Table 2, one can expect the same order of time resolution in fast-time measurements using either of them. Secondly, a fast-timing set-up includes an other stage of detection with a non neglectable intrinsic time resolution, contribution of which therefore will be reduced when dealing with large amount of particles (statistical reduction of the noise contribution).

Table 2. Comparison of main properties of common scintillators (data from [12] for BaF_2, from [13] for other crystals).

	$LaBr_3$	NaI(Tl)	BaF_2	BGO
Light Yield LY (ph./keV)	61	41	1.8	9
Decay time τ (ns)	16	250	0.7	300
F.O.M (τ/LY) [10]	0.26	6	0.38	33
Energy res. @662keV	2.7	5.6	11.4	9

3. Characteristics of Photomultipliers for fast-timing with $LaBr_3$

In scintillators incident gamma energy is converted in photons and not in electrons as in semi-conductors. A stage of conversion is then required in order to deal with an electric signal. This is performed by photomultipliers, which have non neglectable contribution to time resolution in fast-timing set-up.

A photomultiplier tube (PMT) consists in three different parts, gathered in one "tube". The conversion of photons to electrons, realized by photoelectric effect at the photocathode, has a limited quantum efficiency (QE), from 20% to 35% at 380 nm – the wavelength of $LaBr_3$ scintillation light –. The conversion is followed by electron collection which, due to geometrical and electron-optical factors of focusing electrodes, is only partial. Collected electrons are then transmitted to the multiplication part. It consists in several electrodes : a first dynode, accepting the largest amount of collected electrons; similar dynodes optimized for electron transmission between them; and the last dynode drawn to fit with the anode. The energy needed for secondary electron emission, in which one incident electron gives rise to δ (in mean) secondary electrons with lower energy, is provided (re-established) by an inter dynode potential.

Due to the different operations taking place in a PMT, the optimization of such a detector for applications requiring good energy and time resolutions, such as fast-timing, is very delicate and represents a difficult compromise. If the energy resolution is reduced by increasing the total number of electrons (statistics effect) at all operation, this is not true for the time resolution. Indeed, timing properties of a PMT are enhanced when most electrons have spent similar (short) times in the tube (i.e similar trajectories and speeds) and could be degraded if one enlarges the photocathode (or the collection efficiency, or the number of dynodes) without redrawing other components. As only two firms are making PMT development (Hamamatsu and Electron Tubes – Photonis PMT departement closed in July 2009), only few PMTs can be used for a fast-timing set-up.

Up to now we used, the XP2020 family PMTs [11]. Among them, the XP20D0 is used for $LaBr_3$. It has a borosilicate entry window and about 30% at 380 nm. The XP2020Q with a quartz window allowing UV light and about a QE 25% at 220 nm is more appropriate for BaF_2. In these PMTs, the improved collection between photocathode and first dynode is associated with a new anode conception: an additional screening grid has been located in front of it [14], reducing the time jitter (time resolution). The time resolution measured with a light emitting diode for XP20D0 is between 220±13 ps and 260±13 ps [11], of the same order (250±1) for XP2020Q [15]. As already mentioned the intrinsic PMT time resolution is non negligible but for a scintillator+PMT association, the statistical effects due to the number of photoelectrons reduces the total time resolution. Indeed, it has been shown in ref.[11, 14] for $LaBr_3$ and BaF_2 respectively, that the product of total time resolution with the square root of the photoelectron number is constant, the value of the constant depending of the scintillator type (BaF_2, $LaBr_3$..). The total time resolution at 1.33MeV for a (ø25x15mm) BaF_2/XP2020Q couple is 127 ps±5 (with 2300±80 Photoelectrons in the peak) [14], very near the 140 ps obtained for a (ø25x25mm)$LaBr_3$/XP20D0 (with 17600±500 Phe)[11].

4. Configuration of the fast timing set-up

Two detectors are enough to define start and stop signals. Therefore, for levels populated in β decay, the start can be provided from β detection (for levels directly fed by β decay) or from γ-detection (with energy conditions). The stop can only be provided by the depopulating γ-ray detection. We developed then a fast timing set-up with three detectors: one plastic Pilot U scintillator (associated with a XP2020 PMT) for β-detection and two scintillating crystals (BaF_2 and $LaBr_3$ or two $LaBr_3$) for γ. The BaF_2 scintillator is associated with a XP2020Q PMT and the $LaBr_3$ are equiped with XP20D0 ones.

For each scintillator+PMT couple, the energy information entering the acquisition is made from the signal of the PMT last dynode, after amplification. The time information (a 'top' pulse) is built from the PMT anode signal using an analogical Constant Fraction Discriminator (CFD) module. In addition to the three 'reals' (E,T) channels, three others channels are defined for the time differences to be built : detector 1 for start with detector 2 as stop; detector 1 for start with detector 3 as stop; and detector 2 for start with detector 3 as stop. Indeed, the very good digitization step (400 ps) of the digital COMET card [16] entering the Orsay acquisition system, Narval [17], is too large compared to the expected time resolutions. The time differences signals have then to be obtained analogically before entering the acquisition system. Downstream CFDs, Time to Amplitude Converter (TAC) modules are used to deliver 'E'-like shaped signals in which the amplitudes are related to the time difference bewteen their two inputs (CFD outputs). Each analogical TAC output signal is then plugged as 'E', and its logical output is entered as 'T', in a virtual channel. In the following this 'E' information will be called TAC.

A large number of measurements, with different adjustements of the electronics, with different electronic modules and different detectors are needed to optimize such a fast timing set-up. For this reason, in the Narval acquisition user's program, we developed a new C module [18] allowing to get, on-line, multi-gated spectra. Many energy conditions (gates) can be imposed allowing for many different 'start' and 'stop' definitions in a same run, and half-lives for many levels to be observed simultaneously. For each possible combination of two gates, a double-gated TAC spectrum is built, incremented by 3(at least)-fold events consisting in the two selected energies (from two detectors), and the corresponding TAC information (between the two latter detectors). Any of such a spectrum can be displayed on the acquisition control window.

5. Optimization measurements with ^{60}Co source:

We performed many systematic measurements with a three detector configuration that allows time resolution of each detection channel to be obtained in a same run [19]. These measurements were performed with a ^{60}Co radioactive source, presenting only prompt concidences. We define gates set on the compton background as well as on the full energy peak in order to cover a large range of energies for the start and stop definitions. We first tested different electronic modules

Journal of Physics: Conference Series **205** (2010) 012044 doi:10.1088/1742-6596/205/1/012044

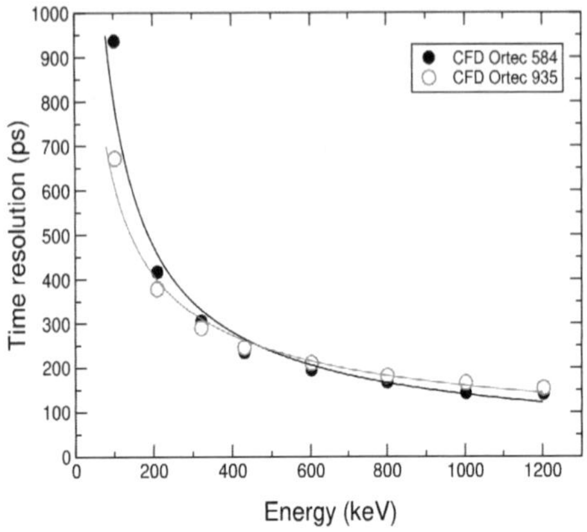

Figure 1. Time resolutions obtained for ⌀25,5x30 mm $LaBr_3$ crystal coupled to a XP20D0 PMT for two different CFDs as a function of energy.

(CFD, TAC) with different adjustments (High voltage applied on the PMT; Walk, external delay of the CFD). In fig.1 one can see the impressive time resolution improvement (200 ps less) at low energy (below 200keV) of a ⌀25,5x30 mm $LaBr_3$+XP20D0 PMT couple brought by the use of a 935 Ortec (empty squares) instead of a 584 Ortec (full square) CFD module.

For different scintillator+PMT detection couples we measured, as a function of the energy, energy and time resolution, absolute efficiency, linearity for several $LaBr_3$ scintillators. Indeed, the time resolution is strongly connected to crystal geometry and dimensions (see for example the recent work of ref.[20]). This is shown in fig.2 where are reported the time resolution curves obtained for two $LaBr_3$ cylindrical crystals with the same diameter (25,5 mm) but different lengthes (10 mm and 30 mm). For comparison, the time resolution curve obtained for the ⌀25,5x30 mm BaF_2 is also reported. A better time resolution (in mean 100 ps of improvement) is obtained with the smallest crystal for all the energy range, but it is more pronounced at low energy (around 200 ps). At the opposite, the peak to total ratio (see top spectrum of fig.3) is degraded in the small crystal due to bad high energy detection as compared to the one with the large one (bottom spectrum).

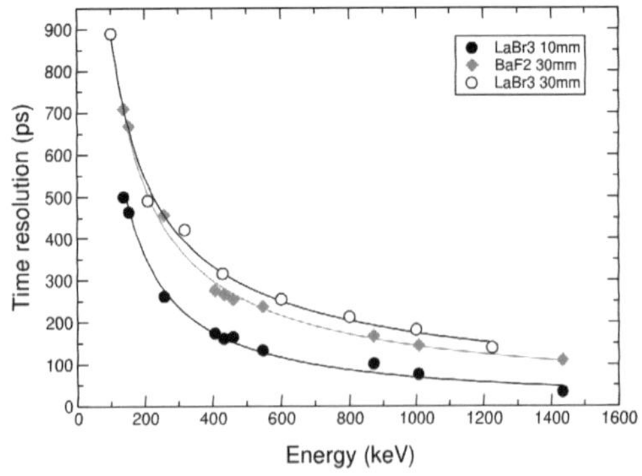

Figure 2. Time resolution obtained with a ⌀25,5x30 mm $LaBr_3$ (full line) and with a ⌀25,5x10 mm $LaBr_3$ (square), both coupled to XP20D0 PMT. For comparison is also reported the time resolution measured with a ⌀25,5x30 mm BaF_2+XP2020Q (square)

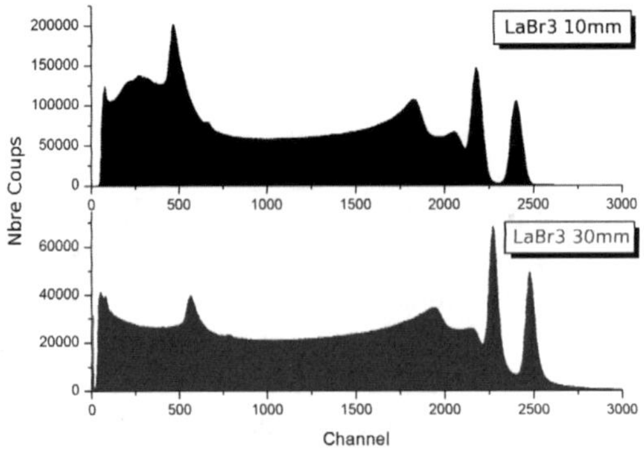

Figure 3. Energy spectra (^{60}Co source) obtained with the ⌀25,5x30 mm $LaBr_3$ (top spectrum) and with the ⌀25,5x10 mm $LaBr_3$ (bottom spectrum), both coupled to XP20D0 PMT.

6. Measurements at low intensity at ALTO:

In order to test on-line the performances of our fast timing set-up, we requested to use, at low intensity, the beam of the ALTO facility for measuring half-lives in $^{137-139}$Cs levels fed by the β^--decay of $^{137-139}$Xe. Indeed, if half-lives are known for levels up to nearly 450keV in ^{133}Cs, up to 250keV in ^{135}Cs, and, on the other side, up to 200keV in ^{141}Cs, none are known in the in between $^{137-139}$Cs isotopes. One half-life measurement has been carried out in ^{137}Cs using only two fast detectors (a plastic scintillator and a NaI crystal) [21], which lead to an upper limit of $T_{1/2} \leq 0.1$ ns for the first excited state. In the $^{133-139}$Cs isotopes the two first excited states are both $5/2^+$ states. The first one is originating from a $d_{5/2}$ quasiparticle excitation, the second from the coupling of 2_1^+ from the even core and one quasi-particle excitation from $1g_{7/2}$ – that leads to the $7/2^+$ ground state in all these isotopes. The decay of the $5/2_1^+$ to the ground state is then an l-forbiden transition but the half-lives for this state were expected to be less than or equal to 100 ps and of the order of 500 ps in $^{137-139}$Cs, respectively [22].

We performed the measurements last spring with the experimental set-up shown in fig.4. One can recognize in the top picture of fig.4, downwards, the tape station, the collection point, and the measurement point surrounded by the four detectors for the fast-timing setup: three scintillators, as previous, and one Ge that we added in order to increase our selectivity. The bottom picture is a zoom of this measurement point and shows counterclockwise : the ⌀25,4x3mm Pilot U, ⌀25,4x10 mm $LaBr_3$, and ⌀25,4x30 mm BaF_2), and the Ge detectors.

Figure 4. The 4-detector Fast-timing set-up used to measure half-lives at the ALTO beam line.

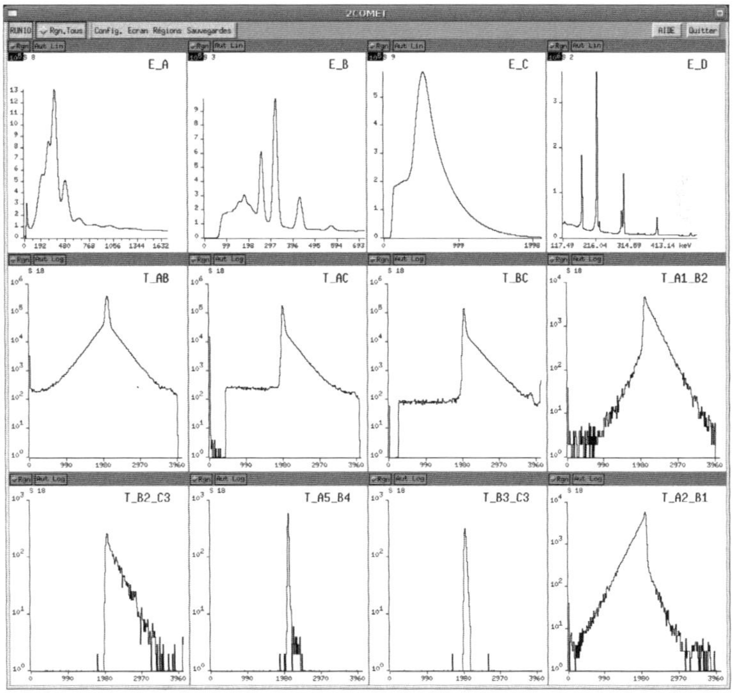

Figure 5. Picture of the Narval acquisition Spectrum Window. Direct Energy, TAC, as well as doubly gated TAC spectra obtained during the ^{139}Xe → ^{139}Cs run, are displayed.

Journal of Physics: Conference Series **205** (2010) 012044 doi:10.1088/1742-6596/205/1/012044

As already mentioned, due to the implementation in the acquisition system Narval of multigating technique, we were able to observe on-line, during the experiment, the presence of some half-lives. A picture of the main window of the acquisition is presented Fig.5. On the first row one sees the energy spectra of the four detectors (A means BaF_2, B $LaBr_3$, C Pilot U and D Ge) and, on the second row the three total TAC spectra (no gates) (T_AB for the TAC between BaF_2 and $LaBr_3$, T_AC between BaF_2 and Pilot U, and between T_BC for Pilot U and $LaBr_3$). The following spectra are doubly gated TAC spectra. Their names T_Ai_Bj allow to recognize which gate (number between 1 to 5) was set on which detector (A to D) energy. The presence of a measurable half-life is easily recognizable in the T_A1_B2 (or T_A2_B1) spectrum, and in the T_B2_C3 one. They exhibit wide TAC, as compared for example to the narrow distribution observed in T_A5_B4 or T_B3_C3.

A preliminary analysis of the data allow us to attribute to the $5/2_1^+$ in ^{139}Cs the half-life of the T_A1_B2 spectrum and to estimate it, by the slope method, around 2.2ns. The Analysis is still in progress but the results are very promissing, half-lives below the time resolutions are already identified, lower than or of the order of 100 ps ($\pm$10 ps) for the levels at 289, 393 and 515 keV in ^{139}Cs.

Conclusions

A fast timing set-up has been developed for ALTO experiment. Its sensitivity in the range of few tens of picosecond to few nanosecond range has been recently established by the measurements of half-lives of $^{137-139}$Cs levels, that were populated in the decay of $^{137-139}$Xe isotopes produced at Alto by photofission.

Aknowledgement

This work was supported by the french Ministère des Affaires étrangères et européennes (ECO-NET program).

References

[1] S Essabaa, J Arianer, P Ausset, O Bajeat, J. P. Baronick *et al.* 2003 *Nucl. Inst. and Method.* B **204** 780
[2] B Roussière 2004 *Presentation to the ALTO Workshop* `http://ipnweb.in2p3.fr/tandem-alto/alto/conferences/semin_pdf/roussiere.pdf`
[3] M A Cardona, D Hojman, B Roussière, I Deloncle *et al.* 2006 *Proposal to the Tandem Pac*
[4] H Mach, R L Gill and M Moszynski 1989 *Nucl. Inst. and Method.* A **280** 49
[5] E V D. Van Loef *et al.*, 2000 *Applied Physics Letter* **77** 1467
[6] J Kiener Geant simulation 2009 *private communication*
[7] C W Beausang et al. 1992 *Nucl. Inst. and Method.* A **204** 37
[8] R F Post and L I Schiff 1950 *Phys. Rev* **80** 1113
[9] A G Wright 2006 *Nucl. Phys.* B**150** 239
[10] Saint Gobain 2007 Technical Note
[11] M Moszynski *et al.* 2006 *Nucl. Inst. and Method.* B **567** 31
[12] Lawrence Berkeley Laboratory Scintillator Database *http://scintillator.lbl.gov/*
[13] O Sellès 2006 *Thèse de Doctorat de l'Université Paris VI*
[14] M Moszynski *et al.*, 2004 *Nucl. Sci. IEEE Trans.* **51** 1701
[15] M Moszynski *et al.*, 2005 *IEEE Trans. Nucl. Sci.* **52** 1027
[16] J Le Bris *et al.*, 2004 *Internal Report IPNO 05-04*
[17] J Le Bris, R Sellem, J C Artiges, J F Clavelin, S Du *et al.* 2006 *COMET-NARVAL ACQUISITION notice Rapport de recherche* 25 and http://hal.in2p3.fr/in2p3-00107358/fr/
[18] I Deloncle *to be published*
[19] J Pouthas *et al.* 1977 *Nucl. Inst. and Method.* **145** 445
[20] G Hull*et al.* 2008 *Nucl. Inst. and Method.* A **588** 384
[21] E Monnand *et al.*, 1975 *J. Physique* **36** 1.
[22] B Roussière *et al.* 2008 *Proposal to the Tandem Pac*

Installation of the AGATA Demonstrator and commissioning experiments at LNL

Francesco Recchia
INFN Sezione di Padova, Padova, Italy
E-mail: recchia@pd.infn.it

Abstract. The first phase of the AGATA project, namely the AGATA Demonstrator Array, consists of a subset of 5 triple clusters of the final array and will be used to demonstrate the feasibility of the γ-ray tracking. The performance of the instrument has been estimated up to now only through Monte Carlo simulations and indirect measurements. The first installation is presently ongoing ay the Laboratori Nazionali di Legnaro, Italy, where it has replaced the CLARA array at the target position of the PRISMA magnetic spectrometer. In the present contribution, the details of the installation will be reviewed and preliminary results from the first in-beam commissioning test will be given.

A new generation of γ-ray spectrometer has been developed, based on the γ-ray tracking concept. The performance foreseen for a full 4π spectrometer exceeds the previous generation of escape-suppressed germanium arrays by 2-3 orders of magnitudes. Thanks to the position sensitivity obtained exploiting this technology it will be possible to obtain an unprecedented absolute efficiency approaching 50% while preserving the superb resolution of germanium detectors even in high recoil velocity ($\beta \approx 0.5$) Doppler corrected spectra. This performance is obtained through the use in real time of pulse shape analysis (PSA) and γ-ray tracking algorithms.

During the initial phase of the project, a subset of the array, composed of five triple clusters and known as the AGATA Demonstrator, will operate at the Laboratori Nazionali di Legnaro. The initial goal of the campaign at LNL is to prove that indeed PSA and γ-ray tracking can be successfully performed in real time. The validation of the γ-ray tracking at LNL will occur on the most demanding conditions achievable in a low-energy stable-beam facility, i.e. with reactions with velocities of the γ-emitting products up to $\beta \approx 10\%$ and relatively high intensity beams. Once this is achieved, the AGATA Demonstrator will be used in coupled operation with the PRISMA magnetic spectrometer [1] to perform spectroscopic studies of moderately neutron-rich nuclei populated by grazing reactions as multi-nucleon transfer or deep inelastic collisions with the stable beams delivered by the Tandem-ALPI and the PIAVE-ALPI accelerator complex. Nevertheless, the coupling of the AGATA Demonstrator with complementary detectors, other than PRISMA, opens experimental possibilities beyond the aforementioned reactions, with direct, Coulomb excitation as well as fusion-evaporation reactions.

The compact arrangement of five triple clusters of AGATA is the optimal geometry of the array for the experimental activity of the Demonstrator foreseen at LNL. The detectors will be placed in front of the PRISMA spectrometer input aperture, therefore distributing the active volume in the best positions regarding Doppler broadening, considering that the nuclei of interest will be detected by PRISMA. The photopeak efficiency of the AGATA Demonstrator placed at the nominal 23.5 cm target-to-detector distance is roughly 3%. Given the low solid angle coverage, the Demonstrator can be also used at shorter target-to-detector distances, with an increase in photopeak efficiency (the value is approximately 7% when the detectors are moved by 10 cm closer to the target position) and without significant losses in the resolution and peak-to-total performance [2].

PRISMA is a large acceptance magnetic spectrometer designed to work with grazing reactions with the heavy ion beams provided by the LNL accelerator complex. The basic characteristics of PRISMA are described in ref. [1]. For the following discussion it is relevant to mention that PRISMA uses ion-tracking position-sensitive detectors to achieve the mass resolution. The tracking detectors provide the basic information to obtain the trajectory and velocity of the reaction products. According to the Monte Carlo simulations [3], up to velocities of approximately v/c=10%, the intrinsic AGATA detector resolution is almost fully recovered if the recoil velocity module is measured with a relative precision better than 1%, and if the recoil velocity direction is measured with a precision better than 1°. These values are actually well within the possibilities of PRISMA.

As mentioned before, the AGATA Demonstrator at LNL is strongly constrained by the experimental campaign coupled to PRISMA. The different elements of the mechanics and infrastructures are such that the coupling and experimentation with both setups are possible. In the following, the different elements of the infrastructure will be described in more detail.

The setup is intended to measure coincidences between the γ-rays detected by the AGATA Demonstrator and the reaction products detected by PRISMA. The AGATA Demonstrator is installed on a mobile platform that will rotate together with PRISMA in such a way that reaction products, detected in the spectrometer focal plane in coincidence with the γ-rays, will have a forward trajectory with respect to the array in order to benefit from the lowest Doppler broadening. The detectors are hosted into a shell made out of 15 elementary AGATA flanges. The shell is positioned on a trolley which can slide on the same platform, rigidly linked to PRISMA. It is thus possible to easily modify the target-detector distance and to access the scattering chamber and to the instrumentation placed closed to the target. The whole support structure has minimal impact on the rotation of PRISMA. Taking into account the rest of the mechanical structure (beam line, scattering chamber), the angular range 41° to 110° is possible (with the Demonstrator placed at the closest distance from the target), while for the largest distance from the target the possible range is 37° to 110°. If one or more of the detectors are removed, PRISMA can be positioned at smaller angles, respectively $15° - 32°$ and $14° - 29°$.

The sliding seal scattering chamber previously used with CLARA was replaced with a new lower absorption chamber. The final part of the beam line as well was replaced by a specially designed telescopic beam line.

The digitizers and the Detector Support System (autofill and power supplies) will be hosted in racks mounted on the same platform used previously for the front-end electronics and power supplies of CLARA, which is rigidly linked to the structure of PRISMA. The 75 m long optical fibres connecting the digitizers to the pre-processing electronics are taken through the basement up to the Pre-Processing racks sitting in the AGATA-PRISMA control room. The racks for the pre-processing electronics are water cooled and fully protected for thermal and acoustic noise insulation. The pre-processing racks are connected to the PSA computer farm, placed in the main computer room of the Tandem building, via 15 m long fibres. The disk storage for AGATA is instead placed in another building, at the location of the TIER-2 centre for the CMS and ALICE experiments, and is connected to the computing farm via optical fibres.

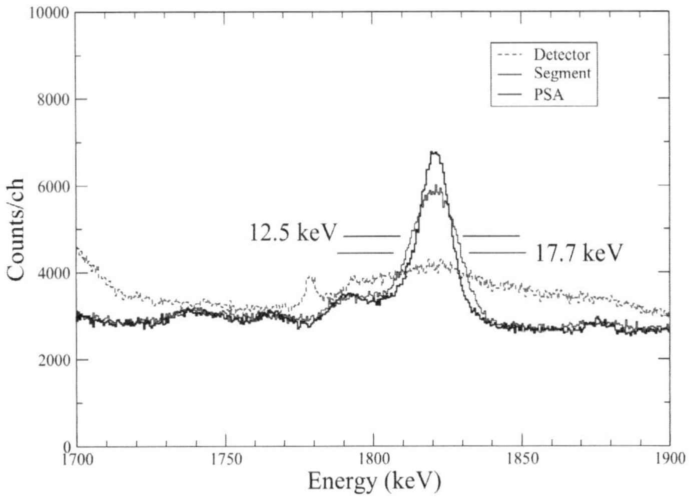

Figure 1. Comparison of Doppler-corrected spectra obtained under different conditions. See text for details.

The first in-beam commissioning experiment

Following the installation of the required infrastructures, an in-beam commissioning test was performed early in 2009. Besides providing the opportunity to verify the correct functioning of the several parts composing the front-end electronics and data acquisition chain of the AGATA Demonstrator, the main goal of the experiment was to measure the overall position resolution provided by the pulse shape analysis algorithms currently implemented in the system. As discussed thoroughly in [4], this parameter, which is critical in determining the overall performance of the array, can be determined with an in-beam measurement by comparing the effective energy resolution of the detectors (taking into account the proper Doppler correction) at several target-detector distances.

During this test, a beam of ^{30}Si with an energy of 70 MeV was fired onto a ^{12}C target, 200 μg/cm^2 thick. The prompt γ radiation was detected with the first AGATA asymmetric triple-cluster detector (ATC1), positioned as close as possible to 90° with respect to the beam direction. The detector was operated with a partial version of the AGATA Detector Support System, including the autofill system and the low-voltage power supply. The high-voltage was provided instead with a standard SY527 system by CAEN.

A full AGATA system was used to collect data to disk, including AGATA digitizers, Global Trigger and Synchronization, pre-processing electronics and a Narval system [5] running on several nodes. The DAQ chain was started using the Cracow GUI, communicating with the Narval system via a Run Control server. Dedicated Narval actors were performing pulse shape analysis and γ-ray tracking in real time. As this was the first time in which the full system was running, the original digitized signals were stored to disk for later replay. Because of this, the overall counting rate was limited to approximately 1.5 kHz per crystal, corresponding to approximately 1 GB/min per crystal.

Through a preliminary data analysis it was possible to spot and to correct for some inconsistencies in the cabling and in the positioning of the detector. The Doppler-corrected spectra for the 1823 keV line of ^{40}K, namely the pn evaporation channel, are shown in Figure 1. Since no ancillary devices to measure the recoil vector velocity on an event-by-event basis were used, Doppler correction was performed by assuming an average recoil velocity. Using

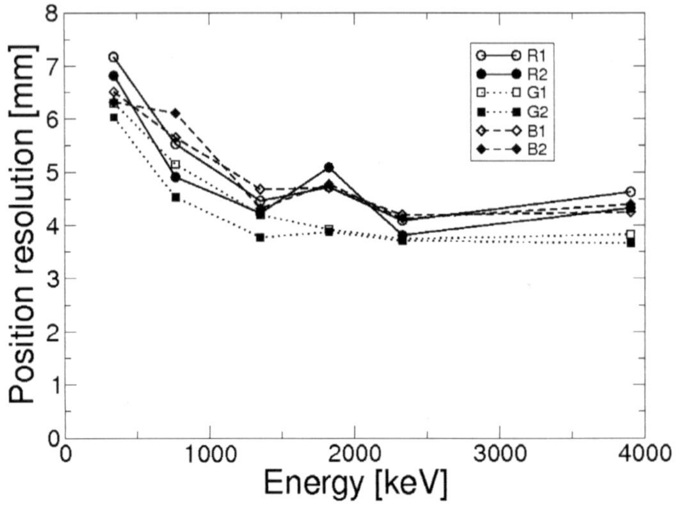

Figure 2. Position resoulution as a function of the energy of γ-ray. The comparison is performed separately for each gamma-ray detector and two estimations are given corresponding to different datasets collected.

the crystal centre-of-gravity to perform Doppler correction, no peak can be clearly identified corresponding to the 1823 keV transition. This case corresponds to discarding the information from the segments, in other words, to treat the detectors as "conventional" ones. The peak FWHM is instead 17.7 keV using the segment centre-of-gravity to perform Doppler correction. A peak FWHM of 12.5 keV is obtained when a full PSA algorithm is applied, in this case a grid search algorithm [6] using a signal basis calculated with the JASS code [7]. This last value is compatible with estimates considering realistic values for the dispersion in recoil velocity module and direction that are the dominationg contributions to the Doppler broadening.

As during previous measurements [4] the main part of the uncertainty on the estimate of the position resolution comes from the difficulty in taking into proper account within the simulation all of the sources of Doppler broadening, a different strategy has been followed here in order to extract the position resolution directly from the experimental data in a model-independent way. This was possible because the data were acquired with the detector in two positions, the former having the detector as close as possible to the target, and the latter having the detector placed farther from the target. Hence, in this experiment, the estimate of the position resolution is reduced to a comparison between the two datasets, since the difference in energy resolution between the Doppler-corrected spectra gathered in the two positions is due to the solid angle subtended by the PSA voxel, i.e. the angular spread on the initial direction of the photon corresponding to the position resolution. From the comparison of the various datasets acquired it is possible to obtain an estimation of the position resolution as a function of the γ-ray energy as depicted in Figure 2.

The estimation of the position resolution that was obtained from this test-beam experiment suggests that the performance is of the order of 4 - 4.5 mm for photons with an energy above MeV. According to Monte Carlo simulations, such value is adequate to reach the design performance of the AGATA array in terms of efficiency and Doppler correction capabilities.

References

[1] A. Latina et al., Nucl. Phys. **A734** (2004), E1.
[2] http://agata.pd.infn.it/documents/simulations/demonstrator.html

[3] F. Recchia et al., LNL-INFN(REP)-202/2004 (2004), 160.
[4] F. Recchia, *In-beam test and imaging capabilities of the AGATA prototype detector*, PhD Thesis, University of Padova (2008).
[5] X. Grave et al, 14th IEEE-NPSS (2005) 1.
[6] R. Venturelli and D. Bazzacco, LNL-INFN(REP)-204/2005 (2005), 220.
[7] M. Schlarb et al., GSI Report 2009-1 (2009), 232.

Journal of Physics: Conference Series **205** (2010) 012046 doi:10.1088/1742-6596/205/1/012046

Nuclear reaction studies with particle-gamma coincidences using the *Saci-Perere* spectrometer

J R B Oliveira[1], V A B Zagatto[1], D Pereira[1], J Lubian[2], P R P Allegro[1], L C Chamon[1], E W Cybulska[1], R Linares[1], N H Medina[1], R V Ribas, E S Rossi Jr[1,3], W A Seale[1], C P Silva[1], D L Toufen[4], M A G Silveira[5], G S Zahn[6], F A Genezini[6], L Gasques[7], J M B Shorto[2]

[1]Instituto de Física da Universidade de São Paulo, São Paulo, SP, Brazil
[2]Instituto de Física da Universidade Federal Fluminense, Niterói, RJ, Brazil
[3]Centro Universitário FIEO - UNIFIEO, Osasco, SP, Brazil
[4]Instituto Federal de Educação, Ciência e Tecnologia, Guarulhos, SP, Brazil
[5]Centro Universitário da FEI, São Bernardo do Campo, SP, Brazil
[6]Instituto de Pesquisas Energéticas e Nucleares, São Paulo, SP, Brazil
[7]Centro de Física Nuclear da Universidade de Lisboa, Lisboa, Portugal

Abstract. The *Saci-Perere* spectrometer of the University of São Paulo has been configured to perform particle-gamma coincidence measurements in order to study nuclear reaction mechanisms. The motivation of this type of measurement comes from the recent development of nuclear reaction models based on the São Paulo potential with the inclusion of an imaginary part with no adjustable parameters. New preliminary data on the $^{18}O+^{110}Pd$ transitional system are presented, and apparent similarities to weakly bound cases (*e.g.* $^{7}Li + ^{120}Sn$) are briefly discussed.

E-mail: `zero@if.usp.br`

1. Introduction

The *Saci-Perere* (acronym for *Sistema Ancilar de Cintiladores - Pequeno Espectrômetro de Radiação Eletromagnética com Rejeição de Espalhamento*) is a gamma-spectrometer located at the Pelletron 8MV Tandem of the University of São Paulo. It consists of four HPGe detectors equipped with BGO Compton suppressors, with a total photopeak efficiency close to 0.5% (at $E_\gamma = 1.3$ MeV), and an ancillary system of 11 $\Delta E - E$ plastic phoswich scintillators, with a total solid angle coverage of 76% of 4π [1]. It was designed and used for nuclear structures studies (e.g. [2]) in fusion-evaporation reactions where p and α charged particles are detected, mainly. For nuclear reaction studies, however, the beam-like particles should be detected, and the solid angle range of each detector should be limited, to reduce both the count-rate and the scattering angle integration range [3]. This has been accomplished with a set of collimators placed on the face of each scintillator. With this setup, the $^{18}O+^{110}Pd$ reaction has been measured at various beam energies (between 46 and 60 MeV) in two separate experiments. The preliminary results show that there is an enhancement of the cross sections in comparison to a "standard" calculation with the potential proposed in [4], similar to the results encountered in weakly bound systems such as $^{7}Li + ^{120}Sn$ [5, 6]. This had already been observed in the $^{18}O+^{110}Pd$ transitional system in inelastic excitation functions at a very backward angle with an annular

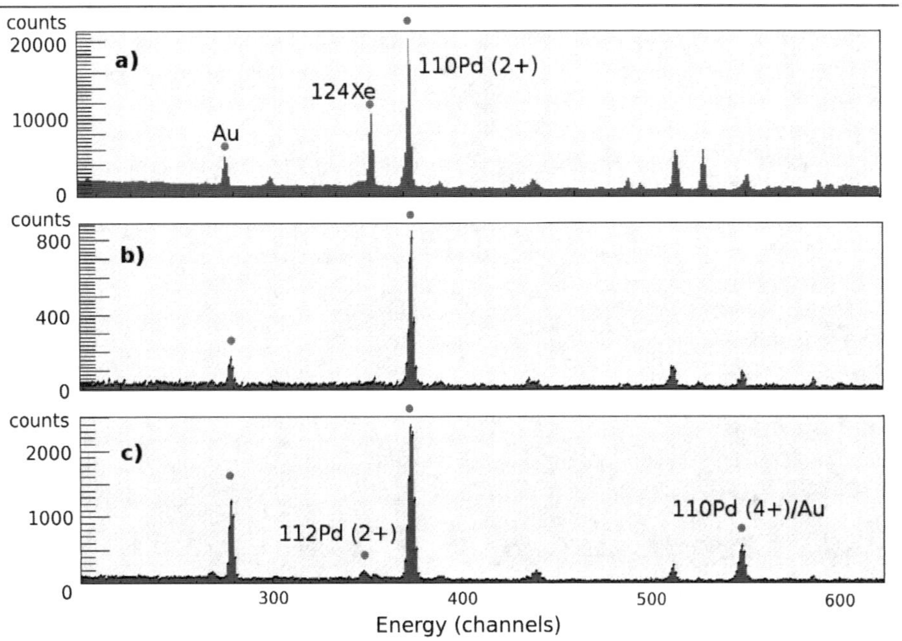

Figure 1. Gamma-ray spectra of the $^{18}O+^{110}Pd$ reaction at $< E >= 55.8$ MeV. a) Total projection. b) In coincidence with particles at 40°. c) In coincidence with particles at 140°.

Si detector [7]. In the present work, additional information has been obtained with regards to the angular distributions for this system. Part of these results has been presented also in the CNR*09 (Second International Workshop on Compound Nuclear Reactions and Related Topics) [8], where the theoretical models are presented in more detail.

2. The $^{18}O+^{110}Pd$ experiments

The $^{18}O+^{110}Pd$ was investigated with the *Saci-Perere* spectrometer in two separate experiments. In the first one, only two HPGe and 7 phoswich detectors were in operation, while on the second one, the full system was used (one of the phoswich detectors, however, presented low gain in the spectra and its analysis has been postponed). The target consisted of a 0.84 mg/cm^2 ^{110}Pd foil with a gold (Au) evaporated layer of 1.49 mg/cm^2 thickness. The beam entered by the Pd layer in the first experiment and the Au layer in the second one, in order to explore the vacuum de-orientation effect as a function of the recoil velocity.

The collimator system consisted of 0.3 mm-thick Al plates, with 0.5-3.0 mm diameter holes (increasing with angle), with 11 average (nominal) angles (with relation to the beam direction) varying from 30° to 140° in 10° steps (skipping 90°). Including the 3 mm diameter beam spot, the scattering angle range of each detector varies between 6° and 12°.

Figures 1 and 2 present sample spectra at a beam energy of 55.8 and 54.1 MeV, respectively (average laboratory energies, considering the energy loss of about 2 MeV through the target), obtained in these experiments. The ^{124}Xe line comes from total fusion followed by evaporation of neutrons, and is present only at the higher energies and in the singles spectra. The gamma-decay transition from the 4^+ of ^{110}Pd is contaminated with an Au line. The 2n transfer related lines (^{112}Pd) are observed only at large scattering angles.

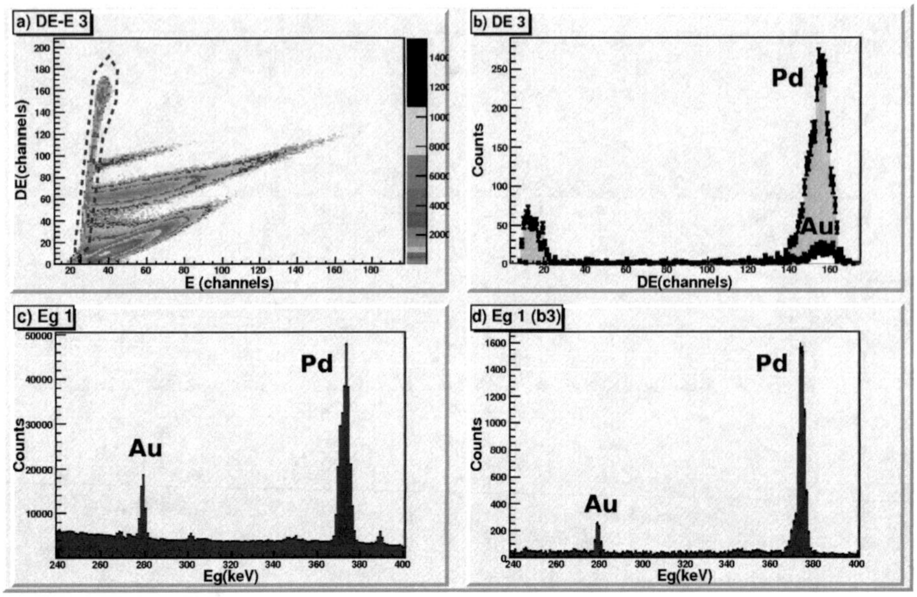

Figure 2. Sample spectra of the $^{18}O+^{110}Pd$ experiment obtained with the Saci-Perere spectrometer. a) Typical plastic scintillator $\Delta E - E$ bi-parametric spectrum, at $\theta_{lab} = 50°$- the dashed contour indicates the cut corresponding to the particles which stop within the ΔE detector thickness (0.1 mm), such as beam-like particles; b) ΔE-projection spectra selected from inside the cut and gated on the $2^+ \rightarrow 0^+$ transition (374 keV) of ^{110}Pd, or on the $\frac{5}{2}^+ \rightarrow \frac{3}{2}^+$ transition (279 keV) of ^{197}Au, as indicated in the figure (the Compton and random coincidences related background has been subtracted). c) Singles gamma spectrum at $\theta_{lab} = 110°$. d) Gamma spectrum of the same detector as in c), in coincidence with particles inside the cut indicated in a), *i. e.*, beam-like particles, essentially.

3. Data normalization

In order to extract the inelastic and transfer to excited states scattering cross sections from particle-gamma measurements, one has to consider, besides the corrections from upper level feeding, the gamma-ray angular distribution in correlation with the beam like particle scattering angle. These effects can be calculated from the population of magnetic sub-states in a nuclear reaction code. The gamma-ray angular anisotropy is attenuated due to the finite angular coverage of the detectors and to the hyperfine interaction when the recoiling nuclei decays in vacuum (vacuum de-alignment). For energies sufficiently below the barrier, these corrections have been successfully applied for many years in Coulomb excitation codes such as GOSIA [9]. Also, for near 180° scattering angle (as in [7]), the gamma-ray angular distribution is independent of the reaction mechanism.

For energies close to the barrier, the Coulomb excitation gamma-ray angular distribution $W(\theta_\gamma)$ can be used as a first order approximation, and, in addition, it turns out that it is weakly dependent on energy. Under this assumption, the experimental yields Y can be related to the cross section using a low energy reference (E^{ref}) measurement (below or at least very near the barrier) to normalize data at each angle, and a low scattering angle (θ^{ref}) reference (where the theoretical Coulomb cross section can be considered reliable at all energies):

$Y(E, \theta_p, \theta_\gamma) \approx I(E)N\frac{d^2\sigma}{d\Omega_p d\Omega_\gamma}\Delta\Omega_p(\theta_p)\Delta\Omega_\gamma(\theta_\gamma)$, where $I(E)$ is the beam intensity of the measurement performed at energy E, and N the number of target nuclei. Under the assumption

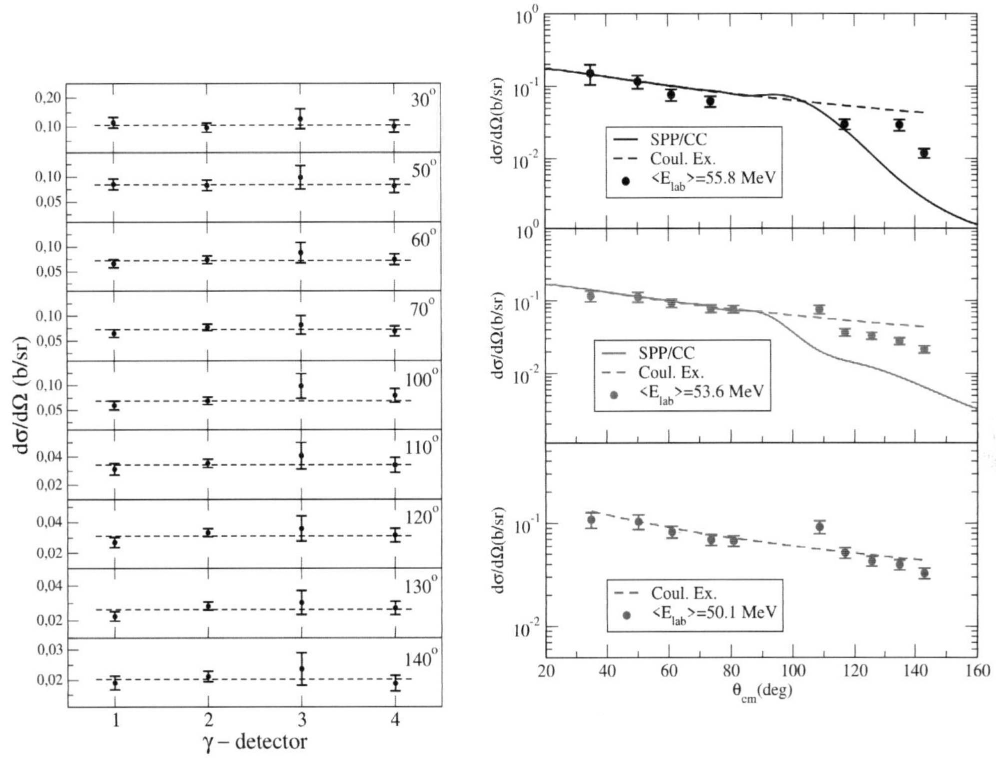

Figure 3. Angular distributions of the inelastic excitation of the 2^+ states of ^{110}Pd (^{18}O +^{110}Pd reaction). *Left*: Comparison of results among gamma-detectors at different angles (θ, φ): 1-$(101°, 0°)$; 2-$(37°, 180°)$; 3-$(101°, 36°)$; 4-$(37°, 144°)$. The ^{18}O scattering angle is indicated at the top right of each graph. The average is indicated as a dashed line. *Right*: average cross section obtained at three different energies, compared to the SPP/CC (see next section) calculation with potential U_1 (solid lines) and to Coulomb excitation (dashed lines).

that $\frac{d^2\sigma}{d\Omega_p d\Omega_\gamma} \approx \frac{d\sigma}{d\Omega_p}(E,\theta_p)W(\theta_p,\theta_\gamma)$, and defining $R_Y(E,\theta_\gamma) = \frac{Y(E,\theta_p)}{Y(E^{ref},\theta_p)} \div \frac{Y(E,\theta_p^{ref})}{Y(E^{ref},\theta_p^{ref})}$ and $R_\sigma = \frac{d\sigma}{d\Omega_p}(E,\theta_p^{ref}).\frac{\frac{d\sigma}{d\Omega_p}(E^{ref},\theta_p)}{\frac{d\sigma}{d\Omega_p}(E^{ref},\theta_p^{ref})}$, one obtains: $\frac{d\sigma}{d\Omega_p}(E,\theta_p) = R_y R_\sigma$, independently of $W(\theta_\gamma)$.

Figure 3 presents the results of the measurements with the above (preliminary) normalization, using $E^{ref} = 46$ MeV and $\theta^{ref} = 40°$. It is shown, at left, that the results are consistently independent of the gamma detector angles relative to the beam and to the particle detector (at 53.6 MeV average beam energy). Similar result is obtained at 55.8 MeV (not shown), however, in this case only two gamma-detectors were used in the experiment (at polar angles 37° and 101°, both in the horizontal plane). At right, the angular distributions obtained from the average of all available gamma detectors are presented at three different energies.

4. The theoretical models

It is well established [10] that the São Paulo Potential (SPP) [11, 12, 13] provides a good description of elastic and quasi-elastic processes near the barrier, when used in a Coupled

Channels (CC) calculation with the inclusion of the main channels. The SPP is theoretically founded on the Pauli non-locality, which arises from quantum exchange effects. The SPP, in the local equivalent version, is given by:

$$V_{LE}^{SP}(R,E) = V_F(R)e^{-\frac{4v^2(R)}{c^2}} \quad (1)$$

where $V_F(R)$ is the double folding [14] potential, and the term $e^{-\frac{4v^2(R)}{c^2}}$ is the local velocity ($v(R)$) dependent correction arising from Pauli non-locality. More recently, a new generation of SPP/CC calculations has been developed [4] with the aim of describing reactions at energies for which dissipative processes, such as deep inelastic collisions (DIC), are important. This has been done with the introduction of an imaginary potential with the same shape of the SPP:

$$U_1(R) = (1 + 0.6i)V_{LE}^{SP}(R) \quad (2)$$

which has been applied successfully to a significant number of systems. For the weakly bound cases, however, modifications of this potential are proposed [5]. Sakuragy, Yahiro, and Kamimura [15] have shown, with the method of coupled discretized continuum channels (CDCC), that the dynamic polarization potential due to the ^{6}Li breakup is strongly repulsive, and effectively reduces the intensity of the real double-folding potential (in a single-channel calculation) by a factor of 0.5-0.6. Indeed, a reasonable description for the ^{7}Li $+^{120}$Sn system [6, 5, 8] is obtained with the following potential:

$$U_2(R) = (0.6 + 0.6i)V_{LE}^{SP}(R) \quad (3)$$

It is expected that other weakly bound systems might also be well described with the use of this potential (see also [16]). It has to be noticed, however, that the above approach could not be applied to a collision with a very high-Z target, where the long-range Coulomb-breakup plays an important role. The inclusion of the Coulomb break-up in the calculations is under development.

5. Discussion

As mentioned above, there has been a previous particle-gamma measurement of the ^{18}O+^{110}Pd system [7]. In that experiment, besides quasi-elastic cross sections obtained from separate particle singles spectra runs, particle-gamma measurements were performed with an annular particle detector in coincidence with two gamma-ray detectors. The excitation functions extracted for the inelastic and transfer to excited states are well described, below and near the barrier, with a coupled channels calculation employing the São Paulo Potential as the bare interaction (Eq. 1). At somewhat higher energies, however, the experimental cross sections are relatively enhanced and a better agreement with the data is again accomplished [6, 5, 8] by the use of the potential U_2 as in the ^{7}Li case. This suggests that again a dynamic polarization potential which decreases the effective real potential is in operation. This could be due to an increased level density, at somewhat high excitation energies, of the transitional nucleus of ^{110}Pd.

The present angular distribution measurements confirm the tendency of enhancement of experimental cross sections relative to the theoretical results with potential U_1. These results are preliminary. The data are still under analysis. The calculations with the potential U_2 have not been performed yet. Detailed calculations of the particle-gamma angular correlation, considering the nuclear potential, are necessary for a more precise and reliable comparison.

6. Conclusions

Particle-gamma angular distributions can be measured with the *Saci-Perere* spectrometer system. Preliminary angular distributions have been obtained for the ^{18}O+^{110}Pd reaction. The results confirm the tendency observed in [7] of an enhancement of the cross sections at large scattering angles which could be related to the increase in the density of states in transitional regions.

Acknowledgments
We acknowledge support by the agencies FAPESP: Fundação de Amparo à Pesquisa do Estado de São Paulo, CNPq: Conselho Nacional de Desenvolvimento Científico e Tecnológico, and CAPES: Coordenação de Aperfeiçoamento de Pessoal de Nível Superior (Brazil), and FCT: Fundação para a Ciência e a Tecnologia (Portugal).

References

[1] Alcántara-Núñez J A, Oliveira J R B *et al.* 2003, Nucl. Inst. Meth. **A 497**, 429
[2] Alcántara-Núñez J A, Oliveira J R B *et al.* 2003, Phys. Rev. C **69**, 024317
[3] Oliveira J R B *et al.*, 2009, AIP Conf. Proc. **1139**, 172
[4] Pereira D, Lubian J, Oliveira J R B, Sousa D P and Chamon L C, 2009, Phys. Lett. **B 670**, 330
[5] Pereira D *et al.*, 2008, AIP Conf. Proc. **1098**, 139
[6] Souza D P *et al.*, AIP Conf. Proc. **1139**, 106, and submitted to Nucl. Phys. A
[7] Pereira *D et al.*, 2006, Phys. Rev. C **74**, 034608
[8] *Proceedings of the CNR*09*, to be published in Euro. Phys. Jou. Web of Conferences, 2009.
[9] Czosnyka T, Cline D, Wu C Y, 1983 Bull. Amer. Phys. Soc. **28**, 745
[10] Pereira D *et al.*, 2009, Nucl. Phys. **A826**, 211 (and references therein).
[11] Cândido Ribeiro M A C *et al.*, 1997, Phys. Rev. Lett. **78**, 3270
[12] Chamon L C *et al.*, 1997 Phys. Rev. Lett. **79**, 5218
[13] Chamon L C *et al.*, 2002, Phys. Rev. C **66**, 014610
[14] Satchler G and Love W, 1979, Phys. Rep. **55**, 183
[15] Sakuragi Y, Yahiro M, Kamimura M, 1983, Prog. Teor. Phys. **70**, 1047
[16] Otomar D R *et al.*, 2009, Phys. Rev. C **80**, 034614

Journal of Physics: Conference Series **205** (2010) 012047 doi:10.1088/1742-6596/205/1/012047

Astrophysical reaction rates for ^{6}He and ^{9}Be production by electromagnetic radiative capture and four-body recombination

R. de Diego[1], E. Garrido[1], D.V. Fedorov[2], and A.S. Jensen[2]

[1]Instituto de Estructura de la Materia, CSIC, Serrano 123, E-28006 Madrid, Spain

[2]Department of Physics and Astronomy, University of Aarhus, DK-8000 Aarhus C, Denmark

E-mail: e.garrido@iem.cfmac.csic.es

Abstract. The triple alpha reaction $\alpha+\alpha+\alpha \rightarrow {}^{12}\text{C}+\gamma$ is considered to be the most relevant one in stars at the helium burning stage, since it permits to bridge the A=5 and A=8 instability gaps. However, at higher temperatures and neutron rich environments, other reactions can also play a relevant role. In this work we investigate the astrophysical reaction and production rates for the two-particle radiative capture processes $\alpha+n+n \rightarrow {}^6\text{He}+\gamma$ and $\alpha+\alpha+n \rightarrow {}^9\text{Be}+\gamma$. The hyperspherical adiabatic expansion method is used. With this method no assumption is made about the capture mechanism. The four-body recombination reactions $\alpha+\alpha+n+n \rightarrow {}^6\text{He}+\alpha$, $\alpha+n+n+n \rightarrow {}^6\text{He}+n$, $\alpha+\alpha+n+n \rightarrow {}^9\text{Be}+n$ and $\alpha+\alpha+\alpha+n \rightarrow {}^9\text{Be}+\alpha$ are also investigated as a possible alternative as a source of ^{6}He and ^{9}Be.

1. Introduction

Due to the lack of neutrons in the helium burning red giants, the triple alpha reaction $\alpha+\alpha+\alpha \rightarrow^{12} \text{C}+\gamma$ is considered to be the most relevant one at this stage of the star life. This reaction is the way such that the A=5 and A=8 instability gaps are bridged.

Nevertheless, other different scenarios are also possible. Stars more massive than about ten times the mass of the sun are expected to end its life as a Type-II supernova explosion, leading to a neutron star as a residue. The gravitational collapse of the star produces neutrinos and antineutrinos heating the external layers of the star, which are expelled in the form of neutrino wind. This rapidly expanding high-temperature and low density region is the so called "hot bubble", which is in fact a rapidly expanding matter with a significant neutron excess where temperatures of about (7$\sim$10) GK and nuclear statistical equilibrium favors also the presence of α-particles. This is the most likely environment where the rapid neutron capture (r-process) nucleosynthesis can happen [1].

Among these processes, two relevant ones are $\alpha+\alpha+n \rightarrow^9 \text{Be}+\gamma$ and $\alpha+n+n \rightarrow^6 \text{He}+\gamma$. They are an alternative way to overtake the A=5 and A=8 gaps. The triple α reaction is too slow at the temperature-density conditions in the hot bubble, and the reactions mentioned above play a crucial role. Determination of their reaction rates is essential for the initial condition of the r-process, namely, the amounts of seed elements and remaining neutrons. Calculation of these reaction and production rates are usually made assuming that the reactions take place in a two-step process (sequential capture) [2, 3].

The purpose of this work is two-fold. First, we investigate how simultaneous inclusion of sequential and direct capture modifies previous results already available for the radiative capture reactions $\alpha + n + n \rightarrow^6$ He $+ \gamma$ and $\alpha + \alpha + n \rightarrow^9$ Be $+ \gamma$. This will be done by use of the hyperspherical adiabatic expansion method [4]. Second, we shall estimate the production and reaction rates for the four-body recombination reactions $\alpha+\alpha+n+n \rightarrow^6$ He$+\alpha$, $\alpha + n + n + n \rightarrow^6$ He $+ n$, $\alpha + \alpha + n + n \rightarrow^9$ Be $+ n$ and $\alpha + \alpha + \alpha + n \rightarrow^9$ Be $+ \alpha$ which in principle compete with the radiative capture as a source of ^{6}He and ^{9}Be. Under the hot bubble conditions these reactions could become the dominating ones and contribute to the bridging of the A=5 and A=8 gaps.

2. The method

The production rate in some stellar environment for a reaction leading from an N-particle system to a final M-particle state ($A_1 + A_2 + \cdots + A_N \rightarrow B_1 + B_2 + \cdots + B_M$) gives the number of reactions taking place per unit time and unit volume in that star, and it can be written as

$$P^T = \int dEB(E,T)P(E), \tag{1}$$

where $P(E)$ is the production rate at a given kinetic energy E in the N-body center of mass, and $B(E,T)$ is the Maxwell-Boltzmann distribution giving the probability for finding the N particles with that precise relative kinetic energy [5]. This distribution takes the form:

$$B(E,T) = \frac{1}{\Gamma(\frac{3N-3}{2})}\frac{1}{K_BT}\left(\frac{E}{K_BT}\right)^{\frac{3N-5}{2}} e^{-\frac{E}{K_BT}}, \tag{2}$$

where K_B is the Boltzmann constant and T is the temperature of the star.

The presence of the exponential in the previous expression implies that, for a given temperature T, the only relevant energies are the ones smaller than no more that a few times K_BT. Typical temperatures in the stars (i.e., in the core of the sun) are of the order of 10^7 K, which leads to $K_BT \approx 0.001$ MeV. Even more, for the temperatures estimated in the hot bubble (7$\sim$10 GK) one has that $K_BT < 1$ MeV. Therefore, in the stellar medium only very low relative kinetic energies are relevant.

The total production rate at a given energy ($P(E)$) is the product of the so called *reaction rate* and the densities n_i of the N nuclei ($i = 1, \cdots, N$) involved in the initial state. These densities are usually written as

$$n_i = \rho N_A \frac{X_i}{A_i}, \tag{3}$$

where N_A is the Avogrado number, A_i and X_i are the mass number and mass abundance of the nucleus i, and ρ is the density of the star [5]. The density is, together with the temperature, the crucial property of the star determining the production rate. In fact, $P(E)$ is proportional to ρ^N, meaning that, for a sufficiently large ρ, processes involving more particles could play a role.

Finally, the reaction rate ($R(E)$) is given by the Fermi's golden rule integrated over all the possible momenta for the final products of the reaction. Assuming M particles in the final state, $R(E)$ is written as:

$$R(E) = \int \frac{2\pi}{\hbar} \left|\langle \Psi_i(E)|W|\Psi_f(E_f)\rangle\right|^2 \delta(E - E_f)\frac{d^3p_1}{(2\pi)^3}\cdots\frac{d^3p_M}{(2\pi)^3}, \tag{4}$$

where Ψ_i and Ψ_f are the initial and final wave functions, $\vec{p}_1, \cdots, \vec{p}_M$ are the momenta of the final nuclei, and W represents the interaction. When only two particles are involved in the initial state, the reaction rate is the cross section of the process times the relative velocity between the two particles.

Obviously, the matrix element contained in the integrand of Eq.(4) is the same for a given reaction and for the inverse process. It is then possible to relate the reaction rates (and therefore the production rates) corresponding to both processes. This means that the production rate for a reaction leading to two particles in the final state can be written in terms of the cross section of the inverse process.

2.1. Three-body radiative capture

Let us consider here the particular case of the radiative capture reaction $\alpha + n + n \rightarrow ^6$ He $+ \gamma$. This a pure electromagnetic process where only the bound ^{6}He nucleus and a photon are found in the final state. Following the discussion above, the corresponding production rate can be written in terms of the photo-dissociation cross section (σ_γ) of ^{6}He. To be precise, this production rate takes the form:

$$P_{\alpha,2n}(\rho,T) = n_\alpha n_n^2 \frac{\hbar^3}{c^2}\left(\frac{m_\alpha + 2m_n}{m_\alpha m_n^2}\right)^{\frac{3}{2}} \frac{2\pi}{(K_B T)^3} e^{-\frac{Q}{K_B T}} \int_{|Q|}^{\infty} E^2 \sigma_\gamma(E) e^{-\frac{E}{K_B T}} dE \tag{5}$$

where $Q = m_{^6\text{He}} - m_\alpha - 2m_n$, and $m_{^6\text{He}}$, m_α, and m_n are the masses of ^{6}He, the α particle, and the neutron, respectively. A similar expression could be written for the reaction $\alpha + \alpha + n \rightarrow ^9$ Be$+\gamma$.

The photodisintegration cross section $\sigma_\gamma(E)$ is usually expanded in terms of electric and magnetic multipoles, each of them given by a well known expression in terms of the strength function of the reaction [6]. For the particular case of the reactions $\alpha + n + n \rightarrow ^6$ He $+ \gamma$ and $\alpha + \alpha + n \rightarrow ^9$ Be $+ \gamma$, the electric dipole term ($E1$) dominates. For three identical charged bosons, like for instance in the $\alpha + \alpha + \alpha \rightarrow ^{12}$ C $+ \gamma$ reaction, it can be seen that the electric quadrupole term is the leading contribution [7].

2.2. Four-body recombination

Similarly, for a the reaction like $\alpha + n + n + n \rightarrow ^6$ He $+ n$, where one neutron takes the excess of energy released when the remaining particles combine into a bound state, it is possible to express the corresponding production rate in terms of the cross section $\sigma_n(E)$ for the inverse process (cross section for neutron breakup of ^{6}He). In particular, the production rate takes the form:

$$P_{\alpha,3n}(\rho,T) = n_\alpha n_n^3 \mu_{n^6\text{He}} \left(\frac{m_\alpha + 3m_n}{m_\alpha m_n^3}\right)^{\frac{3}{2}} \frac{\hbar^6 (2\pi)^{\frac{5}{2}}}{(K_B T)^{\frac{9}{2}}} e^{-\frac{Q}{K_B T}} \int_{|Q|}^{\infty} E \sigma_n(E) e^{-\frac{E}{K_B T}} dE. \tag{6}$$

Calculation of $\sigma_n(E)$ requires a proper description of the four-body initial and final states. In this work we describe them as a three-body state ($\alpha + n + n$), whose wave function is accurately computed, plus a free neutron relative to the center of mass of the previous three-body system. In this way, the transition amplitude can be written as the sum of three terms, each of them containing the interaction between the spectator neutron and each of the constituents of the $\alpha + n + n$ system. More precisely, each of these three terms takes the form [8]:

$$T^{(i=1,2,3)} = \langle \Psi_c(\boldsymbol{r}_{jk}, \boldsymbol{r}_{jk,i}) | e^{i\boldsymbol{q}\cdot\boldsymbol{r}_{jk,i}} | \Psi_b(\boldsymbol{r}_{jk}, \boldsymbol{r}_{jk,i}) \rangle \langle e^{i\boldsymbol{P}'\cdot\boldsymbol{r}_{ni}} | V_{ni} | e^{i\boldsymbol{P}\cdot\boldsymbol{r}_{ni}} \rangle, \tag{7}$$

where Ψ_c and Ψ_b are the continuum and bound three-body wave functions of the $\alpha + n + n$ system, $\boldsymbol{q}$ is the momentum transfer to the three-body center of mass, and $\langle e^{i\boldsymbol{P}'\cdot\boldsymbol{r}_{ni}} | V_{ni} | e^{i\boldsymbol{P}\cdot\boldsymbol{r}_{ni}} \rangle$ is the Fourier transform of the two-body interaction between the neutron and constituent i in the $\alpha + n + n$ system.

After squaring the sum of three terms in Eq.(7), one can then estimate the value of $|\langle\Psi_i(E)|W|\Psi_f(E_f)\rangle|^2$, which after insertion in Eq.(4) permits to obtain the reaction rate.

Expressions equivalent to the ones above could also be obtained for the other four-body recombination reactions considered in this work: $\alpha+\alpha+n+n \rightarrow^6$ He$+\alpha$, $\alpha+\alpha+n+n \rightarrow^9$ Be$+n$ and $\alpha+\alpha+\alpha+n \rightarrow^9$ Be $+ \alpha$.

2.3. Three-body wave functions

As described above, calculation of the three-body radiative capture and four-body recombination reaction rates require knowledge of continuum and bound three-body wave functions. In this work they are obtained by solving the Faddeev equations in coordinate space. This is done by using the hyperspherical adiabatic expansion method, as described in [4].

For the particular case of the three-body continuum wave functions, they have been computed by imposing a box boundary condition. This procedure is automatically discretizing the continuum spectrum. Inclusion of all the discrete continuum states amounts to include all the possible capture mechanisms: direct capture, where intermediate two-body states are not populated, sequential capture, where intermediate two-body states are populated before capturing the third particle, and resonant capture, where a three-body resonance is populated. This point can be visualized by means of the complex scaling method, which, after the scaling transformation, permits to see how each of the discrete continuum states is associated to one of the possible mechanisms [9].

In all the cases the only nuclear interactions involved are the neutron-neutron, neutron-α, and α-α interactions. For the first two potentials we have used the ones given in Eqs.(57) and (58) of [10]. For the neutron-neutron interaction, they are simple gaussian potentials adjusted to reproduce the experimental s-wave and p-wave scattering lengths and effective ranges. For the neutron-α potential gaussian central and spin-orbit terms are adjusted to reproduce the phase shifts for s, p, and d waves up to 20 MeV. For the α-α we use the modified Ali-Bodmer potential [11] specified in [12]. The Coulomb repulsion has been introduced assuming a gaussian charge distribution reproducing the experimental charge radius of the α-particle.

3. Production of ^{6}He and ^{9}Be by radiative capture

The reaction rates (5) for these reactions are obtained from the photodissociation cross section of the inverse process. Calculation of this cross section requires knowledge of $d\mathcal{B}/dE$, where $\mathcal{B}$ is the transition probability from the ground state to each of the box discretized continuum states. For the particular case of the two reactions considered here, the dominant contribution to $d\mathcal{B}/dE$ comes from the electric dipole term.

Once the photodissociation cross sections are obtained, we can compute the reaction rates according to Eq.(5). In Fig.2 we show as a function of the temperature the (energy weighted) reaction rate for the $\alpha+n+n \rightarrow {}^6He+\gamma$ reaction. The thick solid lines are the results obtained in this work. The higher and lower curves correspond to the electric dipole and electric quadrupole contributions, respectively. As seen in the figure the dipole term dominates by roughly four orders of magnitude. This procedure includes all the possible capture mechanisms. As seen in the figure, the quadrupole result agrees with previous estimates by Görres et al.[2] (dot-dashed) and Fowler et al. (dotted). For the dipole contribution our reaction rate is about one order of magnitude higher than Görres et al., Efros et al., and Barlett et al [3]. The reason is that in these calculations a fully sequential capture process is assumed. In fact, in [3] they also estimated the dipole reaction rate including the contribution from dineutron capture. This estimate (dotted line) is above our calculation.

For the $\alpha+\alpha+n \rightarrow {}^9Be+\gamma$ reaction, the dominant contribution to the electric dipole reaction rate comes from the $1/2^+$ states, which produce a peak below 1 GK related to the low-lying resonant state seen in the photodissociation cross section [13]. Only for temperatures beyond 5

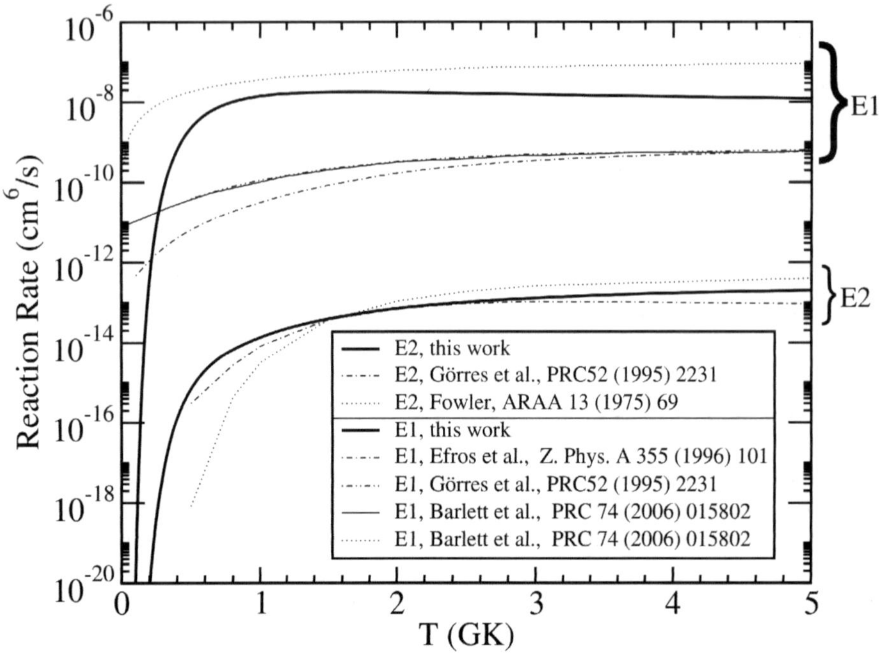

Figure 1. Dipole and quadrupole reaction rates for $\alpha+n+n\rightarrow{}^6$He+$\gamma$. The thick solid lines are the calculations corresponding to this work. The dipole curve is about four orders of magnitude above the quadrupole curve. The different thin curves are the results found in the references given in the figure legend.

GK the $5/2^+$ contributions begins to dominate. This is due to the presence of a $5/2^+$ resonance in ^{9}Be at about 1.5 MeV above threshold. The $3/2^+$ contributions always plays a minor role.

4. Four-body recombination reactions and production rates

The production rate gives the velocity at which a certain nucleus (^{6}He or ^{9}Be in our case) is created in a given environment (number of reactions per unit time and unit volume). Direct comparison between different sources of a given reaction product is then possible. In general this can not be done from the reaction rates, since when the processes involve a different number of particles the reaction rates have different units.

Once the reaction rates have been computed, the corresponding production rates can be easily obtained by multiplying them by the densities of the nuclei involved in the reaction. These densities are given as in Eq.(3). This implies that the production rates depend on the mass density ρ, the abundances of the different elements, and of course on the temperature. Furthermore, it is clear that while the production rate for the two-particle radiative capture processes goes like ρ^3, the one for the four-body recombination reactions goes like ρ^4. Environments with large density favors then the production of ^{6}He or ^{9}Be through four-body recombination processes.

In Fig.2 we show the production rates for the different radiative capture and four-body recombination processes considered in this work. To permit an easier comparison between them we have normalized them to the production rate for the $\alpha+n+n\rightarrow{}^6\text{He}+\gamma$ process. The left, central, and right panels correspond to mass abundances of the α particles of X_α=0.1, 0.5, and 0.9, respectively. They are plotted as a function of the temperature. In the calculation we have used a mass density of ρ=15000 g/cm^3, which amounts to a neutron density of the order of $n_n \sim 10^{27}$ neutrons/cm^3. Larger values of ρ would favor the production rates through the

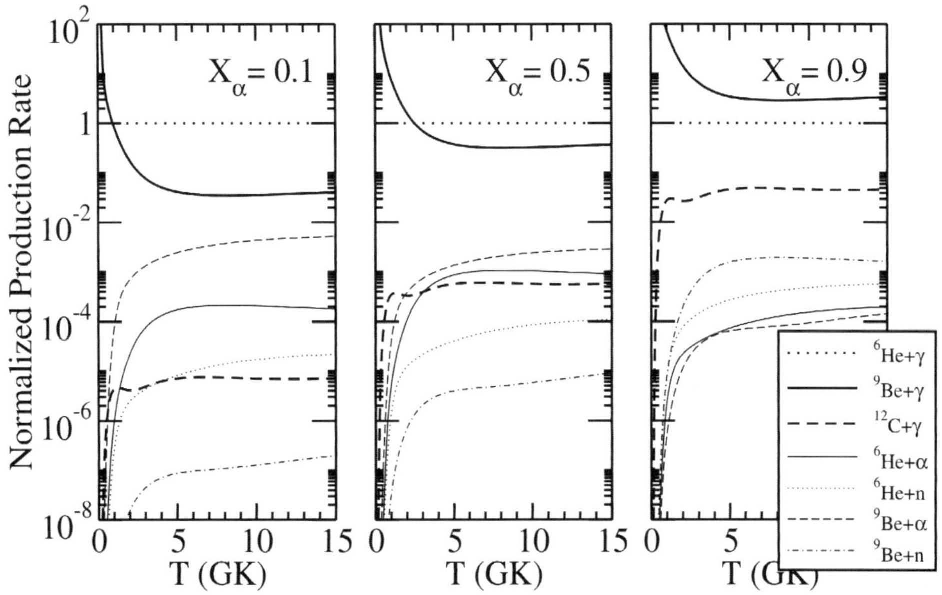

Figure 2. Normalized production rates for electromagnetic and four-body processes as a function of the temperature for α-abundances X_α=0.1, (left), 0.5 (middle), and 0.9 (right). The thick dashed line is the production rate for the triple α process computed as in Eq.(5). The production rate for the radiative capture process α+n+n $\rightarrow$^{6}He+γ is taken as reference.

four-body recombination reactions.

In the figure, the thick curves correspond to the radiative processes. Obviously, the horizontal dotted line at 1 refers to the $\alpha + n + n \rightarrow {}^6\text{He} + \gamma$ reaction. The thick solid line represents the $\alpha + \alpha + n \rightarrow {}^9\text{Be} + \gamma$ process, and the thick dashed line is the production rate for the triple α process computed as in Eq.(5). The production of ^{9}Be is dominating in all the cases for low temperatures. This is consistent with the fact that the reaction rate for the process leading to ^{9}Be$+\gamma$ is, at low temperatures, orders of magnitude bigger than the one leading to ^{6}He$+\gamma$ (this agrees with the result quoted in [2], where for a temperature of 0.5 GK the reaction rate for the $\alpha + \alpha + n \rightarrow {}^9\text{Be} + \gamma$ process is computed to be four orders of magnitude bigger than the one for the $\alpha + n + n \rightarrow {}^6\text{He} + \gamma$ reaction). For higher temperatures, as seen in Fig.2, the reaction rate for production of ^{9}Be is comparable to the one obtained for the production of ^{6}He, and the dominance of one process or another strongly depends on the abundance of neutrons and alphas. In fact, as seen in the figure, a small presence of α's makes the $\alpha + n + n \rightarrow {}^6\text{He} + \gamma$ reaction to be the dominant electromagnetic reaction, while when the abundance of α-particles increases the production rate for the $\alpha + \alpha + n \rightarrow {}^9\text{Be} + \gamma$ process progressively increases as well, and eventually become the dominant electromagnetic process in the whole range of temperatures. As also seen in the figure, a neutron density of about 10^{27} neutrons/cm^3 (as used in the three panels of Fig.2) is enough to make the production of ^{9}Be (of $E1$ character) dominant over the triple α reaction (of $E2$ character). For an environment with a neutron density orders of magnitude smaller (as in the core of the red giants) the triple α process appears however as the dominant one, and therefore as the responsible for the bridging of the A=5 and A=8 instability gaps.

The thin lines in the figure are the production rates corresponding the four-body recombination processes. The dominance of one four-body process over another strongly depends on the abundance of α's and neutrons. It is important to remember that an increase in the mass density by a certain factor, enhances by the same factor the relative four-body recombination production rate. In a hot bubble the neutron density is estimated to range between 10^{20} and 10^{30} neutrons/cm^3. The calculation shown in the figure has used $n_n \sim 10^{27}$ neutrons/cm^3. Therefore, for neutron densities in the higher limit of the estimated values, and for temperatures as the ones estimated for a hot bubble (units of GK), the four-body recombination reactions could easily be very relevant.

5. Summary and conclusions

The main goal when investigating nuclear reactions with astrophysical interest is to estimate their production rate, which gives the velocity (number of reactions per unit time and unit volume) at which the products of the reaction are created. Typical temperatures in the stellar medium are at most of a few GK. This fact implies that the reactions taking place in the interior of a star occur at very low energies (at the nuclear scale). Relative kinetic energies are often clearly smaller than 1 MeV.

The production rates depend critically on the star temperature and mass density. More precisely, the density appears as ρ^N, where N is the number of particles involved in the reaction. This means, that for a sufficiently high mass density, the factor ρ^N could compensate the smaller probability for the N particles to react simultaneously, and therefore become a relevant process.

In this work we have found that four-body recombination reactions can compete with three-body radiative processes provided that the temperature and density conditions are the appropriate. In particular they can play a role as a source of ^{6}He and ^{9}Be, and therefore be relevant in the process of bridging the A=5 and A=8 instability gaps.

In this work we have investigated the reaction and production rates for the radiative processes $\alpha + n + n \rightarrow^6$ He $+ \gamma$ and $\alpha + \alpha + n \rightarrow^9$ Be $+ \gamma$, as well as for the four-body recombination reactions $\alpha + \alpha + n + n \rightarrow^6$ He $+ \alpha$, $\alpha + n + n + n \rightarrow^6$ He $+ n$, $\alpha + \alpha + n + n \rightarrow^9$ Be $+ n$ and $\alpha + \alpha + \alpha + n \rightarrow^9$ Be $+ \alpha$. No particular assumptions about the reaction mechanism has been made. Discretization of the continuum spectrum by using a box boundary condition permits simultaneous inclusion of sequential, direct, and resonant processes. Contrary to what very often is assumed, we have found that the assumption of a sequential process (through intermediate two-body states) can lead to inaccurate results.

6. Acknowledgments

This work was partly supported by funds provided by DGI of MEC (Spain) under contract No. FIS2008-01301. One of us (R.D.) acknowledges support by a predoctoral I3P fellowship from CSIC and the European Social Fund.

References

[1] Meyer B S 1994 Annu. *Rev. Astron. Astrophys.* **32** 153
[2] Görres J, Herndl H, Thompson I J and Wiescher M 1995 *Phys. Rev.* C **52** 2231
[3] Bartlett A, Görres J, Mathews G J, Otsuki K, Wiescher M, Frekers D, Mengoni A and Tostevin J 2006 *Phys. Rev.* C **74** 015802
[4] Nielsen E, Fedorov D V, Jensen A S and Garrido E 2001 *Phys. Rep.* **347** 373
[5] Fowler W A, et al. 1967 *Annu. Rev. Astron. Astrophys.* **5** 525
[6] Forssén C, Shul'gina N B, Zhukov M V 2003 *Phys. Rev.* C **67** 045801
[7] de Diego R, Garrido E, Jensen A S and Fedorov D V 2008 *Phys. Rev.* C **77** 024001
[8] Garrido E, Fedorov D V and Jensen A S 2001 *Nucl. Phys.* A **695** 109
[9] Álvarez-Rodríguez R, Jensen A S, Fedorov D V, Fynbo H O U and Garrido E 2007 *Phys. Rev. Lett.* **99** 072503
[10] Garrido E and Moya de Guerra E 1999 *Nucl. Phys.* A **650** 387
[11] Ali S and Bodmer A R 1966 *Nucl. Phys.* **80** 99
[12] Fedorov D V, Garrido E and Jensen A S 2003 *Few-body Syst.* **33** 153
[13] Sumiyoshi K, Utsunomiya H, Goko S and Kajino T 2002 *Nucl. Phys.* A **709** 467

Journal of Physics: Conference Series **205** (2010) 012048 doi:10.1088/1742-6596/205/1/012048

The Trojan Horse Method as an indirect approach for nuclear astrophysics studies

A Tumino[1,2], C Spitaleri[1,3], S Cherubini[1,3], M La Cognata[1,3], L Lamia[1,3], RG Pizzone[1,3], SMR Puglia[1,3], GG Rapisarda[1,3], S Romano[1,3], ML Sergi[1,3]

[1]Laboratori Nazionali del Sud - INFN, Catania, Italy
[2]Universita degli Studi di Enna "Kore", Enna, Italy
[3]Dipartimento di Metodologie Fisiche e Chimiche per l'Ingegneria - Università di Catania, Catania, Italy

E-mail: `tumino@lns.infn.it`

Abstract. The Trojan Horse method (THM) is a powerful indirect technique that provides a successful alternative path to determine the bare nucleus astrophysical S(E) factor for rearrangement reactions down to astrophysical energies. This is done by measuring the cross section for a suitable three body process in the quasi-free kinematics regime. Prescriptions and basic features will be presented together with some applications to demonstrate how THM works.

1. Introduction

Despite the Coulomb force is extremely weaker than the nuclear one, its in uence cannot be always disregarded. For example, it takes considerable energy to force nuclei to fuse, even those of the lightest element, hydrogen. This is because all nuclei have a positive charge, and as like charges repel, nuclei strongly resist being put too close together. This is a critical point in nuclear astrophysics, that deals with explaining the nucleosynthesis in the universe via nuclear reactions. At the relevant temperatures all reactions are strongly inhibited because of the Coulomb repulsion, responsible for the exponential decrease of the cross section (E) at those temperatures. For this reason, the behavior of (E) at low energy is usually extrapolated [?] from higher energies (usually E>100 keV) using the de nition of the astrophysical S(E)-factor

$$S(E) = E \quad (E)\exp(2 \quad \eta), \tag{1}$$

where η is the Coulomb parameter of the colliding nuclei, $\exp(2 \quad \eta)$ is the inverse of the Gamow factor, which essentially removes the dominant energy dependence of (E) due to the barrier penetration. Although the S(E)-factor allows for an easier extrapolation, large uncertainties to (E) may be introduced, for instance, because of the presence of unexpected resonances.

A second relevant source of uncertainty in extrapolating the S(E)-factor down to zero energy is its enhancement due to the electron screening e ect. In the extrapolation of the cross section using Eq.1, it is assumed that the Coulomb potential of the target nucleus and the projectile is that resulting from bare nuclei. However, as regards nuclear reactions studied in the laboratory, the target nuclei and the projectiles are usually in the form of neutral atoms or molecules and ions, respectively. Because of the electron clouds surrounding the interacting nuclei, the projectile e ectively sees a reduced Coulomb barrier. This, in turn, leads to a higher cross section for screened nuclei, $_s(E)$, compared to the cross section one would get in the case of bare nuclei $_b(E)$ [1, 2]. Therefore the enhancement factor, de ned by the following relation

$$f_{\text{lab}}(E) = \ _s(E)/ \ _b(E) \approx \exp(\ \eta U_e/E) \,, \tag{2}$$

where U_e is the so-called "electron screening potential" [1, 2], must be taken into account to determine the bare nucleus cross section.
In the astrophysical environment, the cross section under plasma conditions $_{pl}$(E) is related to the bare nucleus cross section by a similar enhancement factor:

$$f_{pl}(E) = \ _{pl}(E)/ \ _b(E) \approx \exp(\ \eta U_{pl}/E) \tag{3}$$

that can be calculated from the knowledge of the plasma screening potential U_{pl}, depending on important properties of the plasma such as the Debye-Hückel radius. A measurement of U_e, which is needed to calculate $_s$(E) from Eq. 2, would also help to better understand U_{pl}.
However the U_e values deduced in fusion reactions are quite larger in all cases than the upper limit, given by the adiabatic model as the di erence between the electron binding energies of the separate atoms in the entrance channel and that of the composite atom [3, 2]. This disagreement in laboratory experiments is not justi ed yet, and it does not help in understanding the e ects under astrophysical conditions.
A week point in the laboratory approach - and thus in the deduced U_e value - is the need for an assumption about the energy dependence of $_b$(E) at ultra-low energies. In this framework, the THM overcomes all these difficulties providing a successful alternative path to determine $_b$(E).

2. Basic features of the Trojan Horse Method

The THM [4, 5, 6, 7] replaces the astrophysical $A + x \rightarrow c + C$ two-body reaction by a suitable $A + a \rightarrow c + C + s$ three-body process, establishing a relation between the two reactions by using nuclear reaction theories. The three-body process is chosen in such a way that the involved target a (or equivalently the projectile) has a wave function with a large amplitude for a $x \quad s$ cluster con guration, x being the target (or equivalently the projectile) of the two-body reaction. In a selected part of the three-body phase space where the other cluster s remains spectator to the process, the three-body reaction can be regarded as an o -shell two-body reaction (actually half-o -shell, the two body reaction being virtual only in the entrance channel), usually referred to as a quasi-free reaction. Since the three-body process occurs at an energy above the Coulomb barrier, the main feature is the actual suppression of both the Coulomb barrier and the screening e ects in the o -shell two-body cross section. Nevertheless, the quasi-free $A + x$ process can occur even at very low sub-Coulomb energies thanks to the key role of the $x \quad s$ binding energy in compensating for the $A + a$ relative motion. This is indeed a di erent approach to the THM compared to the original idea of ref. [8], where the initial velocity of the projectile A is compensated for by the Fermi motion of x. In that framework a quite large momentum of the order of 200 MeV/c or more is needed. But the relative yield of the experimental momentum distribution at such momenta can be very small, in particular for a l=0 inter-cluster motion (for example p-n motion inside ^{2}H or α-d motion inside ^{6}Li). This would make very critical the separation from other competitive reaction mechanisms. Moreover, the theoretical description of the tails of the momentum distribution is a hard task, their shape being very sensitive to it. In our approach to the THM the inter-cluster motion is only needed to x the accessible astrophysical energy region within a chosen cuto in momentum distribution, usually of the order of few tens of MeV/c. In this framework, the so called "quasi-free two-body energy" is given by:

$$E_{QF} = \frac{m_x}{m_x + m_A} E_A \quad B_{x-s}. \tag{4}$$

where E_A represents the beam energy, m_x and m_A are the masses of x and A particles respectively, and B_{x-s} is the binding energy for the $x \;\; s$ system. Then, a cuto in the momentum distribution, which is related to the Fermi motion of s inside the Trojan-horse a, xes the range of energies around the "quasi-free two-body energy" accessible in the astrophysical relevant reaction.
In the Impulse Approximation, based essentially on the assumption that the interaction of the spectator with particles C and c is reduced, the three body-cross cross section can be factorized as:

$$\frac{d^3}{dE_c d\Omega_c d\Omega_C} \propto \left[KF \, |\varphi_a(\mathbf{p}_{sx})|^2\right] \left(\frac{d}{d\Omega_{c.m.}}\right)^{\mathrm{HOES}} \tag{5}$$

where KF is a kinematical factor containing the nal state phase-space factor. It is a function of the masses, momenta and angles of the outgoing particles [6]; $\varphi_a(\mathbf{p}_{sx})$ is proportional to the Fourier transform of the radial wave function $\chi(\mathbf{r})$ for the $x \;\; s$ inter-cluster relative motion; $(d \;\, /d\Omega_{c.m.})^{\mathrm{HOES}}$ is the half-o -energy-shell (HOES) di erential cross section for the binary reaction at the center of mass energy $E_{c.m.}$ given in post-collision prescription by

$$E_{c.m.} = E_{cC} \quad Q_{2b} \, . \tag{6}$$

Here, Q_{2b} is the Q-value of the binary reaction and E_{cC} is the relative energy of the outgoing particles c and C.

The factorization of Eq.5 is strictly valid in Plane Wave or in the so called Modi ed Plane Wave [9], this changing only its absolute magnitude but not the energy dependence of the two-body cross section.

3. Experimental prescriptions

In a typical THM experiment the decay products (c and C) of the virtual two-body reaction of interest are detected and identi ed by means of telescopes (silicon detector or ionization chamber as E step and position sensitive detector as E step) placed at the so called quasi free angles. After the selection of the reaction channel, the most critic point is to disentangle the quasi free mechanism from other reaction mechanisms feeding the same particles in the nal state, e.g. sequential decay and direct break-up. An observable which turns out to be very sensitive to the reaction mechanism is the shape of the experimental momentum distribution of the spectator. In order to reconstruct the experimental p_s distribution, the energy sharing method [10] is applied for each pair of coincidence QF angles: the quasi-free coincidence yield with a cuto in relative energy of few hundreds of keV at most is divided by the kinematic factor, providing a quantity which is proportional to the product of the momentum distribution for the neutron with the di erential two-body cross section (see Eq.5). The cuto is chosen in such a way that the di erential two-body cross section in this range can be considered almost constant. Thus the quantity de ned above represents essentially the momentum distribution for the spectator that has to be compared with the the Fourier transform of the $x \;\; s$ bound state wave function. An example of experimental momentum distribution for the neutron inside ^{2}H is reported in Fig.1 together with the theoretical behaviour (solid line). The points drawn as full and open circles, belong to two di erent experimental runs. As expected, they have the same trend. Further data analysis is limited to the data lying in the region where the agreement between the two distributions exists. However, usually a window of few tens of MeV/c is chosen not to move too far from E_{QF}.
Therefore, it is possible to derive the HOES $((d \;\, /d\Omega_{cm})^{\mathrm{HOES}}$ from the three-body coincidence yield by simply inverting Eq.5. In a nal step, the HOES cross section has to be related to the relevant on-energy-shell (OES) cross section by applying the corresponding corrections. It was demonstrated that there is no Coulomb barrier in the two-body amplitude extracted from

the TH reaction [11] and this is due to the virtuality of particle x. This seems to be the only consequence of o -energy-shell e ects as suggested by the agreement between HOES and OES cross-sections for the ^{6}Li(n,α)^{3}H reaction [12]. If one looks at the angular distributions no correction is needed because once the energy is xed, it would mean to introduce simply a scaling factor. Thus, the OES data are directly comparable with HOES ones projected onto the emission angle of C (or c) in the C c center of mass system, $\theta_{c.m.}$, as given by [13]:

$$\theta_{c.m.} = \arccos \frac{(\mathbf{v}_{\mathrm{A}} \quad \mathbf{v}_{\mathbf{x}}) \ (\mathbf{v}_{\mathbf{c}} \quad \mathbf{v}_{\mathrm{C}})}{|\mathbf{v}_{\mathrm{A}} \quad \mathbf{v}_{\mathbf{x}}||\mathbf{v}_{\mathbf{c}} \quad \mathbf{v}_{\mathrm{C}}|} \tag{7}$$

where the vectors $\mathbf{v}_{\mathrm{A}}, \mathbf{v}_{\mathbf{c}}, \mathbf{v}_{\mathbf{c}}, \mathbf{v}_{\mathrm{C}}$ are the velocities of projectile, transferred particle and outgoing nuclei respectively. These quantities are calculated from their corresponding momenta in the lab-system, where the momentum of the transferred particle is equal and opposite to that of spectator when the quasi-free break-up takes place in the target, otherwise a little di erent formula has to be used [13]. If HOES data are projected onto the E_{Cc} axis, Coulomb suppression has to be introduced before comparison with OES data. In a heuristic approach this is done by means of the penetrability factor:

$$P_l(k_{Ax}R) = \frac{1}{G_l^2(k_{Ax}R) + F_l^2(k_{Ax}R)} \tag{8}$$

with F_l and G_l regular and irregular Coulomb wave functions. This procedure does not allow us to extract the absolute value of the two-body cross section. However, this is not a real problem since the absolute magnitude can be derived from a scaling to the direct data available at higher energies.

4. Applications

The THM was applied several times to rearrangement reactions connected with fundamental astrophysical problems [14]. A list of reactions studied by means of the THM is given in Table 1 together with the relevant references.

Low-energy cross sections for reactions producing or destroying lithium isotopes are fundamental for a number of still not completely solved astrophysical problems, e.g. understanding Big Bang nucleosynthesis and the "Lithium depletion" in the Sun or in other galactic stars. An experimental program was undertaken to study p-capture reactions on ^{6,7}Li, main responsible for their destruction [19, 20, 5, 21]. The extracted astrophysical S(E) factors were compared with those from direct measurements and found in fair agreement in the energy range where screening e ects are negligible. Since the THM provides the bare S(E) factor, improved information on the screening potential U_e was obtained from comparison with shielded direct data. The extracted values of U_e=(450$\pm$100)eV for ^{6}Li and U_e=(330$\pm$30)eV for ^{7}Li con rmed the isotopic independence of U_e, although they were found to be much larger than the adiabatic limit (176 eV). The S(E) factor for the ^{6}Li(p,α)^{3}He reaction is shown in g. 2 (black dots). Direct data from [22, 23] are also shown as open symbols. The solid line represents a polynomial t to the THM data as reported in [5, 12]. The implications of these THM results for the "Lithium problem" have been discussed in [24, 25], showing that $_b$(E) for these reactions is now well determined and that the solution of the problem of light element destruction in stars lies elsewhere.
The ^{15}N(p,α)^{12}C, ^{18}O(p,α)^{15}N reactions are involved in the production path of ^{19}F in AGB stars, whose chemical evolution is strongly in uenced by ^{19}F abundance. Available data from observations provide evidence that uorine is deeply produced within the interiors of these stars during helium ashes, and its abundance, enhanced by large factors with respect to the solar one, cannot be matched by the models. Among the primary sources of uncertainty in predicting uorine abundances in AGB stars are the adopted reaction rates [26]. The astrophysical S(E)

Journal of Physics: Conference Series **205** (2010) 012048 doi:10.1088/1742-6596/205/1/012048

Table 1. Two-body reactions studied via the THM

	Direct reaction	TH reaction	E_{inc} (MeV)	Q (MeV)	TH_{nucl}	ref
[1]	^{7}Li(p,α)^{4}He	^{7}Li(d,α α)n	19,22	15.122	d = (p$\otimes$n)	[20]
[2]	^{6}Li(d,α)^{4}He	^{6}Li(^{6}Li,α α)^{4}He	5	22.372	^{6}Li = ($\alpha\otimes$d)	[4]
[3]	^{6}Li(p,α)^{3}He	^{2}H(^{6}Li,α ^{3}He)n	14, 25	1.795	d = (p$\otimes$n)	[5]
[4]	^{11}B(p,α)^{8}Be	^{2}H(^{11}B,^{8}Be α)n	27	6.36	d = (p$\otimes$n)	[6]
[5]	^{10}B(p,α)^{7}Be	^{2}H(^{10}B,^{7}Be α)n	27	-1.079	d = (p$\otimes$n)	[15]
[6]	^{9}Be(p,α)^{6}Li	^{2}H(^{9}Be,^{6}Li α)n	22	-0.099	d = (p$\otimes$n)	[16]
[7]	^{6}Li(n,α)^{3}H	^{2}H(^{6}Li,α^{3}H)p	14	4.78	d = (p$\otimes$n)	[12]
[8]	^{2}H(^{3}He,p)^{4}He	^{6}Li(^{3}He,p α)^{4}He	5, 6	16.88	^{3}He = (p$\otimes$n)	[18]
[9]	^{2}H(d,p)^{3}H	^{2}H(^{6}Li,t p)^{4}He	14	2.59	^{6}Li = ($\alpha\otimes$d)	[17]
[10]	^{15}N(p,α)^{12}C	^{2}H(^{15}N,α^{12}C)n	60	2.74	d = (p$\otimes$n)	[7]
[11]	^{18}O(p,α)^{15}N	^{2}H(^{18}O,α^{15}N)n	54	1.76	d = (p$\otimes$n)	[7]
[12]	^{1}H(p,p)^{1}H	^{2}H(p,pp)n	5,6	0	d = (p$\otimes$n)	[31, 32]
[13]	^{17}O(p,α)^{14}N	^{2}H(^{17}O,α^{15}N)n	41	1.19	d = (p$\otimes$n)	[33]
[14]	^{2}H(d,p)^{3}H	^{2}H(^{3}He,t p)^{1}H	17,18	4.03	^{3}He = (p$\otimes$n)	
[15]	^{2}H(d,n)^{3}He	^{2}H(^{3}He,^{3}He n)^{1}H	18	3.27	^{3}He = (p$\otimes$n)	
[16]	^{19}F(p,α)^{16}O	^{2}H(^{19}F,α^{16}O)n	14	8.11	d = (p$\otimes$n)	

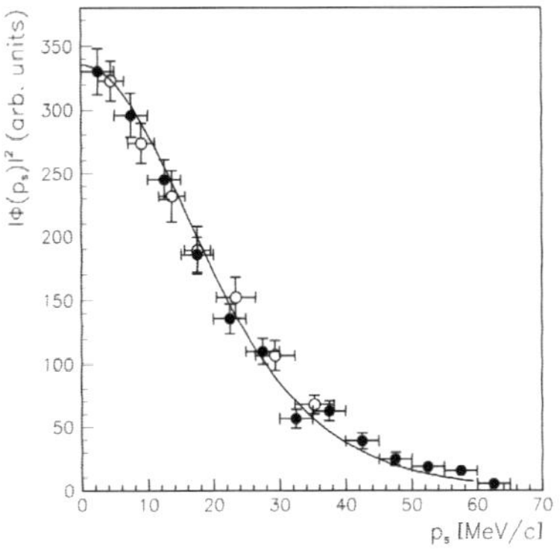

Figure 1. Experimental momentum distribution for the neutron inside d (full and open circles). The solid line represents the shape given by the square of the n p Hulthen bound state wave function in momentum space.

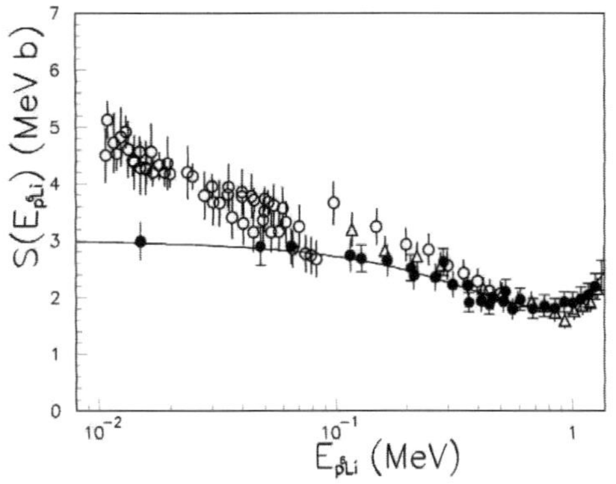

Figure 2. THM bare nucleus S(E) factor for the ^{6}Li(p,α)^{3}He (black dots) [21] compared with direct data (open symbols) from [22]. The solid line is the result of a polynomial t to the THM data.

factor for the ^{15}N(p,α)^{12}C has been recently measured by means of the THM down to 20 keV (to be compared to 73 keV in direct measurements), thus covering for the rst time the whole

Gamow window (94 ± 66 keV for a temperature of 10^8 K) without the complications due to the electron screening. A calculation called Modi ed R-matrix approach [7] was applied. It accounts for the virtuality of the proton in the entrance channel of the relevant two body reaction by replacing the entry resonance widths with the form factors. The calculations include the parameters associated with the rst three resonances in the $^{15}N + p$ system at $E_{^{15}Np} = 0.312$, 0.9624 and 1.0014 MeV, and the subthreshold state at $E_{^{15}Np} =$ 2.53 MeV. For details concerning the formal expression of the reaction amplitude in terms of the form factors see [7]. The calculated S(0) value of the astrophysical factor ranges from 65.0 to 70.0 MeVb for a channel radius $r_0 = 5.0$ to 6.0 fm, in agreement with the estimate derived from a Breit-Wigner extrapolation [27] on the direct data.

As regards the $^{18}O(p,\alpha)^{15}N$ reaction, its reaction rate strongly depends on the strengths of three low-lying resonances corresponding to the 8.014 ,8.084 and 8.138 ^{19}F states (20, 90 and 144 keV in the ^{18}O-p relative energy axis respectively). While the nuclear parameters for the 8.138 state are quite well known, the strengths of the rst two states are very much uncertain. This experiment has provided the strengths of both resonances. The Modi ed R-matrix approach was applied in order to extract this information, after measuring the relative strengths of the nearby known resonances. Its uncertainty was reduced almost by a factor 10 [28] with respect to that in the previous value adopted in the NACRE compilation [29]. The resulting o -shell excitation function is reported in Fig.3 for $\theta_{c.m.}$ of 55°, in the 0-250 keV energy range relevant for astrophysics. A Gaussian t was used to disentangle the contribution of each peak to the overall coincidence yield. The solid line is the sum of the three Gaussian functions drawn as dashed, dotted, and dashed-dotted lines, and the straight line (dashed-dotted line) is to provide for the nonresonant behaviour. A reaction rate about 35% larger than the Nacre value has been estimated. The astrophysical consequences are being investigated. Recently we have extended

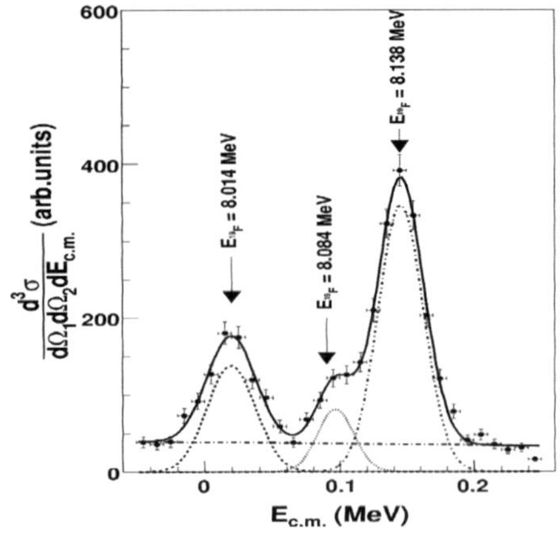

Figure 3. Excitation function for the $^{18}O(p,\alpha)^{15}N$ reaction (black dots) [7] See the text for the meaning of the di erent lines.

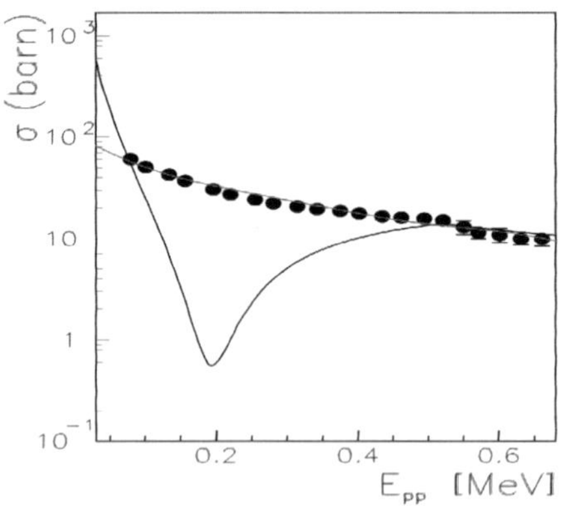

Figure 4. THM two-body cross section (black dots) vs. pp relative energy [31]. Solid line represents the theoretical OES p-p cross section [30]. The dashed-dotted line is the calculated HOES cross section [31].

the application of the THM to scattering processes. In particular, we have address the study of the $p + p$ elastic scattering through the $^2H(p,pp)n$ reaction [31]. The aim of the experiment was to investigate the suppression of the Coulomb amplitude also for scattering. The extracted p p HOES cross section is presented in g.5 as a function of E_{pp} (black dots) and compared to the free p-p cross section (solid line) [30] where the deep minimum due to the interference between Coulomb and nuclear amplitude dominates the excitation function. The dashed-dotted line represents the calculated HOES p p cross section [31] where the Coulomb amplitude is strongly suppressed compared to the nuclear one. We observe a striking disagreement between the THM (HOES) and the free p p (OES) cross sections throughout the region of the interference minimum, which is missing in the THM data. Instead, the calculated HOES p p nicely ts the THM data. In order to strengthen this result, the THM p p cross section was compared with the OES n n, p n and pure nuclear p p ones [30] as reported in [31]. A good agreement shows up providing compelling evidence of the validity of the THM method for elastic scattering. Moreover, another issue has been readdressed by looking at the energy dependence of the HOES n n cross section. Surprisingly, it fairly reproduces the behaviour of the n n and pure nuclear p p cross sections. This result is in agreement with the one reported in [12] since it excludes the existence of o -energy-shell e ects when the Coulomb barrier is absent. These results are extensively explained and displayed in [32]. Thus, through a mechanism di erent from that of nuclear rearrangement reactions, the present work strongly sustains the THM basic feature, namely the suppression of Coulomb e ects in the two-body cross section at sub-Coulomb energies. This appears to be a universal e ect whether we consider binary elastic or rearrangement processes. In conclusions, all these applications put on rmer grounds the applicability of the THM in nuclear astrophysics as well as in all physics contexts where it can be important to investigate nuclear e ects at low energies.

5. Bibliography

[1] H.J. Assenbaum et al. 1987 *Z . Phys.* A **327 461**
[2] F. Strieder et al. 2001 *Naturwissenschaften* **88 461**
[3] G. Fiorentini et al. 1995 *Z. Phys.* A **350 289**
[4] C. Spitaleri et al. 2001 *Phys. Rev.* C **63 005801**
[5] A. Tumino et al. 2003 *Phys. Rev.* C **67 065803**
[6] C. Spitaleri et al. 2004 *Phys. Rev.* C **63 055806**
[7] M. La Cognata et al. *Phys. Rev.* C **76 065804**
[8] G. Baur 1986 *Phys. Lett.* B **178 135**
[9] S. Typel and G. Baur 2003 *Ann. Phys.* **305 228**
[10] J. Kasagi et al. 1975 *Nucl. Phys.* A **239 233**
[11] A.M. Mukhamedzhanov et al. 2006 *Eur. Phys. J.* A **27 205**
[12] A. Tumino et al. 2005 *Eur. Phys. J.* A **25 649**
[13] M. Jain et al. 1970 *Nucl. Phys.* A **153 49**
[14] C.J. Copi et al 1995 *Science* **267 192**
[15] L. Lamia et al. 2008 *Il Nuovo Cimento* **31 N.4 423**
[16] Qun-Gang Wen et al. 2008 *Phys. Rev.* C **78 035805**
[17] A. Rinollo et al. 2005 *Nucl. Phys.* A **758 146**
[18] M. La Cognata et al 2006 *Eur. Phys. J.* A **27 249**
[19] C. Spitaleri et al. 1999 *Phys. Rev.* C **60 055802**
[20] M. Lattuada *et al.*, Ap.J. **562**, 1076
[21] A. Tumino et al. 2004 *Prog. Theor. Phys. Suppl.* **154 341**
[22] S. Engstler et al. 1992 *Z. Phys.* A **342 471**
[23] A.J. Elwyn et al. 1979 *Phys. Rev.* C **20 1984**
[24] R.G. Pizzone et al. 2003 *A & A* **398 423**
[25] R.G. Pizzone, A. Tumino et al. *A & A* **438 779**
[26] M. Lugaro et al. 2004 *Ap. J.* **615 934**
[27] A. Redder et al. 1982 *Z. Phys.* A **305 325**
[28] M. La Cognata et al. 2008 *Phys. Rev. Lett.* **101 152501**
[29] C. Angulo et al. 1999 *Nucl. Phys.* A **656 3**
[30] J.D. Jackson and J.M. Blatt 1950 *Rev. Mod. Phys.* **22 77**
[31] A. Tumino et al. 2007 *Phys. Rev. Lett.* **98 252502**
[32] A. Tumino et al. 2008 *Phys. Rev.* C **78 064001**
[33] M.L. Sergi et al., submitted to *Astrophysical Journal Letters*

Journal of Physics: Conference Series **205** (2010) 012049 doi:10.1088/1742-6596/205/1/012049

The ^{10}B(n,α_0)^{7}Li and ^{10}B(n,$\alpha_1\gamma$)^{7}Li alpha-particle angular distributions, branching ratios and cross-sections measurements for E_n < 3 MeV

Hambsch F J[1], Ruskov I[2] and Vidali M[1]

[1]EC-DG-JRC-Institute of Reference Materials and Measurements, Retieseweg 111, B-2440, Geel, Belgium
[2]Institute for Nuclear Research and Nuclear Energy of Bulgarian Academy of Sciences Tzarigradsko chaussee blvd. 72, BG-1784, Sofia, Bulgaria

franz-josef.hambsch@ec.europa.eu

Abstract. The ^{10}B(n,α_0)^{7}Li and ^{10}B(n,$\alpha_1\gamma$)^{7}Li angular distributions have been measured at the GELINA time-of-flight spectrometer in the incident neutron energy range from 0.1 keV to 1 MeV by means of a twin Frisch-grid ionization chamber. With this type of detector it is possible to measure the angular distribution of the charged reaction fragments in a close to 2x2π solid angle with ~100% efficiency and a clear separation of both reaction channels: emission to the ^{7}Li ground state (α_0) or to its first excited state (α_1). A strong angular anisotropy was observed at ~ 520 keV. In order to extend the energy range up to 2.5-3 MeV and to measure, also, the reaction cross sections, a double twin Frisch-grid ionization chamber was constructed. It is loaded with two very thin 94% ^{10}B-enriched samples, mounted back-to-back with ^{235}U samples on the common cathodes. New data acquisition, visualization and analysis software is used in a new set of long-term measurements, which are still going on.

1. Introduction

At the last standards evaluation of the IAEA [1], the ^{10}B(n,α)^{7}Li {^{10}B(n,α_0) + ^{10}B(n,$\alpha_1\gamma$)} and ^{10}B(n,$\alpha_1\gamma$) standards got a large attention due to its improved ENDF/B-VII [2] database and its extension from 250 keV up to 1 MeV incident neutron kinetic energy.

For the ENDF/B-VII standards evaluation, a combined procedure similar to that used for the ENDF/B-VI was implemented. The ^{10}B(n,α) standards were evaluated using, also, the GMAP code (GMA code [3], modified by adding the Chiba-Smith option [4] to handle Peelle's Pertinent Puzzle problems), within a combining procedure, using input from the RAC [5] (Tsinghua Univ.) and EDA [6] (LANL) R-matrix analyses, and a thermal constants evaluation. The R-matrix analyses used all the available data, including total and scattering cross-section. The only boron data for direct use in the GMAP code were the ratio measurements. Because the ^{10}B(n,α) and ^{10}B(n,$\alpha_1\gamma$) cross-sections obtained from the RAC and EDA analyses were not identical, the cross-sections from both analyses were averaged (without weights) and used as the R-matrix input to GMAP.

This way, the difficulties met in the previous evaluation were overcome via improvements of the evaluation procedure and using new experimental data sets, mainly those on the α-particle angular distributions [7] and ^{10}B(n,α_0) / ^{10}B(n,$\alpha_1\gamma$) branching ratios [8], obtained at the EC-JRC-IRMM, for incident neutron energies $E_n \leq 1$ MeV.

For the first time, also, the full angular distributions for both alpha-particle emission channels (α_0-^{7}Li ground state transition and α_1-^{7}Li* first excited state transition) were used in a multi-level, multi-channel R-matrix calculation [9].

The experimental setup, method and obtained results have been published in [7, 8] and in the literature cited therein. Here we focus our discussion on the improvements made on the existing experimental setup and measuring conditions. This was done in order to measure the α-particle angular distributions and obtain the ^{10}B(n,α_0), ^{10}B(n,$\alpha_1\gamma$) and ^{10}B(n,α) differential cross sections $d\sigma_{n,\alpha 0}/d\Omega$, $d\sigma_{n,\alpha 1}$, $/d\Omega$, $d\sigma_{n,\alpha}/d\Omega$, the angle integrated cross sections $\sigma(\alpha_0)$, $\sigma(\alpha_1)$, $\sigma(\alpha)$ and their branching ratios $BR \equiv \sigma(\alpha_0) / \sigma(\alpha_1)$ for $E_n \leq 2.5$-3 MeV.

Measurements of the differential cross-section $d\sigma_{n,\alpha}/d\Omega$ in the fast neutron energy range E_n = 1.5-5.6 MeV were made by Zhang *et al.* [10] and Giorginis and Khryatchkov [11] using Frisch-grid ionization chambers. A large discrepancy was observed between the two measurements. This deviation was attributed to the so-called particle leaking effect, a kinematic effect resulting from forward emission of both reaction products with respect to the direction of the incoming neutron [12]. If it is not taken into account in the analysis, the resulting cross section will not be accurate.

The shape of the ^{10}B(n,$\alpha_1\gamma$) cross-section (relative to a "black" detector) for E_n = 0.2 - 4.0 MeV was measured by Schrack *et al.* [13] in a NIST/ORNL collaboration. The cross-section obtained was normalized to the ENDF/B-VI evaluation over the region of E_n = 0.2 - 1.0 MeV, and agreed with this evaluation below 1.5 MeV. However, it differs from it by about 40% for $E_n \geq 1.5$ MeV. Additional shape measurements relative to H(n,n), in the neutron energy region E_n = 10 keV - 1 MeV [14], led to a cross-section ~ 5 % lower than ENDF/B-VI for E_n = 120 keV - 1 MeV. In the same NIST/ORNL collaboration, the ^{10}B total cross-section was measured for E_n = 20 keV - 20 MeV. The values agree with the ENDF/B-VI evaluation above E_n ~ 2 MeV, but are lower as much as 4 % for E_n = 600 keV - 2 MeV and higher as much as 5 % for E_n < 600 keV.

At GELINA Märten *et al.* [15] measured the ^{10}B(n,$\alpha_1\gamma$) cross section relative to the ^{235}U(n,f) and carbon standard cross sections in the incident neutron energy interval from 320 keV to 2.8 MeV. Their data are in reasonable agreement with those of Schrack *et al.*

With the present experiment we will try to extend the incident neutron energy range, so far reported in [7] and [8], up to about 2.5 - 3 MeV and, additionally, to measure the reaction cross section, relative to the ^{235}U(n,f) standard, in order to compare with the data of [15] and to have an overlap with the measurements performed at the Van de Graaff accelerator [11].

A brief description of the new experimental setup is given elsewhere [16]. Here more details about the measurements are given.

2. The experimental setup

The GELINA facility of JRC-IRMM [17] is utilised as a white spectrum pulsed neutron source at which the time-of-flight (TOF) method is used to determine the energy of the interacting neutrons. Among the pulsed white spectrum neutron sources available in the world, GELINA is the one with the best time resolution. It is a multiuser facility serving 12 completely independent flight-paths with a length of up to 400 m.

The experimental setup is commissioned at the FP16 experimental station, situated at a distance of about 60 m from the GELINA neutron producing target (figure 1).

This distance was chosen by several reasons: to use the un-moderated neutron beam, obtain better time (energy) resolution and to reduce the intensity of the very strong γ-flash from the neutron producing target. The γ-flash was additionally attenuated by a number of in-beam filters (bismuth and lead), situated at a distance of ~ 1 m in the front of the chamber (figure 1). The neutron resonance absorption in these filters is used for time-to-energy conversion of the recorded TOF spectra.

As charge particle detector a double twin Frisch-grid gas ionization chamber (IC) was used. With this type of detector-spectrometer it is possible to register with an efficiency of ~100% the kinetic energy and the direction of irradiation of the charged particles (α-particles and fission fragments in this case), covering nearly 2x2π solid angle.

Some parameters of the IC are shown in table 1. It is constructed from 2 ionization chambers of the same type, as the one used in the previous measurements [7].

Figure 1. The ionization chamber loaded with the boron and uranium targets at FP16/60m, in-beam filters ~ 1 m in the front of it, signal collecting electronics and (n+γ) beam profile with a diameter ~ 115 mm, recorded remotely by a scintillation counter at ~ 1.5 m behind the chamber during the adjustment of the collimation system.

Table 1. Ionization chamber parameters.

component	vessel	cathodes	grids		anodes/shields	separators
type	cylinder	annular disc	wires	ring	foil	cylinder
material		stainless steel			Al	ceramics
thickness (mm)	1	1	0.11	1	0.05	4
pitch (mm)			1.00			
inner ø (mm)	228	100		152		6
outer ø (mm)	230	178				10
length (mm)	364					6, 10, 30

The only difference here is the distance between the common cathode and grids, which was chosen to be 40 mm. This way all the α-particles with kinetic energies up to ~ 4.5 MeV (from spontaneous α-decay of ^{235}U and from the ^{10}B(n,α) reaction) will spend their kinetic energy for ionization of the atoms of the mixed 95% Ar + 5% CO_2 gas media at a pressure of ~ 86 kPa before reaching the grids. The gas is refreshing (~ 30 ml/min) via a continuous flow through the chamber.

The samples are mounted in a back-to-back geometry (uranium/boron, boron/uranium) on the cathodes, perpendicular to the up/down neutron stream (backward/forward direction). The main characteristics of the samples are listed in table 2.

Because of different diameters of the uranium and boron samples, the beam was collimated to a diameter of ~ 95 mm at the entrance window of the IC, to assure that all 4 samples are fully covered by the beam. In fixed target geometry the homogeneity of the beam is required. It was checked by exposing photo-plates and remote scanning by a fast (neutron+gamma) scintillation detector (figure 1).

Table 2. Physical characteristics of the boron and uranium samples.

characteristics		boron			uranium	
twin chamber	sample	density ($\mu g/cm^2$)	weight (μg)	sample	density ($\mu g/cm^2$)	weight (μg)
1	Nº 3	14.5±0.8	804±44	Nº 2	193.1±2.9	3071±46
2	Nº 4	15.7±0.8	870±44	Nº 1	193.7±2.9	3081±46
support		stainless steel			stainless steel	
thickness (mm)		0.3			0.5	
diameter (mm)		100			70	
chemical form		B_4C			UF_4	
purity		98%		Sp. α-activity:	3.913±0.016 (dps/μg)	
isotope		94% ^{10}B			1.66530% ^{234}U, 97.6630% ^{235}U	
composition		6% ^{11}B			0.14912% ^{236}U, 0.5229% ^{238}U	
deposit ø (mm)		84			45	

The block diagram of the electronic set-up is shown in figure 2. It is the same as that described in [7], except that the present set-up records twice the number of spectra (8 amplitudes + 2 time signals) requiring the use of multiplexer modules (MMPM) developed at JRC-IRMM. The signals from the Frisch-grids were used to trigger the timing information of the neutron interactions with the sample nuclei. By setting appropriate low-level thresholds in the analog-to-digital converters and constant-fraction discriminators, the number of events from the spontaneous alpha-decay of the uranium samples was significantly reduced.

All the 16 outputs of the 4 MMPM modules were fed to the first slot of the MPI8100 data acquisition system and stored via the LISA software [18] (running in PV-WAVE environment) in list-mode on the hard disk for further off-line analysis. Recently, new software for data acquisition and analysis, named GENDARC [19], has been developed at IRMM and is used in this ongoing experiment, too. Some typical 2-dimentional spectra are shown in figures 3, 4 and 5.

The 1- and 2- dimensional visualization capability of the program allows to control the data acquisition and to do some primary analysis on the collected events. The reaction events, from the boron/uranium chamber, as function of the anode amplitude (~ fragment kinetic energy) and sum (anode+grid) amplitude (~ fragment kinetic energy and its angle of irradiation) are shown as 2-dimensional distributions for $^{235}U(n,f)$ (figure 3, left) and for $^{10}B(n,\alpha)$ (figure 3, right). In figure 4 the same distributions are shown for the uranium/boron chamber. In the case of the $^{10}B(n,\alpha)$ reaction, the incident neutron momentum transfer to the reaction fragments is very well seen.

3. Conclusions

A new double twin Frisch-grid gas-flow ionization chamber was constructed for measuring the $^{10}B(n,\alpha)$ reaction channel α-particle angular distributions, cross sections and branching ratios up to 2.5-3 MeV incident neutron energy, relative to the $^{235}U(n,f)$ reaction. The chamber is loaded with very thin 94% ^{10}B samples, mounted in a back-to-back geometry with the ^{235}U reference samples on its cathodes. It is installed at the FP16/60m of the GELINA TOF facility, where an air-conditioning system assures the long-term stability of the electronic modules. A new data acquisition, visualization and analysis software was tested and used in the new measurements, which are still going on.

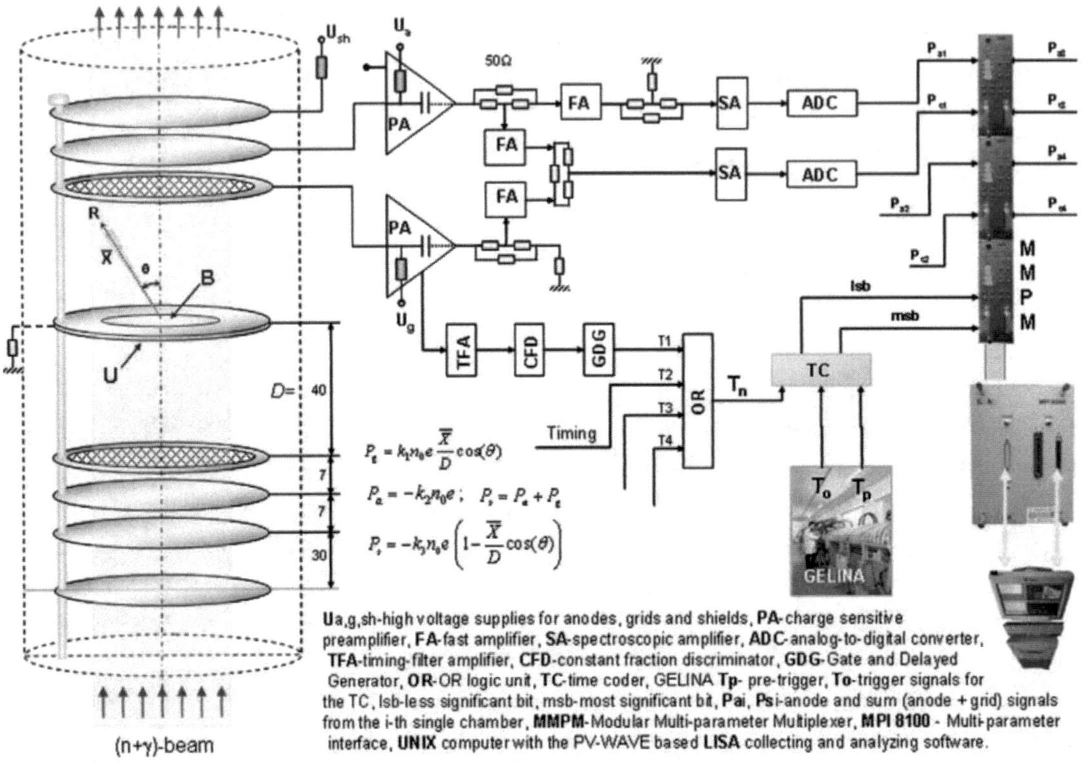

Figure 2. The scheme of the twin ionization chamber and associated electronics. The anode signal is proportional to the energy of the ionizing charged particle, the (anode+grid) signal, in addition, contains the information about the angle of irradiation.

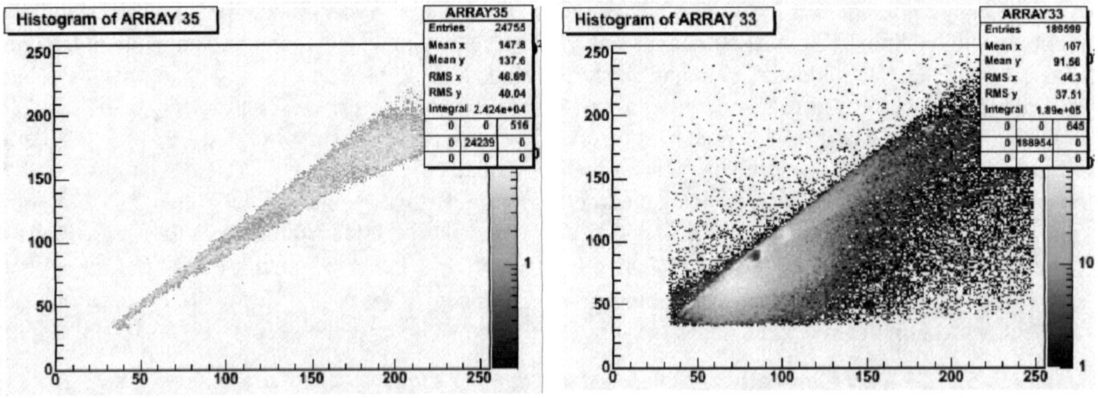

Figure 3. Fission fragment events (anode+grid pulse-height, anode pulse-height) spectra from $^{235}U(n,f)$ (left) and $^{10}B(n,\alpha)$ (right) reactions in the boron/uranium chamber.

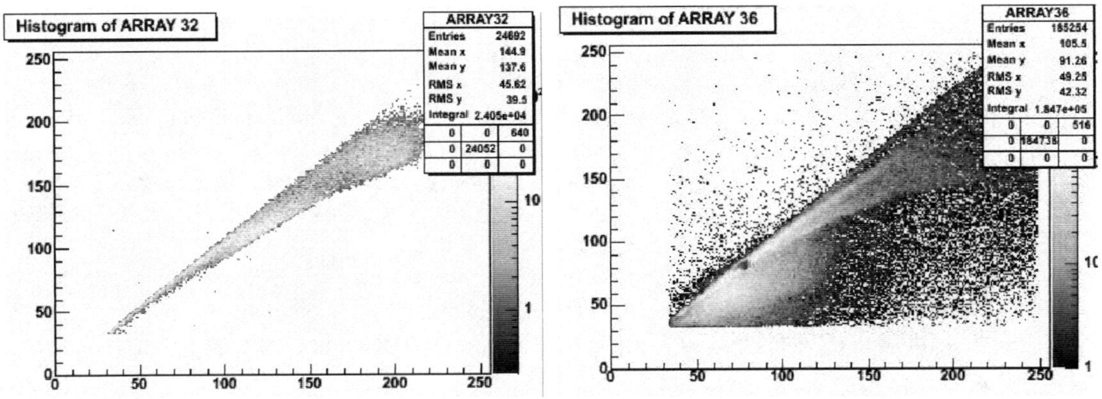

Figure 4. Fission fragment events (anode+grid pulse-height, anode pulse-height) spectra from ^{235}U(n,f) (left) and ^{10}B(n,α) (right) reactions in the uranium/boron chamber.

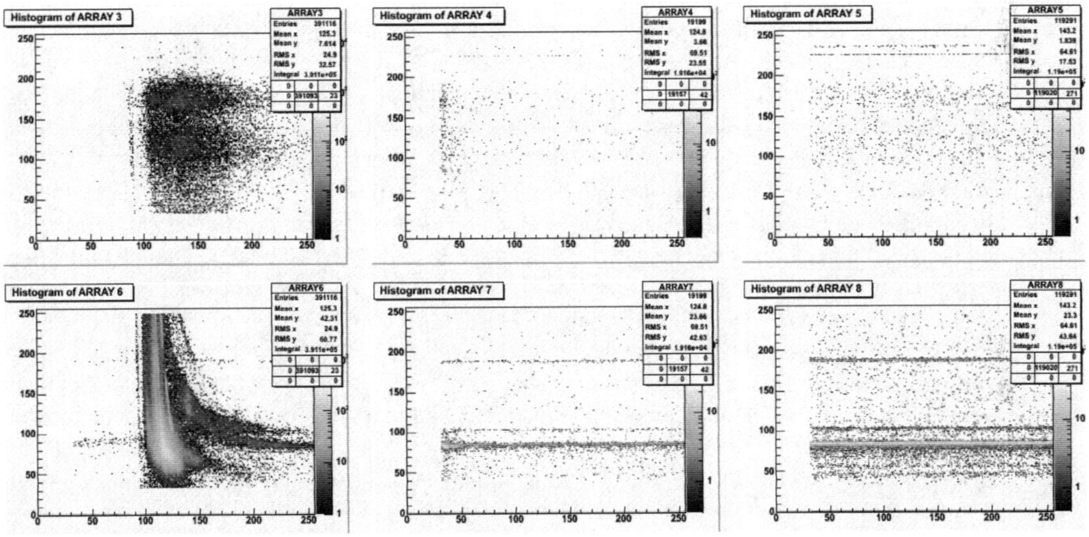

Figure 5. Fission fragment events (pulse-height, TOF) spectra from ^{235}U(n,f) (top raw) and ^{10}B(n,α) (bottom raw) reactions in the uranium/boron chamber.

Acknowledgments
The authors are indebted to the operational staff of GELINA, headed by W. Mondelaers. We would, also, like to acknowledge the very skillful preparation of the thin boron samples by J. Van Gestel from the Sample Preparation group at JRC-IRMM. One of us (I. R.) would like to thank the European Commission for the support to perform this research at the EC-JRC-IRMM.

References
[1] IAEA CRP on *Improvement of the Standard Cross Sections for Light Elements* http://www.iaea.org/programmes/ripc/nd/crps/standcrosect_of_lightelem.htm
[2] *International Evaluation of Neutron Cross Section Standards* 2007 (Vienna: IAEA) http://www-pub.iaea.org/MTCD/publications/PDF/Pub1291_web.pdf
[3] Poenitz W P 1981 Data interpretation, objective, evaluation procedures and mathematical technique for the evaluation of energy-dependent ratio, shape and cross-section data *Proceedings of the Conference on Nuclear Data Evaluation Methods and Procedures* **I** (BNLNCS-51363) p 249

[4] Chiba S and Smith D L 1991 *A Suggested Procedure for Resolving an Anomaly in Least Squares Data Analysis Known as "Peelle's Pertinent Puzzle" and the General Implications for Nuclear Data Evaluation* (ANL/NDM-121)

[5] Chen Z P 1995 Reduced R-matrix analysis for ^{17}O system *Atomic Energy Science and Technology* **29** p 366

[6] Hale G M 1981 Use of R-matrix methods for light element evaluations *Proceedings of the Conference on Nuclear Data Evaluation Methods and Procedures* **II** (BNL-NCS-5136) p 509

[7] Hambsch F J and Ruskov I 2009 The ^{10}B (n,α_0)^{7}Li and ^{10}B (n,$\alpha_1\gamma$)^{7}Li α-particle angular distributions for E_n < 1MeV *Nucl. Sci. Eng.* **163** p 1

[8] Hambsch F J and Ruskov I 2007 The ^{10}B(n, α_o)/^{10}B(n, $\alpha_1\gamma$) branching ratio *Nucl. Sci. Eng.* **156** p 103

[9] Hale G M 2007 *Private communication*

[10] Zhang G, Tang G, Chen J, Zhang S, Shi Z, Yuan J, Chen Z, Gledenov Yu M, Sedysheva M and Khuukhenkhuu G 2002 Differential cross-section measurement for the ^{10}B(n,α)^{7}Li reaction *Nucl. Sci. Eng.* **142** p 203

[11] Giorginis G and Khryatchkov V 2006 The cross-section of the ^{10}B(n, α)^{7}Li reaction measured in the MeV energy range *Nucl. Instr. Meth.* **A562** p 737

[12] Giorginis G and Khriatchkov V 2005 The effect of particle leaking and its implication for measurements of the (n,α) reaction on light elements by using ionization chambers *Nucl. Instr. Meth.* **A538** p 550

[13] Schrack R A, Wasson O A, Larson D C, Dickens J K and Todd J H 1993 The ^{10}B(n,$\alpha_1\gamma$)^{7}Li cross section between 0.2 and 4.0 MeV *Nucl. Sci. Eng.* **114** p 352

[14] Schrack R A, Wasson O A, Larson D C, Dickens J K and Todd J H 1994 The ^{10}B(n,$\alpha_1\gamma$)^{7}Li cross section from 10 keV to 1 MeV *Proceedings of the International Conference on Nuclear Data for Science and Technology* (Gatlinburg TN) p 43

[15] Märten H 1997 Private communications through H. Weigmann

[16] Hambsch F J and Ruskov I 2008 *Neutron Physics Unit Scientific Report 2007* ed. S. Oberstedt and P. Rullhusen (Luxembourg: Office for Official Publications of the European Communities) p 53

[17] Benussan A and Salome J M 1978 Gelina: a modern accelerator for high-resolution neutron time-of-flight experiments *Nucl. Instr. Meth.* **155** p 11

[18] Oberstedt A and Hambsch F J 1994 LISA - a powerful program package for list-mode and spectral data analysis *Nucl. Instr. Meth.* **A340** p 379

[19] Fabry I, Hambsch F J, Oberstedt S, Zeynalov Sh and Kornilov N 2008 *Neutron Physics Unit Scientific Report 2007* ed. S. Oberstedt and P. Rullhusen (Luxembourg: Office for Official Publications of the European Communities) p 86

Journal of Physics: Conference Series **205** (2010) 012050 doi:10.1088/1742-6596/205/1/012050

Measurement of prompt gamma-ray emission in fission

R. Borcea[1], R. Billnert[2], W. Geerts[1], F-J. Hambsch[1], R. Jaime Tornin[1], J. Karlsson[2], A. Oberstedt[2], S. Oberstedt[1]

[1] European Commission, JRC-IRMM, B-2440 Geel, Belgium
[2] Institutionen för Naturveteskap, Örebro Universitet, S-70182, Örebro, Sweden

Corresponding author: Ruxandra.Borcea@ec.europa.eu

Abstract

In this work we present results from the characterization of both $LaCl_3$:Ce scintillation and polycrystalline diamond detectors to be used for measurements of prompt gamma radiation emitted in fission. The properties of these new detector types, such as excellent timing resolution and improved energy resolution will allow to obtain fission data with higher accuracy than those available today.

Introduction

A particular challenge for the modelling of new generation reactor neutron kinetics is the calculation of the γ-heat deposition e. g. in steel and ceramics reflectors without UO_2 blankets, which is required to be known with an uncertainty as low as 7.5% [1]. The comparison of various benchmark experiments with calculated γ-heating shows a systematic underestimate ranging from 10 to 28% for the main fuel isotopes ^{235}U and ^{239}Pu. This is attributed to deficiencies in γ-ray production data in evaluated nuclear data files. Requests for new measurements of prompt γ-emission in the reactions ^{235}U(n,f) and ^{239}Pu(n,f) have been formulated and included in the Nuclear Data High Priority Request List of the Nuclear Energy Agency [2].

Data found in modern nuclear-data libraries all date back to experiments performed in the early 1970's. In those experiments NaI scintillation detectors were used as γ-ray spectrometer with an ionisation chamber as fission trigger. A major difficulty in measuring the competition between neutron and γ-ray emission during fission fragment de-excitation is the suppression of background γ-rays induced by prompt fission neutrons in the γ-detector. A common method is to distinguish between γ-rays and neutrons by their respective different time-of-flight, which however is limited by the timing resolution of the detector (not better than 5 ns for NaI).

The characterization of recently developed Cerium-doped Lanthanum Chloride crystal scintillation detectors is the subject of an ongoing feasibility study at Institute for Reference Materials and Measurements, a Joint Research Centre of the European Commission. Three $LaCl_3$:Ce detectors have been characterized in terms of energy resolution, efficiency and timing resolution. The study, made in perspective of measuring prompt fission gamma-rays, has shown that the detectors are suitable for this purpose, having a much better timing resolution and efficiency than the NaI crystals previously used.

In the framework of a parallel project, a new highly time-resolving detector type based on polycrystalline CVD diamond has been introduced and tested. These detectors will be used in the two-arm time-of-flight spectrometer VERDI for the investigation of prompt and delayed neutron emission in neutron-induced fission that is presently being constructed and tested at IRMM. Among the advantages of this detector material is its very high radiation tolerance compared to others (like e. g. Si) and its excellent timing properties. Using the fission trigger provided by the diamond detector in combination with the $LaCl_3$:Ce gamma detectors, measurements of fission fragments and prompt fission gamma-rays are possible with much higher accuracy.

1. Characterization of $LaCl_3$:Ce detectors

The detectors are cylindrical, with a size of 1.5"x1.5", and are doped with 5% Cerium. The crystals are attached to a photomultiplier of type Photonis XP2550/FB, and were purchased from Scionix Holland. A photograph of the detector is shown in Fig. 1.
The following features were investigated: the intrinsic activity, the energy resolution, the intrinsic efficiency and the timing resolution.

Fig.1. $LaCl_3$:Ce scintillation detector.

1.1. Intrinsic activity

It is known from previous studies [3] that this detector material possesses an intrinsic activity due to the decay of ^{227}Ac (and all the daughters of its decay chain) and ^{138}La. A characteristic spectrum illustrating this feature is shown in Fig. 2.

As can be seen, there are two possibilities of discarding the peaks coming from the intrinsic radioactivity: subtracting a scaled background spectrum from the measured spectrum or using a coincidence condition. The specific intrinsic activity of these particular detectors amounts to1.3 Bq cm^{-3}, in reasonable agreement with previously reported results.

1.2. Energy resolution

The energy resolution of the $LaCl_3$:Ce detectors was investigated using standard calibration γ-sources up to an energy of 2.6 MeV and up to about 7 MeV by measuring γ-rays from excited states of ^{16}O, produced in the reaction ^{19}F(n,α)^{16}O*. The FWHM of the peaks from the measured energy spectra is plotted as a function of γ energy in Fig. 3. The results are satisfactory since the energy resolution for these detectors is 40% better than for NaI(Tl) scintillation detectors.

Journal of Physics: Conference Series **205** (2010) 012050 doi:10.1088/1742-6596/205/1/012050

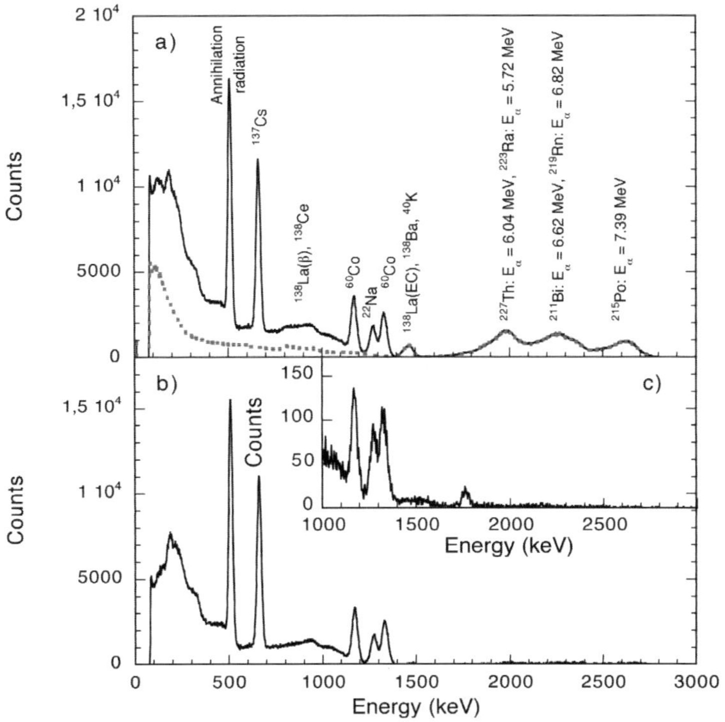

Fig. 2. Characteristic spectra taken with a $LaCl_3$:Ce detector. a). The black spectrum shows peaks from several γ-sources as well as the peaks caused by the intrinsic activity, as indicated on the peaks. The red spectrum is taken without any radioactive source and shows only the intrinsic activity peaks.
b). Background-subtracted spectrum (scaled).
c). Coincidence spectrum of two $LaCl_3$:Ce detectors.

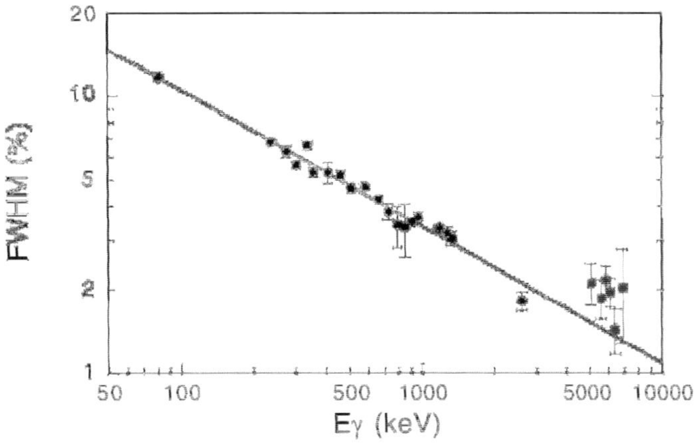

Fig. 3. Energy resolution measured for energies ranging from 80 keV to 7 MeV. The grey dots represent the energies from the decay of excited states in ^{16}O.

1.3. Intrinsic efficiency

The intrinsic efficiency was determined using several calibrated standard sources and compared with the efficiency of a NaI(Tl) detector of same size [4] and also with values obtained previously for other $LaCl_3$:Ce detectors [5]. The results are shown in Fig. 4. It can be observed that the efficiency of these detectors is 53% better than the one of NaI(Tl) detectors of the same size over the entire energy range.

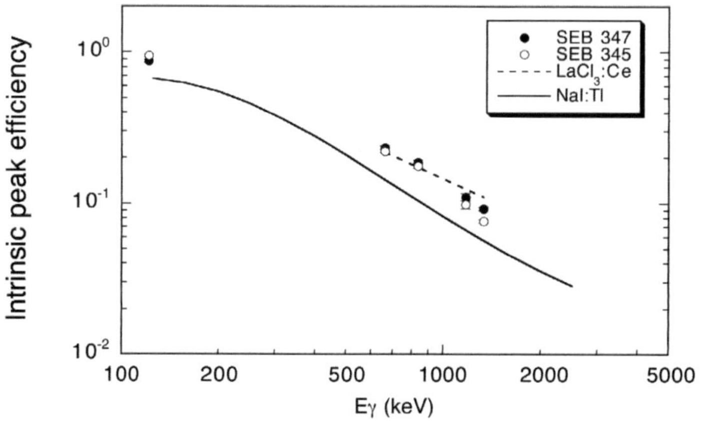

Fig. 4. Intrinsic peak efficiency of two $LaCl_3$:Ce detectors measured in this work: full and empty dots; the dashed line represents results obtained in Ref. [4]. The full line represents the intrinsic efficiency for NaI(Tl) detectors of same size.

1.4. Timing resolution

In order to determine the timing resolution, two $LaCl_3$:Ce detectors were used to measure coincident gamma-rays emitted from ^{22}Na and ^{60}Co sources. A timing spectrum measured in this way is shown in Fig. 5. The analysis of the obtained spectrum results in a timing resolution of 890 ps, which corresponds to an intrinsic timing resolution of 630 ps for each detector.

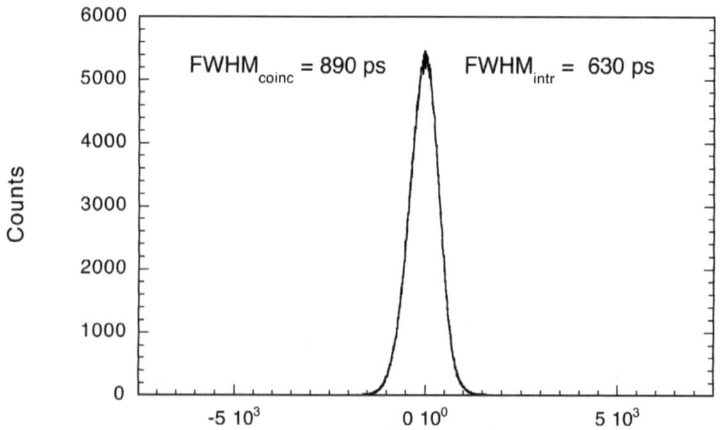

Fig. 5. Timing spectrum measured using two $LaCl_3$:Ce detectors.

2. Characterization of polycrystalline diamond detectors

The use of artificial Polycrystalline and single-crystal Chemical Vapour Deposited (CVD) diamond material is being constantly extended to different research areas due to its excellent timing characteristics, radiation hardness and easy operation. In nuclear physics, diamond detectors are used mostly in high-energy experiments as beam monitors and tracking devices, replacing the traditionally employed Si detectors, because they survive in high radiation environments, have very low leakage current and need no cooling [6].

The detector used in this work is a 100 μm thick polycrystalline CVD diamond film. The active area is 10 mm x 10 mm (Fig. 6). Electrical contact is made by a gold pad deposited on the growth side of the film.

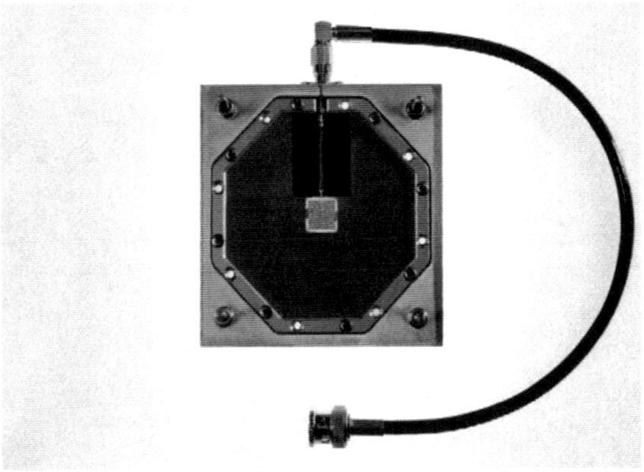

Fig. 6. Photograph of the pcCVD detector used in this work.

In order to estimate the intrinsic timing resolution of the CVD diamond detector, the following experimental set-up was used: two detectors were placed in a test reaction chamber; a spontaneous fission ^{252}Cf source with an activity of 250 fissions/s was attached to one of the detectors. The other diamond detector was placed at a distance of 9.5 cm from the first one. The pressure in the chamber was $2.5 \cdot 10^{-5}$ mbar. The detectors were biased to180 V.

Over the whole duration of the experiment, the radiation dose amounted to $2.6 \cdot 10^{9}$ fission fragments, corresponding to $8 \cdot 10^{10}$ alpha particles and $8 \cdot 10^{9}$ neutrons, respectively. In these conditions the output signals of the detectors remained unchanged which shows their high radiation tolerance and stability. The measured time-of flight spectrum was analysed and a Monte Carlo simulation was performed for the determination of the timing resolution. The time spectrum shown in Fig. 7 corresponds to an intrinsic timing resolution of 280 ps for each CVD detector.

Journal of Physics: Conference Series **205** (2010) 012050 doi:10.1088/1742-6596/205/1/012050

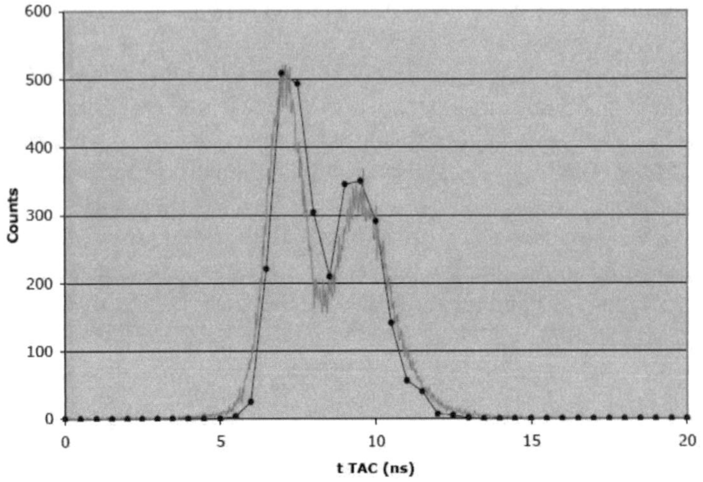

Fig. 7. Comparison of the experimental time-of-flight spectrum for the 2 CVD detectors with a Monte Carlo simulation.

3. Conclusions and outlook

Two different types of detectors were characterized in this work. The $LaCl_3$:Ce detectors intended for measuring gamma-rays were investigated and found to have 40% better energy resolution and 53% better intrinsic efficiency compared to the previously used NaI(Tl) detectors, and also better timing resolution. The intrinsic activity present in the crystals can be suppressed by either imposing a coincidence condition or by subtracting a scaled background spectrum.

The polycrystalline diamond detectors have shown very high resistance to radiation and stability. Also, they are easy to operate and have very good timing resolution.

Therefore, we conclude that both detectors are very suitable for their intended purpose and it should be possible indeed to achieve better quality data on prompt gamma emission in fission by using $LaCl_3$:Ce detectors as gamma-detectors and pcCVD diamond detectors as fission trigger. An experiment on thermal neutron-induced fission of ^{235}U is scheduled for early 2010 in the framework of the transnational access programme EFNUDAT.

References

[1] G. Rimpault, Proc. Workshop on Nuclear Data Needs for Generation IV, April 2005 (Editor P. Rullhusen) Antwerp, Belgium, World Scientific, ISBN 981-256-830-1 (2006) 46
[2] Nuclear Data High Priority Request List of the NEA (Req. ID: H3, H4)
[3] A. Owens et al., Nucl. Instr. Meth. A574 (2007) 110
[4] G. Gilmore, Practical Gamma-Ray spectroscopy, John Wiley & Sons, ISBN 978-0-470-86196-7 (2008)
[5] B. Ayaz Maierhofer and T. A. DeVol, Nucl. Instr. Meth. A578 (2007) 410
[6] R. S. Wallny, Nucl. Instrum. and Meth in Phys. Res. A 582 (2007) 824 – 828

Journal of Physics: Conference Series 205 (2010) 012051 doi:10.1088/1742-6596/205/1/012051

Investigations of a Resonance Structure of Neutron Cross-Sections using the Multiplicity Spectrometry Technique

Zh V Mezentseva[1], Yu V Grigoriev[2] and N B Janeva[3]

[1]Joint Institute for Nuclear Research, JINR, Dubna, Russia, 141980
[2]SSC RF Physics-Energy Institute, Obninsk, Russia, 249033
[3]Institute for Nuclear Research and Nuclear Energy, Sofia, Bulgaria, BG-1784

E-mail: mzv@nf.jinr.ru

Abstract. It is known that uncertainties in the calculation of characteristics of power nuclear reactors such as critical sizes, power density field, reactor life time etc are caused mainly (about 40 %) by uncertainties in the knowledge of neutron constants of reactor materials. Allowed errors on neutron constants for all basic reactor materials and fission products are about 1-5 %. Required errors on constructional materials should be in the range of 5-10 % for σ_t, σ_s, σ_γ and from 2 to 12 % for self-shielding factors. The required accuracy on neutron constants for the majority of reactor materials is not achieved up to now and consequently it is necessary to continue experimental research in this direction. For example, such an accuracy could be achieved using the multiplicity spectrometry technique of the gamma-rays. The description of this method, its features and several experimental results will be presented.

1. Introduction

The information on nuclear data is needed for tunning of different theoretical models of nuclear reactions and a structure of nuclei. More precise definition is important for realization of different transmutation projects and for increasing of a nuclear safety of energetic atomic reactors. Modern calculations of fast reactors require a detailed knowledge of nuclear data, especially of neutron characteristics of fissile and constructional materials. Also new accurate measurements of the neutron cross-sections are needed for construction of the nuclear reactors of type BREST, BN-800 etc [1]. These trends and objectives are closely related to the improvement of estimated nuclear data of basic world data libraries. The world requirements of precision on nuclear data are 1-5 % for basic reactor and fissile materials, 3-7 % for constructional materials and 5-10 % for fission products.

2. The Multiplicity Spectrometry technique

It is possible to achieve precisions mentioned above on neutron cross-sections using the multiplicity spectrometry technique. The multiplicity spectrometry of gamma-quanta and neutrons, emitted by exited nuclei, has been developed to provide wide investigations of interactions of neutrons with nuclei [2]. In contrast to separate experiments in which the same parameters are determined independently from each others, an experiment with such a setup will provide considerably more complete information together with separate quantities and spectra and will be possible to obtain all possible correlations between them. It also allows to obtain neutron cross-sections and their ratios with a high accuracy and to study different channels of decays of excited nuclei.

Examining the probability of some nuclear process as a function of these parameters we obtain the corresponding spectra e.g. the multiplicity spectrum /A(ν)/. The multiplicity spectrum is a distribution of number of particles emitted in a process of de-excitation namely gamma-quanta. It illustrates a probability of emitting one particle, two or more. Here we are talking about the total multiplicity of neutrons and gamma-quanta $\nu=\nu_n+\nu_\gamma$, emitted as a result of interactions of slow neutrons with nuclei. In this case, the multiplicity spectrum A(ν) consists of well-separated maxima, corresponding to scattering, radiative capture and fission processes /see figure1/.

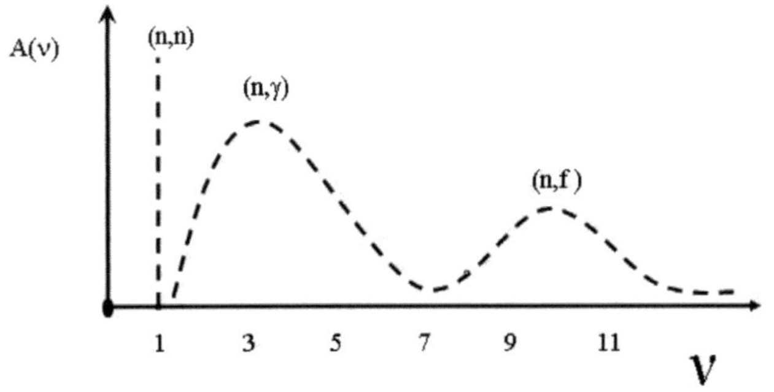

Figure1. A sketch of the multiplicity spectrum with measurements of scattering, radiative capture and fission processes.

In the scattering only a single neutron is detected, so that the A(ν) is everywhere equal to zero except at the point ν=1 so the first maximum corresponds to the scattering process. Approximately 4 gamma-quanta are formed with the radiative capture, i.e. the capture $A_\gamma(\nu)$ is concentrated in the region ν~4. For fission the total number of gamma-quanta and neutrons is ~10. Thus the third maximum corresponds to the fission multiplicity spectrum $A_f(\nu)$.

Such multiplicity spectra could be registered by a 4π-multisectional detector with an efficiency close to 100 %. A sample under investigation is placed in the center of this detector. A number of detected k-fold coincidencies of signals form different sections of the detector gives directly the multiplicity spectra A(ν)=N(k). For this purpose a number of sections in the detector should be larger than a maximum number of emitted particles while an efficiency of detection of each particle by the detector is considered to be close to 100 %.

3. Detector systems.

At the Laboratory of Neutron Physics of the Joint Institute for Nuclear Research over many the technique base has been established. As a result we have two multisectional detectors which have been used for measurements of partial cross-sections and their ratios for many materials. The IBR-30 reactor has served as a neutron source [3]. During the measurements it has been operated at a mean power level of 10 kW with a neutron pulse width of 4 μs and a pulse repetition rate of 100 Hz.

3.1. The multisectional scintillation detector on the base of NaI(Tl) crystals.

The multisectional detector "ROMASHKA" consists of 16 scintillation sections made from NaI(Tl) (figure2). Each section of the detector is optically connected with its photomultiplier. A total volume of the detector is 36 l. A geometrical efficiency is about 90 %. A registration efficiency of single gamma-quanta at the energy of 1 MeV is about 85 % in the absence of its registration in several sections of the detector [4]. Using such a setup it is possible to measure the multi-parameter spectrum of a dependency of a number of events on the time-of-flight of neutrons (or a number of time channel), on multiplicity coincidence of signal in the sections, on energy-release in each section of the detector and in the whole detector.

This spectrometer has the following important features:

(a) the geometrical arrangement,

(b) the actual performance of the individual detector elements,

(c) the associated electronics and data acquisition system, and

(d) the response of the system to input γ-cascades including the effect of crystal-to-crystal scattering and the response to neutrons.

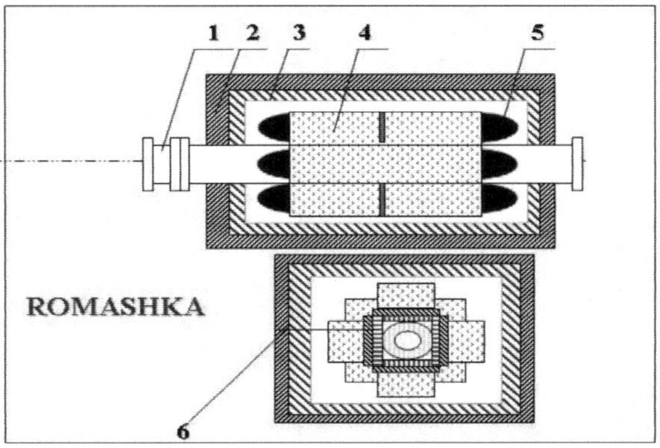

Figure2. The multisectional scintillation detector "Romashka" with 16 crystals of Nai(Tl)
1-vaccum tube, 2-lead shielding, 3- paraffin shielding, 4-NaI(Tl) crystals, 5-photomultiplier, 6-boron converter

3.2. The multisectional liquid scintillation detector "Parus".

Another experimental setup allows to go down in energy to 2 eV (figure3). The detector setup consists of two tanks. The total volume is 80 l. Each tank is divided into 8 sections separated from each other by the aluminum glareshield. Each section is viewed by a photomultiplier. A mixture of 100 g of scintillation additions, methilborate and tholuol is used as a scintillation liquid. The registration efficiency of gamma-quanta is approximately 30 %, the energy resolution for gamma-quanta at the energy of 662 keV (Cs line) is about 30 %.

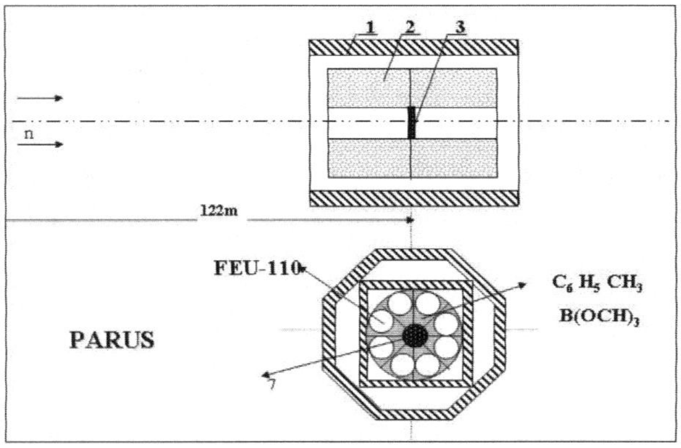

Figure 3. The multisectional liquid scintillation detector "Parus"
1- lead shielding, 2- scintillator, 3-investigated sample

Journal of Physics: Conference Series **205** (2010) 012051 doi:10.1088/1742-6596/205/1/012051

4. Some experimental results.

4.1. Molybdenum.
Molybdenum belongs to reactor constructional material with a high melting temperature 2620^{0}C. The following reaction

$$^{98}Mo(n,\gamma)^{99}Mo \xrightarrow{\beta^-} {}^{99m}Tc \quad (1)$$

is of interest for reactor construction, where Mo is widely used like a constructional material owing to its high resistance to high temperatures and to relatively low neutron capture cross-section. It is important that a product of the reaction mentioned above is ^{99m}Tc radionuclide having ideal characteristics for its usage in radiopharmaceutical preparations for the diagnostic and medical treatment of different organs ($T_{1/2}$=6.006 h, E=140.5 keV, yield of gamma-quanta is 89.6 %, absence of β-rays)[5]. Also isotopes of Mo are accumulated in a large amount at a burning of a nuclear fuel and on 10-20 % determine a "parasitic" capture of neutrons from all fission products in fast reactors. This fact influences the duration of a campaign of nuclear reactors [6].

To obtain neutron cross-sections for Mo we have carried out measurements of the time-of-flight spectra using the multisectional scintillation detector "Romashka" on the basis of NaI(Tl) crystals (already described in subsection 3.1). During the measurements the Romashka setup has been installed at a distance of 500 m from the active zone of the IBR-30 reactor. The NaI(Tl) detector has operated in the multiplicity coincidence spectrometry conditions. The metal disk of Mo with a thickness of 0.4 mm or 0.00281 at/b and with a diameter of 80 mm has been used as a radiator-sample [7]. The multiplicity coincidences spectra have been registered from the 1st up to 15th multiplicities.

After background components had been subtracted and the time-of-flight spectra had been normalized, one determined the quantity pulses in the energy groups of the BNAB constants system [8] over the energy region from 20 eV up to 10 keV. Analogous measurements were realized for thin ^{238}U sample with a thickness of 0.25 mm or 0.00121 at/b and with a diameter of 80 mm and used as a "standard" for determination of group capture cross-sections for Mo. The neutron capture cross - section for Mo can be determined by the following relations:

$$<\sigma_C>_i^{Mo} = \frac{S^U}{S^{Mo}} \cdot \frac{(nd)^U}{(nd)^{Mo}} \cdot \frac{N_i^{Mo}}{M^{Mo}} \cdot \frac{M^U}{N_i^U} \cdot <\sigma_C>_i^U \quad (2)$$

where: M^U and M^{Mo} - monitor coefficients for the U and Mo, S^U and S^{Mo} areas , n^U and n^{Mo} - thickness of radiator samples, d – diameters of the samples, N_i-counts per i-th channel.

After combining the energy intervals the ^{98}Mo group capture cross-sections were determined according to formula (2). The group capture cross-sections for the ^{98}Nb and ^{238}U were obtained by the calculation with the help GRUCON code [9] on the basis of the estimated data libraries. A comparison of the experimental group capture cross-sections of Mo with other available experimental data and with the estimated values is shown in figure 4.

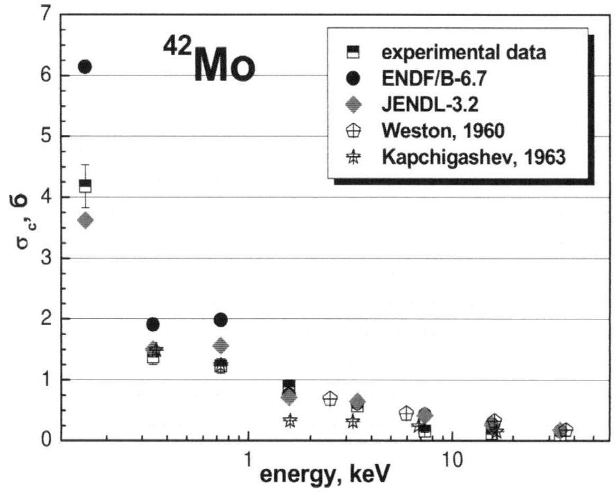

Figure 4. The group capture cross-sections for ^{98}Mo.

As one can see from figure 4 the experimental data are in good agreement not only with the estimated ones [10,11] but also with the other experimental results [12,13].

4.2. Thorium.

^{232}Th as well as ^{238}U is a basic reactor material. Thorium is not a fissile nucleus but it absorbs slow neutrons. The following decay chain leads to the formation of ^{233}U.

$$n + {}^{238}U \to {}^{239}U \underset{\beta^-,\ 23\ \text{min}}{\to} {}^{239}Np \underset{\beta^-,\ 2.3\ \text{days}}{\to} {}^{239}Pu \quad (3)$$

$$n + {}^{232}Th \to {}^{233}Th \underset{\beta^-,\ 22\ \text{min}}{\to} {}^{233}Pa \underset{\beta^-,\ 27\ \text{days}}{\to} {}^{233}U$$

^{233}U is a perfect nuclear fuel, supporting chain reaction. Each neutron absorbed by ^{239}Pu or by ^{235}U gives 2.03-2.08 new neutrons and by ^{233}U-2.37. If they use the thorium cycle in nuclear reactor then Pu and other transuranium elements are accumulated much less in comparison with the uranium fuel cycle.

It is well known that the precise measurement of the capture cross-sections for Th is impeded by the high gamma background arising from its natural decay chain. Commonly, this gamma background is reduced by chemically removing the daughter products responsible for this background or by raising the pulse discrimination level. In the first case the measurements have to be carried out very quickly after a preparation of the sample because the daughter products build up again. The second method has a disadvantage that some of the gamma-rays resulting from the neutron capture are discriminated out which leads to a deterioration of the signal-to-background ratio and loss of experimental information.

To avoid such difficulties we have used the multiplicity spectrometry technique. The neutron capture cross-sections of ^{232}Th were measured in the neutron energy range from 2 eV to 10 keV by using the neutron time-of-flight (tof) method at the IBR-30 pulsed neutron booster of JINR [14]. The gamma rays originating from neutron capture were detected by 16 sections of liquid scintillator located at a distance of 121 m from the neutron source (see subsection 3.2).

With this detector we have performed measurements of tof spectra for thin metal sample of Th (0.2 mm or 0.00665 at/b) and powder of U_3O_8 (0.000498 at/b) with 99.9 % of ^{238}U. We have carried out a detailed analysis, including subtraction of background component, definition of intervals for energy group in the energy range from 2 eV to 10 keV and for low resolved resonance. The capture cross-sections were determined in compliance with the known cross-section of ^{238}U by analogy with ^{98}Mo using the expression (2). These capture cross-sections are presented in figure 5 in comparison with the evaluated data. As one can see from this Figure the experimental results agree very well with the evaluated ones within the experimental errors.

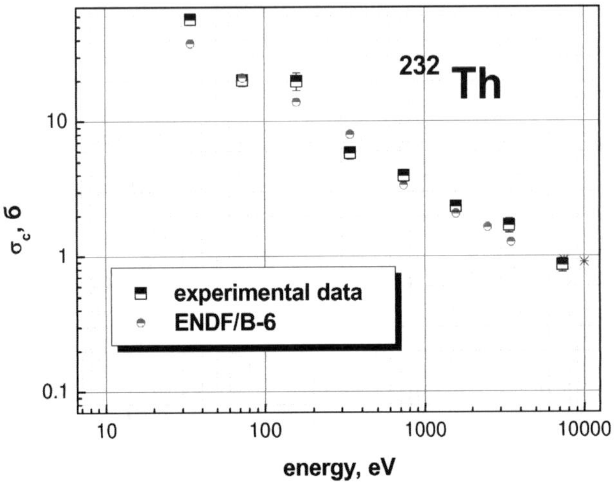

Figure 5. The group capture cross-sections for ^{232}Th.

Also we have extracted capture cross-sections for several low resolved resonances. A comparison of experimental data with the evaluated ones is shown in figure 6.

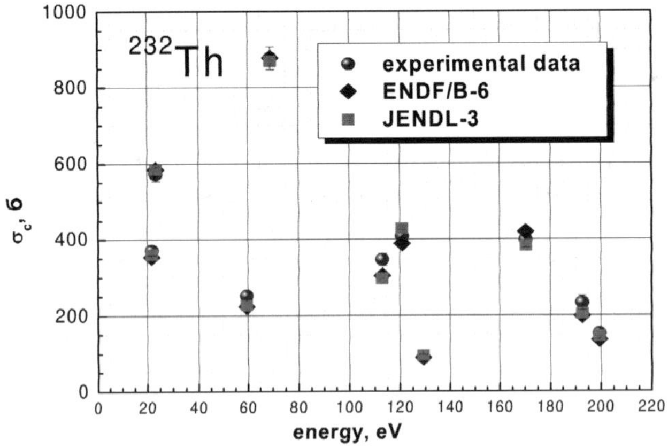

Figure 6. The capture cross-sections for ^{232}Th for low resolved resonances.

This comparison points to very good agreement between two sets of data.

5. Conclusion.

The presented experimental material demonstrates great possibilities of the gamma-ray multiplicity spectrometry for the investigation of exited fissionable and constructional nuclei. In future we are planning to build a similar multisectional detector system with a larger number of crystals and to measure neutron cross-sections for most important materials using the IREN neutron source at FLNP JINR [15].

References

[1] Glazov AG, Leonov BN, Orlov BB, Sila-Novitsky AG, Smirnov BS et al, 2007, The reactor Brest and at-station nuclear fuel cycle, *Nuclear Energy*, **103** 15 (in Russian)

[2] Muradyan G.V, 1981, The Multiplicity spectrometry, *Atomic Energy*, **50 (6)** 394

[3] Frank IM, 1972, Investigation and application in the scientifique investigations of the pulse reactor IBR, *PEPAN* **2(4)** (in Russian)

[4] Georgiev GP, Grigoriev YuV et al., 1992, A setup for measuring neutron cross-sections and radiation multiplcity from neutron-nucleus interaction", *NIM* **A313** 256

[5] Trofimov YuN, Nemilov YuA, 1984, Neutron capture cross-section of ^{98}Mo at the neutron energy of 0.3-2 meV, *VANT, series: Neutron constants and parameters* **3(57)**, 15 (in Russian)

[6] Bednyakov SM, Dulin VA, Manturov GN, Mozhaev VK, 1984, Invetigations of absorption of neutrons by fission products in critical assembling *"VANT, series: Nucler-reactor data*, **1**.62 (in Russian)

[7] Grigoriev YuV et al., 2003 , Investigation of Nb, Mo, Rh, W neutron cross-sections in the resonance energy range, *ISINN XI Dubna* 167

[8] Abagyan LP, Bazazyants OA, Bodarenko II, Nikolaev MN, 1964, *Group constants for the computation on nuclear reactors*, "Atomizdat", Moscow (in Russian)

[9] Sinitsa VV, 1984, Program for calculating the group constants using the library of evaluated neutron data , *VANT, series: Nuclear constants* **5(59)**, 34 (in Russian)

[10] Rose PF, Dunford CK ed. , 1988, ENDF-102 Data Formats and Procedures for the Evaluated Nuclear Data File, *Report BNL,* Upton, New York, 11973, USA.

[11] Shibata K. et al., 1990, Japanese Evaluated Nuclear Data Library Version JENDL-3,*Rep. JAERI 1319*

[12] Weston LW, Seth KK, Bilpuch EG, Weston HW, 1960, Netron Capture Cross Sections in the keV Region. Part II. Spin--Orbit Coupling and the Optical Model, *Atomic Physics* **10** 477

[13] Kapchigashev SV, Popov YuP, 1963, Capture Cross Sections in Construction Materials for Neutrons with Energies up to 50 keV, *Atomic Energy*, **15** 120

[14] Mezentseva ZhV et al, ,2003, Neutron Cross - Section Measurements of ^{232}Th in the Energy Region 10 eV-10 keV, *Journal of the Korean Physical Society* **43 (5),** 704

[15] Furman W et al., April 20-23 2003, Intense resonance neutron source (IREN) - new pulsed source for nuclear physical and applied investigations, *11th International Conference on Nuclear Engineering* Tokyo, JAPAN, ICONE 11 - 36318

Journal of Physics: Conference Series **205** (2010) 012052 doi:10.1088/1742-6596/205/1/012052

Isotope production rates in a thick lead target in the reaction d+Pb at 2,52 GeV

O Yordanov[1], S Batzev[1], L Kostov[1], Ch Stoyanov[1], P Zhivkov[1], J Adam[2], V D Kovalenko[2], M I Krivopustov[2], A A Solnyshkin[2], V M Tsoupko-Sitnikov[2], K Katovsky[3] and M Majerle[4]

[1] Institute for Nuclear Research and Nuclear Energy, 72 Tzarigradsko chaussee, Blvd. BG – 1784 Sofia, Bulgaria
[2] Joint Institute for Nuclear Research, Joliot-Curie 6, 141980 Dubna, Moscow region, Russia
[3] Czech Technical University, Zikova 4, 166 36, Prague 6, Czech Republic
[4] Nuclear Research Institute Rez plc, Husinec – Rez 130, 250 68 Rez, Czech Republic

E-mail: orliny@inrne.bas.bg

Abstract. The "Energy Plus Transmutation"-setup consists of a thick cylindrical lead target of 456 mm length and diameter of 84 mm, which is surrounded by a thick layer containing 206,4 kg of natural uranium. This small-sized target-blanket-assembly resembles the geometry and the constituents of an industry-sized subcritical system for power production and nuclear waste transmutation. The irradiations of the target-blanket-assembly are performed using deuteron beams of variable energies, delivered by the Nuclotron accelerator at the Joint Institute for Nuclear Research in Dubna. Using precision gamma-spectroscopy, the amounts of isotopes of bismuth, lead and thallium, produced in the lead target during the irradiation with a deuteron beam of 2,52 GeV [1,2] have been determined. The collected data on the isotope production are compared with the predictions of nuclear-reaction- and particle-transport-codes.

1. The "Energy+Transmutation" experimental program

In an accelerator-driven system a high-intensity, relativistic particle beam- most common of protons, is impinged on a thick target, made of a heavy material like lead or tungsten. The target nuclei, hit by the primary beam, undergo spallation reactions, in which in average a large amount of fast neutrons is released. Depending on their energies and among the wide variety of the possible reaction channels, the released neutrons are expected in particular to induce fission, or to be captured in nuclei from the material, surrounding the target. The neutron-induced reactions on long-living radioactive nuclides are expected to produce short-living ones, or to induce the fission of fissile nuclides.
In this respect the accelerator-driven systems show a technological possibility to meet the ecological and security problems, caused by the long-term radioactivity and the presence of fissile material, accumulated in the spent fuel of today's nuclear power plants.
The transformation of hazardous nuclear materials by means of nuclear reactions is also referred to as transmutation.

Journal of Physics: Conference Series **205** (2010) 012052 doi:10.1088/1742-6596/205/1/012052

The setup used in the experiments within the "Energy + Transmutation"-program consists of a thick lead target of 456 mm length and diameter of 84 mm, which is surrounded by an array of aluminium cans, containing ^{238}U.
The light particles, mostly neutrons and protons, which are produced in the interaction of the primary beam with the lead target, are passing through thick layers of matter including the target itself and the massive uranium blanket. The particle transport through the setup is determined by diverse reaction mechanisms as scattering-, capture- knock-out-, fission- and fragmentation reactions.

2. The irradiation with 2,52 GeV deuteron beam

During the presently discussed experiment, the setup was irradiated with a deuteron beam of 2,52 GeV, a total amount of 6,5.1012 particles was impinged on the target.
In order to understand the physical behaviour of the setup a large set of probes, including nuclear track detectors, samples for neutron activation analysis and samples containing ^{238}Pu, ^{239}Pu, ^{129}I, and ^{237}Np were placed at different positions within, and on the surface of the setup.
As important, as the detailed measurement of the neutron distributions in energy and intensity within and around the setup, are the experimental data on the interaction of the primary beam with the target. In order to collect experimental data also from the target of the setup, a set of samples, made of lead, as the target itself, have been placed on the surfaces of the target sections, slightly away from the symmetry axis of the setup.

3. Positioning and γ-spectroscopic survey of the lead samples

The interaction of the primary beam with the target was monitored with a set of six lead samples, placed in the vicinity of the symmetry axis of the setup.
The samples have masses of 609 mg, 590 mg, 574 mg, 599 mg, 574 mg and 605 mg, equal diameters of 15mm, and are composed of natural lead. It is expected, that the residual nuclei will be created in fragmentation-, proton-induced-, and neutron-induced reactions.
The six samples of lead were positioned during the experiment 2 cm away from the symmetry axis of the setup, as follows: samples 1 and 5 on the entrance- and exit sites of the target, samples 2, 3, 4 were placed in the slits between the sections of the lead target, the sample "0" was positioned away from the setup, behind the aluminium beam monitor.
After the end of the irradiation the samples have been removed from the setup and transported to the experimental area of the Laboratory of Nuclear Problems of JINR for γ-spectroscopy measurements.
In order to collect data on short- and long-living isotopes, the γ-spectra have been measured sequentially, whereas the time for collecting data was initially set to 30 min, with a time between two measurements in the order of hours. The last spectra of each probe where collected within six hours, with about one week of time between the last two subsequent measurements. A similar time schedule was applied also for the γ-spectroscopy of the other activation samples.
The γ-spectra of the lead samples were collected using an Ortec HPGe-detector coupled with an Ortec-Spectrum Master MCA. No γγ-coincidences have been taken into account.

4. Isotope production in reactions of the primary beam

The determination of the amount of the isotopes produced in the samples during the irradiation proceeds in two steps. The first step is the analysis of the γ-spectra and the unambiguous assignment of the observed γ-lines to the corresponding decaying isotopes. The identification of the decaying nuclide is based on the energy of the detected transition and the beta-decay halflife, the latter being estimated after analyzing a sequence of spectra, in which the observed γ-line is present.
In the second step, based on the known beta-decay halflifes of the identified nuclides and the measured number of decays, one calculates the amount of the decaying nuclide of interest, contained in the sample at the end of the irradiation. The final result has to be corrected with respect to

absorption of the outgoing γ-rays within the volume of the sample, beta-decays in the course of the irradiation, summing effects due to coincident detection of γ-transitions within a cascade.
The R-value for the production of a certain nuclide in one lead sample is defined as:

$$R_{A,Z} = \frac{N_{A,Z}}{Y_{total} \cdot N_{nat.Pb}} \qquad (1)$$

$R_{A,Z}$	production rate of (A,Z)-nuclides
$N_{A,Z}$	total number of (A,Z)-nuclei produced during the irradiation
Y_{total}	total number of projectiles (beam integral)
$N_{nat.Pb}$	number of Pb-atoms in the sample

4.1. Bismuth isotopes

The γ-transitions, assigned to the beta-decays of ^{203}Bi, ^{204}Bi, ^{205}Bi and ^{206}Bi were observed. The beta-decay-halflife of ^{202}Bi and all lighter Bi-isotopes including their isomers are of 1,72 h and less, therefore their decays were not observed during this experiment.
All beta-decay-halflifes of Bi-isotopes, heavier than ^{206}Bi are in the order of years, consequently if even being produced, the activity of ^{207}Bi and ^{208}Bi is expected to be far below the observation limits of this measurement, ^{209}Bi is a stable isotope. The isotope productions of bismuth isotopes are shown on figure 1.

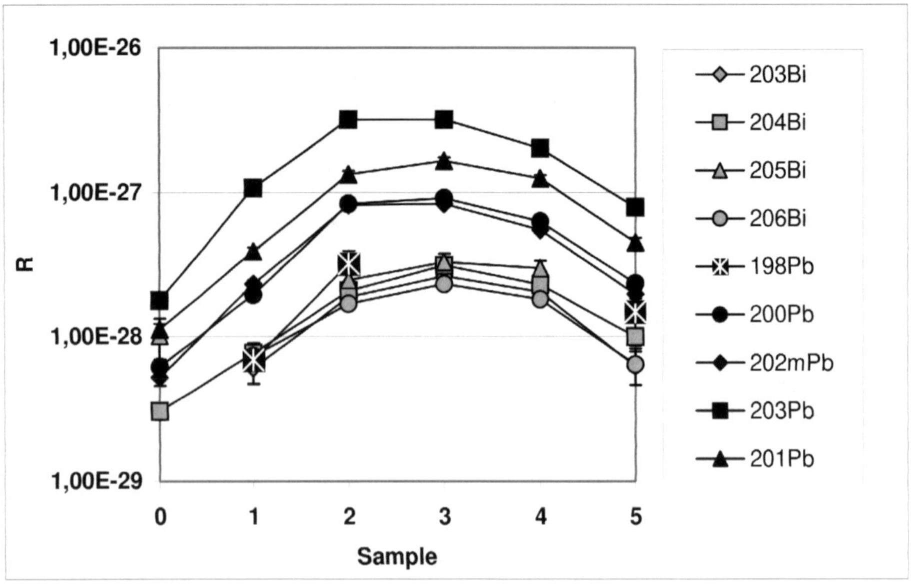

Figure 1. Production rates of isotopes of bismuth and lead, determined in the six lead samples.

4.2. Lead isotopes

The isotope production rates of lead isotopes are shown on figure 1.
The lightest observed isotope of lead is ^{198}Pb with a beta-decay halflife of 2,40 h. The lead isotopes lighter than ^{198}Pb have beta-decay halflifes, which are below one hour, thus their decays were not

Journal of Physics: Conference Series **205** (2010) 012052 doi:10.1088/1742-6596/205/1/012052

observed. The 209,3 keV transition, which coincides with the beta decay of ^{198}Pb was observed only in the spectra of the samples "1", "2" and "5".
The decays of the isotopes ^{200}Pb with T1/2=21 h and ^{201}Pb with T1/2= 9,33 h were observed in all samples.
The isotope ^{202}Pb has a decay halflife of 5.25E4 years. The only γ-transitions could be observed in coincidence with the decay of the isomer 202mPb having a lifetime of 3,53 h.
The decay of the isomeric state (Ip = 9-, E = 2169.83 keV) proceeds together with the emission of γ-rays of 960.67 keV (92%), 786.99 keV (50%), 657.49 keV (32%), 422.18 keV (86%). The measured number of transitions of these energies will reveal the production rate of the isomer ^{202m}Pb, but not the production rate of ^{202}Pb itself. Thus the obtained result is expected to be much lower if compared to the production rates of the neighbouring isotopes ^{201}Pb and ^{203}Pb. The percentage of the produced ^{202m}Pb depends on the isomeric ratio of the reaction, e.g. how many of the residual nuclei have been produced in states with spins greater then 9 relatively to the total ^{202}Pb production.
The calculation of the isomeric ratio of ^{202}Pb is beyond the scope of the present analysis, however it may serve as experimental result for the population of isomers in particle-induced reactions. The heaviest isotope of lead, whose decay was observed in the accumulated γ-spectra, is ^{203}Pb, with a beta-decay halflife of 51,873 h.

4.3. Thallium isotopes

The thallium isotopes, whose decays are observed in the γ-spectra of the lead samples, are of two origins. A part of the thallium nuclides are formed as residual nuclei in the reactions of the primary beam or its residues with the lead nuclei of the samples. Additionally thallium isotopes are produced in the beta-decays of the lead isotopes obtained in the course of the irradiation . Therefore in order to determine the isotopic production-rates of thallium, the isotopic production rates of lead have to be known in advance. In the cases, where
numbers on the production of lead isotopes were not available, the corresponding results for the production of thallium isotopes can be given only as cumulative values.
The isotope ^{198}Tl has a halflife of 5,3 (5) h. The isomers ^{198m}Tl and ^{198m2}Tl have lifetimes of 1,85 h, and 32,1 ms respectively.
^{198m}Tl and ^{198m}2Tl can be produced only in reactions, induced by the primary beam, since the isomeric states are not populated in the beta decay of 198Pb. ^{198m}Tl undergoes with 54% probability beta decay, the internal transition branch amounts 46%, while ^{198m2}Tl decays only by internal transitions, The intensities of the accompanying γ-transitions are different for the beta-decays of ^{198}Tl and ^{198m}Tl. The 587 keV transition has an intensity of 0,20(4) % for the decaying ^{198}Tl and an intensity of 52% for the decaying ^{198m}Tl. Thus the 587 keV γ-line is expected contain mostly decay-events of ^{198m}Tl.
The decay of ^{198}Tl was analyzed using the 675,88 keV transition, having 11% intensity, while this transition is not observed in the decay of ^{198m}Tl.
The decays of ^{198}Tl and the isomer ^{198m}Tl are accompanied also by the strong 411.8 keV transition. The analysis of the 411,8 keV transition has shown, that a part of the intensity originates from the decay of ^{198}Au. Therefore the 411,8 keV transition was excluded from the analysis.
^{199}Tl has a decay halflife of 7,42 (8) h. Since no decays of ^{199}Pb (T1/2=90m) and ^{199m}Pb (T1/2=12,2m) were observed, the production rate of ^{199}Pb remained unknown. Therefore the production rates of ^{199}Tl are provided as cumulative values.
^{200}Tl has a decay halflife of 26,1 (1) h. The beta-decay of ^{200}Tl is accompanied by numerous γ-transitions, the strongest of which – 367,943 keV (Ig=83%), 1205,717keV (Ig=29,9%), 828.32 keV (Ig=10,8%), 579,278keV (Ig=13,8%) were analyzed.
The production rates of ^{200}Pb in the individual samples were obtained in a previous step of the analysis, the final results for the production rates of ^{200}Tl were calculated as the weighted mean of the individual values, obtained within the analysis of the four γ-transitions.
^{201}Tl has a decay halflife of 72,912 (17) h. Also in this case the production rates of the decaying predecessor ^{201}Pb were previously determined. The strongest γ-transition occurring after the decay of

^{201}Tl has 10% of intensity and an energy of 167.43 keV. The determination of the production rates of ^{201}Tl was done using this transition.
^{202}Tl has a decay halflife of 12,23 (2) d. To the production of ^{202}Tl contribute the decays of ^{202}Pb and the isomer ^{202m}Pb. Since the halflife of ^{202}Pb is 5,25E4 years, no decays of ^{202}Pb have been observed in this experiment. A relevant contribution to the production of ^{202}Tl comes from the decay of ^{202m}Pb, which has a lifetime of 3,53h. ^{202m}Pb undergoes to 9,5% beta decay and to 90,5% internal transitions as well. The determination of the production rates of ^{202}Tl was done using the 439,56 keV transition.
The measured production rates if thallium isotopes are shown on figure 2.

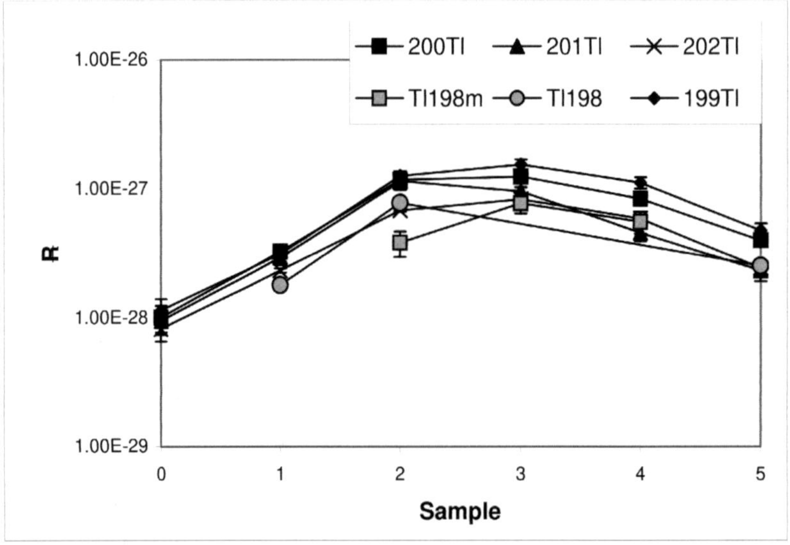

Figure 2. Production rates of isotopes of thallium, determined in the six lead samples.

5. Numerical simulation and comparison with experimental data

The simulations of the experimental data of the measured production rates have been performed using the MCNPX-nuclear-reaction- and particle transport code. The simulation provided the fluxes of the deuterons, protons and neutrons, measured in units of particles/deuteron/cm^2. The production-cross-sections were also calculated within the code, except for reactions of less than 150 MeV energy of the incoming particle. For these cases the production cross sections have been calculated using the THALYS code. The production rate R_{Calc} is calculated as follows:

$$R_{Calc} = \sum_{i,A(Pb)} p_{A(Pb)} \cdot \left(\phi_n(E_i).\sigma^{A(Pb)}_{n,X(A,Z)}(E_i) + \phi_p(E_i).\sigma^{A(Pb)}_{p,X(A,Z)}(E_i) + \phi_d(E_i).\sigma^{A(Pb)}_{d,X(A,Z)}(E_i)\right) 0{,}001.10^{-24} \tag{2}$$

$\phi_n(E_i), \phi_p(E_i), \phi_d(E_i)$: calculated neutron-, proton-, deuteron- flux at certain position and fixed energy, measured in [particles/cm2/incoming deuteron].

$\sigma^{A(Pb)}_{n,X(A,Z)}(E_i), \sigma^{A(Pb)}_{p,X(A,Z)}, (E_i), \sigma^{A(Pb)}_{d,X(A,Z)}(E_i)$: cross sections for the production of the nuclide X(A,Z) in neutron-, proton-, and deuteron-induced reactions on a lead isotope of mass *A(Pb)*, for certain energies, measured in milibarns.

The final result is taken as the sum over the four naturally occurring lead isotopes, weighted by their natural abundances $p_{A(Pb)}$.
Except for the sample positioned outside the setup, the deviations between the calculated and the measured production ratios do not exceed one order of magnitude. The deviations observed in the sample "0" could result either from an incorrect description of the beam interaction with the sample, or from deviations in the calculations of the deuteron-induced reactions on lead.
The calculated and measured production rates of lead isotopes are in nearly perfect agreement for the positions 1 and 2, while for positions 3,4 and 5 a maximal three-fold difference was observed.
The deviations between the measured and calculated values production rates increase for the isotopes of thallium. The lowest deviation by factor of 1,96 between both values of the production rate was found for ^{202}Tl in the sample, placed at position 2.

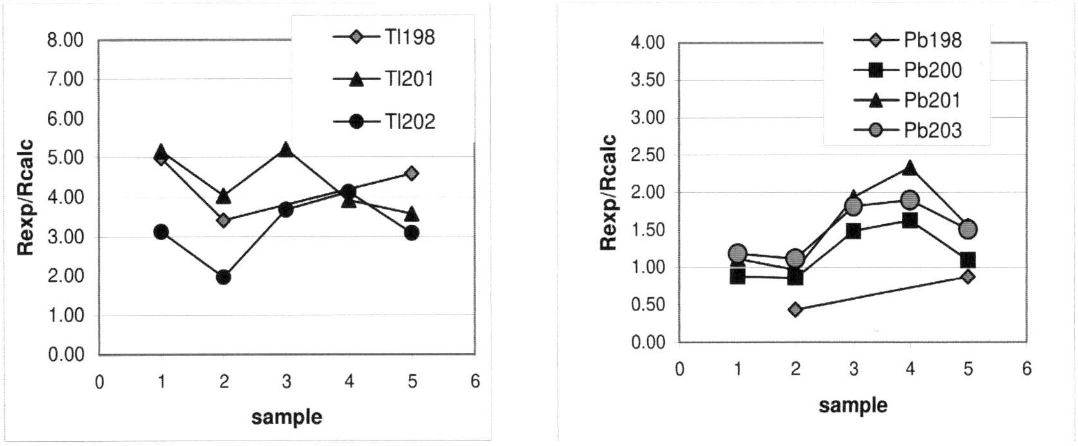

Figure 3. Ratios between the experimental isotope production rates R_{exp} and the calculated isotope production rates R_{calc} of isotopes of thallium and lead in the five lead samples, placed on the surface areas of the target sections .

The analysis of the simulation data has shown, that the neutron-induced reactions dominate the production of the lead- and thallium isotopes in the target. Based on the compared calculated and measured production rates, underestimated cross-sections for the outgoing proton-neutrons-emission-channels could explain the observed deviations for the isotopes of thallium. For the proper understanding of this effect additional simulations, complementary to MCNPX, of the interaction of the primary beam with the lead target are planned. The obtaining of isotope-production data in the lead target at additional energies of the primary beam are desirable.

References

[1] Krivopustov M I et al., About the First Experiment on Investigation of 129I, 237Np, 238Pu and 239Pu Transmutation at the Nuclotron 2.52 GeV Deuteron Beam in Neutron Field Generated in U/Pb-assembly "Energy plus Transmutation", 2007, *JINR Preprint* E1-2007-7, Dubna

[2] Krivopustov M I et al., First results studying the transmutation of 129I, 237Np, 238Pu, and 239Pu in the irradiation of an extended natU/Pb-assembly with 2.52 GeV deuterons , 2009 *Journal of Radioanalytical and Nuclear Chemistry* **279** 567

Physical Start-up of the First Stage of IREN Facility

Belikov O V[1], Belozerov A V[2], Becher Yu[2], Bulycheva Yu[2], Fateev A A[2], Galt A A[1], Kayukov A S[2], Krylov A R[2], Kobetz V V[2], Logachev P V[1], Medvedko A S[1], Meshkov I N[2], Minashkin V F[2], Pavlov V M[1], Petrov V A[2], Pyataev V G[2], Rogov A D[2], Sedyshev P V[2], Shabratov V G[2], Shvec V A[2], Shvetsov V N[2], Skrypnik A V[2], Sumbaev A P[2], Ufimtsev A V[2] and Zamrij V N[2]

[1]Budker Institute of Nuclear Physics Siberian branch of RAS, Novosibirsk, 630090, Russia
[2]Joint Institute for Nuclear Research, Dubna, Moscow region, 141980, Russia

shv@nf.jinr.ru

Abstract. The full-scale scientific research complex IREN will comprise a 200-MeV linear accelerator LUE-200 with a beam power about 10 kW, a subcritical multiplying target, and beam infrastructure with experimental pavilions, as well as technological, control, safety and service systems. The characteristics of the full-scale complex IREN (integral neutron yield 10^{15} n/s and pulse width 0.6 μs) will allow it to rank among the best neutron sources of such class GELINA (Belgium) and ORELA (USA). The realization of the project is conducted in several stages. The first stage includes the construction of the LUE-200 linear accelerator and nonmultiplying target. This will make possible to carry out experiments which require precision neutron spectroscopy in the energy range from fractions of eV to hundreds of eV already at the first stage of IREN. The results of the physical start-up of the first stage of IREN facility at the Frank Laboratory of Neutron Physics of the Joint Institute for Nuclear Research are presented. General scheme and current status of the electron linac are described. Achieved parameters are: pulsed electron beam current – 2.0 A; electron energy – 30 MeV; pulse width – 100 ns; repetition rate – 25 Hz; integral neutron yield $(3\div5)\bullet10^{10}$ n/s.

1. Introduction

Linear electron accelerator LUE-200 [1, 2] was designed at the Budker Institute of Nuclear Physics of Siberian branch of RAS (Novosibirsk) as the driver for intense resonance neutron source IREN [3].

Accelerator consists of the pulsed electron gun, accelerating system, RF power supply system based on 10-cm range klystrons with modulators, beam focusing and transport system including wide aperture magnetic spectrometer and vacuum system. Accelerator is allocated vertically inside 3-floors building. Its designed parameters are listed in table 1. At the beginning the nonmultiplying neutron producing target will be used.

Table 1. LUE-200 designed parameters

Parameter	Value
Maximal electron energy, MeV	200
Pulse beam current, A	1.5 – 2.0
Accelerated electron pulse width, ns	250
Repetition rate, Hz	150
Accelerating rate, MeV/m	35

Beam power, kW	9 – 12

2. IREN facility components

2.1. Electron source

LUE-200 electron source is the 2-electrode electron gun with 12 mm diameter oxide thermocathode. Cathode is fed by 200 kV pulsed transformer. The anode – wall of the vacuum chamber with hole 43 mm in diameter closed with wire frame made of stainless steel. Electron gun provides pulsed electron beam with 8 A peak current, 250-300 ns duration, 50 Hz repetition rate and • 0.01· ·cm·mrad emittance.

2.2. Accelerating system

Accelerating system consists of the RF buncher and 2 short (3 meters long) accelerating sections with high acceleration rate. Accelerating sections were designed and manufactured at BINP (Novosibirsk). Accelerating section represents the circular blinded waveguide with constant impedance fed by RF power. RF power comes from klystrons – RF power amplifiers of 10-cm frequency range (2856 MHz). The prototype of the accelerating system was tested at foreinjector of the VEPP-5 complex of BINP. During the tests the maximal electric field 45 MV/m and average accelerating rate 35 MeV/m were achieved [4].

Full scale accelerator project foresees two accelerating sections. At present one accelerating section is used the second one is replaced with drift section surrounded by two quadruple lenses doublets.

2.3. RF power supply

As the optimal RF power supply for LUE-200 accelerator the SLAC 5045 were chosen providing peak RF power 60 MW at average level 45 kW. RF power compression system is introduced to provide 3-4 times gain in RF power transfer to the accelerating section providing average accelerating rate up to 35 MeV/m. Main components of the RF system of the accelerator are: two-channel high frequency driving generator with possibility to set 180 degrees phase shift between channels and fast phase reversal; two RF preamplifiers for klystrons actuation; two power compression systems; waveguides, couplers etc.

At present due to the absence of SLAC 5045 klystrons only one accelerating section was put into operation with TH2129 (Thomson) klystron providing 20 MW peak power.

2.4. Beam focusing and transport system

Beam focusing and transport system provides beam transporting and trajectory correction on the way from electron gun to the nonmultiplying neutron producing target. From electron gun to the RF buncher 3 short solenoidal magnetic lenses are focusing electron beam. Inside RF buncher and first accelerating section the beam transports due to the continuous solenoidal magnetic field provided with "short" and "long" solenoids enveloping accelerating section.

Beam transportation from the exit of accelerating section to the nonmultiplying target is provided with one separate quadruple lens, four doublets of the wide-aperture quadruple lenses and six correcting magnets distributed along the electron guide.

2.5. Nonmultiplying neutron producing target

Neutron producing target is produced with tungsten based alloy and represents cylinder 40 mm in diameter and 100 mm height spaced within aluminum can 160 mm in diameter and 200 mm height. Distilled water is circulated inside can, providing target cooling and neutron moderation. Water layer thickness in radial direction is 50 mm. Target dimensions and water moderator thickness were optimized with Monte Carlo simulation. Simulation result for neutron flux density at 10 meters flight path is shown at Figure 1.

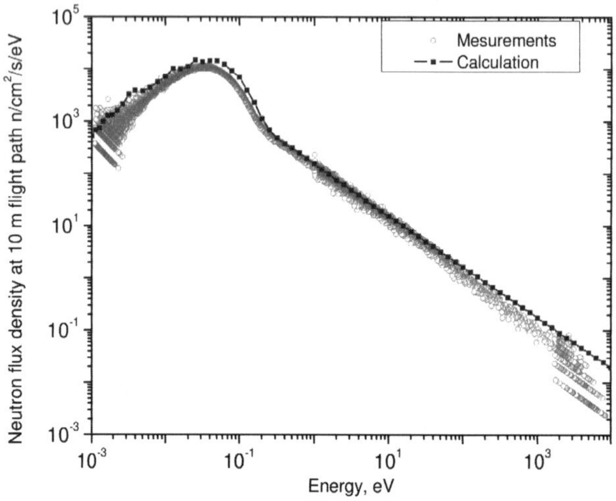

Figure 1. Neutron flux density at 10 m flight path of IREN source. Calculated (solid squares) and measured (open circles).

Accelerated electrons are penetrating inside the can through beryllium window 1 mm thick and hit the top surface of the target cylinder. One thermocouple is placed in the center of the exposed face of the target.

Neutrons are produced in the target with double stage process. At the first stage accelerated electrons stop in the target producing Bremsstrahlung gamma quanta. Energy spectrum of the gamma quanta is limited from above with the maximal energy of the accelerated electrons. At the second stage gamma quanta interacting with tungsten nuclei and neutrons are produced in reactions $^{A}W(\gamma,n)^{A-1}W$, $^{A}W(\gamma,2n)^{A-2}W$, $^{A}W(\gamma,3n)^{A-3}W$.

3. Beam adjustment and transportation

Accelerator startup was carried out by subsequent adjustment of the electron gun, components of beam focusing and transportation system and accelerating system.

During accelerator startup the beam parameters were tested at five diagnostic stations distributed along accelerator: after the electron gun, on the exit from accelerating section, at the end of the drift section, after magnetic spectrometer and in front of the target. Each diagnostic station includes: magnetic induction sensors of the beam current – Rogowski loop; beam viewer – luminescent screen with CCD camera; secondary emission wire frame profilometer. These tools let one to measure beam current and its position in each pulse.

As a result of the optimization of the operational modes of the accelerating and focusing systems at 5 A peak current from electron gun the peak current on the exit from accelerating section reaches 2.2 A and peak current on the target on the level 1.6 A.

4. Measurements of the accelerated electrons spectra

To measure the spectra of the accelerated electrons the broadband magnetostatic analyzer with magnetic field perpendicular to the beam direction was used. Magnet E-core with the variable magnetic rigidity 0.166 – 0.7 T·m is placed after the drift section. Electron beam in the spectrometer magnetic field is turned up to 90 degrees and extracted from the spectrometer vacuum chamber through 50 microns stainless steel window. Thereupon turned electrons hit the position sensitive detector which make it possible to measure electron energy with accuracy $\Delta E/E \leq 5\%$.

Electron spectra were measured at different modes of operation of the electron gun and RF power compression system. Peak current on the exit from the accelerating section varies from 0.7 A up to 2.6 A. It was found that when RF power compression system is off the maximum of the accelerated

Journal of Physics: Conference Series **205** (2010) 012053 doi:10.1088/1742-6596/205/1/012053

electron spectra is between 18 MeV and 20 MeV. When RF power compression system is on the maximum of the accelerated electron spectra is between 27 MeV and 31 MeV. Typical beam spectrum is visualized at Figure 2.

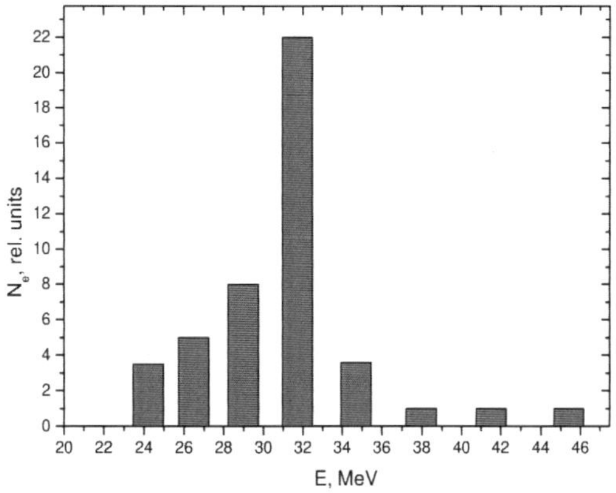

Figure 2. IREN electron beam spectra. Electron gun peak current 4 A, accelerated beam current 1 A, power compression system is on.

5. Estimation of the neutron yield and extracted neutron beam parameters

Actual beam parameters achieved at the middle of 2009 were: average energy of the accelerated electrons – 30 MeV; peak beam current – 1.5 A; electron pulse width 100 ns; repetition rate 25 Hz.

Estimation of the integral neutron yield realized by measurements of the neutron flux density at 10 meters distance from the target by means of the ^{3}He gaseous proportional counter SNM-16 (see Figure 1) and also with the set of activation detectors positioned on the side surface of the target can. The pressure of the ^{3}He in SNM-16 was measured beforehand by the method of transmission of the collimated neutron beam trough the counter and amounts 4.3 bar. The results of the estimation of the neutron yield for abovementioned parameters obtained with two methods are in a good agreement and amount to the value of neutron yield $(3 - 5)\cdot 10^{10}$ n/s. Corresponding neutron flux density at 10 meters from the target (shortest flight path) amounts $(2.4 - 4)\cdot 10^3$ n/cm$^2\cdot$s.

To estimate the parameters of the extracted neutron beam the experiment on transmission of the neutrons through tantalum and indium filter samples 2 mm thick was performed. These elements have strong resonances in total cross sections in the energy region from eV to keV. Neutrons passed through the samples were detected with SNM-16 counter and another ^{3}He detector SHVAPS 38 cm in diameter and 53 cm long with 1 bar ^{3}He pressure. Time of flight spectra were registered by the DAQ system with minimal time channel width 20.83 ns. The results are presented at Figure 3.

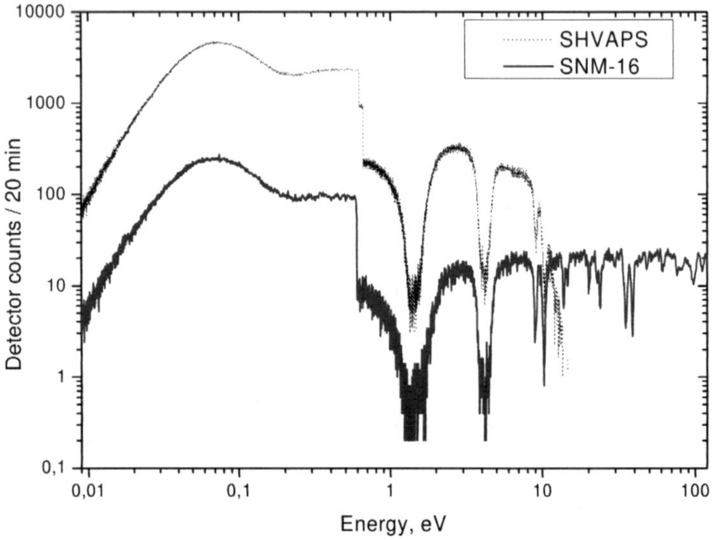

Figure 3. Transmission curves for two filters (Ta + In). Exposition – 20 minutes. Flight path – 9.3 m. Time channel widths in the energy range below 0.5 eV – 10 μs, above 0.5 eV – 2 μs.

To compare the energy resolution of the IREN source with our old IBR-30 + LUE-40 complex the experiment on measurements of the capture gamma quanta from tantalum was performed. Large volume liquid scintillator detector consisting of 6 sections 20 liters each was placed at 60 meters flight path. Tantalum sample was installed inside the detector. Raw experimental data are presented at Figure 4. One can see valuable improvement of the energy resolution in the neutron energy range above 10 eV.

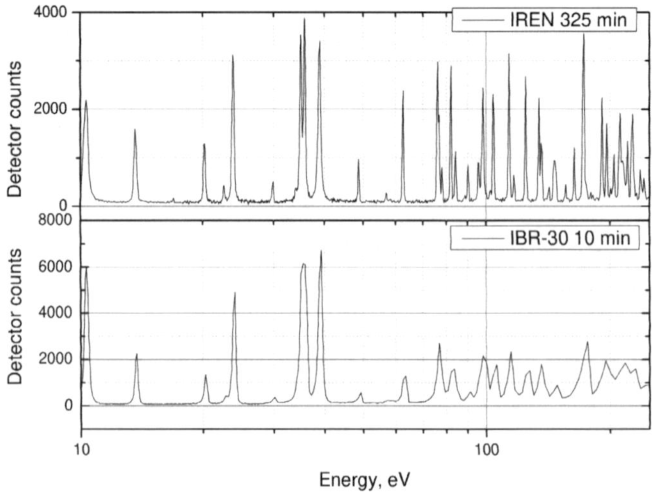

Figure 4. Liquid scintillator detector counts vs neutron energy for IBR-30 and IREN.

6. Conclusion

Physical start up of the first stage of IREN facility was performed at the beginning of 2009. Actual parameters achieved at the middle of 2009 are: average energy of the accelerated electrons – 30 MeV; peak beam current – 1.5 A; electron pulse width 100 ns; repetition rate 25 Hz; neutron yield (3 - 5)$\cdot 10^{10}$ n/s. Extracted beam parameters were estimated within the set of test experiments.

The upgrade of the facility with more powerful klystrons and modulators to reach the designed parameters is planned for the period 2010 – 2015.

References

[1] Novohatsky A et al. Linear accelerator for Intense Resonance Neutron source (IREN) *Proc. of the 2nd Workshop on JINR Tau-Charm Factory* 1994 D1-9-13-459 Dubna p 197

[2] Kaminsky A et al. LUE200 - Driver Linac For Intense Resonant Neutron Spectrometer (IREN) *Proc. of the XVIII International Linear Accelerator Conference* **V 2** 26-30 August 1996 (15 November 1996 Geneva Switzerland CERN 96-07) pp 508-510

[3] Anan'ev V D et al. Intense Resonance Neutron Source (IREN) – New Pulsed Source for Nuclear and Applied Investigations *Particles and Nuclei Letters* 2005 **V 2** No 3(126) pp 11-18

[4] http://arxiv.org/abs/physics/0008100 VEPP-5 Team (Budker INP, Novosibirsk, Russia) Test of accelerating structure for VEPP-5 preinjector *LINAC 2000 Conference* 21-25 August (Monterey California)

Journal of Physics: Conference Series **205** (2010) 012054 doi:10.1088/1742-6596/205/1/012054

Preliminary results of new GELINA data for natural cadmium

I Z Ivanov
Institute for Nuclear Research and Nuclear Energy, 72 Tzarigradsko chaussee, Blvd., BG – 1784 Sofia, Bulgaria

ivanov@inrne.bas.bg

Abstract. The total neutron cross section of cadmium (Cd) is very large in the thermal neutron energy region which determines its special role in reactor technology. The first resonance of ^{113}Cd at 0.178 eV is of high importance in the interpretation of Cd neutron cross section structure in the thermal region. The evaluated data files for Cd isotopes and particularly the resonance parameters do not correspond to the requirements of the advanced reactor technologies. To provide data for a new evaluation, a set of top quality transmission experiments has been designed and performed at the neutron time-of-flight facility GELINA at the Institute for Reference Materials and Measurements (IRMM) in Geel, Belgium. The measurements have been carried out at 26.45 m and 50 m flight paths of GELINA at 50 Hz, 400 Hz and 800 Hz working frequencies of the LINAC, using natural Cd samples with various thicknesses (purity 99.99% in Cd). The experiments at 50 Hz provide high quality data in the thermal energy region, from which we could extract the resonance parameters for the first resonance of ^{113}Cd at 0.178 eV with high accuracy.
Considering the transmission data obtained with various sample thicknesses as a source of additional information for low level cross section we validate a new approach for cross section description in this energy region.

1. Introduction:

Cadmium (Cd) is one of the elements with a large total neutron cross-section in the thermal neutron energy region, which determines its special role in reactor technology. The total neutron cross-section of cadmium is crucial either for the design and development of nuclear reactors and also for the basic study of neutron interaction with nuclei. The first resonance of ^{113}Cd at 0.178 eV is of high importance for the interpretation of Cd neutron cross section structure in the thermal region.
The evaluated data files for Cd isotopes and particularly the resonance parameters do not correspond to the requirements of the advanced reactor technologies. For a variety of nuclides differences between the resonance integral as measured by integral experiments and the values derived from resonance parameters are observed. It has been suggested that the interpretation of the integral measurements could be the source for some of the observed discrepancies. Many of these integral experiments use cadmium foils as shielding against thermal neutrons and it was proposed that these foils are the sources of the observed discrepancies [1]. To clarify the situation and provide data for a new evaluation, a set of top quality transmission experiments has been designed and performed at the neutron time-of-flight facility GELINA, at IRMM in Geel, Belgium.

Journal of Physics: Conference Series **205** (2010) 012054 doi:10.1088/1742-6596/205/1/012054

2. Experimental details

To improve the experimental data base for the determination of the resonance parameters for ^{nat}Cd, high resolution transmission and capture measurements were performed at GELINA [2, 3]. It is a pulsed neutron source providing neutrons in the energy range between 1 meV and 20 MeV. Intense pulsed electron beams with repetition rates up to 800 Hz are accelerated to an energy up to 150 MeV in a linear electron accelerator. The electron pulses are compressed using a specially designed post-acceleration compression magnet to a duration of less than 1 ns [4]. After the compression, these high-energy electrons hit the rotating mercury-cooled uranium target [5]. Colliding with the target, electrons produce Bremsstrahlung in it, and subsequently neutrons by (γ,n) and (γ,f) reactions. To enhance the neutron intensity in the low energy region, two water-filled beryllium containers are used as moderators.

The determination of the neutron energy is based on the Time-Of-Flight (TOF) method. The very short pulse duration of less than 1 ns in combination with the very long flight paths of up to 400 m results in the extremely high energy resolution of the GELINA TOF facility.

Cadmium transmission measurements were built up at 26.45 m and 50 m flight paths of GELINA, which form an angle of +9° and -9 °, respectively, with the normal to the moderator. The experiments were carried out at 50 Hz, 400 Hz and 800 Hz working frequencies of the LINAC, using natural Cd samples enriched to 99.99% in Cd.

A detailed description of the experimental arrangement of the Cd transmission measurements at the 26.45 m flight-path is shown in figure 1. The Cd samples were placed at 9.1 m distance from the neutron producing target in an automatic sample changer, operated by the data acquisition system. Black resonance filters were also mounted on the sample changer. The count rate in the minima of the black resonances was used to determine the time dependence of the background.

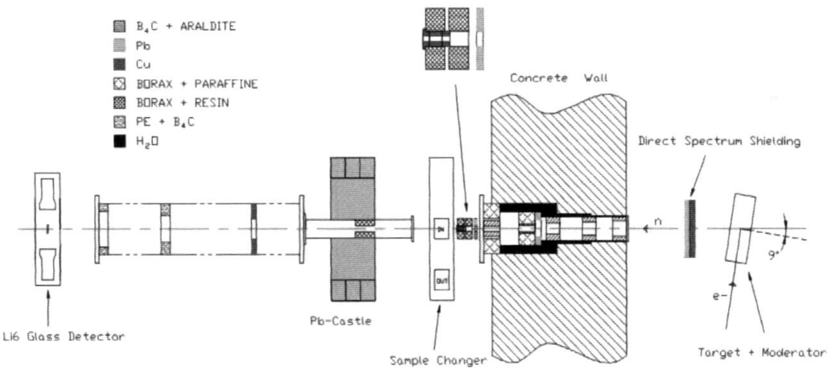

Figure 1 A detailed description of the experimental arrangement of the Cd transmission measurements at GELINA.

After the filters and the sample, the neutron beam was further collimated and finally detected by a Li-glass scintillator (NE 905), enriched to 95 % in ^{6}Li, placed at 26.45 m from the neutron producing target. This scintillator, with an effective diameter of 110 mm and a thickness of 12.7 mm, was placed in an aluminium canning and viewed by two EMI 9823 KQB PhotoMulTipliers (PMT), placed outside the neutron beam and perpendicularly to the neutron beam axis. The aluminium canning was covered with a thin Teflon foil to enhance the light reflection. The temperature at the sample position was continuously monitored. The average temperature is used in the resonance shape analysis program to calculate the Doppler broadening of the resonances.

The anode pulses of both PMT's are transmitted to a constant fraction discriminator to create fast logic signals for a coincidence unit. Whenever a signal from each PMT was present within a 30 ns coincidence window, a coincidence pulse was sent to the Time to Digital Converter (TDC), which determines the TOF of the detected neutron from the time difference between the start signal, given at each electron burst, and the stop signal from the coincidence unit. Selecting only coincident pulses, a significant improvement of the signal-to-noise ratio in the TOF-spectrum was obtained. The time spectra from the BF_3 proportional counters, neutron flux monitors located in the ceiling of the target hall, was also recorded to verify the stability of the accelerator and their count rate used to normalize the sample-in and sample-out measurements to the same total neutron fluence.

The transmission experiments at the 26.45 m flight path were performed at three different working frequencies of GELINA 400 Hz, 800 Hz and 50 Hz.

The measurements at 400 Hz and 800 Hz were carried out with a thick natural Cd sample consisting of plates with a nominal diameter of 50 mm and with a total mass of (441.6460 ± 0.0001) g. The sample thickness was 25.818 mm.

To obtain a quality data in the thermal energy region, four additional Cd transmission measurements were performed at the 26.45 m flight-path at 50 Hz working frequency of GELINA. In the experiments were used four samples. Three were thin natural cadmium metallic foils with nominal diameter of 80 mm, their respective nominal thicknesses were 2.03 mm, 30 μm and 50 μm, and their weights (87.6300 ± 0.0001) g, (1.2814 ± 0.0001) g and (2.0993 ± 0.0001) g respectively. The fourth sample was a solution sample. This sample consisted of Quartz Cell with an internal path-lenght of 4.98 mm, filled with (1.1918 ± 0.0001) g Cd $(NO_3)_2$ dissolved in H_2O. For the sample-out measurements a similar Quartz Cell with an internal pathlenght of 5.21 mm filled only with H_2O was used.

A number of Cd transmission experiments were performed at the 50 m flight-path number 4 of GELINA. The experimental setup for the Cd measurements was the same as the one shown in figure 1. The samples were placed at a 25-m distance from the neutron source. A 6 mm thick NE912 Li-glass scintillator, placed in an aluminium sphere and viewed by a single 125 mm EMI9823-QKB photomultiplier, placed outside the neutron beam, was used to detect the neutrons. A constant fraction discriminator was used to determine time of an event, and only signals lying within the (n,α) peak in the pulse height spectrum were sent to to the Time to Digital Converter (TDC).

The Cd transmission measurements at the 50 m flight path were performed at 400 Hz and 50 Hz working frequency of GELINA.

The 400 Hz transmission experiments were performed with three ^{nat}Cd samples. The first experiment was carried out with a thin natural Cd disk sample with a nominal diameter of 50.00 mm, a thickness of 0.125 mm and weight (1.9878 ± 0.0001) g. In the second experiment it was measured a natural Cd foil sample with a nominal diameter of 100.07mm and a thickness of 2.00 mm. The weight of the sample was (174.2655 ± 0.0001) g. The last measurement at 400 Hz was carried out with a sample consisting of two Cd foils one of which was already used in the previous experiment. The other Cd foil had a nominal diameter of 101.93mm, a thickness of 3.00mm, and weight (272.9500 ± 0.0001) g.

Four more transmission experiments were carried out at the 50 Hz repetition rate of the LINAC. In three of them the same 25.818 mm Cd sample was used as the one already measured at 400 Hz and 800 Hz working frequency at the 26.45 m flight path. The last measurement was done with the 0.125 mm Cd sample, which was already measured at 400 Hz

3. Results and discussion

In the transmission experiments, the neutron time-of-flight (TOF) spectra were measured alternating the sample in and out of the neutron beam, using a sample changer. To avoid systematic uncertainties due to possible slow variations of the beam profile or detector efficiency as a function of time, the alternating sequence of measurements was repeated many times in order to approach an identical incoming average neutron fluence for the "in" and "out" cycles. The dead-time of the whole detection system was measured by registering the time-interval between two events (see table 1). The dead-time correction was well below 1% in the energy region of interest.

The transmission factor $T(t_n)$, as a function of the neutron time-of-flight t_n, was obtained as the ratio of the time-integrated spectra $C_{in}(t_n)$ and $C_{out}(t_n)$, the sample in and out spectra respectively, which have been corrected for the incident neutron flux and the dead-time effects

$$T(t_n)=\frac{C_{in}(t_n)-B_{in}(t_n)}{C_{out}(t_n)-B_{out}(t_n)}. \quad (1)$$

The background spectra $B_{in}(t_n)$ and $B_{out}(t_n)$ were described as a function of the neutron time-of-flight by the formula:

$$B(t_n)=a_1+a_2{t_n}^{a_3}+a_4e^{-a_5t_n}+a_6e^{-a_7t_n}. \quad (2)$$

The parameters were obtained by fitting them to background points determined by the black resonance technique. The background contributions were approximately 1% (see figure 2).

Table 1. Dead-time data obtained in different experiments.

Measurement		Dead time
26.45 m flight path		
	400 Hz	90 ns
	800 Hz	2015 ns
	50 Hz	93.5 ns
50 m flight path		
	400 Hz	2050 ns
	50 Hz	2050 ns

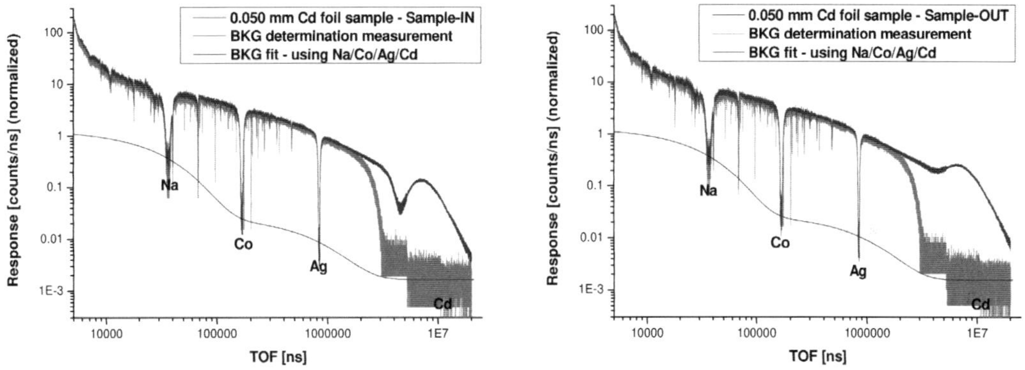

Figure 2. Dead time corrected cadmium TOF spectra and the shape of the background for sample-in and sample–out measurements of the metallic sample with 50 μm thickness.

The transmission measurements at 50 Hz working frequency of GELINA of the two Cd foil samples - 0.030 mm and 0.050 mm, and the 0.5 cm Cd $(NO_3)_2.4H_2O$ solution sample were performed in order to obtain a quality data in the thermal energy region, from which we could extract the resonance parameters for the first well-known resonance in ^{113}Cd at 0.178 eV. In the region of the lowest resonance the full width half maximum of the energy resolution of the experimental setup was of the order of 3 meV, and of the Doppler broadening effect was approximately 30 meV. The natural width for the 0.178 eV resonance is approximately 115 meV, this value is almost completely determined by the capture width. As the broadening is much smaller than the natural line width, the capture width can be extracted from the total width. The resonance area will then give the neutron width. Therefore, by using a shape analysis program both the neutron and the radiation width could be determined from a suitable transmission measurements alone. The free gas model with an effective temperature T_{eff} was used to describe the Doppler effect. The effective temperature is a function of the Debye T_{debye} and room temperature T_{room}:

$$T_{eff}=\frac{3T_{debye}}{8}\coth\left(\frac{3T_{debye}}{8T_{room}}\right). \quad (3)$$

The Debye temperature was taken from the literature as 186 K [6]. The resonance parameters were determined using the program REFIT [7]. This code uses the Reich–Moore approximation of the R-matrix to calculate the nuclear cross sections.
The three sets of experiments were fitted separately and their results compared to each other and to a simultaneous fit to all three data sets (see figure 3). The results are given in table 3. The uncertainties given in table 3 take into account the uncertainties due the counting statistics and the uncertainty of the sample thickness.

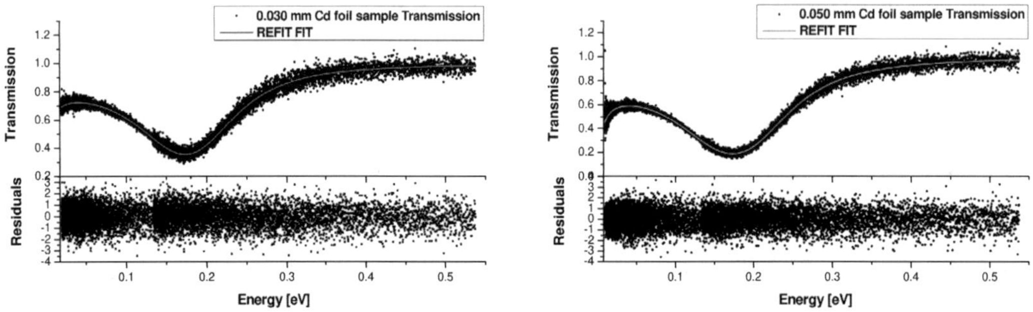

Figure 3. The REFIT fitting results of the transmission data obtained in 0.030 mm and 0.050 mm sample measurements respectively.

The resonance parameters determined in the present measurement have a greater accuracy than the published values (see table 2). The previously published resonance energy and capture width are in reasonable agreement within their quoted uncertainties with the values given in this paper.

Table 2. Comparison of the resonance parameters of the 0.178 eV ^{113}Cd resonance found in the literature.

Experiment:	**sample type**	**Eres (eV)**	**Γγ (meV)**	**Γn (meV)**
R. O. Akyuz et al [8]	metal	0.181 ± 0.003	109 ± 3	0.645 ± 0.025
L. J. Rainwater et al [9]	metal	0.176 ± 0.002	115 ± 2	0.620 ± 0.020
C. N. Brockhouse [10]	metal	0.180 ± 0.003	113 ± 2	0.680 ± 0.020
U. Harz et al [11]	various	0.1783 ± 0.0002	113.5 ± 0.5	0.650 ± 0.005
F. Widder et al [12]	powder	0.1776 ± 0.0006	114.3 ± 0.6	0.618 ± 0.003

Table 3. Resonance parameters fitted to our experimental data.

Sample type	**Sample thickness (at/b)**	**Eres (meV)**	**Γγ (meV)**	**Γn (meV)**
metal	1.36×10^{-4}	178.8 ± 0.2	113.6 ± 0.3	0.642 ± 0.003
metal	2.24×10^{-4}	178.6 ± 0.2	113.4 ± 0.2	0.638 ± 0.003
solution	1.40×10^{-4}	178.5 ± 0.3	114.5 ± 1.0	0.643 ± 0.006
weighted average		178.7 ± 0.1	113.5 ± 0.1	0.640 ± 0.002
Simultaneous fit results		178.6 ± 0.1	113.4 ± 0.2	0.639 ± 0.002
Final results		**178.7 ± 0.1**	**113.5 ± 0.2**	**0.640 ± 0.004**

The derived resonance parameters for the 0.178 eV resonance in ^{113}Cd were validated by integral experiments [13, 14]. The impact of the change of the first resonance is quite significant, with changes in the k_{eff} between 0.02 and 0.05, and the agreement between calculations and measurements is clearly improved [1]. The changes are significant but not quite big enough to bring calculations and experiments in agreement. This might be an indication that higher energy cadmium resonances or the thermal cross sections for other cadmium isotopes should be investigated.

Considering the transmission data obtained with various sample thicknesses as a source of additional information for low-level cross section, we validate a new approach for cross section description in this energy region. The method was proposed by Prof. A. A. Lukyanov [15]. We assume that the application of this method will let us obtain a better description of the cross section in the low energy region (E $\leq$ 1 eV) and the negative levels. In this energy region it is necessary to take into account a number of effects that otherwise would be negligible for energies above 1 eV. For instance, for most of the heavy nuclei the main energy dependence $1/\sqrt{E}$ could be changed accounting of the low-level resonances and the effects of the negative level tails. We take into account the energy dependence of the total resonance width. According to this approach in the analysis of the negative levels, the shift factor Δ has to be proportional to $\sqrt{E}$. In the low energy region, it is very important to study the Doppler Effect with thermal corrections as applying the Exact Free Gas Model with Maxwell-Boltzmann distribution of the thermal motion velocities of the atoms of the target $\vec{u}$:

$$B(\vec{u})d\vec{u} = \frac{1}{\left(\pi u_T^2\right)^{3/2}} e^{-u^2/u_T^2} u^2 du d\vec{\Omega} \tag{4}$$

where $u_T^2 = 2kT/M$, M is the mass of the sample (target) nucleus, k is the Boltzmann's constant,and T is the effective temperature of the sample. For neutrons with velocities $\vec{v}$ in the Lab system, reaction cross section is result of averaging the relative motion velocities of the neutrons $\vec{w} = \vec{v} - \vec{u}$ ($w^2 = v^2 + u^2 - 2uv\cos\theta$) having the same distribution as the one in equation (4). The reaction rate $w\sigma_c(w)$ is presented as:

$$v\sigma_c(v,u_T) = \int d\vec{u}\sigma_c(w) w B(\vec{u}) \tag{5}$$

We obtain as a result:

$$v\sigma_c(v,u_T) = \frac{1}{v\sqrt{\pi}u_T} \int_0^\infty w\sigma_c(w) w dw \left[e^{-(v-w)^2/u_T^2} - e^{-(v+w)^2/u_T^2} \right] \tag{6}$$

We will have for the corresponding energies ($2E = m_n v^2$, $2E' = m_n w^2$) in the center of mass system:

$$\sqrt{E}\sigma_c(E,\delta) = \int_0^\infty \sqrt{E'}\sigma_c(E') S(E',E,\delta) dE' \tag{7}$$

where

$$S(E',E,\delta) = \frac{1}{2\delta\sqrt{\pi E}} \left[e^{-\left(\sqrt{E'}-\sqrt{E}\right)^2/\delta^2} - e^{-\left(\sqrt{E'}+\sqrt{E}\right)^2/\delta^2} \right]$$

$$\delta^2 = kT/A$$

$$\int_0^\infty S(E',E,\delta) dE' = 1$$

This function is normalized to unity, and in the limit of the low energies, where $\sqrt{E'}\sigma_c(E') = const$ it keeps the reaction rate; the cross section does not depend on the temperature.

The analytical method for determination of the integral shown in the equation (7) depends on the choice of the form of the energy dependence of the evaluated resonance cross section $\sigma_c(E)$. It could be presented in various functional forms deduced from the reaction theory: Single Level Breit Wigner (SLBW), Multi Level Breit Wigner (MLBW), Adler Scheme, Reich–Moore approximation of the R –

Matrix theory with a set of evaluated relevant resonance parameters. According to Prof. Hwang's last works [16], all these forms can be combined in one general, universal method of parameterization - Rigorous pole representation of the collision matrix in the momentum domain (as a function of $\sqrt{E}$):

$$\sqrt{E}\sigma_c(E)=\sum_k \mathrm{Re}\frac{(-i)\rho_{kc}}{p_k-\sqrt{E}}=\frac{1}{\sqrt{E}}\sum_k\frac{n_k\alpha_{kc}+\left(m_k-\sqrt{E}\right)\beta_{kc}}{\left(m_k-\sqrt{E}\right)^2+n_k{}^2} \tag{8}$$

where $p_k=m_k-in_k$ and $\rho_{kc}=\alpha_{kc}+i\beta_{kc}$ are energy independent complex constants, which could be determined using evaluated resonance parameters from standard formalisms.

This method of parameterization is very useful for account of the Doppler Effect. In general case, the averaged multi level formulas could be presented as a sum of the same type of integrals for each term presented in equation (8):

As part of our analysis, we determined the nuclear radii for the individual isotopes of Cd. They were defined in the fits from the resonance-potential interference effects in the regions of the larger s-wave resonances. Table 4 contains our values compared to the ones in different evaluated data files.

Table 4. Comparison between the nuclear radii of the individual isotopes of cadmium determined from fitting our data with the ones in the evaluated data files.

Cadmium Isotope	Potential Scattering Radius [fm]			
	ENDF/B - VII	ENDF/B-VI.8 / JEFF 3.1	JENDL 3.3	Preliminary Results
Cd 110	5.790	6.418	6.200	5.800
Cd 111	5.900	5.797	6.200	6.150
Cd 112	5.860	5.814	6.200	6.600
Cd 113	6.300	6.737	6.200	6.500
Cd 114	6.310	6.310	6.200	6.600
Cd 116	6.100	6.500	6.200	6.240

Our analysis is in progress. Cadmium transmission data obtained in the measurements at GELINA have been fitted and analyzed in the energy region from thermal up to about 3 keV. In figure 4 are presented part of the preliminary results.

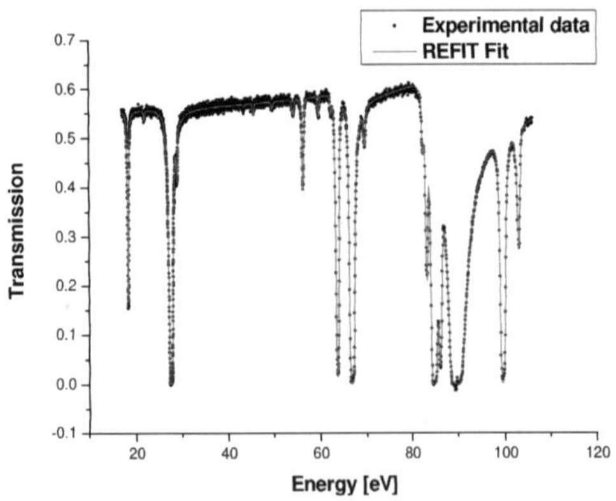

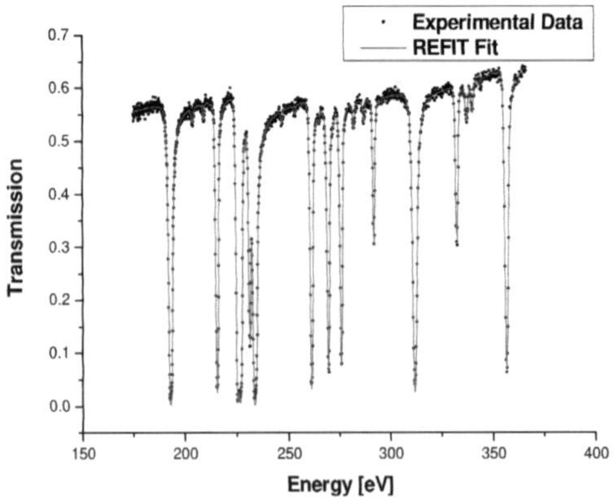

Figure 4. Preliminary results of the analysis of the transmission data of 25 mm thick Cd sample obtained in the experiment at 26.45 m flight path.

4. Conclusions

Cd transmission data have been measured between thermal and 3 keV neutron energy.

We have presented the resonance parameters of the lowest lying resonance of ^{113}Cd as: E_{res} = 178.7 ± 0.1 meV; Γ_γ = 113.5 ± 0.2 meV and Γ_n = 0.640 ± 0.004 meV. They are in general agreement with the available in the literature parameters of this resonance within the given uncertainties. For the neutron width, the values from evaluated data files are 1–2% larger than our result. For the resonance energy and Γ_γ the observed differences are below 1%. Using the parameters recommended in this work gives a better agreement between calculated k_{eff} and measured k_{eff} for the integral experiment by Lloyd et al. [17, 18].

We have reported the determined from us values of nuclear radii for the individual isotopes of Cd and the comparison to the ones in the evaluated data files in the ENDF/B – VI.8, ENDF/B – VII, JEFF – 3.1, JENDL - 3.3 libraries.

In this paper, a scheme of a new approach for cross section description in the low energy region is presented.

Acknowledgements

I would like to express my great gratitude to Prof. Alexander Lukyanov for highly scientific and always-to-the-point comments and suggestions.

I express my thanks to Dr. Peter Siegler and Dr. Stefan Kopecky, my daily IRMM supervisors for the useful scientific dialogs, which significantly improved the quality of our work.

References

[1] Kopecky S, Ivanov I, Moxon M, Schillebeeckx P, Siegler P, Sirakov I 2009 *Nuclear Instruments and Methods in Physics Research B* **267** 2345 – 2350.

[2] Bensussan A, Salome J M 1978 Gelina: a modern accelerator for high-resolution neutron time of flight experiments *Nucl. Inst. Meth.* **155** 11–23.

[3] Salomé J M 1986 GELINA, l'accelerateur lineaire a electrons du BCNM *Physicalia magazine* **8** 261.

[4] Tronc D, Salome J M, Böckhoff K 1985 A new pulse compression system for intense relativistic electron beams *Nucl. Inst. Meth. A* **228** 217-227.

[5] Salome J M, Cools R 1981 Neutron producing target at GELINA *Nucl. Inst. Meth. A* **179** 13-19.

[6] *CRC Handbook of Chemistry and Physics* **52nd ed**. 1971–1972 The Chemical Rubber CO. 18901 Cranwood Parkway Cleveland Ohio 44128.

[7] Moxon M C 1991 REFIT – A least square fitting program for resonance analysis of neutron transmission and capture data *AEA-InTec*-**0470**.

[8] Rainwater L J, Havens W W, Wu C S, Dunning J R 1947 *Phys. Rev.* **71** 65.

[9] Brockhouse B N 1953 *Can. J. Phys*. **31** 432.

[10] Aküz R Ö, Cansoy C, Domanic F 1967 *Nucl. Sci. Eng.* **28** 359.

[11] Widder F, Brunner J, 1968 *Nukleonika* **11** 297.

[12] Harz U, Priesmeyer H G 1974 *Atomkernenergie* **24** 142.

[13] Kim S S 2006 Analysis of high enriched uranyl nitrate solution containing cadmium *Proceedings of PHYSOR-2006* contribution **A071**.

[14] Mosteller R D, MacFarlane R E, Mughabghab S F, Kim S S 2008 Development and testing of a revised ENDF/B-VII capture cross section for 113Cd *Proceedings of PHYSOR-2008* contribution **088**.

[15] Lukyanov A A *Private communications*.

[16] Lukyanov A A Multilevel Parameterization of Neutron Resonance Cross Sections 9-11 October 2006 *WONDER* Cadarache France.

[17] Lloyd R C, Clayton D E, Wilson R E, McBroom R C, Jones R R 1987 *Nucl. Technol.* **79** 82.

[18] NEA Nuclear Science Committee, International Handbook of Evaluated Critical Safety Benchmark Experiments September 2008 *NEA/NSC/DOC(95)03*.

AUTHOR INDEX

AUTHOR INDEX

AUTHOR INDEX

AUTHOR INDEX

Institute of Physics
Dirac House, Temple Back
Bristol BS1 6BE UK

CURRAN ASSOCIATES INC
proceedings.com